Lecture Notes in Mathematics

Edited by A. Dold and B. Eckmann

1273

G.-M. Greuel G. Trautmann (Eds.)

Singularities, Representation of Algebras, and Vector Bundles

Proceedings of a Symposium
held in Lambrecht/Pfalz, Fed. Rep. of Germany,
Dec. 13–17, 1985

Springer-Verlag
Berlin Heidelberg New York London Paris Tokyo

Editors

Gert-Martin Greuel
Günther Trautmann
Fachbereich Mathematik, Universität Kaiserslautern
Erwin-Schrödinger-Straße, 6750 Kaiserslautern, Federal Republic of Germany

Mathematics Subject Classification (1980): 13C05, 13C10, 13D10, 13D15, 13H10, 14B05, 14B07, 14C22, 14C30, 14D20, 14F05, 14H10, 14H15, 14H20, 14H40, 16A46, 16A53, 16A64, 32B30, 32G05, 32G11, 32G13

ISBN 3-540-18263-2 Springer-Verlag Berlin Heidelberg New York
ISBN 0-387-18263-2 Springer-Verlag New York Berlin Heidelberg

Printed in Germany

Printing and binding: Druckhaus Beltz, Hemsbach/Bergstr.
2146/3140-543210

INTRODUCTION

These are the Proceedings of the Symposium "Singularities, Representation of Algebras, and Vectorbundles" held December 13-17, 1985 at the Pfalzakademie in Lambrecht/Pfalz, West Germany.

The purpose of the symposium was to discuss and promote recent developments of the interaction of singularity theory with representation of algebras and vector bundles. The colloquium talks given during the conference initiated intensive and stimulating discussions among the participants of the different areas who usually do not meet at conferences. These discussions led to new insights as well as to new questions concerning the relationship between the three topics of the conference. They partly condensed subsequently in research articles which are - besides the revised texts of oral lectures - presented in this volume. It is the editors' hope that these notes stimulate further development and interaction.
It is nowadays well known that there are close relations between classes of singularities and representation theory by means of the McKay correspondence and representation theory and vector bundles on projective space via the Bernstein-Gelfand-Gelfand construction. On the other hand, these relations can not be considered to be either completely understood or fully exploited.

It became clear during the conference that at least the following can be said about the principal relations between the three areas. The questions and methods of representation theory (as finite and tame representation type, almost split sequences, quivers) have applications to singularities and to vector bundles depending whether one considers modules over complete local rings or graded modules over graded rings. These led in particular to the characterization of the simple singularities in the sense of Arnold by maximal Cohen-

Macaulay modules generalizing the work of Mc Kay, Auslander, Artin and Verdier. Representation theory on the other hand, which had primarily developed its methods for Artinian algebras, starts to investigate algebras of higher dimensions partly because of these applications. There are not only interesting examples within the class of singularities or algebraic varieties, there might be also future research in representation theory, stimulated by the classification of singularities and the highly developed theory of moduli for vector bundles.

Of course, there is the general problem for specialists to understand well enough topics in fields other than their own. In order to overcome this difficulty at least partly during the conference there were three survey talks on the different topics stressing the relationship to the other two: H. Knörrer on "Cohen-Macaulay Modules on Hypersurface Singularities", M.S. Narasimhan on "Moduli of Vector Bundles on Curves", and I. Reiten on "Representation of Algebras and Relations to Singularities". The latter two are presented in this volume, Knörrer's article has already appeared in the proceedings "Representation of Algebras", Lond. Math. Soc., Lect. Notes, Series 116, 147-164, Cambridge Univ. Press. His article surveys the recent development of the interaction of singularity theory and representation theory and is warmly recommended. We are grateful to F.-O. Schreyer for having expanded his oral lecture in order to give a partial survey, together with new results and open problems and conjectures.

Not all the oral lectures are published in these proceedings because they had already been submitted to other journals. On the other hand we are pleased to include others which fit well into the subject of the conference, in particular those which have been initiated by the symposium itself.

With the exception of the survey talks all articles contain original research. The participants served as referees and we owe them much thanks.

ACKNOWLEDGEMENTS

The symposium was made possible by the generous financial support of the Stiftung Volkswagenwerk. Further support was provided by the Kultusministerium des Landes Rheinland-Pfalz and by the Fachbereich Mathematik der Universität Kaiserslautern. We like to thank these institutions for their help.

The pleasant atmosphere of the Pfalzakademie and the beautiful surroundings contributed to the success of the symposium.

G.-M. Greuel

G. Trautmann

TABLE OF CONTENTS

Survey-Talks

Representation of Algebras

Vector Bundles

LIST OF TALKS NOT PRESENTED IN THIS VOLUME

W. Barth
Degeneration of Horrocks-Mumford surfaces

D. Eisenbud
Plane sections of determinantal varieties

A. Holme
On duality for projective varieties

K. Hulek
Complete families of stable vector bundles

M. Maruyama
On the rationality of moduli spaces of vector bundles of rank 2 on $\mathbb{P}_2$

M. Schneider
Linear normality for subvarieties of projective spaces

C.S. Seshadri
Singularities of Schubert varieties

D. Siersma
A class of non-isolated hypersurface singularities

S.A. Strømme
Quot schemes on $\mathbb{P}_1$

LIST OF PARTICIPANTS

M. Auslander
Brandeis

W. Barth
Erlangen

J. Christophersen
Oslo

W. Decker
Kaiserslautern

E. Dieterich
Bielefeld

P. Dowbor
Paderborn

J.M. Drezet
Paris 7

D. Eisenbud
Brandeis

H. Flenner
Göttingen

O. Forster
München

W. Geigle
Paderborn

G.-M. Greuel
Kaiserslautern

D. Happel
Bielefeld

J. Herzog
Essen

A. Holme
Bergen

G. Horrocks
Newcastle

K. Hulek
Bayreuth

Th. de Jong
Leiden

C. Kahn
Hamburg

H. Knörrer
Bonn

H. Kröning
Kaiserslautern

H. Lenzing
Paderborn

M. Maruyama
Kyoto

M.S. Narasimhan
Bombay

J. Le Potier
Paris 7

C.J. Rego
Bombay

I. Reiten
Trondheim

O. Riemenschneider
Hamburg

C.M. Ringel
Bielefeld

K.W. Roggenkamp
Stuttgart

A. Schappert
Kaiserslautern

M. Schlichenmaier
Mannheim

G. Schneider
Kaiserslautern

M. Schneider
Bayreuth

F.-O. Schreyer
Kaiserslautern

C.S. Seshadri
Madras

D. Siersma
Utrecht

W. Singhof
Kaiserslautern

H. Spindler
Göttingen

J.H. Steenbrink
Leiden

D. van Straten
Leiden

St.A. Strømme
Bergen

G. Trautmann
Kaiserslautern

B. Uher
Erlangen

L. Unger
Bielefeld

H. Völlinger
Kaiserslautern

A. Wiedemann
Stuttgart

AUTHORS' ADDRESSES

M. Auslander	Department of Mathematics, Brandeis University, Waltham, Mass. 02257, USA
W. Böhmer	Fachbereich Mathematik der Universität Erwin-Schrödinger-Str., 6750 Kaiserslautern
R.O. Buchweitz	Institut für Mathematik der Universität Welfengarten 1, 3000 Hannover 1
J. Christophersen	Matematisk Institutt, Universitetet i Oslo, P.b. 1053 Blindern, Oslo 3, Norwegen
E. Dieterich	Fakultät für Mathematik der Universität Universitätsstr., 4800 Bielefeld 1
J.M. Drezet	Université de Paris VII, Mathématiques 2, Place Jussieu, 75 230 Paris Cedex 05
D. Eisenbud	Department of Mathematics, Brandeis University, Waltham, Mass. 02254, USA
C. Ellingsrud	Matematisk Institutt, Universitetet i Oslo P.b. 1053 Blindern, Oslo 3, Norwegen
O. Forster	Mathematisches Institut der Universität Theresienstrasse 39, 8000 München 2
W. Geigle	Fachbereich 17 - Mathematik - Informatik der Universität, Warburger Str. 100 4790 Paderborn
D. Happel	Fakultät für Mathematik der Universität Universitätsstr. 4800 Bielefeld 1
J. Herzog	FB Mathematik der Gesamthochschule Universitätsstr. 2, 4300 Essen
H. Knörrer	Mathematisches Institut der Universität Universitätsstr. 1, 4000 Düsseldorf 1
H. Lenzing	Fachbereich 17 - Mathematik - Informatik der Universität, Warburger Str. 100 4790 Paderborn

M.S. Narasimhan	Tata Institute of Fundamental Research Homi Bhabha Road, Bombay 400 005 Indien
C.J. Rego	19, 'Prasanna', Nesbit Road, Bombay 400 010, Indien
I. Reiten	Department of Mathematics, University of Trondheim, Trondheim, Norwegen
K.W. Roggenkamp	Fakultät für Mathematik und Informatik Universität, Pfaffenwaldring 57, 7000 Stuttgart 80
A. Schappert	Fachbereich Mathematik der Universität Erwin-Schrödinger-Str., 6750 Kaiserslautern
F.-O. Schreyer	Fachbereich Mathematik der Universität Erwin-Schrödinger-Str., 6750 Kaiserslautern
C.S. Seshadri	The Institute of Mathematical Sciences Madras 600 113, Indien
J.H.M. Steenbrink	Subfakulteit der Wiskunde, Rijksuniversiteit te Leiden, Wassenaarseweg 80, 23 RA Leiden, Holland
D. van Straten	Subfaculteit der Wiskunde, Rijksuniversiteit te Leiden, Wassenaarseweg 80, 23 RA Leiden, Holland
St.A. Strømme	Matematisk Institutt, Universitetet i Bergen 5014 Bergen, Norwegen
K. Wolffhardt	Mathematisches Institut der Universität Theresienstr. 39, 8000 München 2
J. Wunram	Mathematisches Seminar der Universität Bundesstr. 55, 2000 Hamburg 13
St. Zucker	Department of Mathematics, Johns Hopkins University, Baltimore MD 21218, USA

SURVEY OF VECTOR BUNDLES ON CURVES

M.S. Narasimhan
School of Mathematics
Tata Institute of Fundamental Research
Homi Bhabha Road, Bombay 400 005
INDIA

1. Moduli problem for vector bundles on curves.

Let X be a compact Riemann surface, or what is the same, a smooth projective irreducible algebraic curve over $\mathbb{C}$. It is well known that the set of isomorphism classes of holomorphic (or algebraic) line bundles of degree d has a natural structure of a smooth projective variety; if $d = 0$ we obtain an abelian variety, the Jacobian of X. Moreover any line bundle of degree 0 on X is associated to a (unitary) character of the fundamental group of X [9].

The corresponding 'moduli problem' for (algebraic) vector bundles of higher rank on X was first envisaged by A. Weil in 1938 in a famous paper [30]. Naively formulated, the question is whether there is a natural structure of a variety on the set of isomorphism classes of vector bundles of a given rank and degree on X. However it is easy to see that even 'locally' around certain bundles of rank ≥ 2 one can not have a structure of a variety, for example due to the jump-phenomenon [see 17, p. 126]. This suggests that one can expect a structure of variety only on a suitable subset of the isomorphism classes of vector bundles.

In what follows we shall assume that the genus g of X is ≥ 2. It is well known that any vector bundle on a curve of genus 0 is a direct sum of line bundles. Vector bundles on curves of genus 1 were investigated in detail by M.F. Atiyah [1].

2. Flat bundles and a theorem of A. Weil.

We shall consider in this section a class of vector bundles on X, namely flat bundles, which play an important role in the moduli problem for vector bundles on curves.

Let $\widetilde{X}$ be a universal covering of X and let the fundamental group $\pi = \pi_1(X)$ act on $\widetilde{X}$ on the right. If ρ is a homomorphism of π into the full linear group $GL(r,\mathbb{C})$, we can construct a holomorphic vector bundle $W_\rho = \widetilde{X} \times_\pi \mathbb{C}^r$ on X; the bundle W_ρ is the quotient of $\widetilde{X} \times \mathbb{C}^r$ under the action of π given by :

$$(\widetilde{x},v)\gamma = (\widetilde{x}\gamma,\ \rho(\gamma)^{-1}v),\ \widetilde{x} \in \widetilde{X},\ v \in \mathbb{C}^r,\ \gamma \in \pi .$$

We say that a holomorphic vector bundle V on X arises from a representation (resp. unitary representation) of π, if V is isomorphic to W_ρ where ρ is a homomorphism of π into $GL(r,\mathbb{C})$ (resp. into to the unitary group $U(r)$).

Among other results A. Weil proved in [30]

Theorem 2.1. A vector bundle on X arises from a representation of the fundamental group of X if and only if each of its indecomposable components is of degree zero.

A. Weil also expected that bundles which arise from unitary representations would play an important role.

3. Stable and semi-stable bundles.

The crucial step in the progress of moduli problem for vector bundles on curves was the introduction by David Mumford of the all important notion of stable vector bundles. This concept was motivated by the geometric invariant theory [6].

Definition 3.1. A vector bundle V on X is said to be stable (resp. semi-stable) if for every proper subbundle W of V we have

$$\frac{\deg W}{\operatorname{rank} W} < \frac{\deg V}{\operatorname{rank} V} \quad (\text{resp. } \frac{\deg W}{\operatorname{rank} W} \leq \frac{\deg V}{\operatorname{rank} V}),$$

where $\deg(V) = C_1(V)[X]$, $C_1(V)$ denoting the first chern class of V.

Observe that a semi-stable bundle is automatically stable, if its rank and degree are coprime.

D. Mumford proved [7]

Theorem 3.1. The set of isomorphism classes of stable vector bundles on X of rank r and degree d has a natural structure of a smooth quasi-projective variety of dimension $r^2(g-1)+1$.

4. Stable bundles and unitary bundles.

The following basic theorem was proved by M.S. Narasimhan and C.S. Seshadri [15,16].

Theorem 4.1. A vector bundle on X of degree 0 is stable if and only if it arises from an irreducible unitary representation of the fundamental group of X.

As a consequence one sees that a vector bundle arises from a unitary representation of $\pi_1(X)$ if and only if each of its indecomposable components is of degree 0 and stable. Moreover it is easy to show that two vector bundles arising from unitary representations are isomorphic if and only if the representations are equivalent.

In the same paper a characterisation similar to Theorem 4.1 was also given for stable bundles of arbitrary degree in terms of irreducible unitary representations of a certain Fuchsian group. This result implies that, if $(r,d) = 1$, the space of isomorphism classes of stable bundles of rank r and degree d is compact and is hence a smooth projective variety.

5. The moduli space of semi-stable bundles.

The results stated in § 4 suggest a natural compactification of the space of stable bundles of degree 0, namely the space of equivalence classes of unitary representations (not necessarily irreducible) of a given rank. C.S. Seshadri proved that this compactification is a projective variety [23]. Before stating his result precisely we will introduce an equivalence relation among semi-stable bundles.

Let V be a semi-stable vector bundle on X. Then V has a strictly decreasing filtration by subbundles

$$V = V_0 \supset V_1 \supset \dots \supset V_k = (0)$$

such that for $1 \leq i \leq k$ the bundle $W_i = V_i/V_{i-1}$ is stable and satisfies $\frac{\deg W_i}{\operatorname{rank} W_i} = \frac{\deg V}{\operatorname{rank} V}$. Moreover the bundle $Gr(V) = \bigoplus_{i=1}^{k} (V_i/V_{i-1})$ is uniquely determined by V upto isomorphism (Jordan-Hölder theorem). We say that two semi-stable bundles V_1 and V_2 are S-equivalent if $Gr(V_1)$ is isomorphic to $Gr(V_2)$. Observe that two stable bundles are S-equivalent if and only if they are isomorphic. It is clear, using Theorem 4.1, that the set of equivalence classes of unitary representations is in canonical bijective correspondence with the set of S-equivalence classes of semi-stable vector bundles of degree 0.

C.S. Seshadri, using geometric invariant theory, proved [23]

Theorem 5.1. There is a unique structure of a normal projective (irreducible) variety $U(r,d)$ on the set of S-equivalence classes of semi-stable vector bundles on X of rank r and degree d such that the following property holds : if $\{V_t\}_{t \in T}$ is an algebraic family of semi-stable bundles on X of rank r and degree d parametrised by an algebraic variety T, then the map $T \longrightarrow U(r,d)$, sending $t \in T$ to the S-equivalence class of V_t, is a morphism.

We shall call the variety given by Theorem 5.1 the moduli space of (semi-stable) vector bundles of rank r and degree and denote it by $U(r,d)$.

Theorem 5.1 is also valid for a curve X over an algebraically closed field of arbitrary characteristic [25].

6. Singularities of $U(r,d)$.

The set of singular points of $U(r,d)$ has been determined by M.S. Narasimhan and S. Ramanan [11].

Theorem 6.1. The set of non-singular points of $U(r,d)$ is precisely the set of stable points in $U(r,d)$ except when $g = 2$, $r = 2$ and d even. In this exceptional case $U(r,d)$ is smooth.

Explicit desingularisations of $U(2,0)$ have been given by M.S. Narasimhan-S. Ramanan [14] and by C.S. Seshadri [24].

7. Poincaré bundles.

Let $U_S(r,d)$ denote the open set of stable points in $U(r,d)$.

Definition 7.1. Let Ω be a non-empty open subset of $U_S(r,d)$. A Poincaré family of vector bundles parametrised by Ω is an algebraic vector bundle P on $\Omega \times X$ such that for any $\omega \in \Omega$ the bundle on X obtained by restricting P to $\omega \times X$ is in the isomorphism class ω.

It is not hard to see that when $(r,d) = 1$ there is a Poincaré bundle on $U(r,d) \times X$ [8]. S. Ramanan proved [22]

Theorem 7.1. If r and d are not coprime there is no Poincaré family on X parametrised by any non-empty open subset of $U_S(r,d)$.

The special case of this theorem where $g = 2$, $r = 2$, d even, was proved earlier in [10].

8. The variety $S(r,d)$.

In order to study the varieties $U(r,d)$ let us fix a line bundle L of degree d and consider the subvariety of $U(r,d)$ corresponding to semi-stable bundles V with $\overset{r}{\Lambda} V \simeq L$. We will denote this variety by $S_X^L(r,d)$ or simply by $S(r,d)$. The varieties $S(r,d)$ have been studied intensively by G. Harder, M.S. Narasimhan, P.E. Newstead, S. Ramanan and A. Tjurin, especially in the case $(r,d) = 1$. The results obtained pertain to the computation of numerical invariants like the Betti numbers, questions concerning the rationality of these varieties, the relation between the moduli of curves and the moduli of the varieties $S(r,d)$ and the explicit determination of these varieties in low genus or rank.

9. Betti numbers of $S(r,d)$.

The Betti numbers of $S(r,d)$ were first calculated by P.E. Newstead in the case $r = 2$, $d = 1$, by topological methods using the results of § 4 [18]. Based on these results G. Harder verified the Weil conjecture for $S(2,1)$ in the case of a curve defined over a finite field [4], at a time when the Weil conjecture was not proved in general. Harder observed in turn that the Betti numbers of $S(2,1)$ can be computed by arithmetical methods on the basis of the Weil conjecture.

Harder's method was generalised by him and M.S. Narasimhan to bundles of arbitrary rank [5]. It was shown, in the case $(r,d) = 1$, that the ζ-function of $S(r,d)$ can be calculated from the ζ-function of X. This result, together with Weil conjecture proved by P. Deligne, gives a method for computing the Betti numbers of $S(r,d)$ when $(r,d) = 1$.

As special cases of results proved in [5] we have

Theorem 9.1. 1) The second Betti number of $S(r,d)$ is 1 and the third Betti number is $2g$, when $(r,d) = 1$.

2) Let d and d' be such that $(r,d) = (r,d') = 1$, and $d \not\equiv d' (\mathrm{mod}\ r)$. Then $S(r,d)$ and $S(r,d')$ are not homeomorphic.

Another result proved in [5] is the following :

Theorem 9.2. Let $(r,d) = 1$ and J_r denote the subgroup of $\mathrm{Pic}\,(X)$ consisting of line bundles of order r. Consider the action of J_r on $S(r,d)$ given by

$$(\xi, V) \longmapsto \xi \times V, \ \xi \in J_r, \ V \in S(r,d).$$

Then the induced action of J_r on the real cohomology groups of $S(r,c)$ is trivial.

Another method to compute the Betti numbers of $S(r,d)$ was given recently by M.F. Atiyah and R. Bott [2,29]. They prove also that the integral cohomology is torsion-free, if $(r,d) = 1$.

Using Theorem 8.2 and the Atiyah-Singer fixed point theorem for the action of J_r on $S(r,d)$, M.S. Narasimhan and S. Ramanan computed explicitly the Euler characteristic $\chi(S(r,d),\Omega^p)$ of the sheaf Ω^p [13]. In particular the Euler - characteristic and the index of $S(r,d)$ are zero $[(r,d) = 1]$.

10. Rationality of $S(r,d)$.

It is easy to see that the varieties $S(r,d)$ are unirational [10, p.337]. It is also known that $S(r,d)$ is rational if $d \equiv \pm 1 \ (\mathrm{mod}\ r)$ [21]; in particular $S(2,1)$ is rational. However it is not known whether $S(r,d)$ is rational in general. It is not even known whether $S(2,0)$ is rational when $g \geq 3$.

11. Moduli of curves and moduli of $S(r,d)$.

In this section we shall assume $(r,d) = 1$.

Theorem 11.1. The canonically polarised intermediate jacobian of $S(r,d)$ corresponding to the third cohomology group is naturally isomorphic to the canonically polarised jacobian of X.

This theorem was proved in the case $r = 2$ by D. Mumford and P.E. Newstead [8] and in general by M.S. Narasimhan and S. Ramanan [12]. As a consequence one sees that if X_1 and X_2 are two curves such that the corresponding moduli spaces $S_{X_1}(r,d)$ and $S_{X_2}(r,d)$ are isomorphic, then X_1 and X_2 are isomorphic. This result was also proved by A. Tjurin [26,27].

The following result proved by M.S. Narasimhan and S. Ramanan [12] implies that any small deformation of $S_X(r,d)$ is of the form $S_{X_t}(r,d)$ for a deformation

X_t of X.

Theorem 11.2. 1) The group of automorphisms of $S(r,d)$ is finite and

$$H^i(S(r,d),\Theta) = 0 \quad \text{for} \quad i \neq 1 ,$$

where Θ is the tangent sheaf.

2) $\dim H^1(S(r,d),\Theta) = 3g-3$.

12. Explicit determination of $S(r,d)$ in special cases.

Theorem 12.1. Let $g = 2$. Then

1) $S(2,0)$ isomorphic to $\mathbb{P}^3$.

2) $S(2,1)$ is isomorphic to a smooth intersection of two (smooth) quadrics in $\mathbb{P}^5$.

The proof of Theorem 12.1 given in [11] exploits the surprising connection between the moduli spaces of bundles of rank 2 on a curve of genus 2 and the classical theory of Kummer surfaces and quadratic complexes. The second part of the theorem was also proved by P.E. Newstead [19].

The second part of the theorem was generalised by U.V. Desale and S. Ramanan [3] as follows :

Theorem 12.2. Let X be an hyperelliptic curve of genus $g \geq 2$ and let $\lambda_0,\ldots,\lambda_{2g+1}$ be the branch points of X in $\mathbb{P}^1$. Then $S(2,1)$ is isomorphic to the space of all $(g-2)$ dimensional linear subspaces of $\mathbb{P}^{2g+1}$ which are contained in the intersection of the two quadrics

$$\sum_{i=0}^{2g+1} X_i^2 = \sum_{i=0}^{2g+1} \lambda_i X_i^2 = 0 .$$

BIBLIOGRAPHY

1. M.F. Atiyah : Vector bundles over an elliptic curve. Proc. London Math. Soc. 7 (1957), 414-452.

2. M.F. Atiyah and R. Bott : The Yang-Mills equations on a Riemann surface. Phil. Trans. Royal Soc. Lond. A 308 (1982), 523-621.

3. U.V. Desale and S. Ramanan : Classification of vector bundles of rank 2 on hyper-elliptic curves, Inventiones Math., 38 (1976), 161-185.

4. G. Harder : Eine Bemurkung zu einer Arbeit von P.E. Newstead. Jour. für. Math. 242 (1970), 16-25.

5. G. Harder and M.S. Narasimhan : On the cohomology groups of moduli spaces of vector bundles on curves, Math. Annalen 212 (1975), 215-248.

6. D. Mumford : Geometric Invariant Theory, Springer Verlag, 1965.

7. D. Mumford : Projective invariants of projective structures and applications, Proc. International Congress of Math., (Stockholm 1962), 526-530.

8. D. Mumford and P.E. Newstead : Periods of a moduli space of bundles on curves. Amer. J. Math. 90 (1968), 1200-1208.

9. M.S. Narasimhan : Vector bundles on compact Riemann surfaces, in Complex Analysis and its Applications, Vol. III, International Atomic Energy Agency, Vienna, 1976.

10. M.S. Narasimhan and S. Ramanan : Vector bundles on curves, in Algebraic Geometry, Bombay Colloquium 1968, (O.U.P. 1969), 335-346.

11. M.S. Narasimhan and S. Ramanan : Moduli of vector bundles on a compact Riemann surface, Ann. of Math. 89 (1969). 14-51.

12. M.S. Narasimhan and S. Ramanan : Deformations of the moduli space of vector bundles on an algebraic curve, Ann. of Math. 101 (1975), 391-417.

13. M.S. Narasimhan and S. Ramanan : Generalized prym varieties as fixed points, J. Indian Math. Soc. 39 (1975), 1-19.

14. M.S. Narasimhan and S. Ramanan : Geometry of Hecke Cycles I, In C.P. Ramanujan - A Tribute (T.I.F.R. Studies in Math.) Springer-Verlag, 1978, 291-345.

15. M.S. Narasimhan and C.S. Seshadri : Holomorphic vector bundles on a compact Riemann surface, Math. Annalen 155 (1964), 69-80.

16. M.S. Narasimhan and C.S. Seshadri : Stable and unitary vector bundles on a compact Riemann surface, Ann. of Math. 82 (1965), 540-567.

17. P.E. Newstead : Introduction to moduli problems and orbit spaces, T.I.F.R. Lecture Notes, Springer-Verlag, 1978.

18. P.E. Newstead : Topological properties of some spaces of stable bundles, Topology 6 (1967), 241-262.

19. P.E. Newstead : Stable bundles of rank 2 and odd degree over a curve of genus 2, Topology 7 (1968), 205-215.

20. P.E. Newstead : Characteristic classes of stable bundles of rank 2 over an algebraic curve, Trans. Amer. Math. Soc. 169 (1972), 337-345.

21. P.E. Newstead : Rationality of moduli spaces of stable bundles, Math. Annalen 215 (1975), 251-268, Correction in 249 (1980) 281.

22. S. Ramanan : The moduli spaces of vector bundles over an algebraic curve, Math. Annalen 200 (1973), 69-84.

23. C.S. Seshadri : Space of unitary vector bundles on a compact Riemann surface, Ann. of Math. 85 (1967), 303-336.

24. C.S. Seshadri : Desingularisation of the moduli variety of vector bundles on curves, Proc. Int. Symp. Alg. Geom. Kyoto 1977, Kinokuniya, Tokyo, 1978.

25. C.S. Seshadri : Fibrés vectoriels sur les courbes algébriques (rédigé par J.M. Drezet), Astérisque (96), 1982.

26. A. Tjurin : Analogue of Torelli's theorem for two dimensional bundles over algebraic curves of arbitrary genus Izv. Akad. Nauk. SSSR. Ser. Mat. 30 (1966), 1353-1356, English trans., Math. U.S.S.R. Izv 3 (1969), 1081-1101.

27. A. Tjurin : Analogues of Torelli's theorem for multidimensional vector bundles over an arbitrary algebraic curve, Izv. Akad. Nauk,SSSR Ser Mat. 34 (1970), 338-365, English trans., Math. U.S.S.R. - Izv 4 (1970), 343-370.

28. A. Tjurin : The geometry of moduli of vector bundles, Uspehi Mat. Nauk 29 : 6 (1974) 59-88, English trans., Russian Math. Surveys 29 : 6 (1974), 57-88.

29. J.-L. Verdier and J. Le Potier (Editors) : Module des fibrés stables sur les courbes algébriques, Progress in Maths. vol. 54, Birkhäuser, 1985.

30. A. Weil : Généralisation des functions abéliennes, J. Math. Pures Appl. 17 (1938), 47-87.

FINITE AND COUNTABLE CM-REPRESENTATION TYPE

Frank-Olaf Schreyer
Fachbereich Mathematik der
Universität Kaiserslautern
Erwin-Schrödinger Strasse
D-6750 Kaiserslautern

0.1. A complex analytic hypersurface germ defined by a power series $f \in \mathbb{C}[[x_0,\dots,x_r]]$ is called *simple* c.f. [Arnold], [Wall], if it deforms in only finitely many isomorphism classes of germs. Their classification up to isomorphism is due to Arnold:

$$A_n : x_0^{n+1} + x_1^2 + x_2^2 + \dots + x_r^2 \, , \quad n \geq 1$$

$$D_n : x_0^{n-1} + x_0 x_1^2 + x_2^2 + \dots + x_r^2 \, , \quad n \geq 4$$

$$E_6 : x_0^4 + x_1^3 + x_2^2 + \dots + x_r^2$$

$$E_7 : x_0^3 x_1 + x_1^3 + x_2^2 + \dots + x_r^2$$

$$E_8 : x_0^5 + x_1^3 + x_2^2 + \dots + x_r^2$$

Various other characterizations of these singularities are known see e.g. [Durfee]. For example in dimension $r = 2$, these are precisely those quotient singularities which are hypersurfaces.

0.2. This article reviews a further characterization:

THEOREM [Knörrer 86; Buchweitz-Greuel-Schreyer]

A hypersurface is simple iff it has finite CM-representation type.

i.e. iff there are only finitely many isomorphism classes of indecomposable maximal Cohen-Macaulay modules.

I include a complete proof of this result in this survey article. My emphasis lies on the construction of maximal Cohen-Macaulay modules via matrix factorizations (section 2), from which we deduce Knörrer's periodicity result (section 3). In section 4 we construct for every non-simple hypersurface singularity infinitely many indecomposable maximal Cohen-Macaulay modules.

0.3. The non-isolated singularities

$$A_\infty : x_1^2 + x_2^2 + \dots + x_r^2 = 0$$

$$D_\infty : x_0 x_1^2 + x_2^2 + \dots + x_r^2 = 0$$

can be viewed as "natural limits of the series A_n, D_n of singularities". A_∞ and D_∞ are basic for studying non-isolated singularities c.f. [Siersma].

THEOREM [Buchweitz-Greuel-Schreyer]

A hypersurface has countable CM-representation type iff it is isomorphic to A_∞ or D_∞.

In section 6 we describe all indecomposable maximal Cohen-Macaulay modules on A_n and D_n singularities for $n = 1,2,\dots,\infty$ and their Auslander-Reiten-sequences.

0.4. A complete classification of complex analytic singularities with finite CM-representation type is known only for dimension ≤ 2 (c.f. section 1). Two non-hypersurface examples are given by Auslander and Reiten for dimension 3, no example is known for dimension ≥ 4. However by a result of Auslander they necessarily have an isolated singularity. A list of the known examples is contained in (7.1). In (7.2) I give a few more examples of singularities which have countable representation type.

I do not discuss the McKay correspondence. For this subject (and much of the material covered here) the survey article of [Knörrer 85] gives an excellent overview.

1. FINITE CM-REPRESENTATION TYPE

1.1. Let R be a local Cohen-Macaulay ring. A finitely generated R-module M is *maximal Cohen-Macaulay* (MCM) if depth M = dim R.

We denote by Ind(R) the set of isomorphism classes of indecomposable maximal Cohen-Macaulay R-modules and say that R has *finite* (resp. *countable*) *CM-representation type*, if Ind(R) is a finite (resp. countable) set.

For any finitely generated R-module N the n-th syzygy-module $\Omega_R^n(N)$, i.e. the n-th kernel in a free resolution of N
$0 \to \Omega_R^n(N) \to F_{n-1} \to \dots \to F_1 \to F_0 \to N \to 0$, is maximal Cohen-Macaulay for $n \geq \dim R$. So finite CM-representation type implies that the Grothendieck group of f.g. R-modules is finitely generated.

1.2. The question, which local Cohen-Macaulay rings have finite CM-representation type was first raised by Herzog. He proved:

THEOREM [Herzog]

A Gorenstein local ring with finite CM-representation type is an abstract hypersurface.

A local ring is an abstract hypersurface if it's completion $\hat{R} \cong P/(f)$, where P is a regular complete local ring.

1.3. Moreover Herzog gives an interesting class of examples: Let k be an algebraically closed field and $G \subset GL(2,k)$ a finite group such that char(k) does not divides Ord(G). G acts naturally on $k[[x,y]]$.

THEOREM [Herzog]

The ring of invariants $R = K[[x,y]]^G$ *has finite CM-representation type.*

PROOF:

$k[[x,y]]$ is a finitely generated R-module. We shall prove that each indecomposable MCM R-module is a direct summand of $k[[x,y]]$: The subring $R \subset k[[x,y]]$ is a direct summand via the projection $k[[x,y]] \to R$,
$f \mapsto \frac{1}{\mathrm{ord}(G)} \sum_{g \in G} f^g$. So if M is any MCM over R it is a summand of
$N = M \otimes_R k[[x,y]]$.
N may not be MCM, but the double dual $N^{**} = \mathrm{Hom}_R(\mathrm{Hom}_R(N,R),R)$ is MCM and since MCM-modules over 2-dimensional rings are reflexive, i.e. $M \cong M^{**}$ we obtain M as a direct summand of N^{**}. N^{**} is a $k[[x,y]]$-module; as such N^{**} is free say $N^{**} \cong k[[x,y]]^n$. So by the theorem of Krull-Schmidt [Swan] M is a direct sum of summands of $k[[x,y]]$. □

1.4. If char(k) = 0 the converse to this result is:

THEOREM (Artin-Verdier), [Auslander 86], [Esnault]

Let R be the local ring of a normal analytic surface singularity over an algebraically closed field k of char(k) = 0. If R has finite CM-representation type, then R is a quotient singularity, i.e. there exists a finite group $G \subset GL(2,k)$ such that $R = k[[x,y]]^G$.

REMARK:

If $G \subset SL(2,k)$ is a group without reflections then the multiplicity of $M \in \mathrm{Ind}(R)$ in $k[[x,y]]$ is $\mathrm{rank}_R M$. In particular we have:

$$\sum_{M \in \mathrm{Ind}(R)} (\mathrm{rank}\, M)^2 = [\, k[[x,y]] : R\,] = \mathrm{ord}\,(G).$$

c.f. [Knörrer 82].

These results characterize over an algebraically closed field k of char(k) = 0 all analytic Cohen-Macaulay rings of dimension 2, which have finite CM-representation type.
It is well-known that the hypersurfaces among them are given by quotients by $G \subset SL(2,k)$. These are precisely the simple singularities in dimension 2.

1.5. For dim R = 1 over an algebraically closed field k of char(k) = 0 all analytic rings of finite CM-representation type are classified by

THEOREM [Greuel-Knörrer]

Let R be the local ring of an isolated analytic curve singularity. R has finite CM-representation type iff there exists a simple plane curve singularity S and an embedding $j : R \longrightarrow \overline{S}$ into the normalization $\overline{S}$ of S such that $S \subset j(R) \subset \overline{S}$.

Again the hypersurfaces among them are precisely the simple plane curve singularities. By [Kijek-Steinke] the same result holds for arbitrary characteristic.

1.6. For dim $R \geq 3$ apart from the simple hypersurface singularities only two examples, due to Auslander and Reiten, of rings with finite CM-representation type are known: The cone over the Veronese-surface in $\mathbb{P}^5$ and the cone over the cubic scroll in $\mathbb{P}^4$.

2. MATRIX FACTORIZATIONS

Let (P,m) be a regular local ring, $f \in m-\{0\}$ and $R = P/(f)$ a hypersurface ring. In this section we recall from [Eisenbud] that every maximal Cohen-Macaulay R-module M has an R-free resolution of period 2, which may be obtained from a matrix factorization of f in P.

2.1. A *matrix factorization* (φ,ψ) of f is a pair of morphism $\varphi : F \longrightarrow G$, $\psi : G \longrightarrow F$ between free P-modules F,G such that

$$\varphi \circ \psi = f \cdot id_G , \quad \psi \circ \varphi = f \cdot id_F .$$

A morphism of matrix factorizations is a commutative diagramm

$$\begin{array}{ccccc} G & \xrightarrow{\psi} & F & \xrightarrow{\varphi} & G \\ \alpha \downarrow & & \beta \downarrow & & \alpha \downarrow \\ G' & \xrightarrow{\psi'} & F' & \xrightarrow{\varphi'} & G' \end{array}$$

We call (φ,ψ) *reduced*, if $\varphi(F) \subset mG$ and $\psi(G) \subset mF$. Define

$$\mathbb{M}(\varphi,\psi) := \operatorname{coker} \varphi .$$

2.2. Since f is a non-zero divisor, φ and ψ are injective, rank F = rank G and

$$0 \longrightarrow F \xrightarrow{\varphi} G \longrightarrow \mathbb{M}(\varphi,\psi) \longrightarrow 0$$

is a free resolution of $\mathbb{M}(\varphi,\psi)$ as a P-module. In particular

$$\operatorname{depth} \mathbb{M}(\varphi,\psi) = \dim P-1 = \dim R$$

by the formula of Auslander-Buchsbaum-Serre.

Multiplication by f on this complex is homotopy equivalent to zero:

$$\begin{array}{ccccccccc} 0 & \longrightarrow & F & \xrightarrow{\varphi} & G & \longrightarrow & \mathbb{M}(\varphi,\psi) & \longrightarrow & 0 \\ & & \downarrow f & \swarrow \psi & \downarrow \cdot f & & \downarrow f & & \\ 0 & \longrightarrow & F & \xrightarrow{\varphi} & G & \longrightarrow & \mathbb{M}(\varphi,\psi) & \longrightarrow & 0 \end{array}$$

so $\mathbb{M}(\varphi,\psi)$ is a maximal Cohen-Macaulay module over $R = P/(f)$.

Let $\overline{F} = F \otimes R$, $\overline{G} = G \otimes R$, $\overline{\varphi} : \overline{F} \longrightarrow \overline{G}$ and $\overline{\psi} : \overline{G} \longrightarrow \overline{F}$ denote the restrictions to R. Then

$$\mathbb{F}(\varphi,\psi) : \ \ldots \longrightarrow \overline{F} \xrightarrow{\overline{\varphi}} \overline{G} \xrightarrow{\overline{\psi}} \overline{F} \xrightarrow{\overline{\varphi}} \overline{G}$$

is an R-free resolution of $M(\varphi,\psi)$ of period 2.

2.3. THEOREM [Eisenbud]

Let $R = P/(f)$ be a hypersurface ring. Every maximal Cohen-Macaulay module comes from a matrix factorization of f. Moreover the maps

$$(\varphi,\psi) \longmapsto \mathbb{M}(\varphi,\psi), \quad (\varphi,\psi) \longmapsto \mathbb{F}(\varphi,\psi)$$

induce bijections between the sets of isomorphism classes of

(i) reduced matrix factorizations

(ii) maximal Cohen-Macaulay R-modules without free summands

(iii) minimal 2-periodic acyclic complexes of free R-modules.

PROOF

Any maximal Cohen-Macaulay module M over R has as P-module a free resolution of length 1 on which multiplication with f is homotopy-equivalent to zero. This gives a matrix factorization for M. For the second statement we note that we may choose the presentation

$0 \longrightarrow F \longrightarrow G \longrightarrow M \longrightarrow 0$ of M minimal. Then $\bar{\varphi}$ is minimal. Since coker $\bar{\varphi}$ = ker $\bar{\psi}$, $\bar{\psi}$ is minimal iff M contains no free summand. □

2.4. For any R-module N the n-th syzygy module $\Omega_R^n(N)$ for $n \geq \dim R - \operatorname{depth} N$ is maximal Cohen-Macaulay. Generalizing the construction of a matrix factorization, Eisenbud constructs a R-free resolution of N, hence $\Omega_R^n(N)$, from a P-free resolution $(\mathbb{F}, s_0)$ of N as follows:

THEOREM [Eisenbud]

There are endomorphisms s_α of degree $2\alpha-1$ of $\mathbb{F}$ as graded P-module such that

(i) s_0 is the differential of $\mathbb{F}$

(ii) $s_0 s_1 + s_1 s_0 = f\, id_{\mathbb{F}}$

(iii) $\sum_{\alpha+\beta=\gamma} s_\alpha s_\beta = 0$ for all $\gamma \geq 2$.

Let t be a variable of degree -2 and set

$$\mathbb{D} = Hom_{\text{graded } R\text{-modules}}(R[t], R).$$

Then $\mathbb{D} \otimes_P \mathbb{F}$ equipped with the differential

$$\bar{\partial} = \sum_\alpha t^\alpha \otimes s_\alpha$$

is an R-free resolution of M.

In general however the resulting complex is not minimal.

2.5. EXAMPLE

Suppose $x_1,\dots,x_r \in P$ is a regular sequence and $f = \sum_{i=1}^{r} y_i x_i \in (x_1,\dots,x_r)$. Then $N = P/(x_1,\dots,x_r)$ is an R-module. The Koszul-komplex

$$0 \longrightarrow \overset{r}{\wedge} P^r \longrightarrow \dots \longrightarrow \overset{2}{\wedge} P^r \longrightarrow P^r \longrightarrow P \longrightarrow N \longrightarrow 0$$

is a P-free resolution of N. The differential s_0 is given by contraction with $(x_1,\dots,x_r) \in (P^r)^*$. We may choose s_1 as the wedge-product with $(y_1,\dots,y_r) \in P^r$. Since $s_1^2 = 0$ no higher homotopies are needed. $\partial = s_0 + s_1$ satisfies $\partial^2 = f \cdot \mathrm{id}_{\wedge P^r}$. Hence (∂,∂) is a matrix factorization of f. With

$$\wedge P^r = \overset{even}{\wedge} P^r \oplus \overset{odd}{\wedge} P^r = \bigoplus_{i \equiv 0(2)} \overset{i}{\wedge} P^r \oplus \bigoplus_{i \equiv 1(2)} \overset{i}{\wedge} P^r$$

and

$$\partial^+ = \partial\Big|_{\overset{even}{\wedge} P^r}, \quad \partial^- = \partial\Big|_{\overset{odd}{\wedge} P^r}$$

(∂,∂) decomposes into two summands

$$(\partial,\partial) = (\partial^+,\partial^-) \oplus (\partial^-,\partial^+)$$

which present the even and odd syzygy modules $\Omega_R^n(N)$ for $n \geq r-1$.

2.6. In the 'generic' case $P = k[\![x_1,\dots,x_r, y_1,\dots,y_r]\!]$ and $f = \sum_{i=1}^{r} x_i y_i$ (i.e. f defines a A_1-singularity) (2.5) gives two non-free indecomposable maximal Cohen-Macaulay modules. There are no others cf. (Knörrer), (6. 3).

For an arbitrary regular local k-algebra (P,m) and any presentation

$$f = \sum_{i=1}^{r} g_i h_i$$

the specialization of $(\partial^\pm,\partial^\mp)$ is still a matrix factorization of f. Hence we obtain two maximal Cohen-Macaulay modules over R which we will denote with

$$M^\pm(g_1,\dots,g_r, h_1,\dots,h_r).$$

These matrix factorizations are reduced if

$$g_1,\dots,g_r, h_1,\dots,h_r \in m.$$

This approach allows for an arbitrary regular local ring the construction of infinitely many different maximal Cohen-Macaulay modules (take r larger). What we need is a criterion which guarantees that the isomorphism classes of their indecomposable summands are not all contained in a fixed finite set.

2.7. For a matrix factorization $\varphi : F \longrightarrow G$, $\psi : G \longrightarrow F$ we consider the natural map

$$F \otimes G^* \longrightarrow P, \quad x \otimes \lambda \longmapsto \lambda(\varphi(x)).$$

The image of this map is the ideal $I(\varphi) \subset P$ generated by the entries of the matrix φ. Similarly we consider $I(\psi) = \mathrm{Im}(G \otimes F^* \longrightarrow P)$.

LEMMA [Buchweitz-Greuel-Schreyer]

(i) *$I(\varphi)$ and $I(\psi)$ depend only on the isomorphism class of (φ,ψ). Hence for a reduced matrix factorization $I(\varphi)$, $I(\psi)$ are invariants of $\mathbb{M}(\varphi,\psi)$.*

(ii) *If $(\varphi,\psi) = (\varphi_1,\psi_1) \oplus (\varphi_2,\psi_2)$ then $I(\varphi) = I(\varphi_1) + I(\varphi_2)$, $I(\psi) = I(\psi_1) + I(\psi_2)$.*

(iii) *$f \in I(\varphi) \cdot I(\psi) \subset (I(\varphi)+I(\psi))^2$.*

(iv) *$I(\varphi) + I(\psi) \subset P$ is a proper ideal if neither $\mathbb{M}(\varphi,\psi)$ nor $\mathbb{M}(\psi,\varphi)$ contains a free summand.*

PROOF

For (iii) notice that f is the product of the first column of φ with the first row of ψ. □

2.8. COROLLARY [Buchweitz-Greuel-Schreyer]

If a hypersurface $R = P/(f)$ has finite (countable) CM-representation type, then there are at most finitely (countably) many ideals $I \subset P$ with $f \in I^2$.

PROOF

For every ideal $I = (h_1,\ldots,h_r) \subset m$ with $f \in I^2$ we may write

$$f = \sum_{i=1}^{r} g_i h_i \quad \text{with} \quad g_i \in I.$$

Then (2.6) gives a reduced matrix factorization (φ,ψ) with $I = I(\varphi) + I(\psi)$. By (2.4) and (2.7) at most finitely (countable) many ideals can occur in this way if R has finite (countable) CM-representation type. □

3. KNÖRRER'S PERIODICITY RESULT

3.1. Let $P = k[\![x_0,\dots,x_n]\!]$ be a complete regular local k-algebra over a field k with $char(k) \neq 2$. For a hypersurface $R = P/(f)$ we consider the double covering

$$R' = P[\![z]\!]/(f+z^2)$$

of P branched along (f). MCM's over R and R' are closely related. For M' a MCM over R' the restriction

$$Rest(M') = M'/z.M'$$

is a MCM over R. Conversely any MCM M over R can be viewed as R'-module and the first syzygy-module $\Omega_{R'}(M)$ is a MCM over R':

$$0 \longrightarrow \Omega_{R'}(M) \longrightarrow G' \longrightarrow M \longrightarrow 0$$

with G' a free R'-module of rank $G' = \dim M/mM$.

3.2. THEOREM [Knörrer 86]

Suppose $char(P/m) \neq 2$; let M resp. M' be MCM's over R resp. R'. Then:

(a) $Rest(\Omega_{R'}(M)) \cong M \oplus \Omega_R(M)$

(b) $\Omega_{R'}(Rest(M')) \cong M' \oplus \Omega_{R'}(M')$.

PROOF

It suffices to prove the statement for MCM's M resp. M' without free summands. Suppose (φ,ψ) is a reduced matrix factorization with $M = \mathbb{M}(\varphi,\psi)$. Then

$$\Omega_{R'}(M) = \mathbb{M}\left(\begin{pmatrix} z & \psi \\ \varphi & -z \end{pmatrix}, \begin{pmatrix} z & \psi \\ \varphi & -z \end{pmatrix}\right)$$

and

$$\begin{aligned} Rest(\Omega_{R'}(M)) &= \mathbb{M}\left(\begin{pmatrix} 0 & \psi \\ \varphi & 0 \end{pmatrix}, \begin{pmatrix} 0 & \psi \\ \varphi & 0 \end{pmatrix}\right) \\ &\cong \mathbb{M}(\varphi,\psi) \oplus \mathbb{M}(\psi,\varphi) \\ &= M \oplus \Omega_R M. \end{aligned}$$

For (b) we note that we may regard M' as a free P-module P^r with an P-linear action $z : P^r \longrightarrow P^r$ given by a matrix φ, which satisfies $\varphi^2 = -f.id_{P^r}$. So

$$M' = \mathbb{M}(z-\varphi,\ z+\varphi).$$

This matrix factorization is reduced (i.e. φ is minimal) if M' contains no free summand. We obtain

$$\text{Rest}(M') = \mathbb{M}(-\varphi,\varphi)$$

and

$$\Omega_{R'}(\text{Rest}(M')) = \mathbb{M}\left(\begin{pmatrix} z & \varphi \\ -\varphi & -z \end{pmatrix}, \begin{pmatrix} z & \varphi \\ -\varphi & -z \end{pmatrix}\right)$$

$$\cong \mathbb{M}(z-\varphi, z+\varphi) \oplus \mathbb{M}(z+\varphi, z-\varphi)$$

since 2 is invertible in P. □

3.3. COROLLARY [Knörrer 86]

R has finite (countable) CM-representation type iff R' has finite (countable) CM-representation type.

PROOF

By the theorem of Krull-Schmidt [Swan] and (3.2) the sets

$\{(M,M') \in \text{Ind}(R) \times \text{Ind}(R') \mid M \text{ is a summand of } \text{Rest}(M')\}$

and

$\{(M,M') \in \text{Ind}(R) \times \text{Ind}(R') \mid M' \text{ is a summand of } \Omega_{R'}(M)\}$

are equal. We denote this correspondence by T. Both projections $p_1 : T \to \text{Ind}(R)$, $p_2 : T \to \text{Ind}(R')$ are surjective and have finite fibers. Hence if any one of the sets Ind(R), T, Ind(R') is finite (countable) all three are finite (countable). □

3.4. COROLLARY [Knörrer 86]

For char(k) ≠ 2, simple singularities have finite CM-representation type.

PROOF

For char(k) = 0 we know from (1.3) that simple surface singularities have finite CM-representation type. If char(k) ≠ 2 we know it for simple curve singularities. By (3.3) it follows for all dimensions. □

3.5. If k = P/m is algebraically closed one can describe the correspondence T (3.3) in more detail.

LEMMA [Knörrer 86]

$k = P/m$ algebraically closed, char$(k) \neq 2$. Let M, M' be MCM over R and R'. The following are equivalent

(a) $M \cong \Omega_R M$

(b) $M \cong \mathbb{M}(\varphi,\varphi)$

(c) $M \cong \mathrm{Rest}(M')$ *for a suitable* M'

(d) $\Omega_{R'}(M)$ *is decomposable*

Similarly

(a') $M' \cong \Omega_{R'}(M')$

(b') $M \cong \mathbb{M}\left(\begin{pmatrix} z & \psi \\ \varphi & -z \end{pmatrix}, \begin{pmatrix} z & \psi \\ \varphi & -z \end{pmatrix}\right)$

(c') $M' \cong \Omega_{R'}(M)$ *for a suitable* M

(d') $\mathrm{Rest}(M')$ *is decomposable.* □

3.6. THEOREM [Knörrer 86]

Suppose (P/m) is an algebraically closed field of characteristic $\neq 2$. There is a bijection between maximal Cohen-Macaulay modules over $P/(f)$ and $P[\![z,w]\!]/(f+z^2+w^2)$.

PROOF

By (3.2), (3.5)

$$\mathbb{M}(\varphi,\psi) \longmapsto \mathbb{M}\left(\left(\begin{array}{c|c} z-iw & \psi \\ \varphi & z+iw \end{array}\right)\left(\begin{array}{c|c} z+iw & \psi \\ \varphi & -z+iw \end{array}\right)\right)$$

induces a bijection. □

4. IDEALS I WITH $f \in I^2$

4.1. THEOREM [Knörrer 86, Buchweitz-Greuel-Schreyer]

For an analytic hypersurface $R = P/(f)$ over an algebraically closed field of characteristic $\neq 2$ the following are equivalent:

(i) *R is simple*

(ii) *R has finite CM-representation type*

(iii) *There are at most finitely many ideals $I \subset P$ with $f \in I^2$.*

PROOF

(3.4) gives the first implication. (2.8) the second. We prove (iii) $\Rightarrow$ (i) by induction on $n = \dim R$. □

4.2. If $n = 1$ we use:

LEMMA

$R = k[\![x_0,x_1]\!]/(f)$ is simple iff for all $x,y \in m$ the following holds:

(i) $f \notin (x,y)^4$

(ii) $f \notin (x^3,x^2y^2,xy^4,y^6)$

(iii) $f \notin (x^2)$

PROOF

All curves which satisfy (i), (ii), (iii) are classified in [Barth-Peters-Van de Ven] pp. 61-65. They are simple. □

Hence for every non-simple curve we have with

(i) $I_\lambda = (\lambda_0x+\lambda_1y) + (x,y)^2$; $\lambda = (\lambda_0:\lambda_1) \in \mathbb{P}^1_k$

(ii) $I_\lambda = (x+\lambda y^2, y^3)$ $\qquad \lambda \in \mathbb{A}^1_k$

or

(iii) $I_\lambda = (x,y^\lambda)$ $\qquad \lambda \in \mathbb{N}$

an infinite family of ideals I_λ with $f \in I_\lambda^2$.

4.3. For $\dim R > 1$, we proceed by induction. If mult $R = 2$ we may write

$$f = g(x_0,\ldots,x_{n-1}) + x_n^2$$

by the Weierstraß preparation theorem and Tschirnhausen transformation. Then with f also g is non-simple. By induction hypothesis there are infinitely many ideals $I_\lambda \subset k[\![x_0,\ldots,x_{n-1}]\!]$ with $g \in I_\lambda^2$. Then $I'_\lambda = (x_n)+I_\lambda \cdot P$ gives an infinite family of ideals with $f \in I_\lambda^2$.
If $\text{mult}(R) \geq 3$ we write

$$f = f_3 + \text{higher order terms}$$

with f_3 homogeneous of degree 3. For every point $\lambda \in C = \{(\lambda_0:\ldots:\lambda_n) \in \mathbb{P}^n \mid f_3(\lambda) = 0\}$ we consider the ideal

$$I_\lambda = ((\lambda_i x_j - \lambda_j x_i)_{i,j}) + m^2.$$

Then $f \in I_\lambda^2$. Since $\dim C \geq 1$, C contains infinitely many points and we have the desired result. □

4.4. Analysing the proof of (4.1) we find that in case $k = \mathbb{C}$ we actually constructed uncountable infinitely many MCM's unless

$$f \cong g(x_0,x_1) + x_2^2 + \ldots + x_n^2$$

where g satisfies (4.2.(i),(ii)). If g violates (4.2.(iii)), g is given by

$$g(x_0,x_1) = x_1^2 \quad \text{or} \quad g(x_0,x_1) = x_0 x_1^2.$$

This proves one direction of the following:

<u>THEOREM</u> [Buchweitz-Greuel-Schreyer]

Let R be a complex analytic hypersurface. R has countable infinite CM-represenation type if $R \cong \mathbb{C}[\![x_0,\ldots,x_n]\!]/(f)$ *with*

$$(A_\infty) : f = x_1^2 + \ldots + x_n^2$$

or

$$(D_\infty) : f = x_0 x_1^2 + x_2^2 + \ldots + x_n^2$$

The singularities A_∞ and D_∞ are singular along the line defined by $(x_1,\ldots,x_n)$. They are basic for the study of non-isolated singularities, cf. [Siersma].

5. AUSLANDER-REITEN-SEQUENCES

5.1. Let R be a ring. A map $f : A \longrightarrow B$ of indecomposable R-modules is *irreducible*, if it is not an isomorphism and for any factorization

$$f : A \longrightarrow B, \qquad A \xrightarrow{g} C \xrightarrow{h} B$$

either g is a splittable monomorphism or h is a splittable epimorphism. A short exact sequence

$$0 \longrightarrow A \xrightarrow{f} B \xrightarrow{g} C \longrightarrow 0$$

of R-modules is *left (right) almost split* if it does not split, but for any map $A \longrightarrow A'$ $(C' \longrightarrow C)$ which is not a splittable monomorphism (epimorphism) the pushout (pullback) sequence splits:

$$\begin{array}{ccccccccc} & & & & & & C' & & \\ & & & \swarrow & & & \downarrow & & \\ 0 & \longrightarrow & A & \longrightarrow & B & \longrightarrow & C & \longrightarrow & 0 \\ & & \downarrow & \swarrow & & & & & \\ & & A' & & & & & & \end{array}$$

If A and C are indecomposable then an exact sequence $0 \longrightarrow A \longrightarrow B \longrightarrow C \longrightarrow 0$ is left almost split iff it is right almost. We call such a sequence an *Auslander-Reiten-sequence*. If $B = \bigoplus_i B_i$ is a decomposition into indecomposable summands then the components $f_i : A \longrightarrow B_i$, $(g_i : B_i \longrightarrow C)$ are precisely the irreducible maps with source A (target C).

The set of indecomposable R-modules together with an arrow for each isomorphism class of irreducible maps, from a directed graph, which we call the *Auslander-Reiten quiver* of R.

5.2. THEOREM [Auslander 84]

Let R be a complete local Cohen-Macaulay ring. The following assertions are equivalent:

(i) R has an isolated singularity

(ii) For each $C \in Ind(R)$, $C \ncong R$, there exists a right almost split sequence

(iii) For each $A \in Ind(R)$, $A \ncong \omega_R$, there exists a left almost split sequence.

ω_R denotes the dualizing module.

COROLLARY [Auslander 84]

If R has finite CM-representation type, then R has an isolated singularity.

REMARK

The theorem and its corollary hold more generally for R a not necessarily commutative P-algebra which is finitely generated and free as a P-module, where P denotes a complete regular local ring, and the categorie of left R-modules which are free as P-modules. For a notion of isolated singularity in this case see [Auslander 84].

6. AUSLANDER-REITEN QUIVER OF A-D-SINGULARITIES

In this section we briefly describe the Auslander-Reiten sequences for A-D-singularities cf. [Auslander-Reiten], [Dieterich-Wiedemann], [Knörrer 86], [Buchweitz-Greuel-Schreyer].

6.1. We first treat the case of dimension 1; $R = k[\![x,y]\!]/(f)$, k an algebraically closed field of $\mathrm{char}(k) \neq 2$. Set $y^0 := 1$, $y^\infty := 0$ and $\infty \pm k := \infty$.

$$A_n : f = x^2 + y^{n+1} \quad \text{for} \quad n = 1,2,\dots,\infty.$$

Define matrix factorizations (φ_k,φ_k) hence Cohen-Macaulay modules $M_k := \mathbb{M}(\varphi_k,\varphi_k)$ by

$$\varphi_k := \begin{pmatrix} x & y^k \\ y^{n+1-k} & -x \end{pmatrix}$$

for all k with $0 \leq k \leq n+1$ and extensions

$$\Phi_k := \left(\begin{array}{c|c} \varphi_k & \varepsilon_k \\ \hline 0 & \varphi_k \end{array}\right) \quad \text{with} \quad \varepsilon_k := \begin{pmatrix} 0 & y^{k-1} \\ -y^{n-k} & 0 \end{pmatrix}$$

for all k with $0 < k < n+1$.
Since

$$(\Phi_k, \Phi_k) \cong (\varphi_{k-1}, \varphi_{k-1}) \oplus (\varphi_{k+1}, \varphi_{k+1})$$

we obtain short exact sequences

$$\alpha_k : 0 \longrightarrow M_k \longrightarrow M_{k-1} \oplus M_{k+1} \longrightarrow M_k \longrightarrow 0.$$

Roughly speaking the M_k's are all the indecomposable maximal Cohen-Macaulay modules and the α_k's are all the Auslander-Reiten sequences.
Notice that

$$M_0 \cong R,$$

$$M_k \cong M_{n+1-k} \quad \text{if} \quad n < \infty \quad \text{and}$$

$$M_{(n+1)/2} \cong M^{+}_{(n+1)/2} \oplus M^{-}_{(n+1)/2} \quad \text{if} \quad n \equiv 1(2)$$

$$\text{with} \quad M_{(n\pm1)/2} \cong \mathbb{M}(x \pm iy^{(n+1)/2},\ x \mp iy^{(n+1)/2}),$$

$$M_\infty = 2\, M_{\infty/2} \quad \text{with} \quad M_{\infty/2} := \mathbb{M}(x,x).$$

The Auslander-Reiten quiver of A_n is given by

$$M_0 \rightleftarrows M_1 \rightleftarrows \ldots \rightleftarrows M_{(n-2)/2} \rightleftarrows M_{n/2} \quad , \quad \text{if } n \equiv 0(2),$$

$$M_0 \rightleftarrows M_1 \rightleftarrows \ldots \rightleftarrows M_{(n-1)/2} \begin{array}{l} \rightleftarrows M^{+}_{(n+1)/2} \\ \rightleftarrows M^{-}_{(n+1)/2} \end{array} \quad , \quad \text{if } n \equiv 1(2), \text{ or}$$

$$M_0 \rightleftarrows M_1 \rightleftarrows \ldots \rightleftarrows M_k \rightleftarrows M_{k+1} \rightleftarrows \ldots\ldots M_{\infty/2} \circlearrowleft, \quad \text{if } n = \infty.$$

There is no Auslander-Reiten sequence for $M_{\infty/2}$, c.f. (5.2.). □

<u>$D_n : f = x^2y + y^{n-1}$ for $n = 4,5,\ldots,\infty$</u>

Define matrix factorizations

$$\varphi_k = \begin{pmatrix} x & y^k \\ y^{n-2-k} & -x \end{pmatrix} \qquad \psi_k = \begin{pmatrix} xy & y^{k+1} \\ y^{n-1-k} & -xy \end{pmatrix}$$

$$\theta_k = \begin{pmatrix} x & y^k \\ y^{n-1-k} & -xy \end{pmatrix} \qquad \xi_k = \begin{pmatrix} xy & y^k \\ y^{n-1-k} & -x \end{pmatrix}$$

for $0 \le k \le n-2$, hence Cohen-Macaulay modules

$$M_k^+ = \mathbb{M}(\varphi_k, \psi_k), \quad M_k^- = \mathbb{M}(\psi_k, \varphi_k)$$

$$N_k^+ = \mathbb{M}(\theta_k, \xi_k), \quad N_k^- = \mathbb{M}(\xi_k, \theta_k).$$

The extensions

$$\Phi_k = \left(\begin{array}{c|c} \varphi_k & \varepsilon_k \\ \hline 0 & \psi_k \end{array}\right) \qquad \Psi_k = \left(\begin{array}{c|c} \psi_k & \varepsilon_k \\ \hline 0 & \varphi_k \end{array}\right), \quad 0 \le k \le n-2$$

$$\Theta_k = \left(\begin{array}{c|c} \theta_k & \varepsilon_k \\ \hline 0 & \xi_k \end{array}\right) \qquad \Xi_k = \left(\begin{array}{c|c} \xi_k & \varepsilon_{k-1} \\ \hline 0 & \theta_k \end{array}\right), \quad 0 < k < n-2$$

with

$$\varepsilon_k = \begin{pmatrix} 0 & y^k \\ -y^{n-2-k} & 0 \end{pmatrix}$$

give short exact sequences

$$\alpha_k : 0 \longrightarrow M_k^{\pm} \longrightarrow N_{k+1}^{\pm} \oplus N_k^{\mp} \longrightarrow M_k^{\mp} \longrightarrow 0$$

$$\beta_k : 0 \longrightarrow N_k^{\pm} \longrightarrow M_k^{\pm} \oplus M_{k-1}^{\mp} \longrightarrow N_k^{\mp} \longrightarrow 0 .$$

Roughly speaking, the $M_k^{\pm}$, $N_k^{\pm}$ are all the indecomposable maximal Cohen-Macaulay modules and the $\alpha_k^{\pm}$, $\beta_k^{\pm}$ all the Auslander-Reiten sequences. More precisely:

$$N_0^{\pm} \cong R$$

$$M_0^+ \cong \mathbb{M}(x^2+y^{n-2}, y), \quad M_0^- \cong R \oplus \widetilde{M}_0^-, \text{ with } \widetilde{M}_0^- \cong \mathbb{M}(y, x^2+y^{n-2})$$

$$M_k^{\pm} \cong M_{n-2-k}^{\pm}, \quad N_k^{\pm} \cong N_{n-1-k}^{\pm} \quad \text{if} \quad n < \infty ,$$

and in case $n \equiv 0(2)$

$$M_{(n-2)/2}^+ = M_{(n-2)/2}^{++} \oplus M_{(n-2)/2}^{+-}, \quad M_{(n-2)/2}^- = M_{(n-2)/2}^{--} \oplus M_{(n-2)/2}^{-+}$$

with

$$M^{+\pm}_{(n-2)/2} \cong \mathbb{M}\,(x \pm iy^{(n-2)/2},\ xy \mp iy^{n/2})$$

$$M^{-\pm}_{(n-2)/2} \cong \mathbb{M}\,(xy \mp iy^{n/2},\ x \pm iy^{(n-2)/2})$$

in case $n \equiv 1(2)$

$$N_{(n-1)/2} := N^{+}_{(n-1)/2} = N^{-}_{(n-1)/2}.$$

If $n = \infty$, then

$$M^{\pm} = 2 \cdot M^{\pm}_{\infty/2}$$

with

$$M^{+}_{\infty/2} = \mathbb{M}\,(x, xy), \quad M^{-}_{\infty/2} = \mathbb{M}\,(xy, x).$$

The Auslander-Reiten-quivers are given by

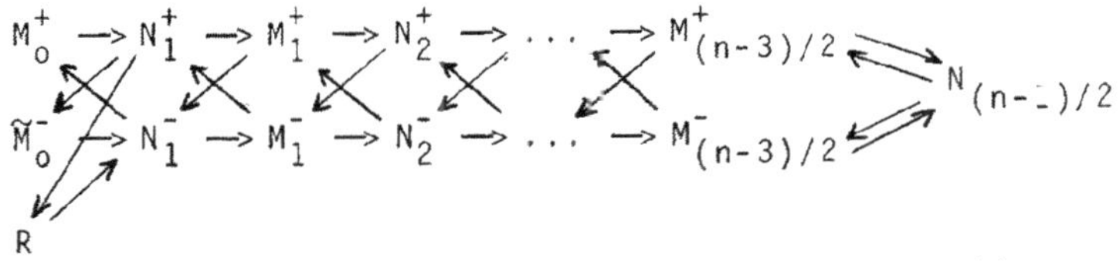

if $n \equiv 1(2)$

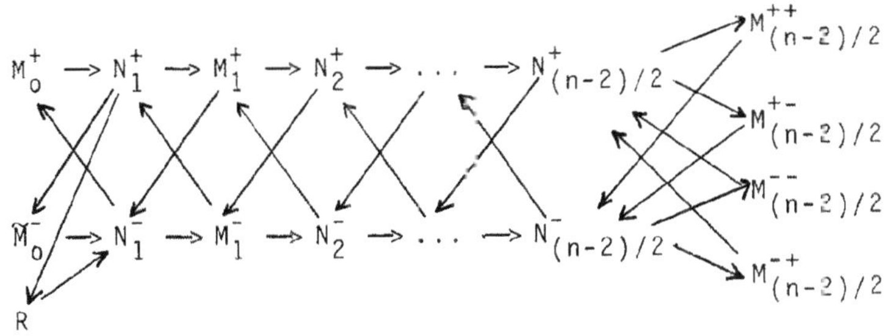

if $n \equiv 0(2)$

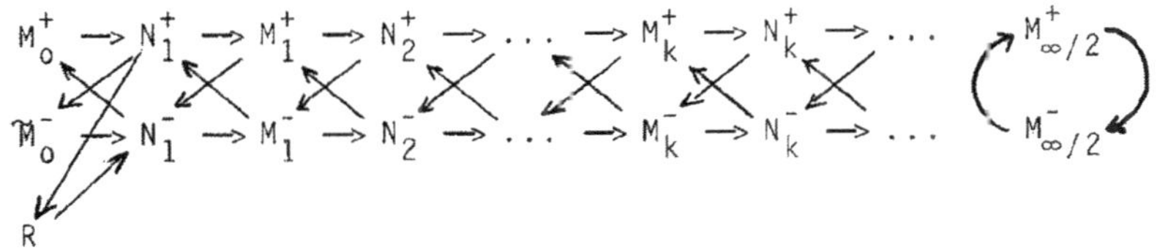

if $n = \infty$.

There are no Auslander-Reiten sequences for $M^{\pm}_{\infty/2}$ c.f. (5.2).

6.2. In case of dimension 2, i.e $R' = k[\![x,y,z]\!]/(z^2+f(x,y))$ the maximal Cohen-Macaulay modules and similarly the Auslander-Reiten sequences are given by (3.5).

$\underline{A_n : z^2 + x^2 + y^{n+1} \text{ for } n = 1,2,\dots,\infty}$.

$$\Omega_{R'}(M_k) \cong M_k^+ \oplus M_k^-$$

and similarly

$$\Omega_{R'}(\alpha_k) \cong \alpha_k^+ \oplus \alpha_k^-.$$

Note that: $M_0^+ \cong M_0^- \cong R$ and

$$M_k^+ \cong M_{n+1-k}^-.$$

The Auslander-Reiten Quiver is given by

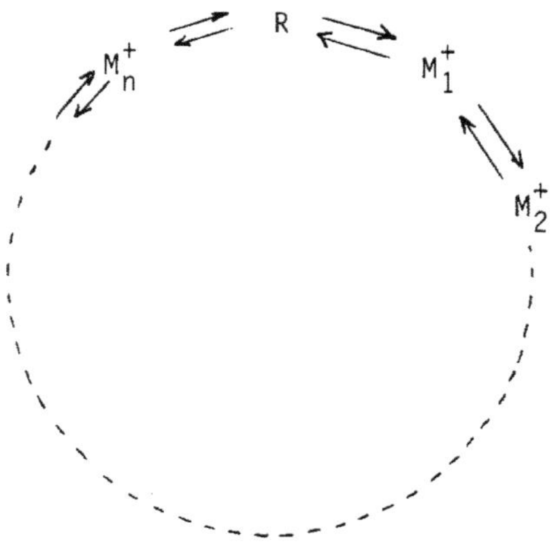

, if $n < \infty$

and

$$\circlearrowleft M^-_{\infty/2} \;\cdots\; \rightleftarrows M_k \rightleftarrows \cdots \rightleftarrows M_1^- \rightleftarrows R \rightleftarrows M_1^+ \rightleftarrows M_2^+ \rightleftarrows \cdots \rightleftarrows M_k^+ \rightleftarrows \;\cdots\; M^+_{\infty/2} \circlearrowright$$

if $n = \infty$. □

$\underline{D_n : z^2 + x^2 y + y^{n-1} \text{ for } n = 4,5,\dots,\infty.}$

$$M_k := \Omega_{R'}(M_k^+) = \Omega_{R'}(M_k^-), \quad 0 \le k \le n-2$$

$$N_k := \Omega_{R'}(N_k^+) = \Omega_{R'}(N_k^-), \quad 1 \le k \le n-1$$

and

$$\Omega_{R'}(N_{(n-1)/2}) = N^+_{(n-1)/2} \oplus N^-_{(n-1)/2} .$$

Note that: $M_k \cong M_{n-2-k}$, $N_k \cong N_{n-1-k}$. Similarly for the $\alpha_k^\pm$, $\beta_k^\pm$. We obtain the Auslander-Reiten-quivers

$$\begin{array}{l} M_0 \rightleftarrows N_1 \rightleftarrows M_1 \rightleftarrows N_2 \rightleftarrows \dots \rightleftarrows M_{n-3/2} \begin{array}{l} \nearrow\!\!\swarrow N^{+}_{(n-1)/2} \\ \searrow\!\!\nwarrow N^{-}_{(n-1)/2} \end{array} \quad , \quad \text{if} \quad n \equiv 1(2) \\ \quad\;\; \downarrow\uparrow \\ \quad\;\;\, R \end{array}$$

$$\begin{array}{l} M_0 \rightleftarrows N_1 \rightleftarrows M_1 \rightleftarrows N_2 \rightleftarrows \dots \rightleftarrows N_{(n-2)/2} \begin{array}{l} \nearrow\!\!\swarrow M^{+}_{(n-2)/2} \\ \searrow\!\!\nwarrow M^{-}_{(n-2)/2} \end{array} \quad , \quad \text{if} \quad n \equiv 0(2), \text{ or} \\ \quad\;\; \downarrow\uparrow \\ \quad\;\;\, R \end{array}$$

$$\begin{array}{l} M_0 \rightleftarrows N_1 \rightleftarrows M_1 \rightleftarrows N_2 \rightleftarrows \dots \rightleftarrows M_k \rightleftarrows N_{k+1} \rightleftarrows M_{k+1} \rightleftarrows \dots \quad M_{\infty/2} \circlearrowleft, \quad \text{if} \quad n = \infty. \\ \quad\;\; \downarrow\uparrow \\ \quad\;\;\, R \end{array}$$

6.3. For higher dimensions $r \geq 3$ the results are similar according to Knörrer's periodicity theorem. The Auslander-Reiten quivers are nearly the same as in dimension 1 respectively 2 if $r = 2k-1$ respectively $r = 2k$. The only difference is that the projective module is replaced by a direct sum of k copies of the ring.

6.4. For dimension 2 the Auslander-Reiten quivers of A_n and D_n look like extended Dynkin-diagramms of type A_n resp. D_n, with the extended vertex corresponding to R. An explanation of this observation due to McKay is given in [Gonzalez-Sprinberg-Verdier], see also [Esnault-Knörrer]. Even in case $n = \infty$ one can discover a diagram of type A_∞ resp. D_∞.

7. OPEN PROBLEMS

7.1. CLASSIFICATION OF FINITE CM-REPRESENTATION TYPE SINGULARITIES (in the complex analytic case)

A Gorenstein ring has finite CM-representation type, iff it is a simple hypersurface singularity by (1.2) and (0.2). The examples of non-hypersurface singularities, which are known to have finite CM-representation type are the following:

dimension 1 [Greuel-Knörrer] , (1.5)

$$D^{s}_{n+1} \; : \; \mathbb{C}[[x,y,z]]/\langle rg \begin{pmatrix} x & y & 0 \\ y & x^n & z \end{pmatrix} \leq 1 \rangle \quad ; \text{ for } n = 1,2,3,\dots$$

$$E^{s}_{2+2n} \; : \; \mathbb{C}[[x,y,z]]/\langle rg \begin{pmatrix} x & y & z \\ y & z & x^n \end{pmatrix} \leq 1 \rangle \quad ; \text{ for } n = 2,3.$$

dimension 2 [Herzog, Artin-Verdier, Esnault, Auslander 84]

Quotient singularities $\mathbb{C}[\![x,y]\!]^G$, $G \subset GL(2,\mathbb{C})$ a finite group without reflections and not contained in $SL(2,\mathbb{C})$. For equations see [Riemenschneider].

dimension 3 [Auslander-Reiten 86]

The cone over the cubic scroll in $\mathbb{P}^4$:

$$\mathbb{C}[\![x_0,x_1,x_2,y_0,y_1]\!]/<rg\begin{pmatrix} x_0 & x_1 & y_0 \\ x_1 & x_2 & y_1 \end{pmatrix} \leq 1>,$$

the cone over the Veronese surface in $\mathbb{P}^5$

$$\mathbb{C}[\![x_0,\ldots,x_5]\!]/<rg\begin{pmatrix} x_0 & x_1 & x_2 \\ x_1 & x_3 & x_4 \\ x_2 & x_4 & x_5 \end{pmatrix} \leq 1>.$$

No example is known for dimension ≥ 4.

Notice that all these examples have minimal multiplicity for their embedding dimension:

$$\text{mult}(R) = \dim_{\mathbb{C}}(m/m^2) - \dim R + 1.$$

Question: Is this list complete?

REMARK

Auslander and Reiten prove that the cones over smooth irreducible projective varieties of minimal multiplicity, which have finite CM-representation type are among this list: The cone over a rational normal scroll has infinite CM-representation type unless it is a rational normal curve $\mathbb{P}^1 \subset \mathbb{P}^r$, the quadric hypersurface in $\mathbb{P}^3$ or the cubic scroll in $\mathbb{P}^4$.

7.2. CLASSIFICATION OF COUNTABLE CM-REPRESENTATION TYPE SINGULARITIES (in the complex analytic case)

We first give a few examples of non-hypersurface singularities with countable CM-representation type.

7.2.1. For dimension 1 we note

PROPOSITION

D^s_∞ : $\mathbb{C}[\![x,y,z]\!]/<rg\begin{pmatrix} x & y & 0 \\ y & 0 & z \end{pmatrix} \leq 1>$ has countable CM-representation type.

PROOF

D_∞^s has not finite CM-representation since it is not an isolated singularity (5.2). To establish countable CM-representation type we notice that $D_\infty = \mathbb{C}[[u,v]]/(u^2v)$ is a subring via $u = x$, $v = y+z$. It suffices to prove that a torsion free D_∞^s-module M decomposes iff it decomposes as D_∞-module. This follows from

$$\mathrm{Hom}_{D_\infty}(M,M) = \mathrm{Hom}_{D_\infty^s}(M,M),$$

i.e. if the diagram

$$\begin{array}{ccc} M & \xrightarrow{\varphi} & M \\ {\scriptstyle y-z}\downarrow & & \downarrow{\scriptstyle y-z} \\ M & \xrightarrow[\varphi]{} & M \end{array}$$

commutes for every $\varphi \in \mathrm{Hom}_{D_\infty}(M,M)$. Since M is torsion free and $(x-y+z) = u+v$ is a non-zero divisor

$$(y-z)(x+y+z) \equiv xy-z^2 \equiv (y+z)(x-(y+z)) = v(u-v)$$

implies the desired result. □

7.2.2. For dimension 2 we note:

PROPOSITION

Quotients of A_∞ and D_∞ by a finite group have at most countable CM-representation type.

PROOF

As Herzog's theorem (1.3). □

EXAMPLES

a) A_∞ : $\mathbb{C}[[u,v,w]]/(uw)$ and $\mathbb{Z}_r$-action induced by $(u,v,w) \longmapsto (\xi_r u, \xi_r v, w)$, $\xi_r = e^{2\pi i/r}$. The invariant ring is generated by

$$u^r, u^{r-1}v, \ldots, v^r, w$$

hence isomorphic to

$$\mathbb{C}[[x_0,\ldots,x_r,w]]/\langle rg\begin{pmatrix} x_0 & x_1 & \cdots & x_{r-1} & 0 \\ x_1 & x_2 & \cdots & x_r & w \end{pmatrix} \leq 1 \rangle .$$

b) D_∞ : $\mathbb{C}[[u,v,w]]/(u^2+v^2w)$ and $\mathbb{Z}_3$-action induced by $(u,v,w) \longmapsto (u,\xi_3 v,\xi_3 w)$. The invariant ring is generated by

$$u, v^3, vw^2, w^3$$

hence isomorphic to

$$\mathbb{C}[[u,x,y,z]]/\langle rg \begin{pmatrix} x & u^2 & y \\ u^2 & y & z \end{pmatrix} \leq 1 \rangle .$$

QUESTION

Do all countable CM-representation type singularities of dimension 2 arise in this way?

7.2.3. CONJECTURE

Countable CM-representation type implies 1-dimensional singular locus.

7.2.4. QUESTION

Are all countable CM-representation type singularities "natural limits" of a "series of singularities" of finite CM-representation type?

For the example 7.2.1 we can consider the series

$$D_n^s : \mathbb{C}[[x,y,z]]/\langle rg \begin{pmatrix} x & y & 0 \\ y & x^n & z \end{pmatrix} \leq 1 \rangle , \quad n \geq 1$$

for the examples 7.2.2 we may take

a) $$\mathbb{C}[[x_0,\ldots,x_r,w]]/\langle rg \begin{pmatrix} x_0 \cdots x_{r-1} & x_r^n \\ x_1 \cdots x_r & w \end{pmatrix} \leq 1 \rangle$$

b) $$\mathbb{C}[[u,x,y,z]]/\langle rg \begin{pmatrix} x & u^2+z^n & y \\ u^2 & y & z \end{pmatrix} \leq 1 \rangle .$$

Their resolution graphs are:

a) 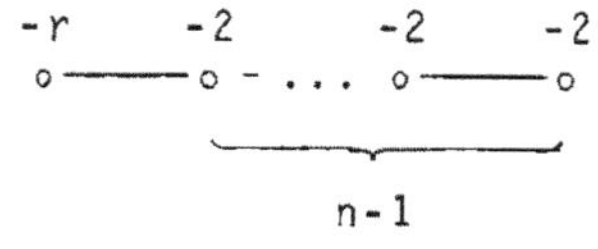

respectively

b) 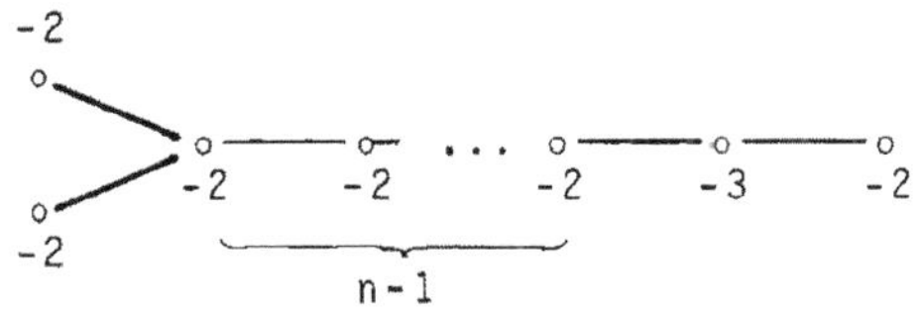

c.f. [Riemenschneider].

PROBLEM

Give a precise meaning to the phrase "natural limit of a series of singularities" and prove that A_∞ and D_∞ have countable CM-representation type from $\lim_{n\to\infty} A_n = A_\infty$ resp. $\lim_{n\to\infty} D_n = D_\infty$.

7.3. FINITE CM-REPRESENTATION TYPE FOR ARBITRARY LOCAL COHEN-MACAULAY RINGS

Let (R,m) be an arbitrary local Cohen-Macaulay k-Algebra.

CONJECTURE

a) R has finite CM-representation type iff the m-adic completion $\hat{R}$ has finite CM-representation type.

b) R has finite CM-representation type iff $R \otimes_k \bar{k}$ has finite CM-representation type, where $\bar{k}$ denotes the algebraic closure of k.

One might ask similar questions in the mixed characteristic case.

7.4. CONSTRUCTION OF MODULI-SPACES FOR MAXIMAL COHEN-MACAULAY MODULES

7.5. CLASSIFICATION OF TAME CM-REPRESENTATION TYPE

I call a CM-singularity of tame CM-representation type, if the dimensions of the components of the moduli spaces for *indecomposable* maximal Cohen-Macaulay modules is bounded.

7.5.1. QUESTION

Is the bound, if it exists, necessarily 1 ?

Dieterich has shown that $\mathbb{C}[[x,y]]/x(x+y)^2(x+ay^2)$ is tame. Schappert proved that a plane curve singularity has 1-dimensional families of torsion free rank 1-modules, iff the singularity is strictly unimodular in the sense of C.T.C. Wall, [Schappert].

7.6. BEHAVIOR UNDER DEFORMATIONS

7.6.1. <u>CONJECTURE</u> [Knörrer]

Finite CM-representation type is an open property in a flat family of rings.

<u>REMARK</u>

In the complex analytic case a confirmative answer to question 7.1 implies the conjecture, since it is known that the singularities of this list deform only in each other. For the hardest part, "quotient singularities deform into quotient singularities" see [Esnault - Viehweg] .

7.6.2. One can pose similar questions for countable (resp. tame) CM-representation type.

7.6.3. Recently Knörrer has shown that the dimension of the moduli-space of torsion free sheaves of a fixed rank on an isolated curve singularity behave semicontinuous under deformations of the curve.

7.6.4. <u>QUESTION</u>

How does the Auslander-Reiten quiver behave under deformations?

7.7. POSSIBLE RANKS FOR INDECOMPOSABLE MAXIMAL COHEN-MACAULAY MODULES

Dieterich proved that for an isolated singularity which has not finite CM-representation type infinitely many ranks are achieved by indecomposable maximal Cohen-Macaulay modules. For hypersurface singularities the rank is bounded from below in terms of the codimension of the singular locus:

7.7.1. <u>CONJECTURE</u> [Buchweitz-Greuel-Schreyer]

Let $R = P/(f)$ be a n-dimensional isolated hypersurface singularity. A maximal Cohen-Macaulay module M is free if

$$\operatorname{rank} M < 2^{[n/2]-2}.$$

REFERENCES

ARNOLD, V.I. (1981): Singularity Theory - Selected Papers; London Math. Soc. Lecture Notes Series 53, Cambridge University Press

ARTIN, M. and VERDIER, J.-L. (1985): Reflexive Modules Over Rational Double Points; Math. Ann. 270, 79-82

AUSLANDER, M. (1986): Almost Split Sequences and Rational Singularities; Trans AMS 293, 511-531

AUSLANDER, M. (1984): Isolated Singularities and Existence of almost Split Sequences; Proceedings of the Fourth International Conference on Representations of Algebras; Ottawa, Springer Lecture Notes 1178, 194-242

AUSLANDER, M. and REITEN, I. (1986): The Cohen-Macaulay type of Cohen-Macaulay rings; Preprint, University of Trondheim

AUSLANDER, M. and REITEN, I. (1986): Almost Split Sequences for Rational Double Points; Preprint, University of Trondheim

BARTH, W., PETERS, C. and VAN DE VEN, A. (1984): Compact Complex Surfaces; Erg. d. Math. Band 4, Springer Verlag

BUCHWEITZ, R.-O., GREUEL, G.-M. and SCHREYER, F.-O. (1986): Maximal Cohen-Macaulay Modules on Hypersurfaces II; to appear in Inv. Math.

DIETERICH, E. (1985): Classification of the Indecomposable Representations of the Cyclic Group of Order Three in a Complete Discrete Valuation Ring of Ramification Degree Four; Preprint Universität Bielefeld

DIETERICH, E. (1986): Reduction of Isolated Singularities; Preprint Universität Bielefeld

DIETERICH, E. and WIEDEMANN, A. (1986): The Auslander-Reiten-Quiver of a Simple Curve Singularity; Trans AMS 294, 455-475

DURFEE, A. (1979): Fifteen Characterizations of Rational Double Points and Simple Critical Points; L'Enseignement Math. 25, 131-163

EISENBUD, D. (1980): Homological Algebra on Complete Intersections, with an Application to Group Representations; Trans AMS 260, 35-64

ESNAULT, H. (1985): Reflexive Modules on Quotient Singularities; J. Reinge Angew. Math. 362, 63-71

ESNAULT, H. and KNÖRRER, H. (1985): Reflexive Modules over Rational Double Points; Math. Ann. 272, 545-548

ESNAULT, H. and VIEHWEG, E. (1985): Two Dimensional Quotient Singularities Deform to Quotient Singularities; Math. Ann. 271, 439-449

GONZALEZ-SPRINBERG, G. and VERDIER, J.-L. (1983): Construction geométrique de la correspondence de McKay; Ann. Sc. École Norm. Sup. 16, 409-449

GREUEL, G.-M. and KNÖRRER, H. (1985): Einfache Kurvensingularitäten und torsionsfreie Moduln; Math. Ann. 270, 417-425

HERZOG, J. (1978): Ringe mit nur endlich vielen Isomorphieklassen von maximalen unzerlegbaren Cohen-Macaulay Moduln; Math. Ann. 233, 21-34

KIJEK, K. and STEINKE, G. (1986): Einfache Kurvensingularitäten in beliebiger Charakteristik; to appear in Arch. d. Math.

KNÖRRER, H. (1982): Group Representations and Resolutions of Rational Double Points; Proc. Conf. Group Theory, Montreal

KNÖRRER, H. (1986): Maximal Cohen-Macaulay Modules on Hypersurface Singularities I; to appear in Inven. Math.

KNÖRRER, H. (1987): Torsionsfreie Moduln bei Deformation von Kurvensingularitäten; this volume

KNÖRRER, H. (1985): Cohen-Macaulay Modules on Hypersurface Singularities; In: Representations of Algebras, London Math. Soc. 116, 147-164

McKAY, J. (1980): Graphs, Singularities and Finite Groups; Proc. Groups, Proc. Symp. Pure Math. 37, 183-186

RIEMENSCHNEIDER, O. (1981): Zweidimensionale Quotientensingularitäten: Gleichungen und Syzygien; Arch. Math. 37, 406-417

SCHAPPERT, A. (1986): A Characterization of Strictly Unimodular Plane Curve Singularities; this volume

SIERSMA, D. (1983): Isolated Line Singularities; Proc. of Symp. in Pure Math. 40, 485-496

SWAN, R. (1968): Algebraic K-Theory ; Lecture Notes in Mathematics 76, Springer Verlag

WALL, C.T.C. (1984): Notes on the Classification of Singularities; Proc. London Math. Soc. 48. 461-513

WUNRAM, J. (1987): Reflexive Modules on Cyclic Quotient Surface Singularities; this volume

WUNRAM, J. (1986): Reflexive Moduln auf zweidimensionalen Quotientensingularitäten. Thesis, Universität Hamburg

FINITE DIMENSIONAL ALGEBRAS AND SINGULARITIES

Idun Reiten

The purpose of this survey is to introduce some of the ideas and results in the theory of finite dimensional algebras to researchers in singularity theory and the theory of vector bundles. We want to show some of the common elements of the theories, and we therefore concentrate on aspects and results from the finite dimensional algebra theory which have so far been extended to or used in the higher dimensional theory. Our topics include finite representation type, almost split sequences and AR-quivers. Important topics in the theory of finite dimensional algebras which have so far not been extended to higher dimensions have been omitted. Also the theory of one-dimensional orders, where there are analogous developments to the finite dimensional algebra theory, is not treated here.

We hope to convey the idea that developments in the theory of finite dimensional algebras can be useful in studying local rings of singularities, and in particular provide another point of view on some of the problems there.

Since we are concentrating on pointing out parallels between aspects of different theories, it will be convenient to fix some notation which will be used throughout the paper.

Let k be an algebraically closed field, and Λ a finite dimensional k-algebra. mod Λ denotes the category of finitely generated left Λ-modules. R denotes a commutative noetherian complete local Cohen-Macaulay domain with maximal ideal $\underline{m}$, of dimension at least 2, with $R/\underline{m} = k \subset R$. Then R contains a complete regular local subring T such that R is a free finitely generated T-module. CM(R) will denote the finitely generated (maximal) Cohen-Macaulay R-modules. We note that there is a common description of CM(R) and mod Λ since CM(R) is the category of finitely generated R-modules which are free T-modules and mod Λ is the category of finitely generated Λ-modules which are free k-modules.

Let further $U = k[X_1,\ldots,X_n]$, $n \geq 2$, be Z-graded by giving the nonzero elements of k degree zero and the X_i degree one. Let S be a Z-graded U-algebra, that is, we have a degree zero homomorphism $U \to S$ whose image is contained in the center of S. We assume that S is a domain and a finitely generated projective U-module. Denote by $CM(\text{gr } S)_0$ the category of finitely generated graded Cohen-Macaulay S-modules with maps of degree zero.

C will denote any one of the categories mod Λ, CM(R), $CM(\text{gr } S)_0$ or the category of vector bundles or coherent sheaves on a nonsingular projective curve. We note that C has the Krull-Schmidt property, that is, every object has a unique (up to isomorphism) decomposition into a finite number of indecomposable objects.

Most of the theory of finite dimensional algebras discussed here is by now classical, and we will sometimes omit references. Instead we refer to the list of references given in [33].

1. Finite representation type.

A k-algebra Λ is said to be of <u>finite representation type</u> if there is only a finite number of isomorphism classes of indecomposable objects in mod Λ. Important problems in the representation theory of finite dimensional algebras have been to investigate when and why Λ is of finite representation type and to classify all indecomposable modules in this case. Analogous problems for R are when is there only a finite number of isomorphism classes of indecomposable objects in CM(R) and what are the modules in this case. We say that R is of <u>finite representation type</u>, or of <u>finite Cohen-Macaulay type</u>, if there is only a finite number of isomorphism classes of indecomposables in CM(R). That CM(R) is the correct module category to look at in this connection is indicated by the nice description of finite Cohen-Macaulay type of hypersurfaces which we state later in this section, as part of a survey of the main results on finite representation type for R. But first we give some examples of results from the finite dimensional algebra theory. There are now methods for deciding whether a given k-algebra is of finite representation type, which we will not go into here.

We discuss the hereditary k-algebras and the k-algebras with $\underline{r}^2 = 0$, where $\underline{r}$ denotes the radical, since in these cases, like in singularity theory, Dynkin diagrams appear, which is not entirely accidental as we shall see in section 5. Also in some cases the problem of proving that R is of infinite representation type will be reduced to the corresponding problem for a k-algebra Λ with $\underline{r}^2 = 0$.

The hereditary k-algebras are up to Morita equivalence the path algebras of (finite) quivers having no oriented cycles. We here recall that a quiver Γ is a set of vertices with a set of arrows between vertices, and the path algebra $k\Gamma$ has the paths in Γ, including the trivial paths (the vertices) as k-basis. The multiplication between the basis elements is given by composing paths when this is possible, and is zero otherwise. If Λ is a k-algebra with $\underline{r}^2 = 0$, there is an associated quiver Γ, such that $\Lambda \simeq k\Gamma/I$, when I is the ideal generated by the paths which are composites of two arrows. The separated quiver Γ' of Γ has vertices x_0, x_1 for every vertex x in Γ, and an arrow $x_0 \xrightarrow{a'} y_1$ for every arrow $x \xrightarrow{a} y$ in Γ. For example,  is the separated quiver for 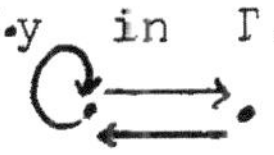 The representation theory for Λ is then closely related to that of the path algebra $k\Gamma'$ of the associated separated quiver. We have the following classical theorem [24].

<u>Theorem 1.1</u>. a) Let Γ be a quiver with no oriented cycles. Then the path algebra $k\Gamma$ is of finite representation type if and only if the underlying graph $|\Gamma|$ of Γ is a finite disjoint union of the Dynkin diagrams A_n, D_n, E_6, E_7, E_8.

b) Let Λ be a k-algebra with $\underline{r}^2 = 0$. Then Λ is of finite representation type if and only if $|\Gamma'|$ is a finite disjoint union of the Dynkin diagrams A_n, D_n, E_6, E_7, E_8, where Γ' denotes the separated quiver.

We now return to the rings R, and give the main results known on finite representation type. We first recall that if R is of finite representation type, then R must be an isolated singularity or regular [6].

1) If $\dim R = 2$, the rings of finite representation type when k is the complex numbers are exactly the fixed rings $k[[X,Y]]^G$, where G is a finite subgroup of $GL(2,k)$ [29] [5] [22]. The Gorenstein rings of finite representation type are exactly the rational double points, in any characteristic [2].

2) If R is a hypersurface in characteristic zero, then R is of finite representation type if and only if R is a simple isolated singularity in the sense of Arnol'd [30] [17]. Actually, the description of the hypersurfaces of finite representation type is the same in any characteristic different from 2 [30] [17]. There are also results similar to [30] in characteristic 2 [38] [40].

3) If R is not a hypersurface, only two examples of finite representation type are known when $\dim R > 2$. These are the fixed ring $k[[X,Y,Z]]^G$, char $k \neq 2$, where the generator of $G = Z_2$ acts by sending each variable to its negative, and $k[[X_0,X_1,X_2,Y_0,Y_1]]/(X_0X_2-X_1^2, X_0Y_1-X_1Y_0, X_1Y_1-X_2Y_0)$ [13].

We note that in all of these cases the indecomposables in CM(R) have also been classified. It is an interesting problem to classify all indecomposables also in cases when R is not of finite representation type. This has been done for the complete local ring of a simple elliptic curve singularity of type $\tilde{E}_8$ [19], and for hypersurfaces with only a countable number of indecomposables [17]. For finite dimensional algebras there has been a lot of work done along these lines, in studying the <u>tame</u> algebras (see [35]).

2. Almost split sequences.

In this section we define almost split sequences and give the relevant existence theorems. We give examples from singularity theory and vector bundles, along with some methods for constructing the almost split sequences. Applications will be given in the next section.

Let as usual $\mathcal{C}$ denote one of our standard categories. We note that for CM(R) and $CM(gr\ S)_0$ we mean exactness as R-modules and S-modules when we talk about exact sequences. We recall that a nonsplit exact sequence $0 \to A \overset{f}{\to} B \overset{g}{\to} C \to 0$ in $\mathcal{C}$ is almost split in $\mathcal{C}$ if (i) A and C are indecomposable, (ii) given $h: X \to C$ with X indecomposable and h not an isomorphism, there is some $t: X \to B$ such that $gt = h$ and (iii) given $h: A \to Y$ with Y indecomposable in $\mathcal{C}$ and h not an isomorphism, there is some $s: B \to Y$ with $sf = h$. It should be noted that (ii) or (i) can be left out from the definition.

Let C be indecomposable in $\mathcal{C}$. A map $g: B \to C$ in $\mathcal{C}$ which is not a split epimorphism is right almost split in $\mathcal{C}$ if any map $h: X \to C$ where X is indecomposable and h is not an isomorphism can be lifted to B. The map $g: B \to C$ is minimal right almost split if in addition for no proper summand B' of B, the restriction map $g': B' \to C$ is right almost split. The notion of (minimal) left almost split is dual (see [7] [8]).

The connection between these concepts is that if $0 \to A \overset{f}{\to} B \overset{g}{\to} C \to 0$ is an exact sequence, then the sequence is almost split if and only if g is minimal right almost split if and only if f is minimal left almost split. We note that if C is an indecomposable projective in $\mathcal{C}$ (that is, $Ext^1(C,X) = 0$ for all X in $\mathcal{C}$) or A is an indecomposable injective (that is, $Ext^1(X,A) = 0$ for all X in $\mathcal{C}$), there cannot be an almost split sequence $0 \to A \to B \to C \to 0$ in $\mathcal{C}$. But it still makes sense to ask if we have minimal right or left almost split maps. Note that in CM(R), R is the only indecomposable projective and the dualizing module $\omega = Hom_T(R,T)$ is the only indecomposable injective in CM(R). We shall say that $\mathcal{C}$ has almost split sequences if for every indecomposable nonprojective C (noninjective A) there is an almost split sequence $0 \to A \to B \to C \to 0$ in $\mathcal{C}$, and that $\mathcal{C}$ has minimal right and left almost split maps if for every indecomposable C in $\mathcal{C}$ there is a minimal right almost split map $B \to C$ and a minimal left almost split map $C \to E$.

We state the following special cases of more general existence theorems. (1) is proved in [7], 2) in [4] for almost split sequences and in [16] for minimal right and left almost split maps, and 2) is also proved in [6]. 3) is proved in [15], 4) in [12], [36]. See [14] for a general theorem containing 1) and 2).)

Theorem 2.1. $\mathcal{C}$ has almost split sequences and minimal left and right almost split maps in the following cases, and the almost split sequences are uniquely determined up to isomorphism by their end terms.

1) mod Λ

2) CM(R) when R is an isolated singularity.

3) $\mathrm{CM}(\mathrm{gr}\ S)_0$ when S is an isolated singularity.

4) The category of vector bundles or coherent sheaves on a nonsingular projective curve.

Before we go on we give some concrete examples of almost split sequences.

(1) Let $\mathcal{C}$ be the vector bundles on $\mathbb{P}^1(k)$. Then the almost split sequences are all of the form $0 \to \mathcal{O}(n-2) \to \mathcal{O}(n-1) \amalg \mathcal{O}(n-1) \to \mathcal{O}(n) \to 0$, where $\mathcal{O}$ denotes the structure sheaf.

(2) Let $R = k[[X,Y,Z]]^{Z_2}$, where the generator of Z_2 sends each variable to its negative. Then $0 \to \Omega^1\omega \to R^3 \to \omega \to 0$ is almost split where $\Omega^1\omega$ is the first syzygy module for ω.

When almost split sequences exist in $\mathcal{C}$, we give some information on how they can be constructed in our various categories $\mathcal{C}$. If $0 \to A \to B \to C \to 0$ is almost split in $\mathcal{C}$, it follows by the uniqueness that C determines A, and we write $A = \tau C$. It is useful to note that we have explicit formulas for τ. For mod Λ, $A = D\,\mathrm{Tr}\,C$, where D denotes the duality $D = \mathrm{Hom}_k(\ ,k)$, and the transpose Tr is defined by $P_1 \to P_0 \to C \to 0$ and $0 \to C^* \to P_0^* \to P_1^* \to \operatorname{Tr} C \to 0$ being exact, where the first sequence is a minimal projective presentation, and $X^* = \operatorname{Hom}_\Lambda(X,\Lambda)$.

If $C = CM(R)$ for an isolated singularity R, of dimension d, then $\tau C = D \operatorname{Tr}_L C$, where $D = \operatorname{Hom}_T(\ ,T)$ and $\operatorname{Tr}_L C = \Omega^d \operatorname{Tr} C$. In particular, we have $\tau C = \Omega^{2-d} C$ if R is Gorenstein. Since for a hypersurface R, $\Omega^2 C \simeq C$ [21], we have $\tau C = C$ if d is even and $A = \Omega C$ if d is odd. If R has dimension 2, then $\tau C = (C \otimes \omega)^{**}$ [5].

For $CM(\operatorname{gr} S)_0$ we have $\tau C = D \operatorname{Tr}_L C(-d)$, where $d = \dim U$ [15]. Using the adjunction formula, we then get that for hypersurfaces $S = k[X_0,\ldots,X_d]/(f)$, $D \operatorname{Tr}_L C = \Omega^{2-d} C(s-1)$, where s is the degree of f, and since $\Omega^2 C = C(-s)$, it is easy to see that if d is even, $\tau C = C(\frac{d}{2} s - d - 1)$. If s is odd, we get $\tau C = \Omega C(\frac{d+1}{2} s - d - 1)$; and $\tau^2 C = C(d(s-2) - 2)$. If S has dimension 2, we have $\tau C = (C \otimes \omega)^{**}$, where $\omega = \operatorname{Hom}_U(S,U)(-d)$ [12].

Correspondingly, we have $\tau(F) = F \otimes \omega$, where ω is the dualizing sheaf, for the category of vector bundles on a nonsingular projective curve [12] [36].

This information on τ is clearly useful in constructing almost split sequences, especially combined with the following result (see [7]).

<u>Proposition 2.2</u>. Let $0 \to A \xrightarrow{f} B \xrightarrow{g} C \to 0$ be a nonsplit exact sequence in $\mathcal{C}$ with indecomposable end terms, and $\tau C = A$. If for any $h: C \to C$ which is not an isomorphism, there is some $t: C \to B$ with $gt = h$, then the sequence is almost split.

The situation is especially easy when $\operatorname{End}(C) = k$, since we then only have to find a nonsplit exact sequence with the given end terms. For example, for $\mathbb{P}^1(k)$ we know that $\omega = \mathcal{O}(-2)$ and since $\operatorname{End}(\mathcal{O}) = k$, and we have an exact sequence $0 \to \mathcal{O}(-2) \to \mathcal{O}(-1) \amalg \mathcal{O}(-1) \to \mathcal{O} \to 0$, it follows directly that this sequence is almost split. Similarly, for an elliptic curve we have $\omega = \mathcal{O}$, and hence if we have a nonsplit exact sequence $0 \to \mathcal{O} \to F \to \mathcal{O} \to 0$, it must be almost split. Such a sequence is given in [3].

For R and S of dimension 2 and for vector bundles on nonsingular projective curves, there is a good way of constructing the almost split sequences from one exact sequence, which has no analogue for finite dimensional algebras. Namely if dim R=2, there is a unique exact sequence $0 \to \omega \xrightarrow{f} E \xrightarrow{g} R \to R/\underline{m} \to 0$ called the fundamental exact sequence, with the property that f is minimal left almost split and g is minimal right almost split. Then if C is indecomposable nonprojective and the characteristic of k does not divide rank C, then $0 \to (\omega \otimes_R C)^{**} \to (E \otimes_R C)^{**} \to C \to 0$ is almost split [5]. There is a similar result for S [12] and for the category of vector bundles. In the last case the result is that if $0 \to \omega \to E \to \mathcal{O} \to 0$ is almost split and F is a vector bundle whose rank is not divisible by char k, then $0 \to \omega \otimes F \to E \otimes F \to F \to 0$ is almost split [12], [36].

We end this section with some remarks on the generality of the existence theorem stated here.

1) For vector bundles (or coherent sheaves) on nonsingular projective varieties of higher dimension we do not have almost split sequences. (An easy argument for this has been given by Schofield.) For a connected projective curve the category of vector bundles has almost split sequences if and only if the curve is Gorenstein [12]. For coherent sheaves, however, it is not hard to see that they exist if and only if the projective curve is nonsingular.

2) There is a close connection between the Serre duality theorem for curves and the isomorphism formulas on which the existence theorems for almost split sequences are based. For a graded isolated singularity S our formula specializes in dimension two to $\mathrm{Ext}^1_S(C, X \otimes \omega(-2)) \simeq \mathrm{Hom}_U(\underline{\mathrm{Hom}}_S(X,C), I_2(-2))$ for X and C in CM(gr S). Here $\underline{\mathrm{Hom}}_S(X,C)$ denotes the ordinary maps from X to C modulo those which factor through projectives and I_2 is the last injective U-module in the minimal graded injective resolution $0 \to U/(X_1,X_2) \to I_0 \to I_1 \to I_2 \to 0$ in $\mathrm{mod}(\mathrm{gr}\ U)_0$. Then a nonzero degree zero element in the socle of

$Ext^1_S(C, C \otimes \omega(-2))$ as End C-module is shown to be almost split. The existence of almost split sequences for vector bundles can be deduced from the graded case. Alternatively a similar argument can be made on the basis of the Serre duality formula, which is analogous, but not equivalent to the above formula, and this is done in the independent proof in [36].

3) If R is not an isolated singularity, but $R_{\underline{p}}$ is Gorenstein for any nonmaximal prime ideal $\underline{p}$ in R, the category $L(R)$ of Cohen-Macaulay modules free outside the maximal ideal still has almost split sequences. In fact, these almost split sequences are also almost split in CM(R) [14]. This last phenomenon was first observed by Schreyer to hold for some specific examples of hypersurfaces, like $k[[x,y]]/(x^2)$.

3. Proving finite representation type.

In proving finite or infinite representation type for R or S, there are two main types of connections with the finite dimensional algebra theory. Either methods from finite dimensional algebras are extended, or the problems are reduced to a problem on finite dimensional algebras, so that the results there can be used. In this section we illustrate both aspects.

Almost split sequences have been used extensively in proving finite representation type for finite dimensional algebras. The criterion we use here for R is like in the case of finite dimensional algebras, not a formal consequence of the existence theorem. For simplicity we here only state the criterion in a form convenient for the non-Gorenstein case [13].

Lemma 3.1. Let R be an isolated singularity, and let $\mathcal{D} = \{R = A_1, \ldots, A_n\}$ be a finite set of nonisomorphic indecomposable objects in CM(R) having the following properties.

1) If $A_i \not\approx R$ is in $\mathcal{D}$, and $0 \to A \to B \to A_i \to 0$ is almost split in CM(R), then all indecomposable summands of $A \amalg B$ are in $\mathcal{D}$. 2) If $A_i \not\approx \omega$, (the dualizing module) then in the almost split sequence $0 \to A_i \to B \to C \to 0$, all indecomposable summands of $B \amalg C$ are in $\mathcal{D}$. Then $\mathcal{D}$ is the set of all indecomposables in CM(R).

We shall give an illustration of how to apply this lemma. At the same time we illustrate a useful relationship between the maps occurring on the left and right hand side in an almost split sequence. We first recall that a map $f: B \to C$ between indecomposables in $\mathcal{C}$ is irreducible if f is not an isomorphism and whenever $f = hg$, then g is a split monomorphism or h is a split epimorphism [8].

Proposition 3.2. Assume that $\mathcal{C}$ has minimal right and left almost split maps. Let $g: B \to C$ be a nonzero map between indecomposables in $\mathcal{C}$. Then $g: B \to C$ is irreducible if and only if there is some map $f: B' \to C$ such that the induced map $B \amalg B' \to C$ is minimal right almost split if and only if there is some map $h: B \to C'$ such that the induced map $B \to C \amalg C'$ is minimal left almost split.

We now give an application of Lemma 3.1 and Proposition 3.2.

Example 3.3. Let $R = k[[X_0,X_1,X_2,Y_0,Y_1]]/(X_0X_2 - X_1^2, X_0Y_1 - X_1Y_0, X_1Y_1 - X_2Y_0)$. Then we can find the five nonisomorphic indecomposable modules $R, A = \omega$, B, C, K in CM(R) where the first four have rank 1 and K has rank 2. We can also show that we have almost split sequences $0 \to R \to A \amalg A \amalg B \to K \to 0$ and $0 \to K \to R \amalg R \amalg C \to A \to 0$, and that $\tau B = C$, $\tau C = B$ (see [13]). We want to show that on the basis of this information we can conclude that there are no other indecomposables in CM(R).

We have almost split sequences $0 \to C \to E \to B \to 0$ and $0 \to B \to F \to C \to 0$. By Proposition 3.2 we have irreducible maps $K \to C$, $C \to A$ and $R \to B$. This shows $K | F$, $R | E$ and

$A|E$. A rank argument now shows $E = R \amalg A$ and $F = K$. Hence we are done by Lemma 3.1.

The fixed ring $k[[X,Y,Z]]^{Z_2}$ mentioned before can also be treated this way [13] , and Solberg has used a similar method for hypersurfaces (in characteristic 2).

There is an analogous criterion for Z-graded isolated singularities S [15]. If S is Z-graded and $R = \hat{S}$ is the ring obtained by completing at the maximal graded ideal in S, it should be true in general that R is of finite representation type if and only if $CM(\operatorname{gr} S)_0$ has only a finite number of indecomposables up to shift. This holds in all cases where $R = \hat{S}$ is known to be of finite representation type (proved in [15] for R as in Ex. 3.3, by Solberg for $S = k[X,Y,Z]^{Z_2}$ and in [30, 38] (see also [40]) for hypersurfaces). We propose the following conjecture, which should be an important step in proving the above: If $0 \to A \to B \to C \to 0$ is almost split in $CM(\operatorname{gr} S)_0$, then the induced exact sequence $0 \to \hat{A} \to \hat{B} \to \hat{C} \to 0$ in CM(R) is almost split, or at least a direct sum of almost split sequences. We note that if Λ is a Z-graded finite dimensional k-algebra, then if $0 \to A \to B \to C \to 0$ is an almost split sequence in the category $\operatorname{mod}(\operatorname{gr} \Lambda)_0$ of finitely generated graded Λ-modules with degree zero maps, then it is also almost split in mod Λ. And it is also known that Λ is of finite representation type if and only if $\operatorname{mod}(\operatorname{gr} \Lambda)_0$ has only a finite number of indecomposable objects up to shift [26].

We now give an example where showing that some R is of infinite representation type is reduced to showing that a certain finite dimensional algebra Λ is of infinite representation type. Such a method is used for the fixed rings $R = k[[X_1,\ldots,X_n]]^G$, for $n \geq 4$, char $k = 0$, where G is a subgroup of GL(n,k) with no pseudoreflections [13]. Writing $S = k[[X_1,\ldots,X_n]]$, $\underline{n} = (X_1,\ldots,X_n)$, G acts also on $S/\underline{n}^2$. Let $\Lambda = (S/\underline{n}^2)G/\langle e\rangle$, where $(S/\underline{n}^2)G$ denotes the skew group ring and e is the sum of the elements in G. Then there is a functor α: mod $\Lambda \to CM(R)$ given by $\alpha(C) = (\Omega^n_{SG}C)^G$, which induces an embedding of isomorphism classes of indecomposable objects. Note that Λ is a k-algebra with $\underline{r}^2 = 0$. One can show that the underlying graph of the separated quiver for Λ is not a disjoint union of Dynkin diagrams, and hence Λ is of infinite representation type by Theorem 1.1.

4. The AR-quiver.

Assume that C has almost split sequences and minimal left and right almost split maps. We associate with C a (possibly infinite) quiver Γ, called the AR-quiver of C, whose vertices are in one-one correspondence with the isomorphism classes [C] of indecomposable objects in C, and there is an arrow from [B] to [C] if and only if there is an irreducible map from B to C. The AR-quiver is equipped with a valuation, that is, a pair of natural numbers (r_{BC}, s_{BC}) associated with the arrow $[B] \to [C]$. Here r_{BC} denotes the multiplicity of B in F when $h: F \to C$ is minimal right almost split, and s_{BC} denotes the multiplicity of C in E when $h: B \to E$ is minimal left almost split. In our categories C in this paper we always have $r_{BC} = s_{BC}$, using that k is algebraically closed. We then say that the valuation is symmetric, and we shall write $\cdot \xrightarrow{2} \cdot$ instead of $\cdot \xrightarrow{(2,2)} \cdot$, or also $\cdot \rightrightarrows \cdot$. Denote for a quiver Γ the set of vertices by Γ_0 and the set of arrows by Γ_1. A translation quiver Γ is a quiver with no double arrows and with an injective map τ from a subset of Γ_0 to Γ_0, such that for every vertex y where τy is defined there is a bijective map σ from the arrows ending in y to the arrows starting in τy. This notion was introduced in [34], with the additional requirement that Γ has no loops. Γ is said to be stable if τ gives a bijection from Γ_0 to Γ_0. A basic type of example of a translation quiver is $Z\Delta$ when Δ is an oriented tree. The vertices are (x,n) for x a vertex of Δ and n an integer, and for every arrow $x \to y$, we have arrows $(x,n) \to (y,n)$ and $(y,n) \to (x,n+1)$ for all

$n \in Z$. The translation τ is defined by $\tau(x,n) = (x,n-1)$, for all x in Δ and n in Z. Since $Z\Delta$ is independent of the orientation, we shall also write $Z\Delta$ when Δ is a (nonoriented) tree. We also have the notion of a <u>valued translation quiver</u>. We shall again for simplicity restrict to the case of symmetric valuation, and we require that a and σa have the same associated number, for each arrow a in Γ.

The existence of minimal right and left almost split maps in C, together with the connection with irreducible maps, has the following direct consequence.

<u>Proposition 4.1</u>. The AR-quiver Γ for C is a locally finite valued translation quiver, with τ the translation induced from the almost split sequences.

Since an irreducible map between two indecomposables can be shown to be either a monomorphism or an epimorphism, there are no loops in the AR-quiver for mod Λ. As we shall see below, this can however happen in general, and this is the reason for allowing loops in the definition of a translation quiver.

We now give some examples of AR-quivers.

(1) If C is the vector bundles on $\mathbb{P}^1(k)$, the AR-quiver is

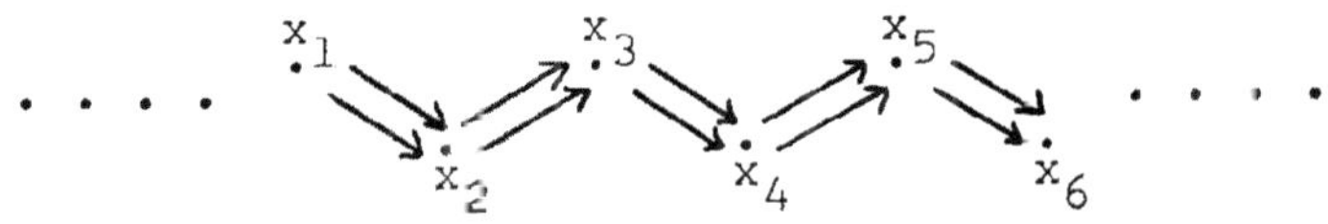

where $\tau x_i = x_{i-2}$ for all i in Z.

(2) If C is the vector bundles on an elliptic curve, every connected component of the AR-quiver is of the form

with τ the identity.

(3) If $R = k[[X,Y]]$, the AR-quiver is 
with τ not defined.

(4) The AR-quiver for $R = k[[X,Y,Z]]^{Z_2}$ is

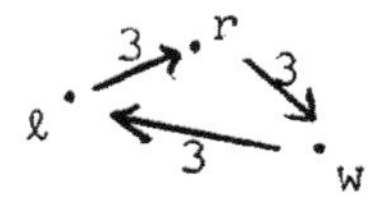

with $\tau w = \ell$ and $\tau\ell = r$.

In addition to giving a way of organizing maps between modules, the AR-quiver (with τ) can in the case of finite representation type be used to compute the Grothendieck group K_0(mod R), that is the free group on the isomorphism classes in mod R, modulo the relations given by exact sequences [9]. The result is that K_0 (mod R) is isomorphic to the free group on the vertices of the AR-quiver modulo the relations $x + \tau x - \sum r_y y$ for every vertex x such that τx is defined, and when the sum runs over all arrows $y \xrightarrow{r_y} x$. This description was first given in the finite dimensional case [18], where one knows the Grothendieck group anyway. It is even more useful for the R since it can be used to compute K_0 (mod R). We also have K_0 (mod R) $= Z \amalg H$, where H can be computed similarly from the AR-quiver with R removed [9]. If R is Gorenstein, this is the <u>stable</u> AR-quiver. Using this, we see that for $R = k[[X,Y,Z]]^{Z_2}$, H is the free group on ℓ, w, modulo the relations $\ell + w$, $3w - \ell$, so that K_0 (mod R) $\cong Z \amalg Z_4$. Since when f determines a hypersurface of finite representation type and char $k \neq 2$, $f + z^2 + u^2$ determines one of finite representation type with the same stable AR-quiver [30], their Grothendieck groups are isomorphic.

There is some interesting combinatorics associated with the AR-quiver, based on using the length function in the finite dimensional case, and this applies equally well to the rank function for CM(R), CM(gr S)$_0$ and vector bundles.

Let $\Gamma = (\Gamma_0, \Gamma_1)$ be a stable valued translation quiver. A function $f\colon \Gamma_0 \to N$, where N is the positive integers is <u>subadditive</u> if for every x in Γ_0, $f(x) + f(\tau x) \geq \sum r_y f(y)$, where the sum ranges over all arrows $y \xrightarrow{r_y} x$. f is <u>additive</u> on Γ if we always have equality. The corresponding related concept for a graph Ω with single edges and possibly with loops, valued with a positive integer assigned to each edge, is that f is subadditive if $2f(x) \geq \sum r_y f(y)$, where the sum ranges over all edges $y \overset{r_y}{\bullet\!\!-\!\!\bullet} x$, and f is additive if we always have equality (see [27], [28]). The basic underlying combinatorical results for the information on the structure of the AR-quiver is given in the following results from [27], [28].

<u>Theorem 4.2</u>. Let Γ be a stable connected translation quiver without loops and with symmetric valuation, and assume that there is a subadditive function f on Γ and a vertex $x \in \Gamma_0$ with $\tau^n x = x$ for some $n > 0$.

a) Then $\Gamma \simeq Z\Delta/G$, where G is a group of automorphisms of $Z\Delta$, and Δ is a valued oriented tree.

b) If f is not additive, then $|\Delta|$ is Dynkin (i.e., here A_n, D_n, E_6, E_7, E_8) or A_∞: 

c) If f is additive, then $|\Delta|$ is extended Dynkin (i.e. here $\tilde{D}_n$, $\tilde{E}_6$, $\tilde{E}_7$, $\tilde{E}_8$ or ) or A_∞,

D_∞: 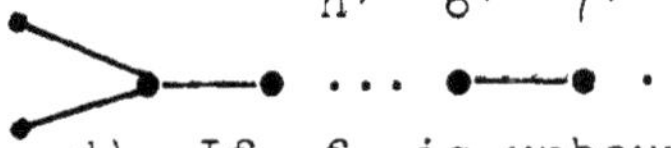, ${}_\infty A_\infty$: .

d) If f is unbounded, then $|\Delta|$ is A_∞.

<u>Theorem 4.3</u>. Let Ω be a connected graph with symmetric valuation and a subadditive function f on Ω.

a) If f is additive, then Ω is one of the finite diagrams $\tilde{A}_n$, $\tilde{D}_n$, $\tilde{E}_6$, $\tilde{E}_7$, $\tilde{E}_8$, , ,

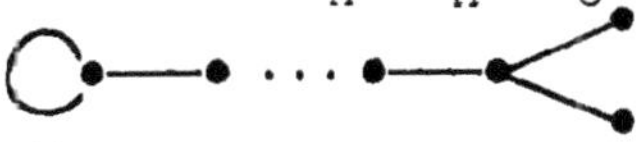 or infinite diagrams A_∞, D_∞, ${}_\infty A_\infty$,

.

b) If f is additive and unbounded on Ω, then Ω is A_∞.

c) If Ω is finite and f not additive, then Ω is Dynkin or 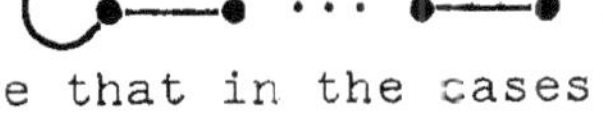.

We note that in the cases of additive functions the functions are known. We just mention that for A_∞ it must be $\underset{n}{\bullet}\!-\!\!-\!\underset{2n}{\bullet}\!-\!\!-\!\underset{3n}{\bullet}\!-\!\!-\!\bullet \cdots$ for a positive integer n.

We now discuss how these combinatorical results are used to get results on the structure of the components of the AR-quiver. The basic observation is that since length for mod Λ and rank in the other cases are additive on exact sequences, we get by defining f on a vertex to be the length or rank of the corresponding module for the AR-quiver, that if x is a vertex such that τx is defined, then $f(x) + f(\tau x) = \sum r_y f(y)$, where the sum ranges over all $y \bullet\!\xrightarrow{r_y}\!\bullet\, x$.

If $\mathcal{D}$ is a component of the AR-quiver, the translation τ does not necessarily give a bijection between the vertices, that is, $\mathcal{D}$ is not necessarily stable. This is, however, the case when $\mathcal{D}$ contains no projectives or injectives, in particular it is always the case for vector bundles and coherent sheaves on a nonsingular projective curve. If R or S is of dimension two, it is possible even if $\mathcal{D}$ contains projectives or injectives, to extend τ to all vertices such that $\mathcal{D}$ becomes a translation quiver. Because of the properties of the fundamental exact sequence $0 \to \omega \to E \to R \to R/\underline{m} \to 0$ discussed before, we can define $\tau R=\omega$.

If $\mathcal{D}$ contains projectives and we are not in dimension 2, we want to remove vertices from the AR-quiver to make τ defined everywhere. Since we want to throw away as few as possible, the situation is nicest when the projectives and injectives coincide, for then it is enough to remove the projectives. In the finite dimensional algebra case these are the selfinjective algebras, and for R and S the Gorenstein rings. Note that if we remove a projective from a component, then f becomes a subadditive function which is not additive. This is especially of interest for finite representation type.

In Theorem 4.2 it is also assumed that τ is periodic on some vertex x. This is necessarily the case for the AR-quiver if we have finite representation type, when the projectives and injectives coincide. We have also seen that if R is a hypersurface, then $\tau^2 = \text{id}$. Theorem 4.2 only applies directly if there is no loop in the AR-quiver. Note however that if for a stable translation quiver $\mathcal{D}$, τ is the identity, then a (sub-) additive function on $\mathcal{D}$ is the same as a (sub-) additive function on the graph Ω obtained from $\mathcal{D}$ by replacing $\cdot\rightleftarrows\cdot$ by $\cdot\!-\!\!-\!\cdot$ and $\circlearrowleft$ by $\bullet\!\bigcirc$. Hence Theorem 4.3 can be applied in this case.

We first look at some consequences for finite representation type. That the stable AR-quiver for a selfinjective algebra is of the form $Z\Delta/G$ with Δ Dynkin follows from Theorem 4.2 (see [27]). This result (with another proof) was the starting point for the classification of selfinjective algebras of finite representation type [34]. For Gorenstein rings R in dimension 2 we have the following (using Theorem 4.3), where the diagrams with loops are excluded by computing K_0 (mod R) [11]. We note that loops may occur in the noncommutative two-dimensional maximal orders of finite representation type classified in [1] (see [11]).

<u>Theorem 4.4</u>. Let R be a nonregular rational double point. Then in the AR-quiver Γ for R arrows occur in pairs, and replacing $\cdot\rightleftarrows\cdot$ by $\cdot\!-\!\!-\!\cdot$ we have an extended Dynkin diagram $\|\Gamma\|$ for Γ, and a Dynkin diagram for the stable quiver Γ_S.

In characteristic zero this result was first proved in [5], by proving an isomorphism with the McKay quiver, whose underlying graph after removing the trivial representation was already known to be isomorphic to the resolution graph for the corresponding singularity. There is now a direct isomorphism from $\|\Gamma\|$ to the resolution graph [23].

For infinite representation type it is useful to know that if Λ is indecomposable, then for every component of the AR-quiver the length is unbounded [4]. Then it follows that if a component $\mathcal{D}$ contains no projectives or injectives and f is periodic on a vertex, then $\mathcal{D} \simeq ZA_\infty/G$. Hence the following recent generalization from artin algebras is very important here [20] (see also [37]).

Theorem 4.5. Assume that R is an isolated singularity of infinite representation type. Then in any connected component of the AR-quiver for R there are indecomposable Cohen-Macaulay modules of arbitrary high rank.

Using the existence of almost split sequences in $CM(gr\ S)_0$, the proofs for Theorem 4.5 modify directly to give the following graded version.

Theorem 4.6. a) If the Z-graded ring S is an isolated singularity, of infinite representation type, then in a connected component of the AR-quiver for S there are indecomposables in $CM(gr\ S)_0$ of arbitrarily high rank.

b) If for a connected nonsingular projective curve there is an infinite number of indecomposable vector bundles up to shift, then in a connected component of the AR-quiver there are indecomposable vector bundles of arbitrary high rank.

We note that these results fail if we drop the assumption of isolated singularity, for example for $R = k[[x,y]]/(x^2)$ [17], but Theorem 4.2 b) c) still applies to L(R) when R_p is Gorenstein for all nonmaximal prime ideals p.

We state the following consequences.

Corollary 4.7. a) Let R be an isolated hypersurface singularity of infinite representation type of dimension $d \geq 2$, and $\mathcal{D}$ a component of the AR-quiver which unless $d = 2$ does not contain R. Then if d is even $\mathcal{D} \simeq ZA_\infty/\langle\tau\rangle$ and if d is odd, $\mathcal{D} \simeq ZA_\infty/\langle\tau\rangle$ or $\mathcal{D} \simeq ZA_\infty/\langle\tau^2\rangle$

b) Let S be a Z-graded isolated hypersurface singularity of dimension $d \geq 2$, given by a homogeneous polynomial f of degree s, of infinite representation type, and let $\mathcal{D}$ be a component of the AR-quiver which unless $d = 2$ does not contain any projectives. Then we have $\mathcal{D} \simeq ZA_\infty$ except when $d = 2$, $s = 3$ when we get $\mathcal{D} \simeq ZA_\infty/\langle\tau\rangle$.

c) For the category of vector bundles on a nonsingular projective curve of genus ≥ 1, each component of the AR-quiver is of the form $ZA_\infty/\langle\tau\rangle$ in genus 1 and ZA_∞ in genus > 1.

Under these assumptions the rank function is additive, so a) follows from the previous theorems and comments [20]. For the graded case in b) we see from our previous formulas for τ that τ is periodic (and of period 1) exactly when $s = 3$, $d = 2$. In the other cases the combinatorics can be applied by using that $f(\tau^2 x) = f(x)$ (or even $f(\tau x) = f(x)$ [21]) for all x in $\mathcal{D}$. In genus 1 we have a hypersurface with $d = 2$ and $s = 3$, so this case follows from b). (This case can also be seen directly from [3], and has also independently been observed in [36].) The rest was first proved in [36]. In our approach here we use that $f(\tau x) = f(x)$, since $\tau(F) = \omega \otimes F$, and as in [36] that no vertex is τ-periodic since $\deg(\omega^n \otimes F) = n(2g-2) + \deg F$, where g denotes genus and deg degree.

5. Dynkin and extended Dynkin diagrams.

We have seen that in some cases there is some common combinatorical explanation for the appearance of Dynkin and extended Dynkin diagrams in the finite dimensional algebra theory and higher dimensional theories via the AR-quiver. In general when the Dynkin diagrams appear there is no organic connection between the rings giving rise to the same (extended) Dynkin diagrams, but we would like to point out that in some instances there is such a connection.

We have seen that there is a Dynkin diagram associated both with a 2-dimensional hypersurface R of finite representation type (that is, a rational double point) and a finite dimensional hereditary algebra of finite representation type. Let R be a rational double point in characteristic zero and $k\Gamma$ a hereditary algebra, associated with the same Dynkin diagram Δ. We shall see that R and $k\Gamma$ are connected through a finite dimensional algebra which can be constructed in a natural way both from R and from $k\Gamma$ (and which directly determines the underlying graph $|\Gamma| = \Delta$ back again). Denote by M the direct sum of one copy of each indecomposable in $CM(R)$, and let $\Sigma = \underline{\mathrm{End}}_R(M)$, the endomorphism ring of M, modulo the maps factoring through projectives. Σ is finite dimensional since R is an isolated singularity. On the other hand there is the construction of a preprojective algebra associated with $k\Gamma$. If we denote by Γ^* the quiver obtained from Γ by adding for every arrow a in Γ an arrow a^* in the opposite direction, the preprojective algebra is the factor algebra $k\Gamma^*/\langle \sum_a aa^*+a^*a \rangle$. It is known to be finite dimensional since $|\Gamma|$ is Dynkin [34] [25]. Using a result of Riedtmann one can show that this preprojective algebra is isomorphic to Σ, thus Σ provides our desired connection. We note that Σ is built both from the indecomposables in $CM(R)$ and those in $\mathrm{mod}\, k\Gamma$, since it is known that the preprojective algebra of $k\Gamma$ as $k\Gamma$-module is the direct sum of one copy of each indecomposable $k\Gamma$-module [34] [25].

There is also a connection between rational double points and finite dimensional hereditary algebras $k\Gamma$ where $|\Gamma|$ is extended Dynkin. These are the <u>tame</u> hereditary algebras, and a complete classification of the indecomposables is known in this case (see [35]). Let R be a rational double point in characteristic zero with AR-quiver Γ. Then R is known to be isomorphic to $k[[X,Y]]^G$, where G is a finite subgroup of $SL(2,k)$. Via this embedding there is also a natural action of G on the path algebra $k\Gamma_1$ of the quiver $\bullet \rightrightarrows \bullet$, which is tame since $\bullet = \bullet$ is extended Dynkin. Then one can show that the skew group ring $(k\Gamma_1)G$ is also tame, and is hence Morita equivalent to the path algebra of a quiver Γ_2, where $|\Gamma_2|$ is a disjoint union of extended Dynkin diagrams. For the separated quiver Γ' of the AR-quiver Γ it is also easily seen that $|\Gamma'|$ is a disjoint union of extended Dynkin diagrams. It can be proved that Γ_2 is isomorphic to (the opposite of) Γ', thus establishing a connection between $R = k[[X,Y]]^G$ and $(k\Gamma_1)G$ (see [10], [31]).

References

1. Artin, M., Maximal orders of global dimension and Krull dimension two, Invent. Math.Vol. 84, Fasc. 1(1986), 195-222.

2. Artin, M. and Verdier, J.L., Reflexive modules over rational double points, Math. Ann. 270 (1985), 71-82.

3. Atiayh, M.F., Vector bundles over an elliptic curve, Proc. London Math. Soc. (3), VII, 27 (1957), 414-452.

4. Auslander, M., Functors and morphisms determined by objects. Applications of morphisms determined by objects. Proc. Conf. on Representation Theory (Philadelphia 1976) (Marcel Dekker, 1978), 1-327.

5. Auslander, M., Rational singularities and almost split sequences, Trans. Amer. Math. Soc., Vol. 293, No. 2 (Feb. 1986), 511-532.

6. Auslander, M., Isolated singularities and almost split sequences, Proc. ICRA IV, Springer Lecture Notes in Math. 1178 (1986), 194-241.

7. Auslander, M. and Reiten, I., Representation theory of Artin algebras III. Almost split sequences, Comm. in Algebra 3 (1975), 239-294.

8. Auslander, M. and Reiten, I., Representation theory of artin algebras IV. Invariants given by almost sequences. Comm. in Algebra 5 (1977), 519-554.

9. Auslander, M. and Reiten, I., Grothendieck groups of algebras and orders, J. Pure and Applied Algebra 39 (1986), 1-51.

10. Auslander, M. and Reiten, I., McKay quivers and extended Dynkin diagrams, Trans. Amer. Math. Soc. Vol. 293, No. 1, Jan. 1986, 293-301.

11. Auslander, M. and Reiten, I., Almost split sequences for rational double points, Trans. Amer. Math. Soc. (to appear).

12. Auslander, M. and Reiten, I., Almost split sequences in dimension 2, Adv. in Math.1987.

13. Auslander, M. and Reiten, I., The Cohen-Macaulay type of Cohen-Macaulay rings. Preprint Trondheim 1986.

14. Auslander, M. and Reiten, I., Almost split sequences for Cohen-Macaulay modules. Preprint Brandeis 1986.

15. Auslander, M. and Reiten, I., Almost split sequences for Z-graded rings, this volume.

16. Auslander, M. and Smalø, S.O., Lattices of orders: Finitely presented functors and preprojective partitions. Trans. Amer. Math. Soc. 273 (1982) 433-445.

17. Buchweitz, R.O., Greuel, G.M. and Schreyer, F.O., Cohen-Macaulay modules on hypersurface singularities II. Kaiserslautern, Preprint no. 109, 1986.

18. Butler, M.C.R., Grothendieck groups and almost split sequences, Proc. Oberwolfach Conf. on Integral Representations and Applications, Springer Lecture Notes in Math. 882 (1981), 357-368.

19. Dieterich, E., Classification of the indecomposable representations of the cyclic group of order three in a complete discrete valuation ring of ramification degree four.

20. Dieterich, E., Reduction of isolated singularities. Preprint Brandeis 1986.

21. Eisenbud, D., Homological Algebra on a complete intersection with an application to group representations. Trans. Amer. Math. Soc. 260 (1980), 35-64.

22. Esnault, H., Reflexive modules on quotient singularities, J. Reine Angew, Mathematik.

23. Esnault, H. and Knörrer, H., Reflexive modules over rational double points, Math. Ann. 272 (1985), 545-548.

24. Gabriel, P., Unzerlegbare Darstellungen I., Manuscripta Math. 6 (1972), 71-103.

25. Gelfand, I.M. and Ponomarev, V.A., Model algebras and representations of graphs, Funkc. anal. i priolož 13 (1979), 1-12.

26. Gordon, R. and Green, E.L., Representation theory of graded artin algebras, J. Algebra 76 (1982) 138-152.

27. Happel, D., Preiser, U. and Ringel, C.M., Vinbergs characterization of Dynkin diagrams using subadditive functions with applications to DTr-periodic modules, Proc. ICRA II, Ottawa 1979, Lecture Notes in Math. 832 (Springer 1980), 280-294.

28. Happel, D., Preiser, U. and Ringel, C.M., Binary polyhedral groups and Eucledian diagrams, Manuscripta Math. 31 (1980), 317-329.

29. Herzog, J., Ringe mit nur endlich vielen Isomorphieklassen von maximalen unzerlegbaren Cohen-Macaulay Modulen, Math. Ann. 233 (1978), 21-34.

30. Knörrer, H., Cohen-Macaulay modules on hypersurface singularities I. Preprint 1986.

31. Lenzing, H., Polyhedral groups and the geometric study of tame hereditary algebras,

32. McKay, J., Graphs, singularities and finite groups, Proc. Sympos. Pure Math. 37, Amer. Math. Soc., Providence, RI (1980), 183-186.

33. Reiten, I., An introduction to the representation theory of artin algebras, Bull. London Math. Soc. 17 (1985) 209-233.

34. Riedtmann, C., Algebren, Darstellungsköcher, Überlagerungen und zurück, Comment. Math. Helv. 55 (1980), 199-224.

35. Ringel, C.M., Tame algebras and integral quadratic forms, Lecture Notes in Math. 1099 (Springer 1984).

36. Schofield, A., Private communication.

37. Yoshino, Y., Brauer-Thrall theorems for maximal Cohen-Macaulay modules. Preprint 1986.

38. Buchweitz, R., Eisenbud, D. and Herzog, J., Cohen-Macaulay modules on quadrics. Preprint 1986.

39. Herzog, J. and Sanders, H., Graded Gorenstein domains and bounded rank Cohen-Macaulay representation type have minimal multiplicity. Preprint Essen 1986.

40. Solberg, Ø., Hypersurface singularities of finite Cohen-Macaulay type. Preprint Trondheim 1986.

Cohen-Macaulay Modules on Quadrics

by

Ragnar-Olaf Buchweitz * (Hannover)
David Eisenbud ** (Brandeis)
Jürgen Herzog (Essen)

with an appendix by
Ragnar-Olaf Buchweitz

Dedicated to Maurice Auslander on the occasion of his 60th birthday

Contents

* Supported by a "Heisenberg-Stipendium", Bu-398/3-1 of the DFG
** Partially supported by the NSF

Abstract This paper analyzes the graded maximal Cohen-Macaulay modules over rings of the form $R = k[x_1,\dots,x_r]/Q$, when Q is a quadratic form defining a regular projective hypersurface, and k is an arbitrary field (the case when k is algebraically closed of characteristic $\neq 2$ is a special case of the theory developed by Knörrer [1986]). For any nonzero quadratic form Q, regular or not, the graded maximal Cohen-Macaulay R-modules define modules over the even Clifford algebra of Q, and we show that this algebra is semi-simple iff Q is regular (this is classical for char $k \neq 2$). As a result of this and other information about the Clifford algebra, we give a detailed account of the Cohen-Macaulay modules when Q is regular, identifying the number of indecomposables (2 or 3, counting R) their ranks, and the relations of duality and syzygy among them.

We also show that when Q is nonsingular every maximal Cohen-Macaulay module over the completion of R at $(x_1,\dots,x_r)$ is the completion of a graded module.

Our approach is via the matrix factorizations of quadratic forms in the sense of Eisenbud [1980], and includes a classification of these. As an appendix we include a sketch of the relevant case of a more general theory of Buchweitz giving equivalences of certain derived categories of modules over the algebras of a Koszul pair, generalizing work of Bernstein-Gelfand-Gelfand.

Introduction:

If R is a local or graded ring, then a maximal Cohen-Macaulay (abbreviated MCM) R-module M is a finitely generated module whose depth is equal to the dimension of R. There has recently been much interest in Cohen-Macaulay rings R which have only finitely many isomorphism classes of indecomposable MCM modules, the "rings of finite CM-type"; see for example Auslander-Reiten [1986]. Two milestones of importance for this paper are 1) Auslander's result [1986] that if R is a local (or graded) ring of finite CM-Type then R has an isolated singularity and 2) Knörrer's results [1986] analyzing the Cohen-Macaulay modules over

$$k[[x_1,\ldots,x_n,y]]/f(x)+y^2$$

or

$$k[[x_1,\ldots,x_n,y,z]]/f(x)+yz$$

in terms of those over

$$k[[x_1,\ldots,x_n]]/f\ .$$

These results suggest that if Q is a quadratic form on a vector-space V over k considered as an element of $S_2(V^*)$, which is regular in the sense that

$$R = k[V^*]/Q$$

has only an isolated singularity, then the MCM modules over R should be particularly tractable; indeed, Knörrers results show that if k is algebraically closed and of characteristic not 2, then there are always 1 or 2 nonfree indecomposable MCM R-modules depending on whether $\dim V$ is odd or even; that they both have the same rank and are syzygies of one another when there are 2; and that writing

$$m = \left[\frac{\dim V}{2}\right] - 1\ ,\quad ([x] \text{ denoting the largest integer smaller or equal to } x)$$

the rank of their direct sum is 2^m (Note that if $R = k[x]/x^2$ or $k[x,y]/xy$, then rank R/x may reasonably be taken to be 1/2 as this formula implies.)

In this paper we derive parallel results for all fields k. (Theorem 3.1 and 3.2). As in Knörrer's case, if $\dim V$ is odd there is a unique nonfree indecomposable MCM R-module; but if $\dim V$ is even, there can be 1 or 2, depending on the discriminant or Arf invariant of Q (Theorem 3.1). Again when there are two modules they have the same rank and are syzygies of one another; but in the general case the rank of their direct sum may be any power of 2 between 2^m and

$2^{\dim V-2}$, depending on the degree of the division algebra which is the Hasse-Witt invariant of Q (except in characteristic 2, where the nullity of the bilinear form associated to Q also plays a role).

To obtain these results we use the following idea to transform the problem to a problem about modules over the even Clifford algebra, $C_o(Q)$:

If M is an MCM R-module, then the free resolution of M is known to be periodic of period 2 (Eisenbud [1980]), and the periodicity isomorphism on the level of $\mathrm{Tor}_*^R(k,M)$ is multiplication by an element $\sigma \in \mathrm{Ext}^2(k,k)$. Thus $\mathrm{Tor}_o^R(k,M) \oplus \mathrm{Tor}_1^R(k,M)$ is a module over the algebra $\mathrm{Ext}_R^*(k,k)/\langle\sigma-1\rangle$, which is isomorphic as a $\mathbb{Z}/2\mathbb{Z}$-graded algebra to the Clifford algebra $C(Q)$. Taking even degree parts, $\mathrm{Tor}_o^R(k,M) = k \otimes_R M$ becomes a module over $C_o(Q)$ in a functorial way. In fact (Theorem 2.1), for any quadratic form $Q \neq 0$ this functor induces an equivalence between the category of modules over $C_o(Q)$ and a certain subcategory of the category of MCM R-modules modulo projectives. If Q is nonsingular, we obtain in this way the whole category of graded MCM R-modules modulo projectives, and even, by Theorem 2.2, all MCM modules over the completion

$$\hat{R} = k[[V^*]]/Q\ .$$

To construct the inverse of the functor $M \to k \otimes_R M$, we use the idea of <u>matrix factorization</u> (Eisenbud [1980]). Recall that a matrix factorization of a nonzerodivisor f of a ring S is a pair of (necessarily) square matrices φ,ψ over S such that $\varphi\psi$ is f times the identity matrix, $f\cdot 1$. It follows at once that $\psi\varphi = f\cdot 1$, too. If φ,ψ is a matrix factorization of f, and S is Cohen-Macaulay, then coker φ is an MCM S/f-module. If S is regular, then every MCM S/f-module arises in this way, and the category of MCM modules mod projectives is equivalent to the category of matrix factorizations (with maps defined up to homotopy in an obvious sense). Thus, we may construct the inverse functor by constructing a matrix factorization of Q from a $C_o(Q)$-module.

If A_o is a $C_o(Q)$-module then $A = C(Q) \otimes_{C_o(Q)} A_o =$
$= A_o \oplus (A_1 := C_1(Q) \otimes_{C_o(Q)} A_o)$ is a $\mathbb{Z}/2\mathbb{Z}$-graded $C(Q)$-module; in fact this construction gives an equivalence of categories iff $C_o(Q)$ is non-zero (Atiyah-Bott-Shapiro [1964]). The module structure on

A yields, via the inclusion $V \to C_0(Q)$, maps

$$V \otimes A_0 \to A_1$$
$$V \otimes A_1 \to A_0$$

whose adjoints

$$A_0 \to A_1 \otimes V^*$$
$$A_1 \to A_0 \otimes V^*$$

define degree 1 maps of $S = k[V^*]$-modules

$$A_0 \otimes S \xrightarrow{\psi} A_1 \otimes S$$
$$A_1 \otimes S \xrightarrow{\varphi} A_0 \otimes S \;;$$

it turns out that φ, ψ is a matrix factorization of Q, and $A_0 \to \operatorname{coker} \varphi$ is the desired inverse to $M \to k \otimes_R M$ in the regular case.

In fact we may conveniently define the functor $M \to k \otimes M$, regarded as a module over the even Clifford algebra, in a completely elementary way using matrix factorizations, and we adapt this method in the text, below.

To use this machine, we need information about $C_0(Q)$ and its modules. This is provided in section 1 below. The case where the characteristic of k is not 2, and even the "non-defective" case in characteristic 2, are classical. Our main - and probably only - new result in this section is that Q is regular in the sense above iff $C_0(Q)$ is semi-simple (Theorem 1.1). We describe the structure of this algebra in the semi-simple case. We also describe the possibilities - which are more limited than one might at first think - over a "global" field, and we show that the Clifford algebra of a generic form in characteristic 0 is a division ring.

Closely related to this paper in several points is the noteworthy paper of Swan [1985]. In particular Swan makes use of the Clifford algebra - in a way closely related to ours - to analyze the K-theory of modules over R when Q satisfies the condition that the hypersurface defined by Q is smooth over k except at the origin. The condition of smoothness - which Swan calls nonsingularity - is somewhat more restrictive than our condition of regularity: a simple example is the form

$$x^2+ty^2$$

over $k_0(t)$, the field of rational functions in t over a field k_0 of characteristic 2 , which is regular but not smooth.

Many open problems about MCM modules over hypersurfaces remain. Some which seem to us particularly interesting are:

1) It seems natural to try to imitate the theory of Knörrer [1986] and consider MCM modules over a ring of the form

$$R = k[x_1,\ldots,x_n,y_1,\ldots,y_m]/f(x)+Q(y)$$

where Q is a quadratic form and k is an arbitrary field. Of course if M is an MCM R-module, then $N = M/(y_1,\ldots,y_m)M$ is an MCM module over

$$R_1 = R/(y) = k[x_1,\ldots,x_n]/f(x) .$$

Further,

$$N \otimes_{R_1} k = M \otimes_R k = (M/(x_1,\ldots,x_n)M) \otimes_{k[y]/Q(y)} k,$$

so $N \otimes k$ is naturally a $C_0(Q)$-module. Can one establish an inverse construction from some compatibility between these two structures on N ?

2) Let Q be a nonsingular quadric on V and $R = k[V^*]/Q$ as above. What R-modules have the indecomposable MCM R-modules as syzygies? For example, if $C_0(Q)$ is a division ring, then the dimension of the unique indecomposable $C_0(Q)$-module (which is $C_0(Q)$ itself!) is $2^{\dim V-1}$, and it follows that the corresponding MCM R-module is the nth syzygy of the residue class field of R (for all $n \geq \dim R$). In general, the nth syzygies of the residue class field always have this number of generators (and the rank $2^{\dim V-2}$), so they are always decomposable if $C_0(Q)$ is not a division ring. In the appendix (Cor. 2.) we give a necessary and sufficient condition for an MCM R-module to be a syzygy module of an R-module of finite length.

3) What is the nature of the matrix factorization of a quadric? More specifically, for Q regular, or for linear factorizations in general we know that matrix factorizations of Q correspond to modules over $C_0(Q)$ as well as to MCM R-modules. If $C_0(Q)$ is simple, so there is just one nonfree indecomposable MCM R-module, then it is easy to see that the corresponding matrix factorization can be chosen in the form $\varphi^2 = Q\cdot 1$. Can φ always be taken to be either symmetric or skew-symmetric? What determines this? Of course if A_0 is the unique indecomposable $C_0(Q)$-module in this case, then A_0 will be iso-

morphic to A_o^* (with action of $C_o(Q)$ via the "main involution"), so we may regard the matrix factorization as involving $\varphi : F^* \to F$ in a fairly natural way. If C_o is not simple, so $C_o(Q) = (k \times k) \otimes_k \overline{C}$, with $\overline{C}$ central simple, then as we show in Theorem 3.2, the first syzygy module N of one of the nonfree indecomposable MCM R-modules M has $k \otimes_R N$ the contagredient $C_o(Q)$-module to $k \otimes_R M$ iff dim V ≡ ≡ 2(mod 4), so at least in this case we may write the minimal free resolution of M naturally as

$$F^* \xrightarrow{\varphi} F \xrightarrow{\psi} F^* \xrightarrow{\varphi} F \to M .$$

Can φ and ψ be taken to be of some special form? In the case dim V ≡ 0(4), $M \cong N^*$, so φ is similar to ψ^*. When can φ and ψ be chosen so that φ <u>is</u> the transpose of ψ ?

1) Regular forms and semi-simple even Clifford algebras

In this section we present the facts about Clifford algebras. Some of the results are new in characteristic 2. The reader primarily interested in Cohen-Macaulay modules may wish to read at least the beginning of section 2, to see how the Clifford algebra arises naturally in a description of linear matrix factorizations, before this section. The text by Jacobson [1980] sect. 4.8 contains an excellent introduction to the theory of Clifford algebras. Other good references are Atiyah-Bott-Shapiro [1964], Lam [1980], and Scharlau [1985].

a) Results

Throughout, we let Q be a nonzero quadratic form on a vectorspace V of dimension r over a field k.

We regard Q as a map $Q : V \to k$ satisfying

(i) $Q(\alpha v) = \alpha^2 Q(v)$ for all $\alpha \in k$,

(ii) $(v,w)_Q = Q(v+w)-Q(v)-Q(w)$ is bilinear.

Quadratic forms may be identified with element of $S_2(V^*)$ and we will henceforth make this identification.

We will write V_{ins} for the nullspace of $(,)_Q$.

The Clifford algebra of Q is the tensor algebra $\otimes V$ modulo the relations generated by those of the form $v\otimes v - Q(v)$ for $v \in V$. Since both terms of the relations are of even degree, $C(Q)$ is $\mathbb{Z}/2\mathbb{Z}$-graded; we write $C_o(Q)$ and $C_1(Q)$ for the even and odd parts. We have $\dim C(Q) = 2^r$, $\dim C_o(Q) = 2^{r-1}$.

We will say that a quadratic form Q is regular if the projective hypersurface it defines in $\mathbb{P}(V^*)$ is regular, in the sense that all its local rings are regular local rings, or, equivalently, if the ring $R = k[V^*]/Q$ becomes regular when localized at any prime other than that generated by V^*. If $\operatorname{char} k \neq 2$, this is equivalent to the usual definition of nonsingularity of a quadratic form.

(The condition that the projective hypersurface defined by Q be smooth over k is equivalent to regularity in characteristics other than 2, but definitely stronger in characteristic 2. Smoothness, in this sense, is equivalent in all characteristics to the condition called nonsingularity by Swan [1985]).

Regularity is a very natural condition from the point of view of the even Clifford algebra:

Theorem 1.1: The following conditions are equivalent:

1) Q is regular
2) $C_o(Q)$ is semi-simple
3) Either the nullspace V_{ins} of $(,)_Q$ is zero, or the restriction of Q to V_{ins} has the form

$$\sum_{i=1}^{n} s_i x_i^2,\ s_1 \neq 0, \quad \text{and} \quad [k(\{(s_i/s_1)^{1/2}\}):k] = 2^{\dim V_{ins}-1}$$

Remarks: 1) If $\operatorname{char} k \neq 2$, then $Q|V_{ins} \equiv 0$, so $V_{ins} = 0$ iff Q is regular.

2) Of course any basis for V_{ins} is orthogonal, in the sense that Q becomes diagonal; and it is easy to see that the condition in 3) is independent of the basis chosen, and is in fact equivalent to the basis-free condition:

$$[k(\{(Q(v)/Q(w))^{1/2} | v,w \in V_{ins}, Q(w) \neq 0\}) : k] = 2^{\dim V_{ins}-1}.$$

3) We will give below a direct but in characteristic 2 somewhat computional proof of the implication 1) $\Rightarrow$ 3). Our original proof was indirect but more conceptual: The equivalence of 3) and 2) is

fairly easy. Given the category equivalence described in section 2, below, the equivalence 1) $\Longleftrightarrow$ 2) follows from the Theorem of M. Auslander [1986] that a local Cohen-Macaulay ring has an isolated singularity if it has finite CM-type.

The main facts we will use about the even Clifford algebra of a regular form are summarized in the next Theorem, where we write

$$\det Q = \Pi s_i, \quad \text{if } Q \sim \Sigma s_i x_i^2, \ \text{char } k \neq 2$$
$$\text{arf } Q = \Pi Q(e_i)Q(f_i), \ \text{if char } k = 2, \ (\ ,\ )_Q \text{ is nongenerate,}$$

and $\{(e_i, f_i)\}$ is a symplectic basis for V with respect to $(\ ,\ )_Q$.

As is well known, $\det Q$ yields an invariant of Q in $k^*/(k^*)^2$, and arf Q an invariant modulo $\{x^2+x \mid x \in k\}$.

Theorem 1.2: If $(\ ,\ )_Q$ is nondegenerate, then $C_o(Q)$ is central simple over k if $\dim V$ is odd, and else the tensor product over k of a central simple algebra with the extension

$$k[x]/x^2-(-1)^{(\dim V)/2}\det Q \quad \text{if char } k \neq 2$$

$$k[x]/x^2+x+\text{arf } Q \qquad \text{if char } k = 2 .$$

If $(\ ,\ )_Q$ is degenerate, but Q is regular, then $C_o(Q)$ is the tensor product of a central simple algebra with the purely inseparable field extension, of degree $2^{\dim V_{ins}-1}$, given in part 3) of Theorem 1.1. In either case, the central simple factor of $C_o(q)$ is Brauer equivalent to a product of quaternion algebras over k.

Remark: Q is called non-defective if $Q|_{V_{ins}} \equiv 0$, as is true for every Q when char $k \neq 2$. The Theorems are new, if at all, only in the defective case; see the proof below.

We will also use two facts about quaternion algebras. Recall that a local field is either $\mathbb{R}$, $\mathbb{C}$, or complete with respect to a non-archimedean discrete valuation and with finite residue field; a global field is a finite algebraic extension of either the rational field, or a field of rational functions over a finite field. The following results are classical. For the convenience of the reader we will include sketches of proofs.

Theorem 1.3: If k is any local or global field, then any tensor product of quaternion algebras over k is Brauer equivalent to a quaternion algebra. In particular, the division algebra corresponding to the central simple factor described in Theorem 1.2 has dimension either 1 or 4 over k.

By contrast, a product of generic quaternion algebras is a division algebra. In terms of the even Clifford algebra, we have:

Theorem 1.4: Let $K = k(s_1,\ldots,s_n)$ be a field of rational functions over a field k. If Q is the quadratic form $\sum_1^n s_i x_i^2$, then $C_c(Q)$ is a central division algebra over an extension field of K as described in 1.2.

b) Proofs and sketches

Proof of Theorem 1.1 and 1.2: First, if Q is non-defective - that is, the restriction of Q to V_{ins} is identically 0 - everything is known: For in Theorem 1.1 condition 3) becomes $V_{ins} = 0$ which by the Jacobian criterion implies nonsingularity, whereas if $V_{ins} \neq 0$, then V_{ins} is contained in the singular locus of $\{Q = 0\}$; thus 1) $\Longleftrightarrow$ 3) in this case. That $V_{ins} = 0$ implies the semi-simplicity of $C_o(Q)$ is classical; see for example Jacobson [1980] section 4.8, which also contains the conclusions of Theorem 1.2 in this non-defective case. On the other hand, even if $V_{ins} \neq 0$, then choosing a non isotropic vector v in V (which is possible since we assumed at the outset of this section that $Q \neq 0$) we see that

$$C_o(Q) \cong C(-Q(v)\cdot Q|_{v^\perp})$$

(as for example in Jacobson Thm. 4.13; the hypotheses there are superfluous for this conclusion) and

$$C(-Q(v)Q|_{v^\perp}) = A \mathbin{\hat{\otimes}} C(-Q(v)Q\,|\,V_{ins}) ,$$

when $\hat{\otimes}$ is the graded tensor product and A is the Clifford algebra of Q restricted to a complement of V_{ins}. Since

$$C(-Q(v)\cdot Q|_{V_{ins}}) = C(0|_{V_{ins}}) \cong \Lambda(V_{ins}),$$

the exterior algebra, we see this is not semi-simple and so 1) $\Longleftrightarrow$ 2) in the non-defective case.

Thus we may henceforward assume

$$\text{char } k = 2$$

$$V_{ins} \neq 0$$

$$Q|_{V_{ins}} \neq 0 .$$

We first prove the equivalence of 1) and 3) in Theorem 1.1. Choose a complement V_{sep} for V_{ins} in V, and a symplectic basis $\{(e_i,f_i)\}$ of V_{sep} with respect to the form $(\ ,\ )_Q$. If y_i, z_i are dual to e_i, f_i, we may write Q as

$$Q = \Sigma y_i z_i + \Sigma r_i y_i^2 + \Sigma t_i z_i^2 + \sum_{j=1}^{n} s_j x_j^2 \quad \text{with} \quad n \geq 1 .$$

Now suppose condition 3) is satisfied. Passing to an affine open cover, we may assume that one of the variables, say x_1 or y_1, is 1. We must show that Q is not contained in the square of any maximal ideal N of $S = k[\underline{x},\underline{y},\underline{z}]$. For this it suffices to show that there is a derivation D of S with $D(Q) \notin N$. Suppose on the contrary that $D(Q) \in N$ for all derivations D. Now the derivations $\frac{\partial}{\partial y_i}$ and $\frac{\partial}{\partial z_i}$ take the values z_i and y_i on Q, respectively, so we need only to consider those primes N containing all y_i, z_i, and Q. In particular, we have no problem with the open set $y_1 = 1$, so we may take $x_1 = 1$.

By condition 3), the $(s_i/s_1)^{1/2}$ are 2-independent over k in the sense of, for example, Matsumura [1980] section 38, and thus the s_i/s_1 are 2-independent over k^2. It follows that there is a derivation D_j of k taking s_i/s_1 to δ_{ij} ("Kroneker δ"), and we may extend D_j to all of S so that $D_j(y_i) = D_j(z_i) = 0$ for all i. Applying D_j to $\frac{1}{s_1} Q$ we get x_j^2 modulo the y_i and z_i, for $j > 1$, so N must contain all these as well. But then N contains 1 (since $x_1 = 1$ by assumption), a contradiction. Thus the hypersurface associated to Q is regular, proving 3) $\Rightarrow$ 1).

For the converse, note that if $n < \dim V_{ins}$, then the projective hypersurface associated to Q is obviously singular, so we may assume $n = \dim V_{ins}$. Suppose that

$$[k(\{(s_i/s_1)^{1/2}\}) : k] < 2^{n-1}$$

we will find a singular maximal ideal containing all the y_i and z_i.

Replacing Q by $\frac{1}{s_1}Q$, we may suppose $s_1 = 1$. Rearranging the s_i, we may assume that

$$k(\{\sqrt{s}_i\}_{i=2,\ldots,n}) = k(\{\sqrt{s}_i\}_{i=2,\ldots,m})$$

for some $m < n$ such that $s_2,\ldots,s_m$ are 2-independent.

To find a singular point on the hypersurface $Q = 0$, we work on the affine open set $x_{m+1} = 1$, and employ the criterion given in Matsumura [1980] Theorem 95, according to which a prime ideal $N \subset S' = k[\{y,z\},x_1,\ldots,x_m,x_{m+2},\ldots,x_n] = k[V^*]/(x_{m+1}-1)$, of height h, containing $Q' = Q|_{x_{m+1}} = 1$ is a singular point of $\operatorname{Spec} S'/Q'$ if there is a set of derivations $\Delta = \{D_i\}$ and a set of elements $\{g_j\} \subset N$ such that

1) $D_iQ' \in N$ for all i.

2) Some $h \times h$ minor of the matrix

$$(D_i g_j)$$

is not contained in N.

First we define Δ. Since $s_2,\ldots,s_m$ are 2-independent over k^2, the $k^2(s_2,\ldots,s_m)$-vectorspace

$$\operatorname{Der}_{k^2} k^2(s_2,\ldots,s_m)$$

has a basis consisting of derivations D_i; $i = 2,\ldots,m$; such that

$$D_i(s_j) = \delta_{ij}$$

and we may extend these first to $\operatorname{Der}_{k^2}(k)$ and then also to derivations of S' which annihilate the variables. We take Δ to be the set

$$\Delta = \{D_2,\ldots,D_m, \frac{\partial}{\partial y_i}, \frac{\partial}{\partial z_j}, \frac{\partial}{\partial x_k}; \text{ all } i,j,k\}.$$

Next we define N, by giving generators. Let N' be the ideal generated by all the y_i and z_i together with $x_{m+2},\ldots,x_n$, and let N'' be the ideal generated by N' and the following polynomials $f_1,\ldots,f_m$:

$$f_1 = x_1^2 + s_{m+1} + \sum_{i=2}^{m} s_i D_i(s_{m+1}) ,$$

and for $i = 2,\ldots,m$,

$$f_i = x_i^2 + D_i(s_{m+1}) .$$

Note that

$$Q' = f_1 + \sum_2^m s_i f_i \mod N' ,$$

so $Q' \in N''$. Also $D_i(Q') \equiv f_i \mod N'$, and so $D_i(Q') \in N''$ for all i.

Clearly N' is prime, but contains neither Q' nor $\Delta Q'$. N'' on the other hand contains both Q' and $\Delta Q'$ but is generally not prime. We will remedy this by adding square roots of some of the f_i to N'' to obtain the ideal N.

Rearranging $x_2,\ldots,x_m$ if necessary, we may suppose that

$$D_2(s_{m+1}),\ldots,D_l(s_{m+1})$$

are 2-independent over k^2 while

$$D_{l+1}(s_{m+1}),\ldots,D_m(s_{m+1})$$

are 2-dependent on them (the case $l = m$ is not excluded), say

$$\sqrt{D_{l+k}(s_{m+1})} = q_{l+k}(\sqrt{D_2(s_{m+1})},\ldots,\sqrt{D_l(s_{m+1})}) \quad \text{for } k = 1,\ldots,m-l$$

where the q_{l+k} are polynomials with coefficients in k.

It follows that for $k = 1,\dots,m-1$,

$f_{l+k} \equiv (x_{l+k}+q_{l+k}(x_2,\dots,x_l))^2$ modulo $N'+(f_2,\dots,f_l)$, and we set

$$g_k = f_k \ , \ k = 2,\dots,l$$

$$g_{l+k} = x_{l+k}+q_{l+k}(x_2,\dots,x_l), \ k = 1,\dots,m-1 \ .$$

Further, if $s_{m+1} + \sum_{i=2}^{m} s_i D_i(s_{m+1})$ is not 2-dependent on $D_2(s_{m+1}),\dots,D_l(s_{m+1})$ we set $g_1 = f_1$, else we write its square roots as $q_1(\sqrt{D_2(s_{m+1})},\dots,\sqrt{D_2(s_{m+1})})$ where q_1 is a polynomial with coefficients in k, and set

$$g_1 = x_1+q_1(x_2,\dots,x_l) \ .$$

We call these case I and case II, respectively.

We take $N = N'+(g_1,\dots,g_m)$. From the construction it is clear that $Q',\Delta Q'$ are in N and N is a maximal ideal of S' which thus has height equal to $\dim V-1$.

Applying $\Delta_I = \{\frac{\partial}{\partial y_i} ; \frac{\partial}{\partial z_j} ; \frac{\partial}{\partial x_{l+k}} ;$ all i,j and $k = 1,\dots,n-1\}$ to $A_I = \{y_i;z_j;g_{l+k};$ all i,j and $k = 1,\dots,n-1\}$ yields an identity matrix of size $\dim V-l-1$ in case I. In case II, $\Delta_{II} = \Delta_I \cup \{\frac{\partial}{\partial x_1}\}$ applied to $A_{II} = A_I \cup \{g_1\}$ yields an identity matrix of size $\dim V-1$. It thus suffices to show that the matrix

$(D_i g_j)$ $i = 2,\dots,m$
$j = 1,\dots,l$ in case I
$j = 2,\dots,l$ in case II

which is actually a matrix with coefficient in $k^2(s_2,\dots,s_m) \subset k \subset S'/N$, has rank equal to l in case I ($l-1$ in case II). Of course $D_i g_j = D_i(g_{j,0})$, where $g_{j,0}$ is the constant term of g_j . Now the $D_2,\dots,D_m$ form a $k^2(s_2,\dots,s_m)$-basis of $\mathrm{Der}_{k^2} k^2(s_2,\dots,s_m)$, and thus linear combinations of them induce all derivations of $k^2(g_{1,0},\dots,g_{l,o})$. The rank of the given matrix $(D_i g_j)$ is consequently the maximal number of 2-independant elements among the $g_{j,o}$, so by construction it is the desired number l in case I (or $l-1$ in case II). Thus $N/(Q)$ represents a singular point of the projective hypersurface $Q = 0$, proving 1) $\Rightarrow$ 3) in Theorem 1.1.

Continuing with the case char $k = 2$, $V_{ins} \neq 0$, $Q|V_{ins} \not\equiv 0$, we analyze $C_o(Q)$; the result will show that 2) and 3) of Theorem 1.1 are equivalent and establish the conclusions of Theorem 1.2.

Since $Q|V_{ins} \not\equiv 0$, we may choose a non-isotropic vector $v_1 \in V_{ins}$. Taking V_1 to be any complement of v_1 in V containing a codimension 1 subspace of V_{ins}, the map

$$C_o(Q) \ni v_1 v \mapsto v \in V_1 \subset C(Q(v_1)Q|V_1)$$

induces an epimorphism of algebras which must be an isomorphism since both sides have dimension $2^{\dim V-1}$, and setting $Q_1 = Q(v_1)Q|V_1$, we need only analyze $C(Q_1)$.

Let $V_{1,ins} = V_{ins} \cap V_1$, which by construction has codimension 1 in V_{ins}, and let V_{sep} be a complement of $V_{1,ins}$ in V_1. We have $C(Q_1) = C(Q_1|V_{sep}) \otimes_k C(Q_1|V_{1,ins})$, (we can ignore the grading on the tensor product because we are in characteristic 2). On the other hand $(\ ,\ )_{Q_1} = Q(v_1)\,(\ ,\ )_{Q|V_1}$, so $Q_1|V_{sep}$ is non-defective and nonsingular; we may now apply the results of Jacobson [1980], section 4.8, exercises, and we see that $C(Q_1|V_{sep})$ is central simple over k, and is equivalent to a product of quaternion algebras.

On the other hand, since $V_{1,ins}$ possesses an orthogonal basis, we see that $C(Q_1|V_{1,ins})$ is commutative. In fact, if V_{ins}^* has the basis $x_1,\dots,x_m$ $(m \geq n)$, $Q|V_{ins}$ has the form

$$Q|V_{ins} = \sum_1^m s_i x_i^2 ,$$

and V_1 is the kernel of the linear form x_1, then

$$Q_1|_{V_{1,ins}} = \sum_2^n s_1 s_i x_i^2 \sim \sum_2^n (s_i/s_1) x_i^2 ,$$

and

$$C(Q_1|_{V_{1,ins}}) \cong k[e_2,\dots,e_m] \big/ (e_2^2-(s_2/s_1),\dots,e_n^2-(s_n/s_1),e_{n+1}^2,\dots,e_m^2),$$

so that $C(Q_1|_{V_{1,ins}})$ is semi-simple(in fact, a field) iff $n = m$ and $s_2/s_1,\dots,s_n/s_1$ are 2-independant, that is, iff condition 3) of Theorem 1.1.is satisfied. This completes the proof of Theorem 1.1 and Theorem 1.2.

Example: The proof of Theorem 1.1, 3) $\Rightarrow$ 1) yields an explicit construction of a singular point of $Q = 0$ if Q is not regular. Consider for example the form

$$Q = x_1^2+sx_2^2+tx_3^2+stx_4^2 ,$$

where s and t are indeterminates over a field k_o of characteristic 2. Then Q is not regular over $k = k_o(s,t)$, since

$$[k(\sqrt{s},\sqrt{t}) : k] = 4 < 2^{\dim V_{ins}-1} = 8 .$$

A singular point N of $Q = 0$ can be found on the affine open set $x_4 = 1$, since s and t are 2-independent over k^2, while $s\cdot t$ depends on s and t. Following the arguments of the proof we see that we are in case II, and that $N = (x_1+x_2x_3, x_2^2+t, x_3^2+s)$ is a singular point of the affine hypersurface $Q' = Q|_{x_4 = 1}$.

Proof of Theorem 1.3 (With thanks to B. Gross and M. Levine): It suffices to show that any product of quaternion algebras over a local or global field is split by a separable quadratic extension of that field; for such an extension is necessarily galois, and the splitting allows us to write an equivalent algebra as 4-dimensional quaternion algebra, associated to the galois 2-cocycle.

The Brauer group of any local field is $\mathbb{Q}/\mathbb{Z}$, which contains a unique element of order 2, so the Proposition is trivial in the local case.

In the global case the fact is well known, see Vignéras [1980], Theorem 3.8, but we include a proof for the reader's convenience.

Let k be a global field, and let A be a central simple k-algebra which is a product of quaternion algebras. By the Hasse principle, a field extension k'/k will split A iff its completion splits A at the finitely many valuations v such that A_v is not split already. Thus it sufficcs to find an extension k'/k which localizes to given extensions at finitely many places. If $\operatorname{char} k \neq 2$, the local extensions have the form $k_v(\sqrt{s_v})$, and it is enough to choose $s \in k$ approximating each s_v (since any unit closely approximating 1 in k_v has a square root.) If $\operatorname{char} k = 2$, the local extensions have the form $k_v[x]/x^2+x+s_v$ for $s_v \in k_v$, and it suffices to find an $s \in k$ which differs from each of the finitely many s_v by an element of

$$P(k_v) = \{x+x^2 \mid x \in k_v\}.$$

But since $t \to t+t^2$ is a continuous homomorphism from $\mathcal{O}_v^+$ to $\mathcal{O}_v^+$, and since $(1+ax)+(1+ax)^2 = ax+a^2x^2$, we see that

$$P(k_v) \supset \{x \in k_v \mid v(x) = 1\} ,$$

so again it suffices to choose s approximating each s_v.

<u>Proof of Theorem 1.4</u>: We have $C_o(Q) \cong C(s_1 \cdot Q|\ker x_1)$ as usual so it suffices to prove the corresponding facts about the whole Clifford algebra.

In $C(Q)$ we have $s_i = x_i^2$, so $C(Q)$ may be viewed as the (non-commutative) field of fractions of the skew polynomial ring $k[x_1,\ldots,x_n]$ with $x_ix_j = -x_jx_i$. A proof that such a skew polynomial ring is a domain, and has a field of fractions, can be found for instance in section 12.2 in the book of P.M. Cohen [1977].

2. Cohen-Macaulay modules and modules over the even Clifford algebra

Let k be a field, let V be a vectorspace of dimension r over k, and let Q be a quadratic form on V, so that $Q \in k[V^*] =: S$. Set $R = S/Q$, and let m be the maximal ideal generated by V^*.

We will say that a graded R-module M is linear if it is an MCM module admitting a graded free presentation of the form

$$R^n(-1) \xrightarrow{\varphi} R^m \to M \to 0,$$

where φ is a matrix of linear forms.

Suppose M is linear, with presentation as above. From the theory of matrix factorizations it follows at once that if M has no free summand, then $m = n$. Since the degree 1 parts R_1 and S_1 of R and S are equal, we may canonically lift φ to a matrix of linear forms over S. If (φ,ψ) is the matrix factorization corresponding to M, then since φ defines a monomorphism of free S-modules, ψ must have only linear entries as well. Thus the syzygies of linear modules are linear modules, up to a shift in degree.

We will say that a module over R_m or $\hat{R}_m$, the localization and completion of R respectively, is linearizable if it is isomorphic to the localization or completion of a linear module.

We next explain the fundamental construction of this paper, which gives an equivalence between linear R-modules and modules over the even Clifford algebra, $C_0(Q)$.

We have already seen that linear R-modules correspond to linear matrix factorizations, that is, pairs of square matrices (φ,ψ) with entries in $S_1 = V^*$ such that $\varphi\psi = \psi\varphi = Q \circ 1$. It is with these matrix factorizations that we will work.

As we remarked in the introduction, the category of modules over $C_0(Q)$ is equivalent to the category of $\mathbb{Z}/2\mathbb{Z}$-graded modules over $C(Q)$ via

$$A_0 \mapsto C(Q) \otimes_{C_0(Q)} A_0$$

$$= A_0 \oplus C_1(Q) \otimes_{C_0(Q)} A_0$$

$$A_0 \oplus A_1 \mapsto A_0.$$

Thus we may work with $\mathbb{Z}/2\mathbb{Z}$-graded $C(Q)$-modules.

If A is a vectorspace, then a $C(Q)$-module structure on A is a linear family of maps.

$$\lambda_v : A \to A, \qquad v \in V,$$

such that $\lambda_v^2 = Q(v)$. If A is $\mathbb{Z}/2\mathbb{Z}$-graded, then each λ_v has two components

$$\varphi_v : A_1 \to A_0$$
$$\psi_v : A_0 \to A_1,$$

and the condition becomes

$$\varphi_v\psi_v = \psi_v\varphi_v = Q(v) \cdot 1.$$

On the other hand, a "linear family of maps induced by V", such as φ_v is the same as a single matrix φ whose entries are linear forms in $S = k[V^*]$; and the conditions $\varphi_v\psi_v = \psi_v\varphi_v = Q(v) \cdot 1$ become

$$\varphi\psi = \psi\varphi = Q \cdot 1$$

as matrices over S. Thus we see that: <u>a linear matrix factorization over S is the same thing as a $\mathbb{Z}/2\mathbb{Z}$-graded $C(Q)$-module.</u> If the matrix factorization (φ,ψ) corresponds to the linear R-module $M = \text{Coker}\,\varphi$, then clearly $A_0 = k \underset{R}{\otimes} M$, and $A_1 = k \underset{R}{\otimes} \Omega^1(M)$, where $\Omega^1(M)$ is the first syzygy-module of M without free summands, form the corresponding $\mathbb{Z}/2\mathbb{Z}$-graded $C(Q)$-module, and $k \otimes M$ is the (underlying vectorspace of) the corresponding $C_0(Q)$-module.

It is not difficult to see that the correspondences just given are the same as the ones described via Ext and Tor in the introduction to this paper; since we will not use this fact explicitly, we will leave it to the reader.

We summarize and complete the properties of this construction:

<u>Theorem 2.1:</u> The functor

$$T : M \to T(M) = M/_{mM}$$

is an equivalence of categories from the category of linear R-modules without free summands, and maps of degree 0, to the category of modules over $C_0(Q)$, the even Clifford algebra of Q. Under this equivalence, the first syzygy module $\Omega^1 M$ corresponds to the C_0-module

$$T(\Omega^1 M) = C_1(Q) \otimes_{C_0(Q)} T(M),$$

and the dual module $M^* = \mathrm{Hom}_R(M,R)$ corresponds to the contragredient

$$T(M^*) = \mathrm{Hom}(C_1(Q) \otimes_{C_o(Q)} T(M),k)$$

regarded as a $C_o(Q)$-module via the main involution of $C_o(Q)$. The dimension of $T(M)$ as a k-vectorspace, which is the minimal number of generators of M, is equal to twice the rank of M (which may be a half-integer if $r = 1$ or 2).

The next two results show that when Q is regular, in the sense of section 1, the equivalence described in Theorem 1 catches the essence of the category of all MCM modules, even over the completion $\hat{R}_m$.

Theorem 2.2: If Q is regular, then every MCM R_m-(or $\hat{R}_m$-) module M is linearizable, so that $M \cong (\mathrm{gr}_m M)_m$ (or $(\widehat{\mathrm{gr}_m M})_m$ respectively).

This says in effect that when Q is regular, linear modules are really arbitrary MCM modules with their gradings normalized. It will be proved by treating arbitrary MCM $\hat{R}_m$-modules as high degree deformations of linear ones, by means of matrix factorizations. Morally speaking, such deformations are trivial because $\mathrm{Ext}^1_R(M,M)$, the tangent space to the versal deformation space, is nonzero in (at most) degree -1 when M is linear, as the following result shows.

To simplify notation, we imitate the notion of Tate cohomology for a finite group. If

$$\dots \to F_1 \to F_o \to F_1 \to F_o \to M \to 0$$

is the periodic R-free resolution of an MCM module M, we may continue it to the right as well, and set

$$\underline{\mathrm{Ext}}^i_R(M,N) = H_i(\mathrm{Hom}(\dots \to F_1 \to F_o \to F_1 \to F_o \to \dots),N)),$$

so that

$$\underline{\mathrm{Ext}}^i_R = \mathrm{Ext}^i_R \quad \text{for } i > 0$$

$\underline{\mathrm{Ext}}^0_R = \underline{\mathrm{Hom}}$, that is homomorphisms modulo those factoring through projectives,

$\underline{\mathrm{Ext}}^{-1}_R =$ kernel of the natural map $N \otimes M^* \to \mathrm{Hom}_R(M,N)$

and

$$\underline{\mathrm{Ext}}^i_R(M,N) = \mathrm{Tor}^R_{-i+1}(N,M^*) \quad \text{for } i < -1$$

Proposition 2.3: Let Q be an arbitrary nonzero quadratic form. If M and N are linear modules such that

$$\mathrm{Ext}^1_R(M,N) \quad \text{and} \quad \mathrm{Ext}^2_R(M,N)$$

have finite length, then

$$\underline{\mathrm{Ext}}^i_R(M,N)$$

is a vectorspace concentrated in degree $-i$; in particular, if Q is regular, this holds for any MCM modules.

Remarks: 1) An interesting singular example to which the Proposition applies is the case

$$R = k[x,y]/x^2 + y^2, \quad \mathrm{char}\, k = 2,$$

$$M = \mathrm{coker}\begin{pmatrix} x & y \\ y & x \end{pmatrix},$$

N any MCM module,

(since M is free on the punctured spectrum of R).

2) The technique of proof will show that for arbitrary Q and linear R-modules M and N, $\underline{\mathrm{Ext}}^i_R(M,N)$ is at least generated in degree $-i$.

Proof of Theorem 2.1: The first statement of the Theorem and the identification of $T(\Omega^1 M)$ are immediate from the construction given before the Theorem.

To identify $T(M^*)$ (where M^* means $\mathrm{Hom}_R(M,R)$) note that since M is an MCM R-module,

$$M^* \cong \mathrm{Ext}^1_S(M,S(-2)),$$

so $M^* = \mathrm{coker}\,\varphi^*$. Thus the matrix factorization corresponding to M^* is (φ^*,ψ^*), and

$$T(M) = (k \otimes_R \Omega^1 M)^*$$

as vectorspaces.

Further, it is clear from the basic construction that each $v \in V$ acts on the $\mathbb{Z}/2\mathbb{Z}$-graded $C(Q)$-module

$$(k \otimes_R \Omega^1 M)^* \oplus (k \otimes_R M)^*$$

by the dual of its action on

$$(k \otimes_R M) \oplus (k \otimes_R \Omega^1 M).$$

Thus the $\mathbb{Z}/2\mathbb{Z}$-graded $C(Q)$-module structure corresponding to M^* is that obtained by regarding $(k \otimes_R \Omega^1 M)^* \oplus (k \otimes_R M)^*$ as a $C(Q)^{op}$-module in the natural way, and then using the isomorphism $C(Q) \to C(Q)^{op}$ which is the identity on $V \subset C(Q)$, that is, the main involution of $C(Q)$. Passing to the $C_0(Q)$-module $(k \otimes_R \Omega^1 M)^* = (C_1(Q) \otimes_{C_0(Q)} TM)^*$, we get the required statement.

It remains to prove the last statement, regarding the rank of an MCM R-module M without free summands. We use the following easy

<u>Lemma 2.4:</u> If k is any field, $Q \in k[x_1,\ldots,x_r]$ any quadratic form, then there exists a regular sequence of length $r - 1$ in $R = k[x_1,\ldots,x_r]/Q$ consisting of elements of degree 1.

<u>Proof:</u> It suffices by induction to find a degree 1 nonzerodivisor, given $r > 1$. Because S is factorial, any degree 1 zerodivisor must be a factor of Q. But $k[x_1,\ldots,x_r]$ has at least 3 distinct linear forms (mod scalars), so not all can be factors.

By the Lemma we may assume that $x_1,\ldots,x_{r-1}$ is a regular sequence modulo Q. Then

$$\begin{aligned} \operatorname{rank}_{k[x_1,\ldots,x_r]/Q} M &= \tfrac{1}{2} \operatorname{rank}_{k[x_1,\ldots,x_{r-1}]} M \\ &= \tfrac{1}{2} \dim_k M/(x_1,\ldots,x_{r-1})M' \end{aligned}$$

which holds because M is free as a $k[x_1,\ldots,x_{r-1}]$-module.

We will complete the proof by using induction on r to show

$$\begin{aligned} &M/(x_1,\ldots,x_{r-1})M \\ &= M/(x_1,\ldots,x_r)M, \end{aligned}$$

which is of course $T(M)$.

If $r = 1$, $R \cong k[x]/(x^2)$, and since M has no free summands, it is annihilated by x.

In the general case, any free summand of $M/x_1M =: \overline{M}$ could be lifted to a free summand of M:

$$\begin{array}{ccc} & & 0 \\ & & \downarrow \\ & & x_1 R \cong R \\ & & \downarrow \\ M & \dashrightarrow\!\!\!\!\rightarrow & R \\ \downarrow & & \downarrow \\ M/x_1 M & \twoheadrightarrow & R/x_1 R \\ & & \downarrow \\ & & 0 \end{array}$$

because $\mathrm{Ext}^1_R(M,R) = 0$, so M/x_1M is a Cohen-Macaulay R/x_1R-module without free summands, and

$$\begin{aligned} M/(x_1,\ldots,x_{r-1})M &= \overline{M}/(x_2,\ldots,x_{r-1})\overline{M} \\ &= \overline{M}/(x_2,\ldots,x_r)\overline{M} \\ &= M/(x_1,\ldots,x_r)M \end{aligned}$$

as required. This completes the proof of Theorem 2.1.

Proof of Proposition 2.3: If Q is regular, then every localization of R at a relevant prime is regular, so M is free on the punctured spectrum, and $\underline{\mathrm{Ext}}^i_R(M,N)$ has finite length for all i. Thus it suffices to prove the first statement. We do an induction using the following well-known change of rings result:

Lemma 2.5: Let R be any graded ring, and let N be a graded R-module. If $x \in R$ is a nonzerodivisor on R and N, and M is a graded R/xR-module, then

$$\mathrm{Ext}^i_R(M,N) \cong \mathrm{Ext}^{i-1}_{R/xR}(M,N/xN\ (1).$$

Proof: The spectral sequence

$$\mathrm{Ext}^i_{R/xR}(M,\mathrm{Ext}^j_R(R/xR,N)) \Rightarrow \mathrm{Ext}^{i+j}_R(M,N)$$

degenerates, since

$$\mathrm{Ext}^*_R(R/xR,N) = \mathrm{Ext}^1_R(R/xR,N) = N/xN(1),$$

to give the desired isomorphisms.

We now continue with the proof of Proposition 2.3. $\underline{\mathrm{Ext}_R^i}(M,N)$ is a quotient of $\mathrm{Hom}_R(\Omega^i M,N)$, when $\Omega^i M$ is the ith syzygy. Since $\Omega^i M$ is generated in degree i, we see that

$$\mathrm{Ext}_R^i(M,N)_p = 0 \quad \text{for} \quad p < -i.$$

We now do induction on dim R. If dim R = 0, then either R is a field (and the result is obvious) or $R \cong k[x]/x^2$. We may clearly assume that M and N are indecomposable and without free summand, whence $M = R/xR$, $N = R/xR$, and the result follows by direct computation.

Now assume dim R > 1. By Lemma 2.4 there is an $x \in R$, which is a nonzerodivisor on R, and thus on M and N. By periodicity, it is enough to prove the Proposition when i is large. From $0 \to M(-1) \xrightarrow{x} M \to M/_{xM} \to 0$ we get the following exact sequence, the first of whose terms we identify by using Lemma 2.5:

$$\begin{array}{l} \to \mathrm{Ext}_R^i(M/xM,N) \to \mathrm{Ext}_R^i(M,N) \xrightarrow{x} \mathrm{Ext}_R^i(M,N)(1) \to \mathrm{Ext}_R^{i+1}(M/xM,N) \to \\ \quad \| \wr \\ \mathrm{Ext}_{R/xR}^{i-1}(M/xM,N/xN)(1) \end{array}$$

It now follows that all $\mathrm{Ext}_{R/x}^i(M/xM,N/xN)$ have finite length, so by induction

$$\mathrm{Ext}_{R/xR}^{i-1}(M/xM,N/xN)(1)$$

is a vectorspace concentrated in degree -i . Since a power of x kills all of $\mathrm{Ext}_R^i(M,N)$, we see that $\mathrm{Ext}_R^i(M,N)_p = 0$ for $p > -i$, as required.

<u>Proof of Theorem 2.2:</u> Write $\hat{R}$ and $\hat{S}$ for the completion of R and S at m. Let M be an MCM $\hat{R}$-module, with minimal free resolution

$$\cdots \xrightarrow{\bar{\psi}} F_1 \xrightarrow{\bar{\varphi}} F_0 \xrightarrow{\bar{\psi}} F_1 \xrightarrow{\bar{\varphi}} F_0 \to M,$$

where $\varphi\psi = \psi\varphi = Q \cdot 1$ in $\hat{S}$, and $\bar{\ }$ denotes reduction mod Q. Write $\varphi = \sum_{i\geq 1} A_i$, $\psi = \sum_{i\geq 1} B_i$, where A_i and B_i are matrices of forms of degree i, in S. Taking homogeneous parts of $\varphi\psi = Q \cdot 1$, we see that

$$B_1A_1 = A_1B_1 = Q \cdot 1,$$

so that $M' := \mathrm{coker}\, A_1$ is a linear module, with resolution

$$\cdots \xrightarrow{\overline{A}_1} F_o \xrightarrow{\overline{B}_1} F_1 \xrightarrow{\overline{A}_1} F_o \dashrightarrow M' .$$

We will produce matrices $s_i : F_0 \to F_0$, $t_i : F_1 \to F_1$, $i \geq 1$, whose elements are forms of degree i such that the matrices

$$A^{(i+1)} := (1 - t_i) \ldots (1 - t_1)A(1 - s_1) \ldots (1 - s_i)$$

satisfy

$$A^{(1)} = A$$

$$A^{(i+1)} - A_1 \equiv 0(\text{mod } m^{i+2}).$$

The matrices

$$\sigma_i = (1 - s_1) \ldots (1 - s_i)$$

$$\tau_i = (1 - t_i) \ldots (1 - t_1)$$

are congruent to 1 modulo m and form Cauchy sequences. Thus setting

$$\sigma = \lim \sigma_i, \quad \tau = \lim \tau_i$$

we see that σ and τ are invertible and

$$A_1 = \tau A \sigma,$$

so $(\text{coker } A_1) \otimes \hat{R}_m = \text{coker } A$, and M is linearizable. Of course it follows from this that the linear module coker A_1 is isomorphic to $\text{gr}_m M$.

Suppose then that for all $j < i$ the maps s_j and t_j have been defined and satisfy the given conditions. Let $A^{(i)}$ be as above, and set

$$B^{(i)} = \sigma_{i-1}^{-1} B \tau_{i-1}^{-1}$$

$$= B_1 + B_2^{(i)} + \ldots ,$$

with $B_j^{(i)}$ a matrix of forms of degree j. We have

$$A^{(i)}B^{(i)} = Q \cdot 1 = A_1 B_1 .$$

Writing $A^{(i)} = A_1 + A_{i+1}^{(i)} + \ldots$, and comparing terms of degree $i + 2$ in these products we get

$$A_1 B_{i+1}^{(i)} + A_{i+1}^{(i)} B_1 = 0.$$

This equation can be interpreted as saying that $A_{i+1}^{(i)}$ and $B_{i+1}^{(i)}$ define the beginning of a map of complexes

$$(*) \qquad \begin{array}{ccccc} \to F_o & \xrightarrow{\overline{B}_1} & F_1 & \xrightarrow{\overline{A}_1} & F_o \\ \Big\downarrow \overline{B}_{i+1}^{(i)} & & \Big\downarrow \overline{A}_{i+1}^{(i)} & & \\ \to F_1 & \xrightarrow{-\overline{A}_1} & F_o & & \end{array}$$

which in turn defines an element of

$$\mathrm{Ext}_R^1(M',M')_{i+1}$$

from a deformation-theoretic point of view, this element is the tangent vector associated to the deformation of M' which is M. Since $i \geq 1$, this group is 0 by Proposition 2.3. Thus the map of complexes in (*) is null-homotopic; that is, there exist maps of degree i, $\overline{s}_i : F_o \to F_o$ and $\overline{t}_i : F_1 \to F_1$ such that $\overline{A}_{i+1}^{(i)} = \overline{s}_i \overline{A}_1 + \overline{A}_1 \overline{t}_i$. Lifting $\overline{s}_i$ and $\overline{t}_i$ back to $S = k[V^*]$ in an arbitrary way, we get

$$A_{i+1}^{(i)} = s_i A_1 + A_1 t_i + u_{i-1} Q,$$

where u_{i-1} is a matrix of forms of degree $i - 1$. Changing s_i to $(s_i + u_{i-1} B_1)$ we get s_i and t_i such that

$$A_{i+1}^{(i)} = s_i A_1 + A_1 t_i,$$

so

$$A^{(i+1)} := (1 - s_i) A^{(i)} (1 - t_i) = A_1 + (\text{terms of degree} \geq i + 2)$$

as required.

3. Cohen-Macaulay modules and properties of quadratic forms

In this section we will derive explicit information about MCM modules over $R = S/Q$ from properties of Q, using the tools of section 2 and the information about Clifford algebras in section 1.

We will preserve the notation introduced at the beginning of section 2, and we assume troughout that Q is regular on V. As usual, we write V_{ins} for the nullspace of $(\, , \,)_Q$. For any R-module M, let $\Omega^1(M)$ be the first syzygy of M.

We begin with the number of MCM modules, and their relationship.

Proposition 3.1: R has either 1 or 2 isomorphism classes of nonfree indecomposable MCM modules. In case there are 2, they have the same rank, and are first syzygies of one another. If $\operatorname{char} k \neq 2$, then there is just 1 if either $\dim V$ is odd or $\dim V$ is even but

$$(-1)^{(\dim V)/2} \det Q$$

is not a square in k. If $\operatorname{char} k = 2$, then there is just 1 iff either Q is defective or Q is non-defective but

$$\operatorname{arf} Q \neq 0.$$

Further, for any MCM R-module M without free summands,

$$M^* \cong \Omega^1(M) \quad \text{if} \quad \dim V \equiv 0 \quad (4)$$
$$M^* \cong M \quad \text{otherwise.}$$

Similar ideas allow us to determine the rank of the MCM modules. We already know when R has only 1 nonfree indecomposable MCM module and when 2, and we know that when there are two they have the same rank. Thus we need only compute the rank of the sum, A, of the 1 or 2 indecomposable, nonfree MCM R-modules.

Recall from Theorem 1.2 that $C_0(Q)$ may be written as a tensor product of a central-simple k-algebra and an algebra which is either a $k \times k$, or a field extension. The central simple factor is of course itself the product of a central division algebra and a full matrix ring. For our purposes here we will split things differently: We write $C_0(Q) = D(Q) \otimes C_0'(Q)$, where $D(Q)$ is a (not necessarily central) division algebra and $C_0'(Q)$ is a full matrix ring over k or a product of such. Note that $D(Q)$ may be identified with the endomorphism ring of (either of) the simple $C_0(Q)$-module(s), and thus, by Theorem 2.1, with the degree 0 endomorphism ring of (one of) the nonfree indecomposable R-module(s).

Recall that the dimension of a division algebra over its center is the square of an integer called the <u>degree</u> of the algebra.

<u>Proposition 3.2:</u> Let D be the division algebra which is the degree 0 part of the endomorphism ring of a nonfree indecomposable MCM module. Set

$$m = \begin{cases} \left[\dfrac{\dim V}{2}\right] - 1 & \text{if char } k \neq 2 \text{ or } Q \text{ is non-defective} \\ \left[\dfrac{\dim V + \dim V_{ins} - 2}{2}\right] - 1 & \text{if } Q \text{ is defective.} \end{cases}$$

If A is the sum of the (1 or 2) distinct indecomposable nonfree MCM modules over R, then rank A is a power of 2 and

$$\text{rank } A = 2^m \deg D.$$

In particular,

$$2^m \mid \text{rank } A \mid 2^{\dim V - 2} .$$

<u>Proof of Proposition 3.1:</u> By Theorem 2.1, the indecomposable nonfree MCM R-modules of some rank n correspond to the irreducible $C_0(Q)$-modules of dimension 2n, so the statements about the number and rank of the MCM R-modules follow directly from the theory of modules over semi-simple rings and Theorem 1.2.

If $C_0(Q)$ is simple, so that there is just 1 indecomposable nonfree MCM R-module, then the other statements of the Proposition are trivial, so we suppose that $C_0(Q)$ is not simple, and its center is then $k \times k$.

To show that the two indecomposable nonfree MCM R-modules are first syzygies of one another, we must, by Theorem 2.1, show that if A_0 is an irreducible $C_0(Q)$-module, then

$$A_1 := C_1(Q) \otimes_{C_0(Q)} A_0 \not\cong A_0.$$

To prove this we will first choose an element of $V \subset C_1(Q)$ which is a unit and does not centralize $k \times k$ = Center $C_0(Q)$. To see that this is possible, we note that $k \times k = k[c]$ where $c \in C_0(Q)$. If the characteristic is not 2 then c may be taken to be Πe_i, where e_i is an orthogonal base of V. If the characteristic is 2, then c may be taken to be $\Sigma e_i f_i$, where $\{e_i, f_i\}$ is a symplectic base of V. In the first case any c_i will do for v; in the second case any $e_i + f_i$ will do.

Conjugation by v now defines a non-trivial automorphism of $k \times k$, which must therefore interchange the two factors. Since $C_1(Q) = vC_0(Q)$, we may think of $C_1(Q) \otimes_{C_0(Q)} A_0$ as $v \otimes A_0$, with the action given by $w \cdot (v \otimes a) = v(v^{-1}wv) \otimes a = v \otimes (v^{-1}wv)a$, that is, as A_0 with the action given via conjugation by v; since conjugation by v interchanges the two simple factors of $C_0(Q)$, it interchanges the two modules.

To prove the last statement of the Proposition, we may assume that $C_0(Q)$ is not simple and M is indecomposable.

If $e_1, e_2 \in k \times k \subset C_0(Q)$ are the two central primitive idempotents, then we may distinguish between the simple $C_0(Q)$-modules by saying which of the e_i annihilates them. Suppose e_1 annihilates $T(M)$. By what was proved above, $T(\Omega^1 M) \not\cong T(M)$, so e_2 annihilates $T(\Omega^1 M)$. Thus $M^* \cong \Omega^1 M$ iff the main involution of $C_0(Q)$ fixes e_2 or, equivalently, acts trivially on the center of $C_0(Q)$. But if $C_0(Q)$ ist not simple, this is true iff rank $Q \equiv 0(4)$ by the results of section 1, concluding the proof.

<u>Proof of Proposition 3.2:</u> Since $C_0(Q)$ has dimension $2^{\dim V - 1}$, it will be a ring of $2^s \times 2^s$ matrices over $D(Q)$ for some s, and the degree of $D(Q)$ will also be a power of 2, say 2^d, as will the dimension of the center (of $D(Q)$ or of $C_0(Q)$), say 2^c. Thus

$$\dim V - 1 = 2s + 2d + c,$$

and the dimension of the sum of the 1 or 2 distinct $C_0(Q)$-modules is

$$2^{s+2d+c}.$$

By the last statement of Theorem 2.1, the rank of A is

$$2^{s+2d+c-1},$$

and putting this together with the fact that the dimension of the center of $C_0(A)$ is

$2^c = 1$ if char $k = 2$ or Q non-defective, dim V odd

$2^c = 2$ if char $k = 2$ or Q non-defective, dim V even

$2^c = 2^{\dim V_{ins}-1}$ if char $= 2$, Q defective,

the first 2 statements follow.

Remark:

Although we were only concerned with quadrics, parts of the result may be extended to hypersurfaces of multiplicity two:

Let $f \in k[[V^*]]$ be an element of order two, not necessarily homogeneous, and denote f_2 its leading form in $gr_{(V^*)}P = k[V^*]$.

Let $R = P/f$ be as usual the hypersurface ring defined by f and $\overline{R} = gr_{(V^*)}R$ its tangent cone. If now M is a MCM over R, given by a reduced matrix-factorization (ϕ,ψ), say, the leading forms $(\overline{\phi},\overline{\psi})$ define a matrix-factorization of f_2 and hence a MCM $\overline{M}$ over $\overline{R}$. Now, if f has multiplicity two, its leading term f_2 is a quadric, denoted Q_f.

It is not hard to see that M and $\overline{M}$ have the same rank. In particular, if Q_f is regular, we may apply Proposition 3.2 to show that over R as over $\overline{R}$ the rank of a MCM without free summands is divisible by an appropriate power of 2 depending only on $C_o(Q_f)$.

4. Explicit examples

Again in this section we maintain the notations $S = k[V^*]$, $R = S/Q$, V_{ins} = nullspace $(\ ,\ )_Q$ (in char. 2) of sections 2, 3, and we continue to assume that Q is nonregular, as in section 3.

We will also adopt the notation of Proposition 3.2: A will denote the sum of the (1 or 2) distinct indecomposable nonfree MCM R-modules, and

$$m = \begin{cases} \left[\dfrac{\dim V}{2}\right] - 1 & \text{if char } k \neq 2 \text{ or } Q \text{ is non-defective} \\ \left[\dfrac{\dim V + \dim V_{ins} - 2}{2}\right] - 1 & \text{if } Q \text{ is defective.} \end{cases}$$

Let us first consider the case of a quadratic form of maximal Witt index. It is given by either

$$Q = \sum_i x_i y_i \qquad \text{in case dim V is even, or}$$

$$Q = \sum_i x_i y_i + \alpha z^2 \qquad \text{in case dim V is odd.}$$

The indecomposable modules over the Clifford-algebra correspond then - as is well known - to the two half-spin representations - for V even-dimensional - or the spin representation - for V odd-dimensional - of the orthogonal group of Q.

Not surprisingly, therefore, the construction of the corresponding MCM's, as given in Buchweitz-Greuel-Schreyer [1986] for example is the "same" as the classical construction of these group representations, a fact now elucidated by Theorem 2.1.

Proposition 3.1 answers rather satisfactorily the question of the number of MCM modules over R. By contrast, Proposition 3.2 refers the problem to another, computing the degree of a division algebra, which is not easy without some technique.

In this section we make some remarks on the situation over various interesting fields. We then treat the case $k = \mathbb{R}$ in somewhat more detail; as we shall show, the mod 8 periodicity associated with the orthogonal groups appears here.

In general we know that

$$2^m \mid \text{rank } A \mid 2^{\dim V - 2},$$

and it follows from Proposition 3.2 that in many cases only the lower estimate is achieved:

Corollary 4.1: If k is finite, algebraically closed, or a rational function field in 1 variable over an algebraically closed field, then

$$\text{rank } A = 2^m.$$

Proof: Classical results assert that every division algebra is commutative over such fields.

More delicate is the situation of global fields such as the rational numbers. Since the Clifford algebra is isomorphic to a product of quaternion algebras we may apply Theorem 1.3 and get:

<u>Corollary 4.2:</u> If k is a local or global field, then

$$\text{rank } A = 2^m \quad \text{or} \quad 2^{m+1},$$

depending on the Hasse-Witt invariant of Q.

Nevertheless it is not hard to construct examples of k and Q for which rank A takes on any of the values allowed by the last statement of Theorem 2. The key observation is that the Clifford algebra of the <u>generic</u> quadratic form is a division algebra - see Theorem 1.4.

<u>Corollary 4.3:</u> Let $k = \mathbb{Q}(s_1,\dots,s_n)$ where the s_i are transcendental over $\mathbb{Q}$. If Q is of the form $Q \sim \sum_0^{n_1} s_i x_i^2 + \sum_1^{n_2} y_i z_i$, so that $\dim V = 1 + n_1 + 2n_2$, then

$$\text{rank } A = 2^{m+n_1};$$

in particular, if $n_2 = 0$ then

$$\text{rank } A = 2^{\dim V - 2}.$$

We turn now to forms over the reals (more generally one could easily treat the forms over any field of characteristic $\neq 2$ which are equivalent to $\sum_i \pm x_i^2$).

Choosing an appropriate basis, we may write

$$Q \sim \sum_{i=1}^{n_1} x_i^2 - \sum_{i=n_1+1}^{n} x_i^2 .$$

Recall that the signature $\sigma(Q)$ of Q is defined to be the number of positive minus the number of negative signs, i. e.

$$\sigma(Q) = 2n_1 - n.$$

Writing as usual $\mathbb{R}$, $\mathbb{C}$, and $\mathbb{H}$ for the reals, complexes, and quaternions, the following table, which follows closely Table 1 of Atiyah, Bott, Shapiro [1964], summarizes what is true:

$\sigma(Q) \bmod 8$	$C_o(Q)$ is a full matrix ring over	If rank $Q = r$: number of indecomposable non-free MCM modules	rank of indecomposable non-free MCM modules
1	$\mathbb{R}$	1	$2^{\frac{r-3}{2}}$
2	$\mathbb{C}$	1	$2^{\frac{r-2}{2}}$
3	$\mathbb{H}$	1	$2^{\frac{r-1}{2}}$
4	$\mathbb{H} \times \mathbb{H}$	2	$2^{\frac{r-2}{2}}$
5	$\mathbb{H}$	1	$2^{\frac{r-1}{2}}$
6	$\mathbb{C}$	1	$2^{\frac{r-2}{2}}$
7	$\mathbb{R}$	1	$2^{\frac{r-3}{2}}$
0	$\mathbb{R} \times \mathbb{R}$	2	$2^{\frac{r-4}{2}}$

Notes on a proof: By virtue of Proposition 3.2, it suffices to establish the correctness of the second column of the table. This is contained in section 4 and Proposition 5.4 of the paper by Atiyah-Bott-Shapiro just cited.

<u>Remark:</u> As Q and $-Q$ have obviously the same MCM's, it is clear that the table only depends on $|\sigma(Q)|$ mod 8.

It is not difficult to explicitly compute the linear matrix factorizations of smallest possible size for a given real quadric Q. By the results of section 2, such a factorization corresponds to an indecomposable $\mathbb{Z}/2\mathbb{Z}$-graded module M over the Clifford algebra $C(Q)$. Choose a basis $e_1,\dots,e_r$ of V, and bases of the even and odd part of M. With respect to the chosen bases let $(a_{ij}^{(k)}) \in Gl(n;k)$ be the matrix describing the multiplication map $M_0 \xrightarrow{e_k} M_1$, and $(b_{ij}^{(k)}) \in Gl(n;k)$ the matrix for $M_1 \xrightarrow{e_k} M_0$. In terms of these, Q has the matrix factorization $Q\cdot 1 = \varphi\psi$ with $\varphi = (\varphi_{ij})$, $\psi = (\psi_{ij})$, where

$$\varphi_{ij} = \sum_{k=1}^{r} a_{ij}^{(k)} x_k \ , \ \psi_{ij} = \sum_{k=1}^{r} b_{ij}^{(k)} x_k$$

We use this method to compute the matrix factorizations of real quadratic forms. By Knörrer's periodicity theorem [1986] it suffices to factorize the forms $Q_n = \sum_{i=1}^{n} x_i^2$. For $n \le 8$, we obtain the following result:

$$Q_n\cdot 1 = \varphi_n \circ \varphi_n^t \ , \quad \text{where}$$

$$\varphi_1 = (x_1),\ \varphi_2 = \begin{pmatrix} x_1 & x_2 \\ x_2 & -x_1 \end{pmatrix}$$

$$\varphi_4 = \begin{bmatrix} x_1 & x_2 & x_3 & x_4 \\ x_2 & -x_1 & -x_4 & x_3 \\ x_3 & x_4 & -x_1 & -x_2 \\ x_4 & -x_3 & x_2 & -x_1 \end{bmatrix}$$

$$\varphi_8 = \left[\begin{array}{cccc|cccc} x_1 & x_2 & x_3 & x_4 & x_5 & x_6 & x_7 & x_8 \\ x_2 & -x_1 & -x_4 & x_3 & -x_6 & x_5 & x_8 & -x_7 \\ x_3 & x_4 & -x_1 & -x_2 & -x_7 & -x_8 & x_5 & x_6 \\ x_4 & -x_3 & x_2 & -x_1 & -x_8 & x_7 & -x_6 & x_5 \\ \hline x_5 & x_6 & x_7 & x_8 & -x_1 & -x_2 & -x_3 & -x_4 \\ x_6 & -x_5 & x_8 & -x_7 & x_2 & -x_1 & x_4 & -x_3 \\ x_7 & -x_8 & -x_5 & x_6 & x_3 & -x_4 & -x_1 & x_2 \\ x_8 & x_7 & -x_6 & -x_5 & x_4 & x_3 & -x_2 & -x_1 \end{array}\right]$$

φ_3 is obtained from φ_4 by setting $x_3 = 0$, and φ_i $(i = 5,6,7)$ from φ_8 by setting $x_j = 0$ for $i < j \leq 8$.

Notice that the entries of these matrices are 0 or $\pm x_i$. Combinatorists call such matrices orthogonal designs. A survey on the theory of orthogonal designs is given in the book of Geramita and Seberry [1979]. It seems hence plausible that the theory of MCM's has also applications to combinatorial questions.

We will use our result on matrix factorizations of real quadrics to give one more proof of the famous theorem of Hurwitz [1898]:

Let $z_1,\dots,z_n$ be real bilinear forms in the variables $x_1,\dots,x_n$ and $y_1,\dots,y_n$ such that

$$z_1^2 + \dots + z_n^2 = (x_1^2 + \dots + x_n^2)(y_1^2 + \dots + y_n^2),$$

then $n = 1,2,4$ or 8.

To relate this theorem with matrix factorizations we consider a finite-dimensional k-vectorspace A together with a bilinear map

$$A \times A \longrightarrow A$$
$$(a,b) \mapsto a \cdot b$$

We call such a pair a k-algebra. The k-algebra A is said to be a

<u>composition algebra</u> if there exists a quadratic form $Q : A \to k$ such that $(\, , \,)_Q$ is nondegenerate and such that

$$Q(a\cdot b) = Q(a)\cdot Q(b) \quad \text{for all} \quad a,b \in A .$$

Recall that if $(\, , \,)_Q$ is nondegenerate then any k-linear map $\varphi : A \to A$ admits a unique k-linear map $\varphi^{ad} : A \to A$, the adjoint of φ, such that

$$(\varphi(a),b)_Q = (a,\varphi^{ad}(b))_Q \quad \text{for all} \quad a,b \in A.$$

<u>Proposition 4.4</u>: Let A be a finite dimensional k-vectorspace, and $Q : A \to k$ a quadratic form for which $(\, , \,)_Q$ is nondegenerate. Consider the following conditions:

1) There exists a k-algebra structure on A such that

$$Q(a\circ b) = Q(a)\circ Q(b) .$$

2) There exists a k-linear map $\varphi : A \to \mathrm{End}_k(A)$ such that

$$Q(a)\cdot id_A = \varphi(a)^{ad} \quad \varphi(a) \quad \text{for all} \quad a \in A.$$

Then 1) implies 2), and if char $k \neq 2$, then 2) implies 1).

First let us derive the Hurwitz theorem from the Proposition and our classification of matrix factorizations of real quadratic forms.

It is clear and well known that a composition of square sums as in the Hurwitz theorem exists exactly for those n for which there exists a composition algebra A over $\mathbb{R}$ of $\mathbb{R}$-dimension n with norm $Q = \sum_{i=1}^{n} x_i^2$. Now Proposition 4.4 implies that composition algebras of dimension n exist exactly for those n for which there exists a matrix factorization $Q\circ 1 = \varphi\varphi^t$, where the size of $\varphi = n = \text{rank } Q$.

The last column of the table on page 90 tells us the rank of an indecomposable nonfree MCM over S/Q (depending on the signature of Q). The minimal size of a matrix factorization of Q is just twice this number.

Checking the table we see that we must have $n \leq 8$, simply because the minimal size of the matrices in a matrix factorization of Q is $\geq 2^{\frac{n-2}{2}}$, and $2^{\frac{n-2}{2}} > n$ for $n > 8$.
Checking the numbers $n \leq 8$, we see that only $n = 1,2,4$ and 8 are possible. That, in fact, these numbers allow a decomposition of square sums as in the Hurwitz theorem follows from Proposition 4.4 and our explicit factorization of $Q = \sum_{i=1}^{n} x_i^2$, $n \leq 8$, where we have seen that $Q \cdot 1 = \varphi\varphi^t$.

Proof of 4.4: 1) $\Rightarrow$ 2): We define $\varphi : A \to \mathrm{End}_k(A)$ by $\varphi(a)(b) = a \circ b$ for all $a,b \in A$. Then for all $a,b_1,b_2 \in A$ we have $(b_1,(\varphi(a)^{ad} \circ \varphi(a))(b_2))_Q = (\varphi(a)(b_1), \varphi(a)(b_2))_Q = (a \circ b_1, a \cdot b_2)_Q = Q(a \cdot b_1 + a \circ b_2) - Q(a \cdot b_1) - Q(a \circ b_2) = Q(a)(Q(b_1+b_2) - Q(b_1) - Q(b_2)) = Q(a) \circ (b_1,b_2)_Q = (b_1, Q(a) \circ b_2)_Q$. Since $(\ ,\ )_Q$ is nondegenerate, this implies that $Q(a)\mathrm{id}_A = \varphi(a)^{ad} \circ \varphi(a)$ for all $a \in A$.

2) $\Rightarrow$ 1): We define $a \circ b = \varphi(a)(b)$ for all $a,b \in A$, then $2Q(a \circ b) = (a \circ b, a \circ b)_Q = (\varphi(a)(b), \varphi(a)(b))_Q = (b, (\varphi(a)^{ad} \circ \varphi(a))(b))_Q = (b, Q(a) \cdot b)_Q = Q(a) \circ (b,b)_Q = 2Q(a) \circ Q(b)$. Since char $k \neq 2$, we may divide by 2 .

References

Atiyah, M. F., Bott, R., and Shapiro, A.: Clifford Modules. Topology 3, Suppl. (1964), 3-38.

Auslander, M.: Isolated singularities and existence of almost split sequences. In Representation Theory II, Groups and Orders, Lecture Notes 1178 (1986), 194-241, Springer-Verlag, New York.

Auslander, M. and Reiten, I.: The Cohen-Macaulay Type of Cohen-Macaulay Rings. Preprint (1986).

Buchweitz, R.-O., Greuel, G.-M., and Schreyer, F.-O.: Cohen-Macaulay modules on hypersurface singularities II. Preprint (1986), to appear in Inventiones.

Cohen, P. M.: Algebra, Vol. 2, John Wiley and Sons, (1977).

Eisenbud, D.: Homological Algebra on a complete intersection, with an application to group representations. Trans. Am. Math. Soc. 260 (1980), 35-64.

Geramita, A. V., and Seberry, J.: Orthogonal designs; quadratic forms and Hadamard matrices. Lecture Notes in Pure and Applied Math. 45 (1979), Marcel Dekker.

Hurwitz, A., Über die Komposition der quadratischen Formen von beliebig vielen Variablen. Nachrichten der k. Gesellschaft der Wissenschaften zu Göttingen (1898), 309-316.

Jacobson, N.: Basic Algebra II, W. H. Freeman and Co., San Francisco, (1980).

Kneser, M.: Bestimmung des Zentrums der Cliffordschen Algebren einer quadratischen Form über einem Körper der Charakteristik 2. J. Reine u. Angew. Math. 193 (1954), 123-125.

Knörrer, H.: Cohen-Macaulay modules on hypersurface singularities I. Preprint (1986), to appear in Inventiones.

Lam, T. Y.: The Algebraic Theory of Quadratic Forms. 2nd printing, with rev. Benjamin, (1980).

Matsumura, H.: Commutative Algebra, 2nd ed., Benjamin Cummings Publ. Co., Reading Mass., (1980).

Scharlau, W.: Quadratic Forms, Springer-Verlag, New York, (1985).

Swan, R.: K-theory of quadric hypersurfaces. Annals of Math. 122 (1985), 113-153.

Vignéras, M.-F.: Arithmétique des Algèbres de Quaternion. Lecture Notes 800 (1980), Springer Verlag, New York.

Appendix: The Comparison Theorem.

by Ragnar-Olaf Buchweitz

Introduction

The aim of this appendix is to put theorem 2.1. of the foregoing paper in its proper conceptual context. In addition, it provides examples for which the theory of maximal Cohen-Macaulay modules over (not necessarily commutative) Gorenstein rings, as developed in [Bu]; has a concrete geometric content.

From our point of view, the theorem, which describes maximal Cohen-Macaulay modules over a quadric in terms of representations of the Clifford-algebra of the associated quadratic form, is just a special application of the generalized Bernstein-Gelfand-Gelfand-correspondence which compares (complexes of) modules over certain graded k-algebras with (complexes of) modules over their associated Yoneda-Ext-algebras. This correspondence was originally established in [BGG] for the case of (graded) polynomial algebras, to obtain a general "monadic" description of complexes of coherent sheaves on the corresponding projective space.

Essentially the same proof extends this correspondence to the class of homogeneous coordinate rings R of complete intersections defined by quadrics.

The distinguishing features of this class of rings R are:

- their Yoneda-Ext-algebra is again noetherian (on both sides). By a theorem of Bøgvad-Halperin [B-H], complete intersections are the only commutative, noetherian rings for which this holds.
- The double Yoneda-Ext-algebra $\mathrm{Ext}^*_{\mathrm{Ext}^*_R(k,k)}(k,k)$ of R coincides with R. Quadratic complete intersections are the only ones with this property as well. (cf. [B-R]).

For this special class of rings, the generalization of the results in [BGG] has been obtained independentely by Kapranov, - to appear in Funkt. Anal. - , as S. I. Gelfand informed us.

Although the above remarks indicate that complete intersections defined by quadrics constitute a natural domain to which the Bernstein-Gelfand-Gelfand-correspondence extends, one may nevertheless drop either restriction of being "commutative" or "noetherian" and still obtain similar results for other classes of graded algebras. This will be explained somewhere else.

To see how the correspondence may be used to study coherent sheaves on projective schemes or maximal Cohen-Macaulay modules, let R denote the (commutative) homogeneous coordinate ring of a projective complete intersection X defined by quadrics, $E^{\cdot} = Ext_R^*(k,k)$ the Ext-algebra over R of the ground field k, - considered as an R-module via the augmentation - , graded by homological degree and with the multiplicative structure defined by the Yoneda-product of extensions.

The main result is the existence of a natural - and explicit - exact equvalence between the triangulated categories $D^b(R.)$ and $D^b(E^{\cdot})$, each denoting the derived category of complexes with bounded cohomology of graded, finitely generated (right) modules over the algebra in question.

Under this equivalence, the full triangulated subcategories $D^b_{perf}(\ .\ )$, consisting of all objects in $D^b(\ .\ )$ isomorphic to a finite, free (="perfect") complex, and $D^b_{art}(\ .\ )$, given by all objects with artinian cohomology, are transformed into each other. Now Serre's description of coherent sheaves of modules on projective schemes tells us that $D^b(R.)/D^b_{art}(R.)$ is canonically equivalent to $D^b(X)$, the derived category of such sheaves of modules on the projective complete intersection X.
On the other hand, the main result in [Bu] states that $D^b(R.)/D^b_{perf}(R.)$ is naturally equivalent to $\underline{MCM}(R.)$, the category of graded maximal Cohen-Macaulay R.-modules modulo projective modules.

To emphasize even more the duality between R. and $E^{\cdot}$, remark that $E^{\cdot}$ is strongly Gorenstein, ([Bu] and proposition 1), so that also $D^b(E^{\cdot})/D^b_{perf}(E^{\cdot})$ might be identified with $\underline{MCM}(E^{\cdot})$, the corresponding category of graded maximal Cohen-Macaulay-modules over $E^{\cdot}$ modulo projectives (in the sense of (loc. cit.)).

Denoting furthermore $\underline{Proj}\ E^{\cdot}$ the quotient of the abelian category of all finitely generated, graded right $E^{\cdot}$-modules by its Serre-subcategory of artinian modules, one has that $D^b(\underline{Proj}\ E^{\cdot})$, its derived category, is naturally equivalent to $D^b(E^{\cdot})/D^b_{art}(E^{\cdot})$.

Hence, the generalized Bernstein-Gelfand-Gelfand correspondence can now be stated as follows:

Theorem 1: With the notations as above, there is a natural exact equivalence, the Bernstein-Gelfand-Gelfand-correspondence, between $D^b(R.)$ and $D^b(E^\cdot)$.
It induces exact equivalences between triangulated categories

$$D^b(\underline{\text{Proj}}\ R.) \overset{\sim}{\leftrightarrow} \underline{\text{MCM}}(E^\cdot) \quad \text{and}$$
$$\underline{\text{MCM}}(R.) \overset{\sim}{\leftrightarrow} D^b(\underline{\text{Proj}}\ E^\cdot) \ .$$

Furthermore, these equivalences are functorial in $(R.,E^\cdot)$. □

To use this theorem efficiently, it is necessary to obtain structural results on $E^\cdot$, and as an example we deduce theorem 2.1. of the foregoing article which motivated this appendix. All that is needed is:

Theorem 2: Assume $R.$ is the homogeneous coordinate ring of a single quadric Q. Then

(i) $\underline{\text{Proj}}\ E^\cdot$ is naturally equivalent to mod-$C^\cdot(Q)$, the category of $\mathbb{Z}/2\mathbb{Z}$-graded, finitely generated right modules over the (full) Clifford-algebra $C^\cdot(Q)$ of Q.

(ii) If Q is not zero, mod-$C^\cdot(Q)$ is equivalent to mod-$C^o(Q)$, the category of finitely generated right modules over the even Clifford-algebra $C^o(Q)$ of Q.

(iii) If $C^o(Q)$ is semi-simple, $D^b(\text{mod-}C^o(Q))$ is abelian and equivalent to grmod-$C^o(Q)$, the category of $\mathbb{Z}$-graded, finitely generated right $C^o(Q)$-modules, $C^o(Q)$ being concentrated in degree zero. □

Here of course only (i) needs a proof, (ii) and (iii) being essentially well-known. (For (ii), see [A-B-S] and for (iii) remark that over a semi-simple ring every complex of modules splits, so that passing to cohomology is an equivalence of categories.)

Combining these two theorems with the description of quadrics which define a semi-simple even Clifford-algebra (theorem 1.1. of the foregoing article), we get the desired result on maximal Cohen-Macaulay modules on a regular quadric:

Corollary 1: Assume that $R.$ is the homogeneous coordinate ring of a regular quadric Q. Then $\underline{\text{MCM}}(R.)$ is an abelian category and as such naturally equivalent to grmod-$C^o(Q)$. □

Considering the algebraic K-groups of the categories involved, one can extend R.G. Swan's result, [Sw], for smooth quadrics to regular ones over a field.

Another application is the following description of those maximal Cohen-Macaulay modules on a quadric which can occur as syzygy-modules of artinian R.-modules.

Corollary 2: With the same assumptions as above, an equi-generated maximal Cohen-Macaulay R.-module occurs as the syzygy-module of some artinian R.-module if and only if the corresponding module over the even Clifford-algebra is free. In particular, an indecomposable maximal Cohen-Macaulay module occurs in this way iff $C^c(Q)$ is a division-ring.

1. The Correspondence

As in the introduction, let $R.$ be the homogeneous coordinate ring of a complete intersection defined by quadrics, $E^{\cdot} = \mathrm{Ext}^{*}_{R.}(k,k)$ its Yoneda-Ext-algebra.

Let $A.$ denote either of these two algebras, (where, contrary to conventions in other sources, raising or lowering of indices will not change signs, i. e. $A_i = E^i$ in case $A.$ represents $E^{\cdot}$!)

Set $\varepsilon = +1$ if $A. = R.$ and $\varepsilon = -1$ if $A. = E^{\cdot}$. If $M. = \bigoplus_i M_i$ is any $\mathbb{Z}$-graded right $A.$-module, for any integer ν we denote $M.(\nu)$ the shifted module with grading $M_i(\nu) = M_{i+\nu}$ and $A.$-module structure determined by

$$m(\nu)a = \varepsilon^{\nu j} ma(\nu) \in M_{i+j}(\nu) ,$$

for $m \in M_{i+\nu}$, $a \in A_j$.

To fix notations on module categories, let Mod-A. be the category of graded, right A.-modules with degree-preserving, A.-linear maps as morphisms, mod-A. its full subcategory of right noetherian modules.

$D^b(A.)$ will denote the full subcategory of the derived category(1) of Mod-A. , whose objects are those complexes of (arbitrary, graded right) A.-modules which have only finitely many non-vanishing cohomology modules, each of which is furthermore in mod-A. .

$D^b_{perf}(A.)$ refers to the full subcategory of complexes isomorphic to finite complexes of graded free A.-modules of finite rank. (Such complexes are also called "perfect".)

$D^b_{art}(A.)$ denotes the full subcategory of $D^b(A.)$ whose objects are all those complexes with right artinian cohomology modules. Both $D^b_{perf}(A.)$ and $D^b_{art}(A.)$ are closed under translation and mapping-cones, whence they inherit a triangulated structure from $D^b(A.)$.

As already remarked, art-A. , the category of right artinian and noetherian A.-modules, is a Serre-subcategory of mod-A. , hence the quotient Proj A. = mod-A./art-A. is again abelian. As in the commu-

(1) For definitions, notations and results on derived or triangulated categories, which are not explicitely given, we refer to [Ver], [BBD] or [Ha]. Remark that in [Ha], $D^b(A.)$ would be denoted $D^b_{\text{mod-A.}}(\text{Mod-A.})$.

tative case, where this is essentially Serre's description of the coherent sheaves on a projective scheme, the quotient $D^b(A.)/D^b_{art}(A.)$ exists as a triangulated category and is naturally equivalent to $D^b(\underline{\text{Proj}}\ A.)$, the derived category with bounded cohomology of <u>Proj</u> A. .

Also, as we will see in a moment, in our case A. is left and right noetherian and of finite injective dimension as either left or right module over itself, hence <u>strongly Gorenstein</u> in the sense of [Bu], and so $D^b(A.)/D^b_{perf}(A.)$ is naturally equivalent to $\underline{\text{MCM}}(A.)$, the category of graded maximal Cohen-Macaulay A.-modules modulo projectives, as a triangulated category. Let us only remark that the equivalence is induced from the functor which to a maximal Cohen-Macaulay module over A. associates the complex with exactly this module as its only non-zero component, placed in degree zero.

We will write T or $-[1]$ as usual for the translation-functor in any triangulated category (to be) encountered. We may combine it with the shift-functor - with which it commutes - and set

$$X[i,j) = T^i X(j)$$

for all $i,j \in \mathbb{Z}$, X in $D^b(A.)$.

In particular, any two complexes X,Y in $D^b(A.)$ define a bigraded Ext-group:

$$\mathrm{Ext}^i_{A.}(X,Y)_j = \mathrm{Hom}_{D^b(A.)}(X,Y[i,j)) \ ; i,j \in \mathbb{Z};$$

and, given a third complex Z in $D^b(A.)$, the <u>Yoneda-product</u>

$$\mathrm{Ext}^p_{A.}(Z,Y)_q \times \mathrm{Ext}^r_{A.}(X,Z)_s \to \mathrm{Ext}^{p+r}_{A.}(X,Y)_{q+s}$$

is defined for all integers p,q,r and s - see [Ver; II.1.2; 2.3]. This product is associative and functorial in the usual way. Thus for any object X in $D^b(A.)$ we get a bigraded algebra $\bigoplus_{p,q} \mathrm{Ext}^p_{A.}(X,X)_q$.

In particular, considering k as an A.-module via the augmentation $A. \to A_0 = k$ and as a complex with k as its only non-zero component placed in degree zero, we recover the usual (bigraded) Yoneda-Ext-algebra of k over A.

We want to emphasize here that $\mathrm{Ext}^p_{A.}(k,k)_q$ as a vectorspace is independent of the sign $\varepsilon = \pm 1$ used in the definition of the shifts but that the multiplicative structure on the whole Ext-algebra depends on this choice!

Leaving abstract homological algebra, let us come back to the algebras in question.

Let $S.$ be a polynomial ring generated by variables of degree 1, and let $R.$ be a quotient of $S.$ by a regular sequence of d quadrics. Set $n = \dim R.$. The properties of $R.$ and $E.$, its Ext-algebra, are summarized in:

Proposition 1: Let $A.$ be either of the augmented, positively graded k-algebras $R.$ and $E^{\cdot}$.

(i) As an associative algebra, $A.$ is generated by its elements in degree one and defined by quadratic relations among these generators. It is left and right noetherian.

(ii) $A.$ is of (the same) finite injective dimension as either left or right module over itself. More precisely, $\operatorname{injdim}_{R.} R. = n = \dim R.$, $\operatorname{injdim}_{E.} E. = d = \operatorname{codim}_{S.} R.$.

(iii) The minimal, graded free resolution of k as an $A.$-module is linear and hence $\operatorname{Ext}^i_{A.}(k,k)_{-j} = 0$ unless $i = j \geq 0$.

(iv) $E^i = \operatorname{Ext}^i_{R.}(k,k)_{-i}$ and $R_i = \operatorname{Ext}^i_{E.}(k,k)_{-i}$ for all i and the multiplicative structure on $E^{\cdot}$ or $R.$ coincides with the Yoneda-product.

(v) $\operatorname{Ext}^i_{R.}(k,R.)_j = 0$ unless $i = \dim R.$, $j = \operatorname{codim}_{S.} R. - \dim R.$,
$\operatorname{Ext}^i_{E.}(k,E^{\cdot})_j = 0$ unless $i = \operatorname{codim} R.$, $j = \dim R. - \operatorname{codim}_{S.} R.$.

Proof: As $R.$ is a complete intersection defined by quadrics, (i) is obvious for $R.$. Furthermore, $R.$ is Gorenstein, whence (ii) in this case. That k has a linear resolution as an $R.$-module follows most easily from the explicit knowledge of that resolution as originally determined by J. Tate; see also [Eis]. Part (v) may be proved by induction on d, the codimension of $R.$ in $S.$: if $d=0$, i.e. $R.=S.$, one has $\operatorname{Ext}^i_{S.}(k,S.)_j = 0$ unless $i = \dim S. = -j$. If the result is assumed for $\tilde{R}.$ of codimension $\tilde{d} \geq 0$ in $S.$, and Q is a non-zero-divisor of degree 2 in $\tilde{R}.$, the exact sequence

$$0 \to \tilde{R}.(-2) \xrightarrow{Q} \tilde{R}. \to \tilde{R}./Q \to 0$$

shows $\operatorname{Ext}^i_{\tilde{R}.}(k,\tilde{R}./Q)_j = \operatorname{Ext}^{i+1}_{\tilde{R}.}(k,\tilde{R}.)_{j-2}$.

As $\mathrm{Ext}^i_{\tilde{R}.}(k,M.)_j = \mathrm{Ext}^i_{\tilde{R}./Q}(k,M.)_j$ for any $\tilde{R}.$-module $M.$ annihilated by Q, (v) is still true for $\tilde{R}./Q$.

In (iv), the statement on $E^{\cdot}$ is tautological. What remains is to see that $R. = \mathrm{Ext}^{\cdot}_{\mathrm{Ext}^*_{R.}(k,k)}(k,k)$.

But this follows already from (iii) in case $A. = R.$ by the work of S. Priddy, [Pr], and C. Löfwall, [Löf]. These papers also show that $E^{\cdot}$ is necessarily generated by E^1, the defining relations being quadric, and that k over $E^{\cdot}$ again has a linear resolution.

That $E^{\cdot}$ is noetherian and of finite injective dimension over itself follows from the fact that $E^{\cdot}$ is indeed a graded, cocommutative Hopf-algebra (as it is the Ext-algebra of a commutative ring), whose graded Lie-algebra of primitive elements is finite-dimensional and (therefore) nilpotent (as $R.$ is a complete intersection). This is explained, for example, in [Av], cf. also [B-R]. That this structural result on $E^{\cdot}$ implies the ring-theoretic properties claimed is then a consequence of the Poincaré-Birkhoff-Witt-theorem for graded Lie-algebras.

It remains to prove (v) for $E^{\cdot}$. This again can be deduced from the Poincaré-Birkhoff-Witt-theorem, given the explicit knowledge of the Lie-algebra of primitives in $E^{\cdot}$. □

Another proof - and a description of the Lie-algebra - will be given below.

The essential properties for the existence of the Bernstein-Gelfand-Gelfand-correspondence are (iii) - with its consequence (iv) - and (v).

Following S. Priddy and C. Löfwall, (loc. cit.), an algebra $A.$ satisfying (iii) is called a Koszul-algebra. As this implies - by [Löf] - that $B^{\cdot} = \mathrm{Ext}^{\cdot}_{A.}(k,k)$ also satisfies (iii), one may rather call $(A.,B^{\cdot})$ a Koszul-pair. It is for such pairs that natural functors $D(A.) \rightleftarrows D(B^{\cdot})$ as in [BGG] can be set up.

A necessary condition for these functors to be inverse equivalences is (v), that is $\mathrm{Ext}^*_{A.}(k,A.)$ and $\mathrm{Ext}^*_{B^{\cdot}}(k,B^{\cdot})$ have to be one-dimensional k-vectorspaces. But for $A.$ commutative, noetherian, this is equivalent to $A.$ being Gorenstein, whence a pair $(A.,B^{\cdot})$ may be called (numerically) Gorenstein, (even without $A.$ commutative).

Hence, the Bernstein-Gelfand-Gelfand correspondence should be considered a property of such Gorenstein-Koszul-pairs.

Now we come to the actual correspondence:

Assume $A.$ is a positively graded k-algebra, (with all A_i finite-dimensional k-vectorspaces), such that $k = A_0$ as a right $A.$-module admits a linear resolution $\mathbb{P}(A.,k) = (P^{\cdot}_{\cdot}, d^{\cdot})$ by finitely generated, graded free $A.$-modules.

Let the shifts on right graded $A.$-modules be defined by using a "commutation-factor" $\varepsilon = \pm 1$ as in the beginning. If $P.^{-i}$ denotes the i-th term in the resolution, (so that $d^i : P.^{i-1} \to P.^i$ increases the complex-degree), it can be identified - non-canonically - as

$$P.^{-i} = T_i \otimes_k A.(-i) ,$$

where $T_i = \mathrm{Tor}_i^{A.}(k,k)_i = H_0^{-i}(\mathbb{P}(A.,k)(i) \otimes_{A.} k)$. (Of course, $P^{-i} = 0$ for $i < 0$.)

More generally, the graded pieces of $\mathbb{P}(A.,k)$ are given by

$$P_{j+i}^{-i} = T_i \otimes_k A_j = \mathrm{Tor}_i^{A.}(k(i),A_j) ,$$

where here and in the sequel the homogeneous components A_j of $A.$ are considered as $A.$-bimodules in degree zero; i. e. A_j is identified with $(A_{\geq j}/A_{\geq j+1})(j)$

Consider now the class $\Theta_j \in \mathrm{Ext}^1_{A.^{op}}(A_j,A_{j+1}(-1))$, represented by the extension of left $A.$-modules

$$0 \to A_{j+1}(-1) \to (A_{\geq j}/A_{\geq j+2})(j) \to A_j \to 0 .$$

Then the restriction of the differential $d^{\cdot}$ in $\mathbb{P}(A.,k)$ to $P_{j+i}^{-i} = \mathrm{Tor}_i^{A.}(k(i),A_j)$ can be recovered from the left action of $\mathrm{Ext}^1_{A.^{op}}(A_j,A_{j+1}(-1))$ on this Tor-group:

$$\mathrm{Ext}^1_{A.^{op}}(A_j,A_{j+1}(-1)) \times \mathrm{Tor}_i^{A.}(k(i),A_j) \to$$

$$\to \mathrm{Tor}_{i-1}^{A.}(k(i),\ A_{j+1}(-1)) \simeq \mathrm{Tor}_{i-1}^{A.}(k(i-1),A_{j+1}) = P_{j+i}^{-i+1} ,$$

in other words, $d^{\cdot}|P_{j+i}^{-i}$ is one of the connecting homomorphisms in the exact sequence obtained by applying $\mathrm{Tor}_*^{A.}(k(i),-)$ to the above short exact sequence, (cf. [ALG X.130; Prop. 7(b)]).

Secondly, we have the natural action

$$\mathrm{Ext}^1_{A.}(k,k(-1)) \times \mathrm{Tor}_i^{A.}(k(i),A_j) \to \mathrm{Tor}_{i-1}^{A.}(k(i-1),A_j)$$

which anti-commutes with Θ_j by [ALG X.129, Prop. 6 (15)] but defines on $\bigoplus_i \mathrm{Tor}_i^{A.}(k(i),A_j) = (\bigoplus_i \mathrm{Tor}_i^{A.}(k,k)_i) \otimes_k A_j$ the structure of a graded left module over $B^{\cdot} = \bigoplus_i \mathrm{Ext}^i_{A.}(k,k)_{-i}$ with its Yoneda-product. Furthermore, $\bigoplus_i \mathrm{Tor}_i^{A.}(k,k)_i \simeq B^{\cdot *}$, the graded k-dual of $B^{\cdot}$, as a left $B^{\cdot}$-module. Hence, forgetting the differential, $\mathbb{P}(A.,k)$ may be considered as the left $B^{\cdot}$-, right A.-module $(B^{\cdot})^* \otimes_k A.$.

Having said all this, let M. be any graded, right A.-module. Then $\bigoplus_i \mathrm{Hom}^{\cdot}_{A.}(\mathbb{P}(A.,k),\ M.[i,-i])$ is a complex of graded k-vectorspaces with j-th term:

$$\begin{aligned}\bigoplus_i \mathrm{Hom}^j_{A.}(\mathbb{P}(A.,k),\ M.[i,-i]) &= \bigoplus_i \mathrm{Hom}_{A.}(T_{j+i} \otimes_k A.(-j),\ M.)\\ &= \bigoplus_i M_j \otimes_k T^*_{j+i}\\ &= M_j \otimes_k \Big(\bigoplus_i \mathrm{Ext}^{i+j}_{A.}(k,k)_{-i-j}\Big) .\end{aligned}$$

But defining now the shifts for B.-modules with the opposite commutation-factor $-\varepsilon$, $B^{\cdot}(j) = \bigoplus_i \mathrm{Ext}^{i+j}_{A.}(k,k)_{-i-j}$ is endowed with the "correct" structure, and this complex becomes even a complex of graded free right B.-modules which we denote

$$\beta^{\cdot}_{\cdot}(M) = M. \otimes_k B^{\cdot}(.) ,$$

with j-th term $$\beta^{\cdot}_j(M) = M_j \otimes_k B^{\cdot}(j) ,$$

(this time the complex-degree as lower index), the differential $\delta^M = \bigoplus_i \mathrm{Hom}^{\cdot}_{A.}(d^{\cdot},M[i,-i])\colon\ M. \otimes_k B^{\cdot}(.) \to M._{+1} \otimes_k B^{\cdot}(. + 1)$ still raising the complex-degree (but preserving $B^{\cdot}$-degree).

Remark 1: Obviously, $\mathrm{Ext}^1_{A.}(k,k(-1)) = \mathrm{Hom}_k(\mathrm{Tor}_1^{A.}(k,k)_1,k)$ is the k-dual of A_1. From the explicit construction of the linear resolution of k over a Koszul-algebra - as given in [Löf] - one gets the following description of δ^M_j . Denote $t : k \to A_1 \otimes_k A_1^* = \mathrm{Hom}_k(A_1,A_1)$ the map sending $1 \in k$ to the identity on A_1 . Denote $\mu_M : M. \otimes_k A_1 \to M._{+1}$ the right action of the 1-forms in A. on M. , $\mu_E : B^1 \otimes_k B^{\cdot} \to B^{\cdot}(1)$ the left multiplication of B^1 on $B^{\cdot}$. Then $\delta^M_j : M_j \otimes_k B^{\cdot}(j) \to M_{j+1} \otimes_k B^{\cdot}(j+1)$ is just the composition

$$\delta^M_j = (\mu_M \otimes_k \mu_B) \cdot (1_{M_j} \otimes_k t \otimes_k 1_{B^{\cdot}(j)}) ,$$

or, in terms of dual bases $\{x_i\} \subseteq A_1$; $\{x_i^*\} \subseteq B^1 = A_1^*$,

$$\delta_j^M(m \otimes b) = \sum_i mx_i \otimes x_i^* b \ .$$

(This is the form of the differential as given in [BGG].) □

The above construction is obviously functorial on Mod-A. and can hence be extended to complexes of such modules by applying it to every term, obtaining a double complex, and finally passing to the total complex. As one easily sees, homotopies of complexes are then carried into such.

Remark that the cohomology of $\beta(M)$ is given by

$$\begin{aligned} H_j^i(M) &= H^j(\mathrm{Hom}_{A.}^{\cdot}(\mathbb{P}(A.,k), M.[i,-i)) \\ &= \mathrm{Ext}_{A.}^{i+j}(k,M)_{-i} \ , \end{aligned}$$

so that $H_j^{\cdot}(M) = \bigoplus_i \mathrm{Ext}_{A.}^{i+j}(k,M)_{-i}$ as a graded right $B^{\cdot}$-module with respect to the natural Yoneda-product of $B^{\cdot} = \mathrm{Ext}_{A.}^{\cdot}(k,k)_{-.}$ on this module.

In particular, we see that $H_j^{\cdot}(k) = 0$ unless $j = 0$ in which case $H_0^{\cdot}(k) = B^{\cdot}$. Also, for $A. = R.$ or $A. = E^{\cdot}$, $H_j^{\cdot}(A.) = 0$ except for $j = \mathrm{injdim}_B.B^{\cdot}$ and

$$H^i_{\mathrm{injdim}\ B^{\cdot}}(A^{\cdot}) = 0 \quad \text{unless} \quad i = \mathrm{injdim}\ A. - \mathrm{injdim}\ B^{\cdot} \ ,$$

in which case the cohomology is just k, by statement (v) of proposition 1. This shows already that always this functor transforms $D^b_{art}(A.)$ into $D^b_{perf}(B^{\cdot})$ and that it transforms $D^b_{perf}(A.)$ into $D^b_{art}(B^{\cdot})$, as soon as $\mathrm{Ext}^*_{A.}(k,A.)$ is one-dimensional over k, (artinian would be enough).

Exchanging the roles of $A.$ and $B^{\cdot}$, let α denote the corresponding functor on $\text{Mod-}B^{\cdot}$, which is possible as, by [Löf], k has also a linear resolution as a $B^{\cdot}$-module. Given a right graded A.-module $M.$, the double complex obtained by applying α term by term to $\beta(M.)$ is just the triply-graded k-vectorspace

$$\alpha\beta(M.) = M. \otimes_k B^*(.) \otimes_k A_-(*) \ ,$$

$\{.,*,-\}$ denoting the three degrees. The two differentials are $d_1 = \delta^M \otimes 1_A$ and $d_2 = 1_M \otimes \delta_B$. Hence, considering the spectral sequence obtained by first passing to cohomology with respect to d_2 yields the E_1-terms

$$E_1 \equiv M. \otimes_k H(B^{\cdot} \otimes_k A.) = M. \otimes_k \mathrm{Ext}_{B^{\cdot}}(k,B^{\cdot}) ,$$

conveniently graded.

We did not indicate the specific degrees as they do not matter. The point is, that _if_ $\mathrm{Ext}^{*}_{B^{\cdot}}(k,B^{\cdot})$ _is one-dimensional_, this spectral sequence obviously collapses and yields the E_∞-term M. as a graded A.-module. In other words, as soon as the spectral sequence converges, the associated total complex of $\alpha\beta(M.)$ has a single cohomology module, namely M. , and hence, in the derived category of A. , represents M. again.

Now to ensure the convergence, it is certainly enough that $\beta(M)$ represents an object in $D^b(B^{\cdot})$, that is, there are only finitely many j such that

$$H^{\cdot}_j(\beta(M)) = \bigoplus_i \mathrm{Ext}^{i+j}_{A.}(k,M.)_{-i} \neq 0$$

and each $H^{\cdot}_j$ is finitely generated over $B^{\cdot}$. But it is known - see [B-R] or [Av, 5.7] and dualize - that for A. = R. and M. a finitely generated R.-module, these two conditions are satisfied.

It follows then that α transforms $D^b(E^{\cdot})$ into $D^b(R.)$, for example by using once again the Poincaré-Birkhoff-Witt theorem for the Hopf-algebra $E^{\cdot}$ and induction on d , the codimension of R. in S. , the case d = 0 being precisely the result in [BGG]. Hence (α,β) is a pair of inverse exact equivalences between $D^b(A.)$ and $D^b(B^{\cdot})$.

As far as functoriality is concerned - which is alluded to above - let R! be another complete intersection defined by quadrics, lying between S. and R. : S. → R! → R. . Set $d' = \mathrm{codim}_{S.} R!$ and $'E^{\cdot} = \mathrm{Ext}^{\cdot}_{R!}(k,k)$. Then any graded right R!-module N. - or more generally an object in $D^b(R!)$ - gives rise to a spectral sequence

$$\mathrm{Ext}^i_{R.}(k,\mathrm{Ext}^j_{R!}(R.,N)) \Rightarrow \mathrm{Ext}^{i+j}_{R!}(k,N.) ,$$

which is in fact a spectral sequence of $E^{\cdot} = \mathrm{Ext}^{\cdot}_{R.}(k,k)$-modules. A priori, the abutment is an $'E^{\cdot}$-module, but it becomes an $E^{\cdot}$-module via the algebra morphism $E^{\cdot} = \mathrm{Ext}^{\cdot}_R(k,k) \to {'E^{\cdot}} = \mathrm{Ext}^{\cdot}_{R'}(k,k)$, which in turn is a boundary map in the above spectral sequence for N. = k.

Conversely, any $E^{\cdot}$-module $L^{\cdot}$ - or a complex of such in $D^b(E^{\cdot})$ - defines a spectral sequence of $R! = \mathrm{Ext}^{\cdot}_{'E^{\cdot}}(k,k)$-modules

$$\mathrm{Ext}^{i}_{'E^\cdot}(k,\mathrm{Ext}^{j}_{E^\cdot}('E^\cdot,L^\cdot)) \Rightarrow \mathrm{Ext}^{i+j}_{E^\cdot}(k,L^\cdot) \ .$$

Writing these spectral sequences in terms of double-complexes in the usual way and interpreting them as statements on exact functors between derived categories, this yields the following diagram of exact functors between triangulated categories

$$\begin{array}{ccccc} & D^b(R^!_\cdot) & \overset{\underline{\rho}'}{\underset{'\underline{\varepsilon}}{\rightleftarrows}} & D^b('E^\cdot) & \\ R\,\mathrm{Hom}_{R^!_\cdot}(R_\cdot,-) & \downarrow \uparrow & \nwarrow \text{"forget"} \nearrow & \downarrow \uparrow & R\,\mathrm{Hom}_{E^\cdot}('E^\cdot,-) \\ & D^b(R_\cdot) & \overset{\underline{\rho}}{\underset{\underline{\varepsilon}}{\rightleftarrows}} & D^b(E^\cdot) & \end{array}$$

whose two squares commute, and where the horizontal pairs $(\underline{\rho}', '\underline{\varepsilon})$ and $(\underline{\rho},\underline{\varepsilon})$ are the equivalences of the Bernstein-Gelfand-Gelfand-correspondence described above.

Taking $R^! = S_\cdot$, in which case $'E^\cdot = \mathrm{Ext}^{\cdot}_{S_\cdot}(k,k)$ is an exterior algebra $\Lambda^\cdot$, one obtains another proof that the equivalences transform $D^b_{perf}(R_\cdot)$ into $D^b_{art}(E^\cdot)$, hence induce equivalences between <u>MCM</u>$(R_\cdot)$ and D^b(<u>Proj</u> $E^\cdot$): By Hilbert's syzygy theorem, $D^b(S_\cdot) = D^b_{perf}(S_\cdot)$ and the image under $R\,\mathrm{Hom}_{S_\cdot}(R_\cdot,-)$ has $D^b_{perf}(R_\cdot)$ as its <u>thick hull</u> (cf. [Bu]). Conversely, $D^b(\Lambda^\cdot) = D^b_{art}(\Lambda^\cdot)$, as $\Lambda^\cdot$ is an artinian k-algebra, and hence it generates $D^b_{art}(E^\cdot)$ under the forgetful functor $D^b(\Lambda^\cdot) \to D^b(E^\cdot)$.
This finishes the sketch of the proof of theorem 1 .

<u>Remark 2:</u> (a) As $R_\cdot$ is Gorenstein, one could as well describe $\underline{\rho}$ in terms of Tor's - which is essentially done in theorem 2.1. of the foregoing article for maximal Cohen-Macaulay modules. For this, consider the two spectral sequences with same limit $\mathbb{E}^\cdot$ for an $R_\cdot$-module $M_\cdot$:

$$\mathrm{Ext}^{i}_{R_\cdot}(\mathrm{Tor}^{R_\cdot}_{j}(k,M_\cdot),R_\cdot) \Rightarrow \mathbb{E}^{i+j}$$

and

$$\mathrm{Ext}^{i}_{R_\cdot}(k,\mathrm{Ext}^{j}_{R_\cdot}(M_\cdot,R_\cdot)) \Rightarrow \mathbb{E}^{i+j} \ .$$

Now the first of these degenerates as $R_\cdot$ is Gorenstein, showing that $\mathrm{Tor}^{R_\cdot}_{j}(k,M_\cdot)^*$, the k-dual, is isomorphic to $\mathbb{E}^{\dim R_\cdot + j}$. If furthermore $M_\cdot$ is MCM , also the second spectral sequence degenerates, so that $\mathrm{Ext}^{i}_{R_\cdot}(k,M^*_\cdot) = \mathrm{Hom}_k(\mathrm{Tor}^{R_\cdot}_{i-\dim R}(k,M_\cdot),k)$, where $M^*_\cdot$ is now the $R_\cdot$-dual of $M_\cdot$. Hence the functors $R\,\mathrm{Hom}_{R_\cdot}(k, R\,\mathrm{Hom}_{R_\cdot}(M_\cdot,R_\cdot))$ and

$\mathrm{Hom}_k(k \overset{\mathbb{L}}{\otimes}_{R.} M.[-\dim R.],k)$ are equivalent - and compatible with the action of $E^{\cdot} = \mathrm{Ext}^{\cdot}_{R.}(k,k)$ on the right.

(b) Both $D^b(R.)$ and $D^b(E^{\cdot})$ are by definition "derived" from abelian categories, hence carry natural t-structures - see [BBD] for the definition - whose "hearts" are mod-R. and mod-$E^{\cdot}$ respectively. As the derived categories are equivalent, each of them carries hence two such structures, which are different:

The cohomology-theory attached to the ordinary t-structure on $D^b(R.)$, for example, is just the usual cohomology of complexes, whereas the "perverse" cohomology given by the equivalence (and with "heart" equivalent to mod-$E^{\cdot}$) is given by ${}^{p}H(X) = H^{\cdot}_{p}(\underline{\varepsilon}(X)) = \bigoplus_i \mathrm{Ext}^{i-p}_{E^{\cdot}}(k,\underline{\varepsilon}(X))_{-i}$, where X is any object in $D^b(R.)$. In particular, a complex X corresponds to a single $E^{\cdot}$-module under $(\underline{\rho},\underline{\varepsilon})$ iff ${}^{p}H(X) = 0$ for $p \neq 0$, that is, iff X is isomorphic to a <u>linear</u> complex of free modules in $D^b(R.)$, (the i-th module being generated in degree i). As a consequence, the intersection of the two "hearts" - on either side - is given by those <u>modules which have linear resolutions</u>.

2. The maximal Cohen-Macaulay modules

We want to investigate $\underline{\mathrm{Proj}}\ (E^{\cdot})$ to obtain more insight into the structure of $\underline{\mathrm{MCM}}\ (R.)$ and to prove theorem 2 and its corollaries.

So far we did not use any representation of $R.$ or $E^{\cdot}$. This will be done now. By assumption, $R.$ is the homomorphic image of $S. = k[V]$, $\deg V = 1$, by an ideal generated by quadrics $W \subseteq S_2^k V$, where both V and W are finite-dimensional k-vectorspaces. Set $d = \dim_k W = \mathrm{codim}_{S.} R.$ and $n = \dim R.$, so that $\dim V = d+n$. Dualizing the inclusion $W \to S_2^k V$ and composing it with the <u>universal quadratic map</u> $\chi : V^* \to (S_2V)^*$, given by $\chi(\lambda)(vw) = \lambda(v)\lambda(w)$ one obtains a <u>k-quadratic</u> map

$$q : V^* \to W^* ,\quad \text{which satisfies the usual conditions:}$$

(i) $q(\alpha\lambda) = \alpha^2 q(\lambda)$ for $\alpha \in k$,

and

(ii) $(\lambda,\mu)_q = q(\lambda+\mu)-q(\lambda)-q(\mu)$
is a bilinear map from $V^* \times V^*$ into W^*.

The triple $L^{\cdot} = (V^*, W^*, q)$ is a simple kind of a <u>graded k-Lie-algebra</u> - as introduced by Milnor-Moore and amended by G. Sjödin and L. Avramov to capture the case $\mathrm{char}\ k = 2$, see [Av]. Here $L^1 = V^*$, $L^2 = W^*$ and $L^i = 0$, for $i \neq 1,2$, the bracket being given by $(\ ,\ )_q$.

Now $E^{\cdot}$ is precisely its <u>graded universal enveloping algebra</u> $U^{\cdot}(L)$, that is, $E^{\cdot}$ admits a presentation $T^k_{\cdot}(V^*) \otimes_k S^k_{\cdot}(W^*)/I$, where
- $T^k_{\cdot}(V^*)$ is the tensor-algebra on V^*, graded by $\deg V^* = 1$,
- $S^k_{\cdot}(W^*)$ is the symmetric algebra on W, with $\deg W^* = 2$,
- the multiplicative structure is given by

  $(x \otimes y).(x' \otimes y') = x \otimes x' \otimes yy'$

  for $x,x' \in T^k_{\cdot}(V^*)$; $y,y' \in S^k_{\cdot}(W^*)$, so that $S^k_{\cdot}(W^*)$ becomes a <u>central</u> subalgebra,

and finally,

- I is the two-sided ideal generated by all elements

  $\lambda \otimes \lambda - q(\lambda)$ for $\lambda \in V^*$.

The Lie-algebra $L^{\cdot}$, which following L. Avramov, [Av], is called the <u>homotopy Lie-algebra</u> of $R.$, is nilpotent in an obvious sense,

$W^* = L^{\geq 2}$ being an abelian ideal in the centre of $L^\cdot$, $V^* = L^\cdot/L^{\geq 2}$ an abelian quotient.

Another interpretation of $L^\cdot$ is obtained from the André-Quillen cohomology of the cotangent-complex. It follows from [Qu] that $L^\cdot = T^\cdot_{k/R.}(k)$, the tangent or André-Quillen cohomology of k over $R.$, that is, $L^\cdot = H^\cdot(\mathrm{Hom}_k(\mathbb{L}_{k/R.},k))$, where $\mathbb{L}_{k/R.}$ is the cotangent-complex of k over $R.$. This is easily seen by direct computation and from (loc. cit.) it follows - in any characteristic, as $R.$ is a complete intersection - that $\mathrm{Ext}^\cdot_{R.}(k,k)$ is precisely the graded enveloping algebra of this graded Lie-algebra, whence the claimed result on the structure of $E^\cdot$.

As to the functoriality of this construction, let us remark only that a k-linear map $\lambda : W_1 \to W$ between two k-vectorspaces W_1 and W defines obvious morphisms of graded Lie-algebras,

$$L^\cdot(\lambda^*) : L^\cdot = (V^*,W^*,q_W) \to L_1^\cdot = (V^*,W_1^*,\lambda^*\cdot q_W),$$

as well as of their graded universal enveloping algebras,

$$U^\cdot(\lambda^*) : E^\cdot = U^\cdot(L^\cdot) \to E_1^\cdot = U^\cdot(L_1).$$

If $i : W_1 \to W$ is a subspace of W, it also defines a complete intersection $R_1 = S./_{(W_1)}$, and $U^\cdot(i^*) : E^\cdot = U^\cdot(L) \to E_1^\cdot = U^\cdot(L_1)$ is the morphism of Ext-Algebras $\mathrm{Ext}^\cdot_R(k,k) \to \mathrm{Ext}^\cdot_{R_1}(k,k)$ referred to earlier.

Two particular cases are worth mentioning:

If $W_1 = 0$, (hence $R_1 = S.$), $L_1^\cdot = V^*$ is the graded abelian Lie-algebra concentrated in degree one and its graded universal enveloping algebra is just $\Lambda^\cdot = \Lambda^\cdot(V^*)$, the exterior k-algebra generated by V^* in degree one. The corresponding morphism $E^\cdot \to \Lambda^\cdot$ is then nothing but the canonical projection $E^\cdot \to E^\cdot \otimes_{S^k(W^*)} k \simeq \Lambda^\cdot$.

If $W_1 = k.Q \subseteq W$ is the subspace generated by a single quadric Q in $W \subseteq S_2^k(V)$, $R_Q = S./_{(Q)}$ is a quadric hypersurface ring with Ext-algebra

$$E_1^\cdot = E_Q^\cdot = T^k_\cdot(V^*)[\sigma]/_{(\lambda \otimes \lambda - \lambda^2(Q).\sigma)},$$

where σ is a central variable of degree two, and λ^2 denotes the linear form on $S_2^k(V)$ given by $\lambda^2(vw) = \lambda(v)\lambda(w)$, i.e. $\lambda^2 = \chi(\lambda)$.

Q can of course also be considered as a linear form on W^* and the corresponding morphism $E^\cdot \to E_Q^\cdot$ is just the projection

$E^\cdot \to E^\cdot \otimes_{S^k(W^*)} k[\sigma]$, $k[\sigma]$ considered a graded $S^k(W^*)$-module via the linear form $Q : W^* \to k.\sigma$.

$E^\cdot_Q$ is nothing but the homogenized Clifford-algebra of the quadratic form defined by Q, in other words, sending σ to 1 defines a morphism of $\mathbb{Z}/_{2\mathbb{Z}}$-graded algebras - as $\deg \sigma = 2$ - :

$$E^\cdot_Q \xrightarrow{\sigma=1} C^\cdot(Q) = E^\cdot \otimes_{S^k=W^*)} k[\sigma]/_{(\sigma-1)} = T^k_\cdot(V^*)/_{(\lambda\otimes\lambda-\lambda^2(Q))}$$

These morphisms allow us to visualize $E^\cdot$ geometrically. An equivalent form of the Poincaré-Birkhoff-Witt theorem for such simple-minded graded Lie-algebras $L^\cdot = (V^*,W^*,q_W)$ can be stated as saying that $E^\cdot$ is a central, flat $S^k_\cdot(W^*)$-algebra with fibre $\Lambda^\cdot = E^\cdot \otimes_{S^k(W^*)} k$ over the "origin" $\{S^k_+(W^*)\}$ in the affine space W^*. Sheafifying $E^\cdot$ to obtain a sheaf of $\mathbb{Z}/_{2\mathbb{Z}}$-graded algebras $\mathcal{E}^\cdot$ on $\mathbb{P}(W^*)$, the underlying projective space of hyperplanes in W^*, the fibre over a k-rational point $W^* \to k$ is the Clifford-algebra $C^\cdot(Q)$ of the corresponding quadric in $W=W^{**}$ defining this point.

In other words, $E^\cdot$ may be considered as the total space of a flat deformation deforming the exterior algebra $\Lambda^\cdot$ into the various Clifford-algebras $C^\cdot(Q)$, $Q \in W$.

Recall now that Proj $(E^\cdot)$ is the category of graded, finitely generated, right $E^\cdot$-modules modulo its Serre-subcategory of artinian modules. As it was said above, $E^\cdot$ is a central, flat $S^k_\cdot(W^*)$-algebra, hence free as a module over this polynomial ring and its rank equals $\dim_k \Lambda. = 2^{\dim V}$. As W^* is concentrated in degree two, any graded $E^\cdot$-module $N^\cdot$ defines two graded $S^k_\cdot(W^*)$-modules,

- $N^\cdot_+$ given by $N^i_+ = N^{2i}$ and
- $N^\cdot_-$ given by $N^i_- = N^{2i+1}$ for all i.

As an $E^\cdot$-module is artinian iff its positive and negative part are artinian $S^k_\cdot(W^*)$-modules, the forgetful-functor defines a functor $(+,-)$: Proj $E^\cdot \to \mathrm{Coh}(\mathbb{P}(W^*))\times\mathrm{Coh}(\mathbb{P}(W^*))$, associating to (the class of) a graded $E^\cdot$-module $N^\cdot$ the sheaves $\tilde{N}^\cdot_+$ and $\tilde{N}^\cdot_-$ on the projective space $\mathbb{P}(W^*)$.

Denoting as usual by $\mathcal{O}(1)$ the Hopf-bundle on $\mathbb{P}(W^*)$ - which is given as a graded module by $S^k_\cdot(W^*)(2)$ as $\deg W^* = 2$ - , the extra-structure on $N^\cdot$, that is the action of V^*, can be recovered as follows:

A pair of coherent sheaves (N_+, N_-) on $\mathbb{P}(W^*)$ comes from a graded $E^{\cdot}$-module $N^{\cdot}$ iff there are $\mathcal{O}_{\mathbb{P}(W^*)}$-linear morphisms

$$\varphi : N_+ \underset{k}{\otimes} V^* \to N_- \quad \text{and}$$
$$\psi : N_- \underset{k}{\otimes} V^* \to N_+(1)$$

such that the following diagrams are commutative:

$$\begin{array}{ccccc} N_+ \underset{k}{\otimes} V^* & & \xrightarrow{\ 1 \otimes q\ } & & N_+ \underset{k}{\otimes} W^* \\ \downarrow{\scriptstyle 1\otimes\tilde{\chi}} & & & & \downarrow{\scriptstyle \text{can.}} \\ N_+ \underset{k}{\otimes} V^* \underset{k}{\otimes} V^* & \xrightarrow[\varphi\otimes 1_{V^*}]{} & N_- \underset{k}{\otimes} V^* & \xrightarrow[\psi]{} & N_+(1) \end{array}$$

and

$$\begin{array}{ccccc} N_- \underset{k}{\otimes} V^* & & \xrightarrow{\ 1 \otimes q\ } & & N_- \underset{k}{\otimes} W^* \\ \downarrow{\scriptstyle 1\otimes\tilde{\chi}} & & & & \downarrow{\scriptstyle \text{can.}} \\ N_- \underset{k}{\otimes} V^* \underset{k}{\otimes} V^* & \xrightarrow[\psi\otimes 1_{V^*}]{} & N_+(1) \underset{k}{\otimes} V^* & \xrightarrow[\varphi(1)]{} & N_-(1) \end{array} .$$

Here $\tilde{\chi}$ denotes the composition of the universal quadratic map $\chi : V^* \to S_2(V)^*$ with the natural inclusion $S_2(V)^* \to V^* \underset{k}{\otimes} V^* = (V \underset{k}{\otimes} V)^*$, given explicitely by $\tilde{\chi}(\lambda) = \lambda \otimes \lambda$ for $\lambda \in V^*$. The map labeled "canonical" is just the tensorproduct of the identity on $N_{+/-}$ with the evaluation of global sections $\mathcal{O} \underset{k}{\otimes} W^* = \mathcal{O} \underset{k}{\otimes} H^0(\mathbb{P}(W^*), \mathcal{O}(1)) \to \mathcal{O}(1)$. Hence:

<u>Proposition 2</u> (i): The category <u>Proj</u> $E^{\cdot}$ is isomorphic to the category of quadruples $(N_+, N_-, \varphi, \psi)$, the $\mathcal{O}$-linear morphisms φ and ψ satisfying the compatibility conditions above. (The description of the morphisms is left to the reader.).
(ii) Under the Bernstein-Gelfand-Gelfand-correspondence, <u>Proj</u> $E^{\cdot}$ corresponds to the subcategory of <u>MCM</u>$(R.)$ given by those modules which admit a linear resolution over $R.$.

<u>Remark 3</u>: The natural t-structure on D^b(<u>Proj</u> $E^{\cdot}$) - see remark 2, (b) - , if interpreted on the equivalent category <u>MCM</u>$(R.)$, has hence the <u>linear modules</u> (generated in degree zero) as its "heart"

<u>Proof</u> (i) is obvious from the above. To prove (ii), let $\tilde{N}$ be an object in <u>Proj</u> $E^{\cdot}$ and represent it by a graded $E^{\cdot}$-module $N^{\cdot}$. By remark 2, (b), the image $\rho(N^{\cdot})$ in $D^b(R.)$ is isomorphic to a

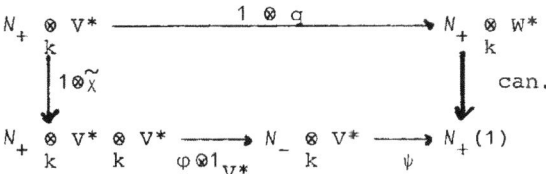

linear complex of free R.-modules. Using [Bu], a maximal Cohen-Macaulay module representing $\underline{\rho}(N^{\cdot})$ in $\underline{MCM}(R.)$ is obtained - up to translation - as a syzygy-module sitting sufficiently far back in the free resolution of $\underline{\rho}(N^{\cdot})$, whence this module - and its translates - have a linear resolution. (A proof using Tate-cohomology, [Bu], can be obtained as follows: Let M. be a MCM representing $\underline{\rho}(N^{\cdot})$. Then, for sufficiently large i and all j, $\underline{\mathrm{Ext}}^i_{R.}(k,M)_{-j} = \underline{\mathrm{Ext}}^i_{R.}(k,\rho(N.))_{-j}$ and $\underline{\mathrm{Ext}}^i_{R.}(k,M.)$ is dual to $\mathrm{Tor}^{R.}_{i-\dim R.}(k,M^*_.)$, which shows that $M^*_.$ - and then also M. - has a linear resolution.)

As $\underline{\varepsilon}(M.)$ represents $N^{\cdot}$, it follows that M. is generated in degree zero (and the i-th term in its resolution by elements of degree i). This establishes the claim. □

In case that W is one-dimensional, hence generated by a single non-zero quadric Q, $\mathbb{P}(W^*)$ is just a point and the pair (N_+,N_-) defines a $\mathbb{Z}/_{2\mathbb{Z}}$-graded module over the full Clifford-algebra $C^{\cdot}(Q)$. Hence in this case, $\underline{\mathrm{Proj}}\ E^{\cdot} = \text{mod-}C^{\cdot}(Q)$ as stated in theorem 2.

It remains to prove corollary 2. If M. is an equi-generated graded MCM over R_Q for some quadric Q, that is, generated by its elements of some fixed degree l, M.(l) is generated in degree zero and has a linear resolution by Proposition 2.3. of the foregoing article.
We may replace M. by M.(l) and can hence assume that M. is already generated in degree zero. Then $\underline{\varepsilon}(M.)$ is (isomorphic in $D^b(E^{\cdot})$ to) a single module $N^{\cdot}$. If now M. is a syzygy-module of some artinian R.-module A., this signifies the existence of a distinguished triangle

$$M.[r-1] \to X^{\cdot}_{\cdot} \to A. \xrightarrow{(1)}$$

in $D^b(R.)$, (the notations as in [BBD]), where $X^{\cdot}_{\cdot}$ is a perfect complex - namely the beginning of a free resolution of A. which exhibits M. as its r-th syzygy, (of course $r \geq \dim R.$, if M. is MCM, but that does not matter).

Applying $\underline{\varepsilon}$ to this triangle and using theorem 1, we obtain a triangle

$$N^{\cdot}[r-1] = \underline{\varepsilon}(M.[r-1]) \to \underline{\varepsilon}(X^{\cdot}_{\cdot}) \to \underline{\varepsilon}(A.) \xrightarrow{(1)}$$

in $D^b(E^{\cdot})$, where now $\underline{\varepsilon}(A.)$ is perfect and $\underline{\varepsilon}(X^{\cdot}_{\cdot})$ is a complex with artinian cohomology. Projecting into $D^b(\underline{\text{Proj}}\ E^{\cdot})$, $\underline{\varepsilon}(X^{\cdot})$ becomes the zero-object, hence the images of $\underline{\varepsilon}(A.)$ and $N^{\cdot}[r]$ become isomorphic. This means that the image of $\underline{\varepsilon}(A.)$ has a single cohomology object, namely the image $\tilde{N}^{\cdot}$ of $N^{\cdot}$ in (complex-)degree $-r$.

Now assume $C^o(Q)$ semi-simple, so that the image of $\underline{\varepsilon}(A.)$ in $D^b(\text{mod-}C^o(Q)) \cong D^b(\underline{\text{Proj}}\ E^{\cdot})$ is (isomorphic to) a finite complex of free $C^o(Q)$-modules, still having a single cohomology module. But $C^o(Q)$ semi-simple implies that this complex splits, whence its only cohomology module is free. (By [Kap; Thm 180] it follows a priori only that the cohomology module is projective having a free complement. But $C^o(Q)$ semi-simple implies that one can cancel.) This proves corollary 2.

References

[ABS]: M. F. Atiyah, R. Bott, A. Shapiro: Clifford Modules, Topology 3 (Suppl.), 3 - 38, (1964)

[ALG]: N. Bourbaki, Algèbre, Masson, Paris 1970 ff

[Av]: L. Avramov: Local algebra and rational homotopy, Astérisque 113/114, 15 - 43, (1984)

[BBD]: A. A. Beilinson, J. Bernstein, P. Deligne: Faisceaux pervers, Astérisque 100, (1983)

[BGG]: I. N. Bernstein, I. M. Gelfand, S. I. Gelfand: Algebraic Bundles over $\mathbb{P}^n$ and Problems of Linear Algebra, Funkt. Anal. 12, No. 3, 66 - 67, (1978) - engl. translation: 212 - 214, (1979) -

[B-H]: R. Bøgvad, S. Halperin: On a conjecture of Roos, pp. 120 - 127, in "Algebra, Algebraic Topology and their Interactions", Proc. Stockholm 1983, ed. by J.-E. Roos, Springer Lect. Notes in Math. 1183, Springer-Verlag Berlin-Heidelberg-New York, 1986

[B-R]: J. Backelin, J.-E. Roos: When is the double Yoneda-Ext--Algebra of a local noetherian ring again noetherian?, pp. 101 - 120, in the same volume as [B-H]

[Bu]: R.-O. Buchweitz: Maximal Cohen-Macaulay modules and Tate--Cohomology over Gorenstein rings; preprint.

[Eis]: D. Eisenbud: Homological Algebra on a Complete Intersection, with an Application to Group Representations. Transactions of the AMS 260, 35-64 (1980)

[Ha]: R. Hartshorne: Residues and Duality, Springer Lecture Notes in Math. 20, Springer-Verlag, Berlin-Heidelberg-New York, 1966

[Kap]: I. Kaplansky: Commutative Rings, rev. edition, Univ. of Chicago Press, Chicago Ill., 1974

[Löf]: C. Löfwall: On the subalgebra generated by one-dimensional elements in the Yoneda-Ext-Algebra, pp. 291 - 339 in the same volume as [B-H]

[Pri]: S. Priddy: Koszul-resolutions, Transactions of the AMS 152, 39 - 60, (1970)

[Qu]: D. Quillen: On the (co-)homology of commutative rings, Proc. Symp. Pure Math. 17, 65 - 87, AMS, Providence, 1970

[Sw]: R. Swan: K-theory of quadric hypersurfaces. Annals of Math. 122 (1985), 113-153

[Ver]: J.-L. Verdier: Catêgories derivées (Etat O), in Sém. de Gêom. Algêbrique 4 1/2, Springer Lecture Notes in Math. 569 (1977)

MONOMIAL CURVES AND OBSTRUCTIONS
ON CYCLIC QUOTIENT SINGULARITIES

Jan Arthur Christophersen
Institute of Mathematics
University of Oslo
Oslo, Norway

0. Introduction

We show how the obstruction space T^2 for isolated singularities is related to the deformations of certain hypersurface sections (§1). This is applied to 2-dimensional cyclic quotient singularities, or what is the same thing: 2-dimensional normal affine toric varieties. For more on quotient singularities see e.g. [Br], [B-K-R], for toric varieties see e.g. [K-K-M-S]. We find a monomial curve C on the cyclic quotient X such that

$$\tau_C - \mu_C = \dim_{\mathbb{C}} T_X^2 + t_C - 1.$$

Here τ is the Tjurina number, μ is the Milnor number and t is the Gorenstein type. τ is computed using a method that works for all affine toric varieties (§2) and μ by a Kouchnirenko type formula for functions on cyclic quotient surfaces (§4).

If r is the minimal codimension of X, i.e. $r = \dim_{\mathbb{C}} \underline{m}_{X,0}/\underline{m}_{X,0}^2 - 2$, then

$$\dim_{\mathbb{C}} T_X^2 = r(r-2)$$

when $r>2$. (Of course T_X^2 is trivial when $r=1$). I have learned through private communication that Jürgen Arndt (Hamburg) has computed this dimension by a different method.

I would like to thank Olav Arnfinn Laudal for constant help and advice. I have also benefited from conversations with A. Sletsjøe,

J. Damon, G.-M. Greuel and the stimulating milieu at Lambrecht. (Supported by the Norwegian Research Council for Science and the Humanities.)

1. Obstructions and deformations of hypersurface sections

With the exception of §2, we shall work over the ground field $\mathbb{C}$. We use the notation T^i_X for $T^i(X/\mathbb{C}, O_X)$ (cotangent complex notation, see [Li-S]) and $H^i(\mathbb{C}, X; O_X)$ (André cohomology notation, see [L]). T^1_X is the space classifying infinitesimal deformations, the tangent space of the miniversal deformation space. T^2_X is the space "in which the obstructions lie". (See [L], [Ri], [S] for the deformation theory involved.)

Lemma 1.1. Let X be an affine scheme with one singular point x and $f \in \underline{m}_{X,x}$, the maximal ideal, such that

(i) $f: X \to \mathbb{C}$ has one critical point x,

(ii) $f \in \mathrm{Ann}(T^2_X)$.

If $Y = f^{-1}(0)$, then

$$\dim_{\mathbb{C}} T^1_Y - e_Y = \dim_{\mathbb{C}} T^2_X$$

where e_Y is the dimension of a smoothing component of the versal deformation space of Y.

By a smoothing we mean a deformation with smooth generic fiber. Thus over a smoothing component the generic fiber is non-singular.

Proof. This is a corollary of "Wahl's conjecture" on the dimension of smoothing components recently proved in [G-L] and [L-P].

Obviously $f: X \to \mathbb{C}$ is a smoothing of Y. From the exact sequence

$$0 \to \mathbb{C}[t] \xrightarrow{\cdot t} \mathbb{C}[t] \to \mathbb{C} \to 0$$

we get a long exact sequence in algebra cohomology

$$0 \to H^0(\mathbb{C}[t],\ X;\ O_X) \xrightarrow{\cdot t} H^0(\mathbb{C}[t],\ X;\ O_X) \to H^0(\mathbb{C},\ Y;\ O_Y)$$
$$\to H^1(\mathbb{C}[t],\ X;\ O_X) \xrightarrow{\cdot t} H^1(\mathbb{C}[t],\ X;\ O_X) \xrightarrow{\alpha} H^1(\mathbb{C},\ Y;\ O_Y)$$
$$\to H^2(\mathbb{C}[t],\ X;\ O_X) \xrightarrow{\cdot t} H^2(\mathbb{C}[t],\ X;\ O_X) \to \cdots$$

where O_X is a $\mathbb{C}[t]$-algebra via f^*. As in the proof of Wahls conjecture $\dim_{\mathbb{C}}(\mathrm{Im}(\alpha))$ equals the dimension of the smoothing component on which $f\colon X \to \mathbb{C}$ "lies", [G-L].

The algebra homomorphisms $\mathbb{C} \to \mathbb{C}[t] \xrightarrow{f^*} O_X$ induce the exact sequence:

$$\cdots \to H^1(\mathbb{C}[t],\ X;\ O_X) \to H^1(\mathbb{C},X;\ O_X) \to H^1(\mathbb{C},\mathbb{C}[t];\ O_X)$$
$$\to H^2(\mathbb{C}[t],\ X;\ O_X) \to H^2(\mathbb{C},X;\ O_X) \to H^2(\mathbb{C},\mathbb{C}[t];\ O_X) \to \cdots$$

Now $H^i(\mathbb{C},\ \mathbb{C}[t];\ O_X)=0$ for $i \geq 1$, so $H^2(\mathbb{C}[t],\ X;\ O_X) \simeq H^2(\mathbb{C},X;\ O_X)$. Since $f^*(t) = f$, we get a short exact sequence

$$0 \to \mathrm{Im}(\alpha) \to T^1_Y \to T^2_X \to 0$$

proving the lemma. □

<u>Lemma 1.2</u>. In the situation of 1.1 assume also that X is a surface with $\mathbb{C}^*$-action and f is homogeneous under this action. Then

$$\dim_{\mathbb{C}} T^1_Y - \mu(f) = \dim_{\mathbb{C}} T^2_X + t_Y - 1$$

where $\mu(f)$ is the Milnor number of f in x and t_Y is the Gorenstein type of Y.

<u>Proof</u>. In [Gr] it is proved that $e = \mu + t - 1$ for quasi-homogeneous isolated curve singularities. □

(As an obvious consequence of 1.2, we see that if $\underline{m}_{X,x} \cdot T^2_X = 0$, then $\tau - \mu$ for quasi-homogeneous functions on X depends only on the surface, generalizing the fact that $\tau = \mu$ if X is smooth.)

2. T^1 for rings over subsemigroups of free abelian semigroups.

In this section k denotes an algebraically closed field. In [La-S] Laudal and Sletsjøe show that for a monoid algebra $k[\Lambda]$, the algebra cohomology groups $H^i(k, k[\Lambda]; M)$, M a $k[\Lambda]$-module, are isomorphic to the cohomology of the monoid with values in M. The algebra cohomology can then be computed using only the monoid structure. We shall use their ideas to find directly a method for computing T^1 for monoid algebras $k[\Lambda]$, when $\Lambda \subseteq Z_0^n$. (Z_0 is the semigroup of non-negative integers.) The main examples are affine toric varieties.

Let Λ be the submonoid of Z_0^n generated minimally by $v_1,\ldots,v_m$, and $k[\Lambda]$ the corresponding monoid algebra. Let $k[Z_0^n] = k[t_1,\ldots,t_n]$. We may assume n minimal so that $\dim k[\Lambda] = \operatorname{rank} \Lambda = n$. Define $\rho\colon Z_0^m \to \Lambda$ by $\rho(a_1,\ldots,a_m) = \sum_{i=1}^m a_i v_i$, and $\rho^*\colon k[x_1,\ldots,x_m] \to k[\Lambda]$ by $x_i \mapsto t^{v_i}$. Then $I = \ker\rho^*$ is generated by $\{x^c - x^d \mid \rho(c)=\rho(d)\}$ ($x^c = x_1^{c_1}\cdots x_m^{c_m}$), [Gi], Thm.7.2. Extend ρ to $Z^m \to Z^n$ and let J be the kernel. J is a free abelian group of rank $m-n = r = \operatorname{codim} k[\Lambda]$.

<u>Lemma 2.1</u>. Let $c,d,a_1,\ldots,a_r,b_1,\ldots,b_r$ (not necessarily distinct) in Z_0^m be such that $c-d = \sum_{i=1}^r (a_i-b_i)$. There exist $\beta_0,\beta_1,\ldots,\beta_r \in Z_0^m$ such that

$$x^{\beta_0}(x^c-x^d) = \sum_{i=1}^r x^{\beta_i}(x^{a_i}-x^{b_i})$$

<u>Proof</u>. The system of equations $\beta_0 + c = \beta_1 + a_1$, $\beta_{i-1} + b_{i-1} = \beta_i + a_i$, $i=2,\ldots,r$ has a solution in Z_0^m. □

Let $\{j_1,\ldots,j_r\}$ be a basis for J and let $(a_i,b_i) = ((a_{i,1},\ldots,a_{i,m}), (b_{i,1},\ldots,b_{i,m}))$ be the unique element in $Z_0^m \times_\Lambda Z_0^m$ such that $a_i - b_i = j_i$ and $a_{i,k} \cdot b_{i,k} = 0$. Then $\{f_i = x^{a_i} - x^{b_i}\}_{i=1}^r$ is a maximal regular sequence for $k[x_1,\ldots,x_m]$ in I. In fact, one can use 3.1 to prove that $\dim k[\Lambda] = \dim k[\underline{x}]/(f_1,\ldots,f_r)$. Since no monomial in $k[x_1,\ldots,x_m]$ can be in I, 3.1 shows that the evaluation map

$$\mathrm{Hom}_{k[\Lambda]}(I/I^2,\ k[\Lambda]) \to \bigoplus_{i=1}^{r} k[\Lambda]$$

defined by $\phi \to (\phi(f_1),\ldots,\phi(f_r))$ is injective. Working in the quotient field $k(\Lambda) \subset k(t_1,\ldots,t_n)$ we get the following description of the image.

<u>Lemma 2.2</u>. $\mathrm{Hom}_{k[\Lambda]}(I/I^2,\ k[\Lambda])$ is isomorphic to

$\{(\phi_1,\ldots,\phi_r) \in \bigoplus_{i=1}^{r} k[\Lambda] \mid \sum_{i=1}^{r} \alpha_i t^{\rho(c)-\rho(a_i)} \cdot \phi_i \in k[\Lambda]$, for all $x^c - x^d$ generating I and $c-d = \sum_{i=1}^{r} \alpha_i j_i$ in J$\}$.

<u>Proof</u>. From 3.1 there exist $\beta_0,\beta_1,\ldots,\beta_r \in Z_0^m$ such that

$$x^{\beta_0}(x^c - x^d) = \sum_{i=1}^{r} \mathrm{sign}(\alpha_i) \cdot x^{\beta_i}(x^{|\alpha_i|a_i} - x^{|\alpha_i|b_i}).$$

Notice that

$$x^{|\alpha_i|a_i} - x^{|\alpha_i|b_i} = \sum_{j=1}^{|\alpha_i|} x^{[(|\alpha_i|-j)a_i + (j-1)b_i]}(x^{a_i} - x^{b_i}),$$

so if $\phi \in \mathrm{Hom}_{k[\Lambda]}(I/I^2,\ k[\Lambda]) = \mathrm{Hom}_{k[\underline{x}]}(I, k[\Lambda])$ then

$$\phi(x^{|\alpha_i|a_i} - x^{|\alpha_i|b_i}) = \sum_{j=1}^{|\alpha_i|} t^{(|\alpha_i|-1)\rho(a_i)} \phi(x^{a_i} - x^{b_i}).$$

Since $\rho(\beta_i) = \rho(\beta_0) + \rho(c) - |\alpha_i|\rho(a_i)$, we get

$$(\star) \quad \phi(x^c - x^d) = \sum_{i=1}^{r} \alpha_i t^{\rho(c)-\rho(a_i)} \phi(x^{a_i} - x^{b_i}) \in k[\Lambda].$$

On the other hand if $(\phi_1,\ldots,\phi_r)$ satisfy the conditions of 3.2 then using 3.1 and (*) one checks that all relations among generators of I are satisfied. □

Using this method one can e.g. compute a basis of T^1 for cyclic quotient singularities getting the equations of the first order deformations. The computation is similar to the one in §6. See also [R], [P2], [La-S] and [B-K-R] for other descriptions of T^1 in this case. For a description of T^1 for monomial curves see [Bu].

3. Cyclic quotient singularities

If G is a finite cyclic subgroup of $GL(2,\mathbb{C})$, let $X = \mathbb{C}^2/G$ be the orbit space. It is a normal algebraic variety $\mathrm{Spec}(\mathbb{C}[x,y]^G)$, where $\mathbb{C}[x,y]^G$ is the invariant ring of the induced action. Since the origin is the only fixed point for the action of G, the corresponding point in X is an isolated singularity, the <u>cyclic quotient singularity</u>.

We may assume that G contains no pseudo-reflections and, since G is abelian, that G is generated by the linear transformation

$$\begin{bmatrix} \zeta_n^q & 0 \\ 0 & \zeta_n \end{bmatrix}$$

where ζ_n is a primitive n'th root of unity, $n = \mathrm{ord}G$, and $0<q<n$, $\gcd(n,q) = 1$. G's induced action on $\mathbb{C}[x,y]$ is generated by $x \to \zeta_n^m x$, $y \to \zeta_n^{-1} y$ where $m = n-q$.

If $\Lambda \subset Z_0^2$ is the semigroup

$$\Lambda = \{(\alpha,\beta) \in Z_0^2 \mid \beta \equiv m\cdot\alpha(n)\}$$

then $\mathbb{C}[x,y]^G$ is the semigroup ring $\mathbb{C}[\Lambda]$. On Λ we have the natural partial order: $\lambda_1 > \lambda_2$ if there exists $\mu\in\Lambda$ such that $\lambda_2+\mu = \lambda_1$. This is just the restriction of the natural partial order on Z_0^2.

Let $\{v_0,\ldots,v_{r+1}\}$ be the minimal elements of $\Lambda\setminus\{0\}$ in this order. Write $v_i = (a_i,b_i)$ and order the indices so that $a_{i+1} > a_i$, $b_{i+1} < b_i$. Then for each $i=1,\ldots,r$ there is a number $e_i \geq 2$ such that $v_{i-1} + v_{i+1} = e_i v_i$. These numbers appear in the continued fraction expansion

$$\frac{n}{m} = e_1 - \cfrac{1}{e_2 - \cfrac{1}{e_3 \ddots - \cfrac{1}{e_r}}}$$

and we could also define v_i by $a_0 = 0$, $a_1 = 1$, $a_{i+1} = e_i a_i - a_{i-1}$, $b_0=n$, $b_1=m$, $b_{i+1} = e_i b_i - b_{i-1}$.

The minimal embedding dimension is therefore $r+2$. If $\mathbb{C}[z_0,\ldots,z_{r+1}] \to \mathbb{C}[\Lambda]$ is the map $z_i \to x^{a_i}y^{b_i}$ then the kernel is generated minimally by $\frac{1}{2}r(r+1)$ polynomials

$$g_{ij} = z_i z_j - z_{i+1} z_{j-1} \prod_{k=i+1}^{j-1} z_k^{e_k-2}$$

for $1 \leq i+1 \leq j-1 \leq r$. ([R]).

From now on we assume $r \geq 2$.

4. The Milnor number of a function on a cyclic quotient.

For the definitions needed below see [K].

<u>Lemma 4.1</u>. Let $(f,0)$ be the germ of an analytic function f on the cyclic quotient singularity $(X,0)$, $X = \mathbb{C}^2/G$ and $n=\mathrm{ord}G$. If

$\pi: \mathbb{C}^2 \to X$ is the natural projection set $\bar{f} = f\circ\pi$. Assume $\bar{f}$ is non-degenerate and "commode" in the sense of Kouchnirenko [K]. Choose s,t minimally such that $x^{s\cdot n}$, $y^{t\cdot n}$ appear as monomials in $\bar{f}$. If A is the area bounded by the Newton polygon of $\bar{f}$ and $S = A/n$, then the Milnor number $\mu(f,0)$ equals $2S-s-t+1$.

Proof. Embed X in $\mathbb{C}^{r+2}$ with $z_i = x^{a_i}y^{b_i}$ as in §3. Using weighted balls

$$B_{\varepsilon,N} = \{x\in\mathbb{C}^{r+2} \mid \sum_{i=0}^{r+1} |z_i|^{2W_i} < \varepsilon,\ \varepsilon>0,\ W_i\cdot(a_i+b_i)=N\}$$

one constructs good representatives ([Lo], chap.3) for f and $\bar{f}$ with Milnor fiber F and $\bar{F}$ such that $F \simeq \bar{F}/G$. Thus $\chi(\bar{F}) = n\cdot\chi(F)$ and $\mu(f,0) = 1 + \frac{\mu(\bar{f},0)-1}{n}$. From [K], $\mu(\bar{f},0) = 2A-sn-tn+1$, so $\mu(f,0) = 2S-s-t+1$. (One checks that since $\bar{f}$ is invariant, $A\equiv 0(n)$) □

The lemma can be generalized to invariant functions for abelian finite subgroups of $GL(d,\mathbb{C})$, see [M]. The assumption "commode" is not essential.

5. Monomial curves on cyclic quotients.

We will now apply 1.2 to the cyclic quotient singularities. We wish to find a hypersurface for which the invariants are easily computed.

Proposition 5.1. With the notation of §3, let p be a positive integer such that $\gcd(p+m,n) = 1$. Then $f=z_0^p - z_{r+1} \in \mathbb{C}[\Lambda]$ satisfies the conditions of 1.2 and $C = \mathrm{Spec}(\mathbb{C}[\Lambda]/(f))$ is a monomial curve (i.e. $\mathbb{C}[\Lambda]/(f) \simeq \mathbb{C}[\Gamma]$ for a semigroup $\Gamma \subset \mathbb{Z}_0$).

The proposition will follow from 5.2 and 5.3, but let's first have a closer look at T_X^2. If $P = \mathbb{C}[z_0,\dots,z_{r+1}]$ and $I = \ker(P \to \mathbb{C}[\Lambda])$, then the relation module $R=R(I)$ is the kernel of the P-homomorphism $P^{\frac{1}{2}r(r+1)} \to I$, $E_{ij} \to g_{ij}$, where E_{ij} is the standard basis of $P^{\frac{1}{2}r(r+1)}$, $1 \leq i+1 \leq j-1 \leq r$. Let $R_0 \subset R$ be the submodule generated by the trivial relations $g_{\alpha,\beta} \cdot g_{ij} - g_{ij} \cdot g_{\alpha,\beta} = 0$. Recall that

$$T_X^2 = H^2(\mathbb{C}, X; \mathbb{C}[\Lambda]) \simeq \operatorname{Hom}_P(R/R_0, \mathbb{C}[\Lambda])/\mathrm{Der}$$

where Der is generated by the derivations

$$D_{\underline{h}}(R_1,\dots,R_{\frac{1}{2}r(r+1)}) = \sum_{i=1}^{\frac{1}{2}r(r+1)} h_i R_i$$

$\underline{h} \in \mathbb{C}[\Lambda]^{\frac{1}{2}r(r+1)}$, ([Li-S], [L]). In our case R is generated by

$$R_{i,j,k} = z_i E_{jk} - z_j E_{ik} + z_{k-1} \prod_{\ell=j+1}^{k-1} z_\ell^{e_\ell - 2} E_{i,j+1}$$

and
$$S_{i,j,k} = z_{i+1} \prod_{\ell=i+1}^{j} z_\ell^{e_\ell - 2} E_{jk} - z_{j+1} E_{ik} + z_k E_{i,j+1}$$

for $0 \leq i < j < k-1 \leq r$. ([R]).

<u>Lemma 5.2</u>. $(z_0, z_{r+1}) \subset \operatorname{Ann}(T_X^2)$.

<u>Proof</u>. The relations among relations:

$$z_{j+1} R_{i,j,k} - z_j S_{i,j,k} \in R_0$$

$$z_0 R_{i,j,k} = z_i R_{0,j,k} - z_j R_{0,i,k} + z_{k-1} \prod_{\ell=j+1}^{k-1} z_\ell^{e_\ell - 2} R_{0,i,j+1}, \quad i \geq 1$$

$$z_{r+1} S_{i,j,k} = z_{i+1} \prod_{\ell=j+1}^{j} z_\ell^{e_\ell - 2} S_{j,k-1,r+1} - z_{j+1} S_{i,k-1,r+1} + z_k S_{i,j,r+1}, \quad k \leq r$$

make $\phi_1, \phi_2 \colon \operatorname{Hom}_P(R/R_0, \mathbb{C}[\Lambda]) \to \mathbb{C}[\Lambda]^{\frac{1}{2}(r-1)r}$, given by

$$\phi_1(\psi) = (\psi(R_{0,1,3}),\dots,\psi(R_{0,j,k}),\dots,\psi(R_{0,r-1,r+1}))$$

$$\phi_1(\psi) = (\psi(S_{0,1,r+1}),\dots,\psi(S_{i,j,r+1}),\dots,\psi(S_{r-2,r-1,r+1})),$$

injective. Let

$$\delta_1 = \left[\frac{\overline{\partial R_{0,j,k}}}{\partial E_{\alpha,\beta}}\right] \quad \begin{matrix} 1 \leq j < k-1 \leq r \\ 1 \leq \alpha+1 < \beta-1 \leq r \end{matrix}$$

and

$$\delta_2 = \left[\frac{\overline{\partial S_{i,j,r+1}}}{\partial E_{\alpha,\beta}}\right] \quad \begin{matrix} 0 \leq i < j \leq r-1 \\ 1 \leq \alpha+1 < \beta-1 \leq r \end{matrix}$$

(If $p \in P$, then $\bar{p}$ is the image in $\mathbb{C}[\Lambda]$). T_X^2 is injectively mapped into $\mathbb{C}[\Lambda]^{\frac{1}{2}(r-1)r}/\mathrm{im}\delta_1$ and $\mathbb{C}[\Lambda]^{\frac{1}{2}(r-1)r}/\mathrm{im}\delta_2$ by ϕ_1 and ϕ_2. For $\alpha \geq 1$, $z_0 = \frac{\overline{\partial R_{0,\alpha,\beta}}}{\partial E_{\alpha,\beta}}$ and $\frac{\overline{\partial R_{0,j,k}}}{\partial E_{\alpha,\beta}} = 0$ for $(j,k) \neq (\alpha,\beta)$. The $\frac{1}{2}(r-1)r$ vectors $(z_0,0,\ldots,0)$, $(0,z_0,0,\ldots,0),\ldots,(0,\ldots,0,z_0)$ are thus in $\mathrm{im}\delta_1$. Similarily the vectors $(z_{r+1},0,\ldots,0),\ldots,(0,\ldots,0,z_{r+1})$ are in $\mathrm{im}\,\delta_2$, proving the lemma. □

<u>Lemma 5.3</u>. If $\rho\colon \Lambda \to Z_0$ is the semigroup homomorphism $\rho(\lambda_1,\lambda_2) = p\lambda_1+\lambda_2$, p is a positive integer with $\gcd(p+m,n) = 1$ and $\mathrm{im}(\rho) = \Gamma$, then the kernel of $\rho^*\colon \mathbb{C}[\Lambda] \to \mathbb{C}[\Gamma]$ is generated by $z_0^p - z_{r+1}$.

<u>Proof</u>. We know that $\ker\rho^*$ is generated by $\{x^\lambda - x^\mu \in \mathbb{C}[\Lambda] \mid \rho(\lambda)=\rho(\mu)\}$ ([Gi] Thm.7.2). Viewing Λ as a subset of $\mathbb{R}^2$, let $[\lambda,\mu]$ be the line segment between λ and μ. We must show that if the slope of $[\lambda,\mu]$ is $-p$ then there is a $g \in \mathbb{C}[\Lambda]$ such that

$$g \cdot (x^{(n,0)} - x^{(0,np)}) = x^\lambda - x^\mu.$$

We may assume that $[\lambda,\mu] \cap \Lambda = \{\lambda,\mu\}$. Write $\lambda = (\lambda_1,\lambda_2)$, $\mu = (\mu_1,\mu_2)$. Then $\mu_2 - \lambda_2 = p(\lambda_1-\mu_1)$, so $(\lambda_1-\mu_1)\cdot(p+m) \equiv 0(n)$. From the assumption $\lambda_1 - \mu_1 \equiv 0(n)$, so $\mu_2 - \lambda_2 \equiv 0(n)$. The lemma is now easily proven. □

6. The invariants of $\mathbb{C}[\Gamma]$.

Let $C = \text{Spec } \mathbb{C}[\Gamma]$ be the curve in 5.1 and choose

$$p = n-m+1 = q+1.$$

From 4.1 the Milnor number is

$$\mu(C,0) = np-p.$$

Since X is a rational singularity and C is a hypersurface in X,

$$t_X = t_C = r$$

([W]). Whats left in formula 1.2 is τ_C.

Proposition 6.1. If C is the curve of 5.1 and $p = q+1$, then

$$\dim_{\mathbb{C}} T^1_C = np - p + r-1 + r(r-2).$$

Before the proof can be given, we must look closer at the semigroup Γ.

Lemma 6.2. (i) If $N \in Z_0$, write $N = s+tn$ with $0 \leq s < n$, $t \geq 0$. Then $N \in \Gamma$ iff $tn \geq qs$.

(ii) If $w_i = \rho(v_i)$, $i=0,\ldots,r+1$ (i.e. $w_{r+1} = pw_0$), then $\{w_0,\ldots,w_r\}$ is a minimal generator set for Γ.

(iii) We have $w_0 = n$, $w_1 = n+1$ and $w_{i+1} = e_i w_i - w_{i-1}$ for $i=1,\ldots,r$.

The proof is left to the reader.

Lemma 6.3. (i) The kernel J of the group homomorphism $w\colon Z^{r+1} \to Z$, $w(s_0,\ldots,s_r) = \sum_{i=1}^{r} s_i w_i$, is generated by

$$j_i = -\varepsilon_{i-1} + e_i\varepsilon_i - \varepsilon_{i+1},\ i=1,\ldots,r-1$$

and

$$j_r = -\varepsilon_{r-1} + e_r\varepsilon_r - p\varepsilon_0,$$

where $\{\varepsilon_i\}_{i=0}^{r}$ is the standard basis of Z^{r+1}.

(ii) The kernel I of the $\mathbb{C}$-algebra homomorphism $w^*: \mathbb{C}[z_0,\ldots,z_r] \to \mathbb{C}[\Gamma]$, $z_i \to t^{w_i}$, is minimally generated by

$$g_{ij} = z_i z_j - z_{i+1} z_{j-1} \prod_{k=i+1}^{j-1} z_k^{e_k - 2}, \quad 1 \leq i+1 \leq j-1 \leq r-1$$

and

$$g_{i,r+1} = z_i z_0^p - z_{i+1} z_r \prod_{i+1}^{r} z_k^{e_k-2}, \quad i=0,\ldots,r-1.$$

The weights of g_{ij} are $w(g_{ij}) = w_i + w_j$.

<u>Lemma 6.4</u>. Let $e_1,\ldots,e_r$ be positive integers, $e_i \geq 2$. Consider the system of r equations

$$\begin{aligned} e_1 x_1 - x_2 &= y_1 \\ -x_1 + e_2 x_2 - x_3 &= y_2 \\ &\vdots \\ -x_{i-1} + e_i x_i - x_{i+1} &= y_i \\ &\vdots \\ -x_{r-1} + e_r x_r &= y_r \end{aligned}$$

in $2r$ variables x_i and y_i. Let

$$\frac{n}{m} = e_1 - \cfrac{1}{e_2 - \cfrac{1}{e_3 \ddots - \cfrac{1}{e_r}}}$$

Define $a_0 = 0$, $a_1 = 1$, $a_{i+1} = e_i a_i - a_{i-1}$ and $b_0 = n$, $b_1 = m$ $b_{i+1} = e_i b_i - b_{i-1}$. Then

$$x_i = \frac{1}{n}\left[b_i\left(\sum_{k=1}^{i} a_k y_k\right) + a_i\left(\sum_{k=i+1}^{r} b_k y_k\right)\right]$$

<u>Proof</u>. The proof is easy when one notices that

$$\frac{a_{i+1}}{a_i} = e_i - \cfrac{1}{e_{i-1} \ddots - \cfrac{1}{e_1}}$$

and that $a_{i+1} b_i - a_i b_{i+1} = n$ for all $i=0,\ldots,r$. □

Proof of 6.3. (i) If $\sum_{k=0}^{r} s_i w_i = 0$, use 6.4 to solve $(s_0,\ldots,s_r)$ $= \sum_{k=1}^{r} \alpha_r j_r$ for $\alpha_i \in Z$. (ii) is obvious from §3. □

We can now apply 2.2. The generator g_{ij} of I corresponds to the element

$$\varepsilon_i + \varepsilon_j - \varepsilon_{i+1} - \varepsilon_{j-1} + \sum_{k=i+1}^{j-1} (e_k - 2)\varepsilon_k = \sum_{k=i+1}^{j-1} j_k$$

in J. So

$$\mathrm{Hom}_P(I, \mathbb{C}[\Gamma]) \simeq \{(\phi_1,\ldots,\phi_r) \in \bigoplus_{i=1}^{r} \mathbb{C}[\Gamma] \mid$$

$$\sum_{\alpha=i+1}^{j-1} t^{w_i + w_j - e_\alpha w_\alpha} \cdot \phi_\alpha \in \mathbb{C}[\Gamma] \text{ for all } i,j \text{ such that}$$

$$1 \leq i+1 \leq j-1 \leq r\}.$$

One checks that the above criterion splits to each summand, i.e.

$$\mathrm{Hom}_P(I, \mathbb{C}[\Gamma]) \simeq \bigoplus_{\alpha=1}^{r} \mathfrak{A}_\alpha$$

where $\mathfrak{A}_\alpha = \{\phi \in \mathbb{C}[\Gamma] \mid t^{w_i + w_j - e_\alpha w_\alpha} \cdot \phi \in \mathbb{C}[\Gamma]$ for all i,j such that $i+1 \leq \alpha \leq j-1\} = \bigcap_{i+1 \leq \alpha \leq j-1} ((t^{e_\alpha w_\alpha}) : (t^{w_i + w_j}))$.

Lemmas 6.5-6.7 are the main steps in the proof of Prop.6.1. We omit computational details.

Lemma 6.5. (i) If $1 \leq i+1 \leq j-1 \leq r$, then $w_i + w_j - (e_\alpha - 1)w_\alpha \in \Gamma$ for $\alpha = i+1,\ldots,j-1$, and $w_i + w_j - w_\alpha \in \Gamma$ for $\alpha = i,\ldots,j$.

(ii) $pw_0 + w_i - (e_\alpha - 1)w_\alpha \in \Gamma$ for $\alpha \in \{1,\ldots r\} - \{i\}$ and $pw_0 + w_i - w_\alpha \in \Gamma$ for all $\alpha = 0,\ldots,r+1$.

(iii) If $h = 0,\ldots,p$ and $\gamma \in \Gamma$, then $h \cdot w_0 - \gamma \in \Gamma$ iff $\gamma = k \cdot w_0$ for a $k = 0,\ldots,h$.

Proof. (i) Continued use of the fact that $w_i + w_j = w_{i+1} + w_{j-1} + \sum_{k=i+1}^{j+1} (e_k - 2) w_k$. (ii) follows from (i) since $pw_0 = (p-1)w_0 + w_0 = w_{r+1}$. For (iii) use 6.2. □

Using 6.5 one can prove

Lemma 6.6. $\mathfrak{A}_\alpha$ is the ideal generated by

$$\{t^w \mid w \in \langle w_{\alpha-1}, w_\alpha, w_{\alpha+1}\rangle \cup \{pw_0 + w_{\alpha+1} - w_i \mid i = \alpha+1, \ldots, r\} \cup \{pw_0 + w_{\alpha-1} - w_i \mid i = 1, \ldots, \alpha-1\}\}.$$

(Notation: $\langle \gamma_1, \ldots, \gamma_k\rangle$ is the semigroup ideal generated by $\{\gamma_1, \ldots, \gamma_k\}$).

Let δ be the matrix $\left[\dfrac{\partial g_{i-1,i+1}}{\partial z_\alpha}\right] =$

$$\begin{bmatrix} t^{-w_2} & e_1 t^{(e_1-1)w_1} & -t^{w_0} & 0 & \cdots 0 \\ 0 & -t^{w_3} & e_2 t^{(e_2-1)w_2} & -t^{w_1} & \cdots 0 \\ & & \ddots & & \\ -pt^{w_{r-1}+(p-1)w_0} & \cdots & & -t^{pw_0} & e_r t^{(e_r-1)w_r} \end{bmatrix}$$

Then $T^1_C \simeq \bigoplus_{\alpha=1}^{r} \mathfrak{A}_\alpha / \mathrm{im}\delta$. If $\{\varepsilon_i\}_{i=1}^r$ is the standard basis of $\mathbb{C}[\Gamma]^r$, then a typical element in $\mathrm{im}\delta$ looks like

$$(*) \qquad \sum_{i=1}^{r-1} (-\phi_{i-1} \cdot t^{w_{i+1}} + e_i \phi_i t^{(e_i-1)w_i} - \phi_{i+1} \cdot t^{w_{i-1}}) \cdot \varepsilon_i$$

$$+ (-\phi_{r-1} t^{pw_0} + e_r \phi_r t^{(e_{r-1})w_r} - p\phi_0 t^{w_{r-1}+(p-1)w_0}) \cdot \varepsilon_r$$

where $\phi_0, \ldots, \phi_r \in \mathbb{C}[\Gamma]$. A computation using 6.4, 6.5, 6.6 and (*) gives

Lemma 6.7. A basis for $T^1_C \simeq \oplus \mathfrak{a}_\alpha / \mathrm{im}\delta$ is

(I) $\{t^{pw_0+w_{\alpha+1}-w_i} \cdot \varepsilon_\alpha \mid \alpha=2,\dots,r-1,\ i=\alpha+1,\dots,r\} \cup$

$\{t^{pw_0+w_{\alpha-1}-w_i} \cdot \varepsilon_\alpha \mid \alpha=3,\dots,r,\ i=1,\dots,\alpha-2\} \cup$

$\{t^{pw_0+w_2-w_i} \cdot \varepsilon_1 \mid i=3,\dots,r\} \cup$

(II) $\{t^{k_\alpha \cdot w_\alpha} \cdot \varepsilon_\alpha \mid \alpha=1,\dots,r,\ k_\alpha = 1,\dots,e_{\alpha-2}\} \cup$

$\{t^{w_{\alpha-1}} \cdot \varepsilon_\alpha \mid \alpha=1,\dots,r-1\} \cup$

$\{t^{w_{\alpha+1}} \cdot \varepsilon_\alpha \mid \alpha=2,\dots,r\} \cup$

(III) $\{t^{w_{r-1}+w} \cdot \varepsilon_r \mid w \in \Gamma - (\langle pw_0, (p+1)w_0 - w_1\rangle$

$\cup \{(p-1)w_0 + (p+1)w_0 - (e_i-1)w_i \mid i=1,\dots,r\})\}$

Proof of 6.1. The basis elements of type I and II sum up to $(r+1)(r-2) + \sum_{i=1}^{r} (e_i-1)$. To count the ones of type III notice the 1-1 correspondance between $\Gamma - \langle pw_0\rangle$ and $\{(\lambda_1,\lambda_2) \in \Lambda \mid \lambda_1 < n,\ \lambda_2 < pn\}$. □

Remark. The basis elements of type III are first order deformations of C in X, see [C].

Adding up the invariants we get:

Theorem. If r is the (minimal) codimension of the cyclic quotient singularity X and $r \geq 2$ then

$$\dim_{\mathbb{C}} T^2_X = r \cdot (r-2).$$

Example. Let X be the affine cone over the embedding of $\mathbb{P}^1$ in $\mathbb{P}^n$ by $\mathcal{O}_{\mathbb{P}^1}(n)$. Then X is the cyclic quotient with ordG=n and q=1. We have $\dim T^1_X = 2n-4$, and for $n > 5$ the formal moduli space

S is geometrically smooth of codimension n-1 ([P2]). Since in this case $r = n-1$

$$\dim T_X^2 = (n-1)(n-3) = (n-1)(\dim T_X^1 - \dim S).$$

In general for cyclic quotients $\dim T^1 = (\sum_{i=1}^{r} e_i)-2$, and there exists an Artin component A of dimension $\sum_{i=1}^{r} (e_i-1)$ ([R]) so

$$\dim T_X^2 = r(\dim T_X^1 - \dim A).$$

References

[B-K-R] K. Behnke, C. Kahn, O. Riemenschneider: "Infinitesimal deformations of quotient surface singularities", Preprint Hamburg 1985.

[Br] E. Brieskorn: "Rationale Singularitäten komplexer Flächen", Invent.math. 4, 336-358 (1967-68).

[Bu] R.-O. Buchweitz: "On deformations of monomial curves", in: Seminaire sur les Singularities des Surfaces, Palaiseau 76-77, Springer Lecture Notes 777.

[C] J.A. Christophersen: "Derivations of simplicial conic algebras", Preprint no 4, 1985, Univ. of Oslo.

[Gi] R. Gilmer: Commutative semigroup rings, Chigago 1984.

[Gr] G.-M. Greuel: "On deformations of curves and a formula of Deligne" in: Algebraic Geometry, Proceedings, La Rabida, 1981, Springer Lecture Notes 961.

[G-L] G.-M. Greuel, E. Looijenga: "The dimension of smoothing components", Duke Math.J. 52, 263-272 (1985).

[K-K-M-S] G. Kempf, F. Knudsen, D. Mumford, B. Saint-Donat: Toroidal Embeddings, Springer Lecture Notes 339.

[K] A.G. Kouchnirenko: "Polyhedra de Newton et nombres de Milnor", Invent. math. 32, 1-32 (1976).

[L] O.A. Laudal: Formal Moduli of Algebraic Structures, Springer Lecture Notes 754.

[L-P] O.A. Laudal, G. Pfister: The Local Moduli Problem. Applications to Isolated Hypersurface Singularities, Preprint no 11, 1986, Univ. of Oslo.

[La-S] O.A. Laudal, A. Sletsjøe: "Cohomology of groups, monoids and their algebras", manuscript 1983.

[Li-S] S. Lichtenbaum, M. Schlessinger: "The cotangent complex of a morphism", Trans. A.M.S. 128, 41-70 (1967).

[Lo] E.J.N. Looijenga: Isolated Singular Points on Complete Intersections, Cambridge 1984.

[M] M.A. Muffett: "Invariant Milnor numbers and Newton polyhedra" Thesis, Univ. of Liverpool.

[P1] H.C. Pinkham: "Deformations of algebraic varieties with G_m action", astérisque 20 (1974).

[P2] H.C. Pinkham: "Deformations of quotient surface singularities", Proc. of Symp. in Pure Math. 30, 65-67 (1977).

[R] O. Riemenschneider: "Deformationen von Quotientensingularitäten (nach zyklischen Gruppen)", Math.Ann. 209, 211-248 (1974).

[Ri] D.S. Rim: "Formal deformation theory", SGA 7(I), Exp V.

[S] M. Schlessinger: "Functors of Artin rings", Trans. A.M.S. 130 (1968).

[W] J.M. Wahl: "Equations defining rational singularities", Ann.Scient. Ec.Norm.Sup., 4e série, t.10, 231-264 (1977).

The Grothendieck group of invariant rings and of simple hypersurface singularities

by

Jürgen Herzog and Herbert Sanders

Let $A = \mathbb{C}[|X_1,\ldots,X_n|]$ be the formal power series ring in n variables over the complex numbers, and let G be a finite group acting linearly on A.

In their paper [1] Auslander and Reiten proved that the invariant ring $R = A^G$ has a finitely generated Grothendieck group $G(R)$ (Chap. 3, Prop. 3.4.) and that the reduced Grothendieck group $\tilde{G}(R) = G(R)/\mathbb{Z}[R]$ is finite if G acts freely on the linear space $L = \sum_{i=1}^{n} \mathbb{C}X_i$ of 1-forms (Chap. 3, Prop. 6.1.).

In this note we want to show that $\tilde{G}(R)$ is finite if G is abelian, without any further assumption on the action of G.

Furthermore we compute the reduced Grothendieck group of all simple hypersurface singularities. A theorem of Knörrer [5] and results of Auslander and Reiten [1] allow to restrict the computation to one-and two-dimensional singularities.

The two-dimensional simple hypersurface singularities are rings of invariants, so one can use the theory of Auslander and Reiten.

Since we are able to compute the Grothendieck group of all one-dimensional complete reduced singularities, we have all the tools to compute the Grothendieck group of simple hypersurface singularities.

To keep the paper self-contained we give the complete proofs of all the theorems, even though the proofs of some of the results can be found in [1].

1. The Grothendieck group of invariant rings

Let R be a noetherian ring and let $M(R)$ be the category of the finitely generated R-modules. An additive function on $M(R)$ is a map $\varphi : M(R) \to A$ into an abelian group such that $\varphi(M) = \varphi(U)+\varphi(N)$ for all exact sequences $0 \to U \to M \to N \to 0$ in $M(R)$. There exists up to isomorphism a uniquely determined abelian group $G(R)$ and an additive function $\alpha : M(R) \to G(R)$, $M \mapsto [M]$ such that any other additive function on $M(R)$ factors through α. $G(R)$ is called the Grothendieck group of R.

If R is a domain, the rank-function is additive on $M(R)$ and we get a split epimorphism $a : G(R) \to \mathbb{Z}$, $[M] \mapsto \text{rank } M$ with the section $\iota : \mathbb{Z} \to G(R)$, $\iota(1) = [R]$. We call $\tilde{G}(R) = G(R)/\mathbb{Z}[R]$ the reduced Grothendieck group. For a domain we have seen that $G(R) \simeq \tilde{G}(R) \oplus \mathbb{Z}$. Since we are mostly dealing with domains we usually restrict ourselves to only consider $\tilde{G}(R)$.

Now let $A = \mathbb{C}[|X_1,\dots,X_n|]$ and G be a finite group acting linearly on A. This means that the group action on A is induced by a linear action of G on the linear $\mathbb{C}$-vectorspace $L = \sum_{i=1}^{n} \mathbb{C}X_i$ of 1-forms. It is well known that the invariant ring $R = A^G$ is a normal Cohen-Macaulay domain.

A finitely generated A-module X is said to be a module with G-action, if X is a module over the twisted group ring $A[G]$, i.e. for all $g \in G$ it holds that $g(a\cdot x) = g(a)\cdot g(x)$ for all $a \in A$ and $x \in X$, and $g(x+y) = g(x)+g(y)$ for all $x,y \in X$.

Example (1.1): Let H be a finitely generated $\mathbb{C}[G]$-module, then $A \otimes_{\mathbb{C}} H$ is a free A-module with G-action, if we let G act diagonally: $g(a \otimes h) = g(a) \otimes g(h)$ for all $g \in G$, $a \in A$ and $h \in H$.

It is clear that the fix-module M^G of an A-module M with G-action is an R-module.

Let $c(G)$ be the class number of G. It is known from classical representation theory that $\mathbb{C}[G]$ decomposes into irreducible modules $\mathbb{C}[G] = \bigoplus_{i=1}^{c(G)} X_i^{n_i}$, where the n_i coincide with the rank of X_i, and that each indecomposable $\mathbb{C}[G]$-module is isomorphic to one of these X_i. For an indecomposable $\mathbb{C}[G]$-module X we set $A_\chi := (A \otimes_{\mathbb{C}} X)^G$, where $\chi : G \to \mathbb{C}$ is the uniquely determined character belonging to X. It is easily seen that A_χ is a maximal Cohen-Macaulay module over R of rank equal to the rank of X.

<u>Proposition (1.2)</u>: Let M be a finitely generated R-module. Then M admits an A_χ-resolution. This means that there exists an exact sequence $0 \to F_n \to \ldots \to F_0 \to M \to 0$, where each F_i is a finite direct sum of certain A_χ.

<u>Corollary (1.3)</u> ([1], Chap. 3, Prop. 3.4.) $G(R)$ is generated by $c(G)$ elements.

<u>Proof of proposition (1.2)</u>: $A \otimes_R M$ is an A-module with G action given by $g(a \otimes m) = g(a) \otimes m$ for all $a \in A$, $m \in M$ and $g \in G$. We claim that for every A-module N with G-action there exists an exact sequence

(*) $0 \to A \otimes_{\mathbb{C}} H_n \to \ldots \to A \otimes_{\mathbb{C}} H_1 \xrightarrow{\partial_1} A \otimes_{\mathbb{C}} H_0 \xrightarrow{\partial_0} N \to 0$

where the H_j are finitely generated $\mathbb{C}[G]$-modules and the

derivations ∂_j are compatible with the G-action (cp.(1.1)).
Applying the exact functor $(\_)^G$ we get the exact sequence
$0 \to (A \otimes_{\mathbb{C}} H_n)^G \to \dots \to (A \otimes_{\mathbb{C}} H_o)^G \to N^G \to 0$ of R-modules. Since every finitely generated $\mathbb{C}[G]$-module is a direct sum of certain X_i we can conclude that the modules $(A \otimes_{\mathbb{C}} H_j)^G$ are direct sums of certain A_χ.
In the case $N = A \otimes_R M$, we have $N^G = M$, so that the assertion follows.

The existence of the sequence (*) is a consequence of the Hilbert-syzygy-theorem and the following observation:
If N is a finitely generated A-module with G-action, then there exists a $\mathbb{C}[G]$-module H with $\mathbb{C}$-vector space dimension equal to the minimal number of generators of N and a G-compatible epimorphism $\partial : A \otimes_{\mathbb{C}} H \to N$ of A-modules.

In fact, consider the G-compatible canonical epimorphism $\varepsilon : N \to N/mN$, where m is the maximal ideal of A.
Since N/mN is a finite dimensional $\mathbb{C}[G]$-module, it is projective as $\mathbb{C}[G]$-module. Therefore ε admits a $\mathbb{C}[G]$-section $j : N/mN \to N$. We let $H := N/mN$ and define $\partial : A \otimes_{\mathbb{C}} H \to N$ by $\partial(a \otimes h) = a \cdot j(h)$ for all $a \in A$ and $h \in H$. Then ∂ is an A-module epimorphism and compatible with the G-action on $A \otimes_{\mathbb{C}} H$ and N. *

We now turn to the case that G is abelian. Then G has only linear characters which form the character group $G^* := \mathrm{Hom}_{\mathbb{Z}}(G,\mathbb{C})$. To each $\chi \in G^*$ there belongs (up to isomorphisms) an R-module A_χ of rank 1. In particular $R = A_\varepsilon$, where ε is the trivial character. As we have seen, the Grothendieck group $G(R)$ is

generated by the elements $[A_\chi]$.
Therefore we get a group epimorphism $\varphi : \mathbb{Z}[G^*] \to G(R)$, $\varphi(\chi) = [A_\chi]$ from the group ring of the character group of G over $\mathbb{Z}$ onto $G(R)$, which induces an epimorphism $\tilde{\varphi} : \mathbb{Z}[G^*]/\mathbb{Z}\varepsilon \to \tilde{G}(R)$.

<u>Theorem (1.4)</u>: $\tilde{G}(R)$ is finite, if G is abelian.

<u>Proof</u>: In order to prove the finiteness we have to find enough relations in the kernel of φ. We may assume that G acts faithfully on the linear space L and after a suitable choice of the variables we may further assume that the action is diagonal, that is $g(X_i) = \chi_i(g)X_i$ for all $g \in G$ and $i = 1,\dots,n$. Then it follows that G^* is generated by the elements $\chi_1,\dots,\chi_n$, since G acts faithfully on L.

Let us first introduce some notations which are useful for this proof. Let $J = \{1,\dots,n\}$ and for $I \subseteq J$ let G_I^* be the subgroup of G^* generated by the characters $\chi_i, i \in J\setminus I$. Set $G_J^* := \{\varepsilon\}$.
For $I \subseteq J$ we further denote the element $\prod_{i\in I} (\varepsilon-\chi_i)$ by ψ_I.
We divide the proof into two parts:

(1) The subgroup $K \subseteq \mathbb{Z}[G^*]$ which is generated by the set $\mathcal{K} = \{\chi\psi_I \mid I \subseteq J,\ \chi \in G^*\setminus G_I^*\}$ is contained in $\ker \varphi$.
(2) $\mathbb{Z}[G^*]/\mathbb{Z}\varepsilon + K$ is finite.

For the proof of (1) we must show that the elements $\chi\psi_I$, $I \subseteq J$, $\chi \in G^*\setminus G_I^*$ are in $\ker \varphi$.

Therefore let $\emptyset \neq I$ and for simplicity assume that $I = \{1,\ldots,s\}$. We consider the Koszul-complex $K. = K.(X_1,\ldots,X_s,A)$ and let $\{e_1,\ldots,e_s\}$ be a free A-basis on which the Koszul-complex is built. Fix $\chi \in G^*$ and define for $a \in A$ and $g \in G$

$$g(a\, e_{i_1} \wedge \ldots \wedge e_{i_t}) = g(a)\chi(g)\ \chi_{i_1}(g)\cdot\ldots\cdot\chi_{i_t}(g) e_{i_1} \wedge \ldots \wedge e_{i_t} \ .$$

This definition makes $K.$ an acyclic complex with G-action. On $H_\chi := H_0(K.) = \mathbb{C}[|X_{s+1},\ldots,X_n|]$ we consider the induced G-action, so that we get the exact sequence

$$\ldots \to \bigoplus_{1\le i_1<\ldots<i_t\le s} A_{\chi_{i_1}\cdot\ldots\cdot\chi_{i_t}\cdot\chi} \to \ldots \to \bigoplus_{i=1}^{s} A_{\chi_i\cdot\chi} \to A_\chi \to H_\chi^G \to 0$$

after applying the functor $(\_)^G$ on $K.$

If we have shown that $H_\chi^G = 0$, for $\chi \notin G_I$, it follows that $\chi\cdot\psi_I \in \ker\varphi$, for $\chi \notin G_I^*$.

Let $F := X_{s+1}^{d_{s+1}}\cdot\ldots\cdot X_n^{d_n} \in H_\chi$ be a monomial and $g \in G$, then the induced action yields

$g(F) = \chi_{s+1}^{d_{s+1}}(g)\cdot\ldots\cdot\chi_n^{d_n}(g)\cdot\chi(g)\ F$, so that $F \in H_\chi^G$ if and only if $\chi^{-1} = \chi_{s+1}^{d_{s+1}}\cdot\ldots\cdot\chi_n^{d_n}$.

Since this is impossible if $\chi \notin G_I^*$, and since H_χ^G is generated by monomials we have shown that $H_\chi^G = 0$ if $\chi \notin G_I^*$.

For the proof of (2) let $U \subseteq \mathbb{C}[G^*]$ be the subspace generated by the elements of K. It suffices to show that $\operatorname{codim}_{\mathbb{C}} U = 1$ and that $\varepsilon \notin U$. For the computations in $\mathbb{C}[G^*]$ we use the $\mathbb{C}$-basis $\dot{G} = \{e_g \mid g \in G\}$ where $e_g := \frac{1}{|G|} \sum_{\chi\in G^*} \chi(g)\cdot\chi$ for $g \in G$. These elements form a complete set of idempotents of the algebra $\mathbb{C}[G^*]$. If we denote by 1 the neutral element of G and if $a = \sum_{\chi\in G^*} x_\chi \chi$ is an element of $\mathbb{C}[G^*]$, then $e_1\cdot a = (\sum_{\chi\in G^*} x_\chi) e_1$. In particular $e_1\cdot\chi = e_1$ for all $\chi \in G^*$, hence $e_1\cdot\varepsilon \neq 0$, but $e_1\cdot a = 0$ for all $a \in K$. This shows that $\varepsilon \notin U$.

It remains to compute the dimension of U. Suppose we have shown that U is the ideal in $\mathbb{C}[G^*]$ generated by the elements ψ_I, $I \in S$, where $S = \{I \subseteq J \mid G_I^* \neq G^*\}$.

Any ideal D in $\mathbb{C}[G^*]$ is a subvectorspace spanned by a subset $\mathcal{D}$ of $\mathring{G}$ and $e_g \in \mathcal{D}$ if and only if $a \cdot e_g \neq 0$ for a generator a of D.

Let $B \subset \mathring{G}$ be the basis belonging to U, then $B = \bigcup_{I\in S} (\bigcap_{i\in I} \mathring{G} \setminus \mathring{G}_i) = \mathring{G} \setminus \bigcap_{I\in S} (\bigcup_{i\in I} \mathring{G}_i)$, where $G_i := \ker \chi_i$ and $\mathring{G}_i = \{e_g \mid g \in G_i\}$.

Since we want to show that $\operatorname{codim}_{\mathbb{C}} U = 1$, it suffices to prove that $\{e_1\} = \bigcap_{I\in S} (\bigcup_{i\in I} \mathring{G}_i)$.

If we use the bijection $G \to \mathring{G}$, $g \mapsto e_g$ and that $G_I^* \neq G^*$ if and only if $\bigcap_{i\notin I} G_i \neq \{1\}$ we must show $\{1\} = \bigcap_{I\in S'} (\bigcup_{i\in I} G_i)$, where $S' = \{I \subseteq J \mid \bigcap_{i\notin I} G_i = \{1\}\}$.

This intersection can be written as a union of sets of the form $\bigcap_{i\in I'} G_i$, where I' is a subset of J that is obtained by picking one element from each $I \in S'$. It is clear that there exists no $I \in S'$ such that $I' = J\setminus I$. By the definition of S' this means that $\bigcap_{i\in I'} G_i = \{1\}$, as we wanted to show.

It remains to prove that U is the ideal $\sum_{I\in S} U_I$ in $\mathbb{C}[G^*]$, where for $I \subseteq J$, U_I is the ideal in $\mathbb{C}[G^*]$ generated by ψ_I.
We set $U_i := U_{\{i\}}$ for $i \in J$.
We show by induction on $t = n - |I|$ that the ideal U_I, $I \in S$, is contained in U. If $t = 0$, then $I = J$ and $G_I^* = \{\varepsilon\}$. Since $\sum_{\chi\in G^*} \chi \cdot \psi_J = 0$ it follows that $\psi_J = - \sum_{\chi\in G^*\setminus\{\varepsilon\}} \chi \cdot \psi_J \in U$.

Now let $t > 0$ and $I \in S$ such that $t = n-|I|$. For simplicity we assume that $I = \{1,\dots,s\}$ where $s = n-t$. It is clear from the definitions of G_I^* and S, that $I \cup \{j\} \in S$ for all $j \in J$. Hence the ideal $V = \sum_{j\in J\setminus I} U_{I\cup\{j\}}$ is contained in U. Let V' be the ideal $\sum_{i\in J\setminus I} U_i$.

Consider the map $\alpha : B \to C$, where $B = \mathbb{C}[G^*]/V$ and $C = \bigoplus_{i=1}^{s} \mathbb{C}[G^*]/U_i \oplus \mathbb{C}[G^*]/V'$, then $\ker\alpha = (\bigcap_{i=1}^{s} U_i \cap V')/V$, so that α is injective, since the product and the intersection of two ideals in $\mathbb{C}[G^*]$ coincide.

For $i \in J$ let $\tilde{U}_i$ be the subgroup of G^*, which is generated by the element χ_i and let $\tilde{V}'$ be the subgroup generated by the elements χ_j, $j \in J\setminus I$. Then C is obviously isomorphic to $\tilde{C} := \bigoplus_{i=1}^{s} \mathbb{C}[G^*/\tilde{U}_i] \oplus \mathbb{C}[G^*/\tilde{V}']$.

For $\chi \in G^*$ we denote its image in $G^*/\tilde{V}'$ by $\overline{\chi}$, while for $a \in \mathbb{C}[G^*]$ we denote the image in B by $[a]$. Then after identifying C with $\tilde{C}$ we get $\alpha[\chi\,\psi_I] = (0,\dots,0,\overline{\chi}\prod_{i=1}^{s}(\overline{\varepsilon}-\overline{\chi}_i))$.

Hence the $\mathbb{C}$-subvectorspace W of B generated by the elements $[\chi\psi_I]$, $\chi \in G^*\setminus G_I^*$ is mapped onto the $\mathbb{C}$-subvectorspace W' in C generated by the elements $(0,\dots,0,\overline{\chi}\prod_{i=1}^{s}(\overline{\varepsilon}-\overline{\chi}_i))$, $\overline{\chi} \in (G^*/\tilde{V}')\setminus\{\overline{\varepsilon}\}$.

As in the case $t = 0$ it follows that the element $(0,\dots,0,\prod_{i=1}^{s}(\overline{\varepsilon}-\overline{\chi}_i)) \in W'$, and since α is injective we get that $[\psi_I] \in W$ and finally we conclude that $\psi_I \in U$.

To finish the proof we have to show that the ideal U_I is contained in U. Thus we have to consider $\chi\psi_I$, where $\chi \in G$.

If $\chi \notin G_I^*$ the element $\chi\psi_I$ clearly belongs to U.

If $\chi \in G_I^*$, then $\chi = \chi_{s+1}^{a_{s+1}}\cdot\ldots\cdot\chi_n^{a_n}$, where $a_j \geq 0$ for $j \in J\setminus I$.

We proceed by induction on $a = \sum_{j\in J\setminus I} a_j$ in order to show that $\chi\psi_I \in U$.

If $a = 0$ we are ready by the calculation above. Now say that $a_j > 0$ for one $j \in J \setminus I$. Then $\chi\psi_I = -\chi_j^{-1}\chi\psi_{I \cup \{j\}} + \chi_j^{-1} \cdot \chi \cdot \psi_I$. The first term on the right hand side belongs to U, since the ideal $U_{I \cup \{j\}}$ is contained in U (by the induction on $t = n - |I|$), while the second one belongs to U by induction on a. *

It would be interesting to know if the relations K in the proof of (1.4) generate the kernel of the map $\varphi : \mathbb{Z}[G^*] \to G(R)$. It seems likely that (1.4) holds for an arbitrary finite group G.

For the next section it will be important that in dimension 2, $\tilde{G}(R)$ can be computed explicitly. This follows easily from

<u>Proposition (1.5)</u>: ([1], Chap. 3, Prop. 2.4.)
Let (R, m, k) be a two-dimensional normal local domain such that $[k] = 0$, then $\tilde{G}(R) \simeq C\ell(R)$ - the class group of R.

<u>Proof</u>: By [2] (Prop. 17 of [3, §4, n°8])
$\tilde{G}(R) \simeq C\ell(R)$ if $[M] = 0$ for the R-modules M such that $M_P = 0$ for all height one prime ideals P in R. But such a module is of finite length, so that $[k] = 0$ guarantees this condition. *

<u>Corollary (1.6)</u>: If G acts faithfully and linearly on $A = \mathbb{C}[|X,Y|]$, then $\tilde{G}(A^G) \simeq (G/H)^*$, where H is the subgroup of G, which is generated by the pseudoreflections on G.

<u>Proof</u>: The Koszul-complex $K. = K.(X,Y,A)$ is an acyclic complex of A^G-modules with $H_0(K.) = k$. It follows that $[k] = 0$. Hence we may apply (1.5) to A^G, and the assertion

follows by the result of Singh ([6], theorem 2), see also [3]. *

2. The Grothendieck group of simple hypersurface singularities

The simple hypersurface singularities are those of the form $R = \mathbb{C}[|X_o,\dots,X_d|]/f(X_o,X_1)+X_2^2+\dots+X_d^2$, where $f(X_o,X_1)$ is one of the following equations.

A_n : $X_o^2 + X_1^{n+1}$, $n = 1,2,3,\dots$

D_n : $X_o^2X_1+X_1^{n-1}$, $n = 4,5,\dots$

E_6 : $X_o^3 + X_1^4$

E_7 : $X_o^3 + X_oX_1^3$

E_8 : $X_o^3 + X_1^5$

The goal of this section is to prove

Theorem (2.1): Let R be a simple hypersurface singularity. Then $\tilde{G}(R)$ is isomorphic to the groups given by the following table.

type of R	R has odd dimension		R has even dimension	
	n even	n odd	n even	n odd
A_n	$\{e\}$	$\mathbb{Z}$	$\mathbb{Z}/(n+1)\mathbb{Z}$	
D_n	$\mathbb{Z}^2$	$\mathbb{Z}$	$(\mathbb{Z}/2\mathbb{Z})^2$	$\mathbb{Z}/4\mathbb{Z}$
E_6	$\{e\}$		$\mathbb{Z}/3\mathbb{Z}$	
E_7	$\mathbb{Z}$		$\mathbb{Z}/2\mathbb{Z}$	
E_8	$\{e\}$		$\{e\}$	

The proof of (2.1) is based on results of Knörrer [5] which say that if $f \in \mathbb{C}[|X_1,\dots,X_n|]$, then the stable Auslander-Reiten quivers of $\mathbb{C}[|X_1,\dots,X_n|]/f$ and $\mathbb{C}[|X_1,\dots,X_n,Y,Z|]/f+Y^2+Z^2$ coincide and that simple hypersurface singularities are of finite Cohen Macaulay representation type.

It follows from [1], Chap. 4, Prop. 3.3. that in this situation the stable Auslander-Reiten quivers of $M(R)$ determine $G(R)$. Therefore we only have to consider the Grothendieck groups of simple hypersurface singularities in dimension one and two. This is the reason for the distinction between odd and even dimension in the above table.

We first compute the Grothendieck group of the even-dimensional simple hypersurface singularities. Therefore we consider these singularities in dimension two. Every such singularity arises as an invariant ring $\mathbb{C}[|X,Y|]^G$, where $G \subseteq SL(2,\mathbb{C})$ is a finite subgroup. In the following table we show the isomorphism types of all finite subgroups of $SL(2,\mathbb{C})$, which were first determined by Klein ([4]). Moreover the table shows the order of these groups and the type of the corresponding invariant ring.

Group G	$\lvert G\rvert$	type of A^G
$\mathbb{Z}/(n+1)\mathbb{Z}$ cyclic group	$n+1$	A_n
$\mathbb{D}_n$ binary dihedral group	$4n$	D_n
$\mathbb{T}$ " tetrahedral group	24	E_6
$\mathbb{O}$ " octahedral group	48	E_7
$\mathbb{I}$ " icosahedral group	120	E_8

Since a subgroup of $SL(2,\mathbb{C})$ has no pseudoreflections it follows from (1.6), that $\tilde{G}(A^G) \simeq Cl(A^G) \simeq G/[G,G]$. The class groups of the rings were computed by Brieskorn [3] . This yields the table (2.1) in even dimension.

In order to compute the Grothendieck group of the odd-dimensional simple hypersurface singularities we have to compute $G(R)$, where $R = \mathbb{C}[|X,Y|]/f(X,Y)$ is the plane curve singularity given by one of the equations on page 11.

It will follow from the next proposition that $\tilde{G}(R) \simeq \mathbb{Z}^{r-1}$, where r is the number of branches of the curve. Hence together with the following result, which generalizes Prop. 2.3 of chap. 3 in [1] , the proof of (2.1) is completed.

Let $R = k[|X_1,\ldots,X_n|]/I$ be a one-dimensional reduced k-algebra (k-field), $\overline{R}$ its integral closure, m the maximal ideal of R and $m_1,\ldots,m_r$ the maximal ideals of $\overline{R}$. We set $f_i = [\overline{R}/m_i : k]$ and $f := \gcd(f_1,\ldots,f_r)$.

<u>Proposition (2.2)</u>: $G(R) \simeq \mathbb{Z}^r \oplus \mathbb{Z}/f\mathbb{Z}$

We introduce some more notations. Let $P_1,\ldots,P_r$ be the minimal primes of R and $R_i = R/P_i$ for $i = 1,\ldots,r$. Then $\overline{R} = \bigoplus_{i=1}^{r} \overline{R}_i$, where $\overline{R}_i$ is the integral closure of R_i.

Since each $\overline{R}_i$ is a discrete valuation ring, the maximal ideal of $\overline{R}_i$ is generated by one element, say x_i. From the exact sequence $0 \to \overline{R}_i \xrightarrow{x_i} \overline{R}_i \to \overline{R}/m_i \to 0$ we conclude that $[\overline{R}/m_i] = 0$ and since $f_i = [\overline{R}/m_i : k]$ we get that $f_i[k] = 0$ for $i = 1,\ldots,r$. Now clearly $f\cdot[k] = 0$.

Let K_i be the quotient field of R_i. For $M \in \mathcal{M}(R)$ we define $a_i(M) := \dim_{K_i}(M \otimes_R K_i)$. Notice that $K_i = R_{P_i}$, to see that a_i is an additive function on $\mathcal{M}(R)$ for $i = 1,\ldots,r$. The first part of the next lemma gives another characterization of the numbers $a_i(M)$.

Lemma (2.3): Let M be an R-module and $(0) = M_0 \subset M_1 \subset \ldots \subset M_t = M$ be a filtration of M so that $M_j/M_{j-1} \simeq R/P$ where $P \in \operatorname{Spec} R$.

a) $a_i(M)$ is the number of the factors isomorphic to R_i in the chosen filtration.

b) Let $a(M) = a+f\mathbb{Z} \in \mathbb{Z}/f\mathbb{Z}$, where a is the number of the factors isomorphic to k in the chosen filtration. Then $a(M)$ does not depend upon the chosen filtration.

Proof of proposition (2.2): Not only the maps $a_i : \mathcal{M}(R) \to \mathbb{Z}$, but also the map $a : \mathcal{M}(R) \to \mathbb{Z}/f\mathbb{Z}$ is an additive function on $\mathcal{M}(R)$, so that we get an induced group homomorphism

$$\rho : G(R) \to \mathbb{Z}^r \oplus \mathbb{Z}/f\mathbb{Z}$$
$$[M] \mapsto (a_1(M),\ldots,a_r(M),a(M)).$$

ρ is surjective, since $\rho([R_i]) = e_i$ for $i = 1,\ldots,r$ and $\rho([k]) = e_{r+1}$, where $e_1,\ldots,e_r$ is the canonical basis of $\mathbb{Z}^r$ and e_{r+1} is the element $(0,\ldots,0,1+f\mathbb{Z}) \in \mathbb{Z}^r \oplus \mathbb{Z}/f\mathbb{Z}$.
It is easy to see that every element of $G(R)$ can be written in the form $[M]-n[R]$, where $n \geq 0$. It follows that $\rho([M]-n[R]) = (a_1(M)-n,\ldots,a_r(M)-n,a(M))$, so that $[M]-n[R]$ is in the kernel of ρ if and only if $a_i(M) = n \geq 0$ for $i = 1,\ldots,r$, and $a(M) = 0 \in \mathbb{Z}/f\mathbb{Z}$.

We induct on n to see that ρ is injective. If $n = 0$ it follows that M has finite length $l(M) \equiv 0 \bmod f$. Using that $[M] = l(M)\cdot[k]$ and that $[k]$ is annihilated by f we get $[M] = 0$. If $n > 0$ there exists a short exact sequence $0 \to R \to M \to N \to 0$ since $a_i(M) > 0$ for $i = 1,\dots,r$. From this sequence we derive that $[M]-n[R] = [N]-(n-1)[R]$, so that the assertion follows by induction. *

<u>Proof of lemma (2.3)</u>

a) follows easily after localisation at a minimal prime of R.

b) We proceed by induction on $n := \sum_{i=1}^{r} a_i(M)$. If $n = 0$ it follows that M has finite length, so that $a(M) = l(M)+f\mathbb{Z}$.

Let $n > 0$ and $(0) = N_0 \subset N_1 \subset \dots \subset N_s = M$ be another filtration of M with factors isomorphic to R/P with $P \in \operatorname{Spec} R$.

We first claim that we can assume that neither N_1 nor M_1 is isomorphic to k.

If $M_1 \simeq k$, choose j minimal with $N_{j+1} \cap M_1 = M_1$. Then $N_j \cap M_1 = 0$.

Consider the filtration $0 \subseteq \overline{N}_1 \subset \overline{N}_2 \subset \dots \subset \overline{M}$, where $\overline{N}_i := (N_i+M_1)/M_1$. It follows that $\overline{N}_i \simeq N_i$ for $i \le j$ and $\overline{N}_i \simeq N_i/M_1$ for $i > j$. Since further $N_{j+1}/N_j \simeq k$ and $\overline{N}_j = \overline{N}_{j+1}$ we can pass to the module $\overline{M} = M/M_1$ to prove the assertion.

Let now $N_1 \simeq R_i$ and $M_1 \simeq R_j$, $i,j \in \{1,\dots,r\}$.

<u>1st case:</u> $N_1 \cap M_1 \neq 0$. It follows that $i = j$. Let $0 \neq x \in N_1 \cap M_1$, then $(x) \simeq R_i$.

By induction we see that $a(M/(x))$ is uniquely determined, so that it remains to prove that $l(N_1/(x))$ and $l(M_1/(x)) \equiv 0 \bmod f$, since then $a(M) = a(M/(x))$ is independent of the choice of the filtration.

We have to compute the length of R_i/y, where y is a non-zerodivisor of R_i. $l(R_i/y) = l(\overline{R}_i/y\overline{R}_i) = \nu \cdot f_i$, if $y = x_i^{\nu}$ and x_i is a generator of the maximal ideal of $\overline{R}_i$.

2nd case: $N_1 \cap M_1 = 0$. By induction we get that $a(M/N_1)$ and $a(M/M_1)$ are well defined. We claim that $a(M/N_1) = a(M/M_1+N_1)$. Then by symmetry we conclude $a(M/N_1) = a(M/M_1)$. This proves that $a(M)$ is independent of the chosen filtration.
To see that $a(M/N_1) = a(M/M_1+N_1)$ consider the filtration $0 \subset N_1 \subset M_1+N_1 \subset \ldots \subset M$ and notice that $(M_1+N_1)/N_1 \simeq$ $\simeq M_1/(N_1 \cap M_1) = M_1 \simeq R_j$. *

References

[1] M. Auslander and I. Reiten, Grothendieck groups of algebras and orders, Journal of Pure and Applied Algebra 39 (1986) 1-51

[2] N. Bourbaki, Algèbre commutative, Chapitre 7, Hermann, Paris, 1965

[3] E. Brieskorn, Rationale Singularitäten komplexer Flächen, Inventiones math. 4 (1968) 336-358

[4] F. Klein, Vorlesungen über das Ikosaeder und die Auflösung der Gleichungen vom fünften Grad, Leipzig (1884)

[5] H. Knörrer, Maximal Cohen-Macaulay modules on hypersurface singularities I, in preparation

[6] B. Singh, Invariants of finite groups acting on local unique factorization domains, Journal of the Indian Math. Soc. 34 (1970) 31-38

Note added:

After having finished this paper we were informed by I. Reiten that she and Maurice Auslander have shown (1985) that $\tilde{G}(R)$ of a ring of invariants R is finite for an arbitrary action of a finite, not necessarily abelian, group. However the result is not yet written up for publication and their proof is essentially different from ours.

We also were informed that the Auslander-Reiten quivers of simple hypersurface singularities have been computed by Dieterich and Wiedemann in dimension 1, and by Auslander in dimension 2. Of course these results allow the computation of the Grothendieck group of simple hypersurface singularities as well.

Torsionsfreie Moduln bei Deformation von Kurvensingularitäten

Horst Knörrer

Die Klassifikation der reduzierten Kurvensingularitäten, über deren lokalem Ring es nur endlich viele Isomorphieklassen torsionsfreier Moduln gibt, hatte auf eine unter Deformation abgeschlossene Klasse von Singularitäten - nämlich diejenigen, die eine ebene einfache Kurvensingularität dominieren - geführt ([Greuel-Knörrer]). In dieser Note soll nun allgemeiner gezeigt werden, daß sich die "Zahl der Parameter" torsionsfreier Moduln von festem Rang über reduzierten Kurvensingularitäten bei Deformation der Singularität nach oben halbstetig verhält.

Der Beweis entstand aus dem Versuch, Argumente, die [Gabriel] im Fall von Artinschen Algebren gab, zu verallgemeinern. Für viele anregende und interessante Diskussionen danke ich G.M. Greuel, C. Rego, F.-O Schreyer und C. Seshadri.

1. Die Zahl der Parameter

Sei K ein algebraisch abgeschlossener Körper und K<t> der Ring der formalen oder - falls K mit einer vollständigen Bewertung versehen ist - konvergenten Potenzreihen in einer Veränderlichen über K. Wir betrachten eine kommutative K<t>-Algebra R ohne nilpotente Elemente, die als K<t>-Modul endlich erzeugt ist; m. a. W., R ist eine direkte Summe von lokalen Ringen reduzierter Kurvensingularitäten. Ferner sei R' ein Ring zwischen R und dem ganzen Abschluß $\tilde{R}$ von R und $I \subset R$ ein R'-Ideal, so daß $\dim_K R/I < \infty$.

Bekanntlich ist jeder torsionsfreie R-Modul vom Rang n isomorph zu einem R-Modul M mit $R^n \subset M \subset \tilde{R}^n$ (siehe etwa [Rego] 1.4). Isomorphismen zwischen solchen Moduln werden durch Automorphismen von $\tilde{R}^n$ induziert, genauer gilt

Lemma:

Seien M_1, M_2 zwei R-Moduln mit $R^n \subset M_i \subset \tilde{R}^n$ und $\dim_K M_1/R^n = \dim_K M_2/R^n$.

Dann sind M_1 und M_2 als R-Moduln isomorph genau dann, wenn es eine Matrix A in $GL(n,\tilde{R})$ gibt, deren Spalten alle in M_1 liegen (die Spalten von A werden aufgefaßt als Elemente von $\tilde{R}^n$) und so daß $A^{-1} \cdot M_1 = M_2$.

Beweis: Nach [Rego] 1.6 sind M_1 und M_2 genau dann isomorph, wenn es $A \in GL(n,\tilde{R})$ gibt mit $A \cdot M_2 = M_1$. Da $R^n \subset M_2$, liegen alle Spalten von A in M_1.

Korollar:
Sind M_1, M_2 Moduln über R mit $R^n \subset M_i \subset R'^n$ und $\dim_K M_1/R^n = \dim_K M_2/R^n$, so sind M_1 und M_2 genau dann isomorph, wenn es $A \in GL(n,R')$ gibt mit $A^{-1} \cdot M_1 = M_2$.

Für jede natürliche Zahl d sei nun $Gr = Gr(d,R')$ die Graßmannvarietät aller d-dimensionalen "K-Untervektorräume" von $(R'/I)^n$, und $B = B(d,R')$ die Menge aller $V \in Gr$, die $(R/I)^n$ enthalten und unter Multiplikation mit R abgeschlossen sind. B ist Zariski-abgeschlossen in Gr. Für jedes $V \in B$ ist das Urbild M_V von V unter der Projektion $R'^n \to (R'/I)^n$ ein torsionsfreier R-Modul mit $R^n \subset M_V \subset R'^n$. Sind $V,V' \in B$, so sind nach dem obigen Korollar die zugehörigen R-Moduln M_V und $M_{V'}$ genau dann isomorph, wenn es $A \in GL(n,R'/I)$ gibt mit $A^{-1} \cdot V = V'$.
$GL(n,R'/I)$ operiert auf Gr, und wir setzen für jedes i

$$B_i(d,R') := \{V \in B(d,R') \mid \dim(GL(n,R'/I) \cdot V \cap B) \leq i\}$$

(dabei sei die Dimension einer konstruierbaren Menge das Maximum der Dimensionen ihrer Komponenten). Es ist klar, daß $B_i(d,R')$ konstruierbar in B ist (*), also macht es Sinn, die "Zahl der Parameter" von torsionsfreien R-Moduln vom Rang n, die sich zwischen R^n und R'^n einbetten lassen, als

$$\mathrm{par}_n(R;R') := \max_{i,d}(\dim B_i(d,R') - i)$$

zu definieren. Man überzeugt sich leicht, daß diese Zahl nicht von der Wahl des Ideals I abhängt. Ferner gilt für jeden Ring R" mit $R' \subset R'' \subset \tilde{R}$

$$\mathrm{par}_n(R;R') \leq \mathrm{par}_n(R,R'') .$$

Da sich jeder torsionsfreie R-Modul vom Rang n zwischen R^n und $\tilde{R}^n$ einbetten läßt, nennen wir

$$\mathrm{par}_n(R) := \mathrm{par}_n(R;\tilde{R})$$

(*) in Abschnitt 2 werden wir zeigen, daß $B_i(d,R')$ abgeschlossen ist.

die "Zahl der Parameter von torsionsfreien R-Moduln vom Rang n". Aus der obigen Konstruktion ergibt sich sofort:

Bemerkung:
Ist $R = R_1 \oplus R_2$ eine direkte Summe von K<t>-Algebren, so ist $\mathrm{par}_n(R) = \mathrm{par}_n(R_1) + \mathrm{par}_n(R_2)$.

2. Halbstetigkeit

Es sei $f: X \to S$ eine flache Familie von reduzierten Kurven über einer glatten Kurve S. Für jedes $s \in S$ sei

$$z_n(s) := \sum_{x \in f^{-1}(s)} \mathrm{par}_n(\mathcal{O}_{f^{-1}(s),x})$$

(man beachte, daß die glatten Punkte von $f^{-1}(s)$ zu dieser Summe nicht beitragen).

Satz:
Die Funktion $S \to \mathbb{N}_0$, $s \mapsto z_n(s)$ ist nach oben halbstetig.

Zum Beweis des Satzes wird die Konstruktion aus Abschnitt 1 simultan für alle Fasern von f durchgeführt. Sei also $g: Y \to X$ die Normalisierung von X und $\mathcal{C} \subset \mathcal{O}_X$ das Konduktorideal, d. h. der Annihilator von $(g_*\mathcal{O}_Y)/\mathcal{O}_X$. Für generisches $s \in S$ ist dann die Einschränkung von g auf $Y_s := (g \circ f)^{-1}(s)$ die Normalisierung von $X_s := f^{-1}(s)$ (vgl. [Buchweitz-Greuel] 4.1.6).

Lemma:
Die Garben $f_*(\mathcal{O}_X/\mathcal{C})$, $f_*(g_*\mathcal{O}_Y/\mathcal{C})$ und $f_*(g_*\mathcal{O}_Y/\mathcal{O}_X)$ sind lokal frei über S.

Beweis: Da die Garben $\mathcal{O}_X/\mathcal{C}$, $g_*\mathcal{O}_Y/\mathcal{C}$ und $g_*\mathcal{O}_Y/\mathcal{O}_X$ auf der kritischen Menge C(f) von f konzentriert sind und $f|_{C(f)}: C(f) \to S$ endlich ist, ist die Sequenz

$$0 \to f_*(\mathcal{O}_X/\mathcal{C}) \to f_*(g_*\mathcal{O}_Y/\mathcal{C}) \to f_*(g_*\mathcal{O}_Y/\mathcal{O}_X) \to 0$$

exakt. Die Tatsache, daß $f_*(g_*\mathcal{O}_Y/\mathcal{O}_X)$ lokal frei über S ist, wird etwa in [Buchweitz-Greuel] 3.2.5, 4.1.4 bewiesen. Es genügt also zu zeigen, daß $f_*(g_*\mathcal{O}_Y/\mathcal{C})$ torsionsfrei über S ist. Dazu seien $s \in S$, $x \in X_s$, $0 \neq a \in \mathcal{O}_{S,s}$ und $\varphi \in g_*\mathcal{O}_{Y,x}$, so daß $a \cdot \varphi \in \mathcal{C}_x$. Dann ist

$a \cdot \varphi \cdot \psi \in \mathcal{O}_{X,x}$ für alle $\psi \in (g_*\mathcal{O}_Y)_x$. Da $f_*(g_*\mathcal{O}_Y/\mathcal{O}_X)$ torsionsfrei über S ist, folgt $\varphi \cdot \psi \in \mathcal{O}_{X,x}$ für alle $\psi \in (g_*\mathcal{O}_Y)_x$, also ist $\varphi \in \mathcal{C}_x$.

Wir bezeichnen mit $\mathcal{V} \to S$ bzw. $\mathcal{W} \to S$ die zu $f_*(g_*\mathcal{O}_Y/\mathcal{C})$ bzw. $f_*(\mathcal{O}_X/\mathcal{C})$ gehörigen Vektorraumbündel auf S. Aus der obigen exakten Sequenz folgt, daß $\mathcal{W}$ ein Unterbündel von $\mathcal{V}$ ist. Für jedes $s \in S$ ist die Faser $\mathcal{V}_s$ von $\mathcal{V}$ über s kanonisch isomorph zu $\bigoplus_{x \in X_s} (g_*\mathcal{O}_{Y_s}/\mathcal{C}_s)_x$, und $\mathcal{W}_s := \mathcal{V}_s \cap \mathcal{W}$ entspricht dem Untervektorraum $\bigoplus_{x \in X_s} (\mathcal{O}_{X_s}/\mathcal{C}_s)_x$ (dabei sei $\mathcal{C}_s := \mathcal{C}|_{X_s}$).
Die Operation der Ringgarbe $f_*(g_*\mathcal{O}_Y/\mathcal{C})$ auf $\mathcal{W}$ entspricht bei Einschränkung der Aktion von $\bigoplus_{x \in X_s} (g_*\mathcal{O}_{Y_s}/\mathcal{C}_s)_x$.

Mit $\mathcal{V}^{\oplus n}$ bezeichnen wir die n-fache Whitneysumme von $\mathcal{V}$, dann ist $\mathcal{W}^{\oplus n}$ ein Unterbündel von $\mathcal{V}^{\oplus n}$. Für jede natürliche Zahl d sei $\pi: \mathcal{Gr}(d) \to S$ das Graßmannbündel der d-dimensionalen Teilräume von $\mathcal{V}^{\oplus n}$ und

$$\mathcal{B} := \mathcal{B}(d) := \{V \in \mathcal{Gr}(d) \quad \mathcal{W}^{\oplus n}_{\pi(V)} \subset V \;,\; f_*(g_*\mathcal{O}_Y/\mathcal{C}) \cdot V \subset V\}$$

$\mathcal{B}$ ist eine Zariski-abgeschlossene Teilmenge von $\mathcal{Gr}(d)$. Auf $\mathcal{V}^{\oplus n}$ und somit auch auf $\mathcal{Gr}(d)$ operiert die Ringgarbe $GL(n, f_*(g_*\mathcal{O}_Y/\mathcal{C}))$, und wir setzen

$$\mathcal{B}_i := \mathcal{B}_i(d) := \{V \in \mathcal{B} \mid \dim GL(n, f_*(g_*\mathcal{O}_Y/\mathcal{C})) \cdot V \cap \mathcal{B} \leq i\}$$

Nach dem oben Gesagten ist klar, daß für jedes $s \in S$

$$\mathrm{par}_n\Big(\bigoplus_{x \in X_s} \mathcal{O}_{X_s,x} \,;\, \bigoplus_{x \in X_s} (g_*\mathcal{O}_{Y_s})_x\Big) = \max_{i,d}(\dim(\mathcal{B}_i(d) \cap \pi^{-1}(s)) - i)$$

Diese Zahl ist stets kleiner oder gleich $z_n(s)$, und für fast alle $s \in S$ gilt Gleichheit. Deshalb genügt es zu zeigen, daß für jedes i,d die Funktion $s \to \dim(\mathcal{B}_i(d) \cap \pi^{-1}(s))$ halbstetig ist. Da $\pi|_{\mathcal{B}}: \mathcal{B} \to S$ eigentlich ist und die Faserdimension bei eigentlichen Abbildungen nach oben halbstetig ist ([Matsumura] 13.E), folgt der Satz nun aus

Lemma:
Für jedes i,d ist $\mathcal{B}_i(d)$ Zariski-abgeschlossen in $\mathcal{B}(d)$

Beweis: Das Vektorraumbündel $\mathcal{V} \to S$ trägt eine multiplikative Struktur $\mathcal{V} \otimes \mathcal{V} \to \mathcal{V}$, die jede Faser zu einem kommutativen Ring macht, denn $\mathcal{V}$ ist das zur Ringgarbe $f_*(g_*\mathcal{O}_Y/\mathcal{C})$ gehörige Bündel. Entsprechend versehen wir $(\mathcal{V}^{\oplus n})^{\oplus n}$ mit der Struktur einer Familie von Matrizenringen, indem wir einem Element $(v_1,\dots,v_n) \in (\mathcal{V}^{\oplus n})^{\oplus n}$ die Matrix mit den Spalten $v_1,\dots,v_n$ zuordnen. Das Bündel $(\mathcal{V}^{\oplus n})^{\oplus n}$ mit dieser Struktur bezeichnen wir auch mit $\mathrm{Mat}(n\times n, \mathcal{V}) \to S$. Darin ist die Menge $GL(n,\mathcal{V})$ aller invertierbaren Matrizen offen, und dicht in jeder Faser. Der Operation von $GL(n,f_*(g_*\mathcal{O}_Y/\mathbb{C}))$ auf $Gr(d)$ entspricht (bis auf die Inversenbildung) der Morphismus

$$\Phi: GL(n,V)\times_S Gr(d) \to Gr(d)$$
$$(A\ ,\ V) \to A^{-1}\cdot V$$

Sei nun $U \subset V^{\oplus n}\times_S Gr(d)$ das universelle Bündel über $Gr(d)$, und $U^{\oplus n} \subset (V^{\oplus n})^{\oplus n}\times_S Gr(d) \simeq Mat(n\times n,V)\times_S Gr(d)$ die n-fache Whitneysumme von U. Die Faser von $U^{\oplus n}$ über $V \in Gr(d)$ ist dann die Menge aller $n\times n$-Matrizen mit Koeffizienten in $\bigoplus_{x\in X_{\pi(V)}} (g_*\mathcal{O}_{Y_{\pi(V)}}/\mathbb{C}_{\pi(V)})_x$, deren Spalten alle in V liegen. $p: E \to B$ sei die Einschränkung dieses Bündels auf B. Wir setzen $E' := E \cap (Gl(n,V)\times_S B)$, dies ist eine Zariski-offene Teilmenge von E. Da für jedes $V \in B$ die Einheitsmatrix in $E'_V := p^{-1}(V)\cap E'$ liegt, ist $p|_{E'}: E' \to B$ surjektiv. Nach dem Lemma aus Abschnitt 1 ist für jedes $V \in B$

$$\Phi(E'_V) = GL(n,f_*(g_*\mathcal{O}_Y/\mathbb{C}))\cdot V \cap B\ .$$

Da p eine glatte Abbildung ist, ist für jedes $i \in \mathbb{N}$ die Menge

$$D_i := \{A \in E' \mid \Phi|_{p^{-1}(p(A))} \text{ hat Rang} \le i \text{ im Punkt } A\}$$

abgeschlossen in E'. Außerdem ist für $V \in B$

$$\dim \Phi(E'_V) = \max_{A\in E'_V} (\text{Rang von } \Phi|_{E'_V} \text{ in } A)\ .$$

Dies zeigt, daß

$$B_i = \{V \in B \mid E'_V \subset D_i\}\ ,$$

also ist $B - B_i = p(E' - D_i)$. Nun ist $p|_{E'}$ flach und von endlichem Typ, also ist $B - B_i$ Zariski-offen in B (vgl. [Hartshorne] III. exercise 9.1).

Literatur:

Buchweitz, R.-O., Greuel, G.-M.: The Milnor number and deformations of complex curve singularities. Inv. Math. 58, 241-281 (1980).

Gabriel, P.: Finite representation type is open. In: Representations of Algebras. Springer Lecture Notes 488, 132-155 (1975).

Greuel, G.-M., Knörrer, H.: Einfache Kurvensingularitäten und torsionsfreie Moduln. Math. Ann. 270, 417-425 (1985).

Hartshorne, R.: Algebraic Geometry. Springer Verlag 1977

Matsumura, H.: Commutative Algebra (2nd edition) Benjamin 1980

Rego, C.: Compactification of the space of vector bundles on a singular curve. Comm. Math. Helvetici 57, 226-236 (1982)

Horst Knörrer
Mathematisches Institut der
Universität
Wegelerstr. 10
D 5300 Bonn 1
BRD

DEFORMATION OF MODULES ON CURVES AND SURFACES

C.J. REGO

Let Z be a smooth variety over an algebraically closed field k. Let T be a sheaf of O_Z modules supported at finitely many points and write $m=h^0(Z,T)$.

Question (*) Does there always exist a flat deformation of T, say over $k[[t]]$, which is supported generically at m distinct points?

We call such a deformation a smoothing of T. Iarrobino has shown by a simple dimension count, that the answer to (*) is negative if $\dim(Z) \geq 3$. When Z is a surface, it was known to Hilbert that for T monogenic a smoothing exists. His argument went as follows. In a neighbourhood of supp(T) we have a presentation

$$O_Z^n \xrightarrow{q} O_Z^{n+1} \longrightarrow O_Z \longrightarrow T \longrightarrow 0$$

Any lift of the matrix q to q_t: $O^n_{Z\times k[[t]]} \longrightarrow O^{n+1}_{Z\times k[[t]]}$ defines a flat deformation of T. One shows that a generic lift of q defines a union of smooth subschemes which is equivalent to (*).

A slicker proof uses the fact that for T monogenic (*) is the same as the irreducibility of $\mathrm{Hilb}^m(Z)$. Since Hilb is connected it suffices to know Hilb is smooth. Now Hilb^m has some smooth points and a simple computation shows that the tangent space is of rank $2m$ at all points. Hence Hilb^m is smooth.

We have the

Theorem(A) : If $\dim(Z) \leq 2$ then every T can be smoothed.

For general T this is equivalent to the irreducibility of $\mathrm{Quot}^m(n,Z)$, the scheme of quotients of O_Z^n of length m. This scheme is not smooth so the arguments for the monogenic case do not generalise. We will prove the stronger

THEOREM(A') : Let X be an integral (possibly singular) curve on a smooth surface Z. Then $\mathrm{Quot}^m(n,X)$ is irreducible for all m, n.

Let N be a torsion free sheaf of rank n on an integral curve X and let U be the smooth locus of X. Note that N is locally free on U. There exists a locally free N_0 and an embedding

$N \longrightarrow N_0$ with (N_0/N) of finite length m. As we will verify later, deforming N_0/N to have support at m distinct points is equivalent to deforming N to a locally free sheaf. This problem is local around the singular points of X and depends only on the isomorphism class of N at each singular point of the curve. Theorem(A') says that for a plane curve singularity, torsion free modules can be made locally free by flat deformation. For non planar curves this is not possible [R-1].

It is easy to construct a total space for torsion free modules of rank n over a one dimensional local domain O containing every isomorphism class. This space is acted on by an algebraic group G with each G orbit defining an isomorphism class. Dividing the space into a union of strata where G operates with equidimensional orbits we can define a generic G quotient for each stratum. Write mod(n,O) for the dimension of the maximal dimensonal G quotient. Greuel and Knörrer have classified O with mod (1,O)=0. It would be interesting to relate mod (n,O) to the other invariants of the singularity O. We offer a lower bound in terms of the multiplicity that is attained in the simplest cases. At the conference H. Knorrer asked if mod(n,O) varies semi continuously in a family. This is true and a proof is sketched below. Nothing substantial seems to be known about mod(n,O). Is it constant under equi- singular deformation ? See note at end of the article.

1. GENERALITIES

Let Y be a scheme over k and $V = O_Y^n$. The functor of $O_Y \otimes O_T$ submodules of $V \otimes O_T$, N_T satisfying "$V \otimes O_T/N_T$ is a locally free O_T module of rank m" is represented by a projective scheme denoted by $\mathrm{Quot}^m(n,Y)$. In the sequel Y will usually be a smooth surface or a curve on a smooth surface. Note that if W is a subscheme of Y we have a closed immersion $\mathrm{Quot}^m(n,W) \longrightarrow \mathrm{Quot}^m(n,Y)$ where $N \in \mathrm{Quot}^m(n,W)$ iff $I_W . V \subset N$ where I_W is the defining ideal of W.

Proposition 1.1 : Theorem (A') implies Theorem (A).

Proof : We want to show that $\mathrm{Quot}^m(n,Z)$ is irreducible for Z a smooth surface. Given two points $P_1, P_2 \in \mathrm{Quot}^m(n,Z)$ it suffices by Theorem(A') to find an integral curve $X \subset Z$ with P_1, P_2 lying on the subscheme $\mathrm{Quot}^m(n,X) \hookrightarrow \mathrm{Quot}^m(n,Z)$. Concretely, this means that any two quotients O_Z^n/M and O_Z^n/N, can be treated as O_X modules for a suitable integral X. To construct X let $Q_1, \ldots Q_s$ and $R_1, \ldots R_t$ be the supports of the two quotients. The annihilator of $O_Z^n/M \oplus O_Z^n/N$ is an ideal I contained in O_Z with O_Z/I supported in the union of the Q_i and R_j. Let B be the semi local ring of the support - which exists as Z is projective. Then all we need to define a suitable X is to find a height one prime $P \subset I$. As B is a U.F.D. any irreducible element in I defines a P and completes the proof.

Proposition 1.2 : Let Z be a smooth surface. Then $\mathrm{Quot}^m(n,Z)$ is singular for $n > 1$, $m > 1$.

Proof : The tangent space at a point corresponding to $N \subset V = O^n_Z$ is canonically identified with Hom (N,V/N). Suppose V/N is supported at m distinct points of Z. We claim V/N defines a smooth point of Quot. To see this first compute the tangent space. Since it is a local question it suffices to fix a local ring O of Z with maximal ideal $\mathfrak{m}$ and suppose $N_0 = \mathfrak{m} \oplus O \oplus O \oplus \ldots O \oplus O \subset O^n = V$. Then $\mathrm{Hom}(N_0, V/N_0)$ has dimension (n + 1). If V/N is supported at m distinct points its tangent space has rank m(n+1) at V/N. We now check that Quot^m has dimension m(n+1) at V/N. Again it suffices to prove that $\mathrm{Quot}^1(n,Z)$ has dimension n+1 at V/N_0. For each point of Z we have a $\mathbb{P}^{n-1} \hookrightarrow \mathrm{Quot}^1$ of quotients supported at that point. As dimension Z=2 we find dim $\mathrm{Quot}^1 \geqslant (n-1)+2 = n+1$. By the tangent space computation dim $\mathrm{Quot}^1 = (n+1)$.

We denote by U^m the smooth open subset of Quot^m defined by quotients supported at m distinct points. To see that Quot^m is singular for $m \geq 2$, $n \geq 2$ we pick a point in the closure of U^m which has a tangent space greater than m(n+1). One such point is defined by the module $N \subset V$ of colength 1 at (m-2) points and of the type $\mathfrak{m} \oplus \mathfrak{m} \oplus O \oplus \ldots O \subset V$ at one point. It is clear how to deform this quotient so that it has support at m points. If x,y are generators of $\mathfrak{m}$ just take the k[[t]] deformation $((x+t,y) + \ldots + O) \subset V \otimes k\,[[t]]$. This shows that V/N is in the closure of U^m. However its tangent space has rank equal to (n+1)(m-2) + 2(n+2) which is greater than m(n+1). This proves the proposition.

Proposition 1.3 : Let X be a smooth irreducible curve; then $\mathrm{Quot}^m(n,X)$ is irreducible. In particular, for any irreducible curve, the open subset of Quot supported at smooth points is irreducible.

Proof : Write $Q = \mathrm{Quot}^m(n,X)$ and recall we have an exact sequence

$$(1.3.1) \qquad 0 \longrightarrow N \longrightarrow O^n_{X \times Q} \longrightarrow H \longrightarrow 0$$

Where N is a rank n vector bundle on X x Q and $p_{2*}H$ is a rank m vector bundle on Q. The determinant defines a map d: $\Lambda^n N \longrightarrow O_{X \times Q}$ with cokernel finite of rank m over Q. Hence we get a morphism p from Q to $\mathrm{Hilb}^m(X) = \mathrm{Quot}^m(1,X)$ with fibres representing quotients which are supported at the "determinantal cycle." Consider the subset $U^m(n,X) \subset \mathrm{Quot}^m(n,X)$ of quotients supported at m distinct points. As in the proof of Proposition (1.1) we see that the fibres of p are m-fold products of $\mathbb{P}^{n-1}$,s. Since $\mathrm{Hilb}^m(X)$ is irreducible and the image of $U^m(n,X)$ is dense open we find $U^m(n,X)$ is irreducible. It remains to prove that $U^m(n,X)$ is dense. But any $N \subset O^n_X$ is locally free as X is smooth so the arguments used in the proof of Proposition (1.2) show that N can be deformed over k [[t]] so that the quotient is supported at m distinct points. This proves the proposition.

Remark (1.4). Since any finite set of points on a smooth irreducible projective variety Z can be joined by a smooth irreducible curve X, given two points on $U^m(n,Z)$ we can find an X with $U^m(n,X)$ containing them. Hence $U^m(n,Z)$ is irreducible.

For the rest of this section O is the singular local ring of an integral curve X. Write $F = O^n$, $\tilde{O}$ for the normalization of O, K the quotient field of $\tilde{O}$, $\tilde{F} = \tilde{O}^n$, δ=length $(\tilde{O}/O)$, $C \subset O$ the conductor of O in $\tilde{O}$. Let N be a torsion free O module of rank n. Write $\tilde{N} = N.\tilde{O} = N \otimes \tilde{O}$/Torsion and as N is torsion free over a PID it is free. Choose n elements in N which generate $\tilde{N}$ over $\tilde{O}$. These define an imbedding $F \subset N$ so that $F.\tilde{O} = N.\tilde{O} = \tilde{F}$. Thus every isomorphism class of O modules is represented by one between F and $\tilde{F}$.

Definition-Proposition (1.5) : The functor of O submodules of $\tilde{F}$ with colength d is denoted by E(d). It is represented by a closed subset of a Grassmanian.

By the above the union of E(d) for $d \leqslant n\cdot\delta$ 'contains' every iso- morphism class. We write E'(d) for the subset of E(d) consisting of $N \subset \tilde{F}$ with $F \subset N$. Note that $G = Gl_n(\tilde{O})$ acts on the E(d) by restriction of the action on $\tilde{O}^n$ to $N \subset \tilde{O}^n$. However G does not act on E'(d).

Remark (1.6) ; The Grassmanian in (1.5) is finite dimensional since given d there is an integer q = q(d) so that if N is an O module with $N \subset \tilde{F}$, length $(\tilde{F}/N)$ = d then $C^q.\tilde{F} \subset N$. See R[1] for more details.

It is not necessary to take O local to define E but note that E depends only on the semi local ring of the singular points. The construction of the scheme E does not extend to families of curves but we shall outline what exactly happens.

Let $f : X_S \longrightarrow S$ = Spec k[[t]] be a family of reduced curves and $\tilde{f} : \tilde{X}_S \longrightarrow S$ be the induced map from the normalization of X_S. To streamline notation write $O = O_{X_S}$, $\tilde{O}$ the normalization of O, $O_0 = O/t.O$, $O_1 = \tilde{O}/t.\tilde{O}$ and $\tilde{O}_1$ = normalization of O_1. Note that $O_0 \subset O_1 \subset \tilde{O}_1$ and $\tilde{O}_1$ is the normalization of O_0. In general (ie. unless the δ of the special and general fibres are equal), $O_1 \neq \tilde{O}_1$. Let C' be the conductor of O and write C for its unmixed component as an O ideal, (i.e. C is the intersection of the isolated primary components of C'). Then one checks that O/C has no t-torsion and hence is a (finite) free O_S module.

Write F, $\tilde{F}$, F_0, F_1, $\tilde{F}_1$ for the the finite free rank n modules over the obvious rings. Then we can construct a sort of E scheme over S by looking at the O modules among the vector spaces in Grass $(\tilde{F}/C,d)$. Notice that a point of this scheme, (which we again denote by E(d)) specializes to a point of the $E(d_1)$ scheme of the special fibre where d_1 = d + length $(\tilde{F}_1/F_1)$. See R[1] for more details.

2. mod(n,O)

In this section we shall sketch a few facts about the moduli of isomorphism classes of O modules. It is easy to check (using the remarks of the previous section and the fact that modules split) that if O has multiplicity 2 then mod (n,O) = 0. Or equivalently, there are only finitely many isomorphism classes of O modules. In particular mod (n,O) does not increase with δ as may be conjectured at a first guess. However the moduli does increase with the multiplicity which is the main point of Proposition 2.1 below.

Using the notation of the previous section observe that for $F \subset N \subset \tilde{F}$, Aut(N) is naturally a subset of $\mathrm{Aut}(\tilde{F}) = Gl_n(\tilde{O})$ and contains $\mathrm{Aut}(F) = Gl_n(O)$. Two modules $N_1, N_2 \in E'(d)$ are isomorphic iff there is a $g \in Gl_n(\tilde{O})$ with $g(N_1) = N_2$. Write $E_c(d)$ for the G closure of E'(d) in E(d). Then the orbit of G through N,G(N), can be viewed as a subscheme of $E_c(d)$ and dim G(N) = dim (G/Aut(N)) is less than or equal to $\dim (Gl_n(\tilde{O})/Gl_n(O)) = \delta \cdot n^2$. To obtain the moduli of isomorphism classes of torsion free modules it is enough to work with $E_c(d)$ by dividing it into a union of strata with equidimensional G orbits. Taking the generic quotient for each stratum and taking the maximum dimension so obtained we arrive at a fairly concrete representation of mod(n,O).

In the following proposition we shall for simplicity assume that n=1 and O is unibranched.

Proposition 2.1 Let O, $\tilde{O}$ be as above but with $\tilde{O}$ local having maximal ideal $(z).\tilde{O}$. Let m = mult(O), s=m - [m/2] and q = [s/2]. Then

$$\mathrm{mod}\,(1,O) \geq q^2$$

Proof : Let $O_1 = k + (z^m).\tilde{O}$ and $O_2 = k + (z^s).\tilde{O}$. Let V be any k vector subspace between O_1 and O_2. Notice that the O_i are subrings of $\tilde{O} = k[[z]]$ and any V as above is closed under multiplication by elements of O_1 (and hence of O). This means V is an O module. In fact one checks easily that V is a ring. Now take N, M both O modules with $O_1 \subset N \subset O_2$ and $O_1 \subset M \subset O_2$. Then if $N \simeq M$ there exists a unit $u \in \tilde{O}$ with u. N = M and since $O \subset N$, $u \in$ M. As M is a ring u^{-1} is in M and so N = M. The estimate follows on computing the dimension of the largest Grassmanian of subspaces of an s dimensional vector space.

Since the schemes E(d) do not sit in a family it is not immediately obvious that mod(n,O) varies semi continuously.

Proposition 2.2 : Let $f : X_S \longrightarrow S = \mathrm{Spec}\, k[[t]]$ be a family of reduced curvs with special fibre X_o and generic fibre X_g Then

$$\mathrm{mod}(n,X_o) \geq \mathrm{mod}\,(n,X_g)$$

Proof : It will be convenient to use the 'relative' E scheme of § 1. There exists a $d \leq \delta \cdot n$ such that the E'(d) of X_g contains an irreducible component W_g which defines the largest moduli of modules on X_g. Using earlier notation we have O, the local ring of the surface X_S, $\tilde{O}$ the normalisation of O, $O_o = O/t.O$ the local ring of the special singularity, $O_g = O\,[t^{-1}]$ the (non local) ring of the generic singularities, $O_1 = \tilde{O}/t.\tilde{O}$ and $\tilde{O}_1$ the normalization of O_1 (and also the normalization of O_c). Write $G' = Gl_n(\tilde{O})$, $G_g = Gl_n(\tilde{O}_g)$, $G_1 = Gl_n(O_1)$, $G = Gl_n(\tilde{O}_1)$. Within the 'relative' E scheme we have a W with generic fibre W_g and specializing to $W_o \subset E'(d_1)$. Since G' does not act on E' it does not act on W but to obtain $\mathrm{mod}(n,O_g)$

we can look at the equivalence relation defined by G_g on W_g. General arguments now show that the G_1 equivalence relation on W_0 gives a quotient greater or equal to mod (n,O_g). But to get the moduli of isomorphism classes of modules in W_0 we have to quotient out by G action and not just G_1.

So we need to verify that if $N,M \subset F_1$ and contains F_0 with N,M isomorphic by an element of $Gl_n(\tilde{O}_1)$ then N and M are equivalent by a $Gl_n(O_1)$ transformation. But this is immediate since if $h \in Gl_n(\tilde{O}_1)$ $(=Gl_n(\tilde{O}_0))$ and $N = h(M)$ then since $F_0 \subset M$ all the columns of h are in N. So in fact $h \in N^n \subset F_1^n = M_n(O_1)$. Since h has invertible determinant h is in $Gl_n(O_1)$.

Remark (2.3) : The main point in the above argument seems to be the choice of the family W in E'(d) and not just in E(d). Readers who find the above proof too sketchy may first read R[1]. See note at end of article.

3. Gorenstein Curves

In this section O will stand for a one dimensional Gorenstein domain. The first result due to D'Souza is a powerful tool in the study of deformations of torsion free modules over O.

D'Souza' Deformation Theorem : Let $A' \longrightarrow A$ be a surjective map of complete local k algebras (with residue field k). Let B' be an A' algebra with $B' \otimes k$ a Gorenstein domain of dimension one. Let $N_{A'}$ be a B' module flat over A' and let $N_A = N_{A'} \otimes A$. Suppose further that $N = N_A \otimes k$ is torsion free over $B' \otimes k$ of rank n and there is an embedding of N_A in $(B' \otimes A)^n$ with $(B' \otimes A)^n/N_A$ a (finite) flat A module. Then the embedding can be lifted to an embedding of $N_{A'}$ in $(B')^n$ so that the diagram

$$\begin{array}{ccc} N_{A'} & \longrightarrow & (B')^n \\ \downarrow & & \downarrow \\ N_A & \longrightarrow & (B' \otimes A)^n \end{array}$$

is commutative.

Proof : See [D-S]

We shall use the notation of the previous section. Take an O module N with $F \subset N \subset \tilde{F}$. Then

$$N^* = \left\{ (n_i^*) \in K^N \;\middle|\; \sum n_i^* \cdot n_i \in O \quad \forall (n_i) \in N \right\}$$

is canonically identified with Hom (N,O). Note that for N as above $N^* \subset F$ and $C\cdot F \subset N^*$. By reflexivity

$$\text{length } (F/N^*) = \text{length } (N/F)$$

(Remark. It is a standard fact that rank one torsion free modules over O are reflexive. For higher rank just use induction on the rank and the vanishing of $Ext^1(N, O)$ for N torsion free).

PROPOSITION 3.1 (a) Every module N can be represented by $C.F \subset N \subset F$
(b) If $N \subset F$, $C\cdot F \not\subset N$ then there is an $N' \cong N$ with $C\cdot F \subset N' \subset F$ satisfying

$$\text{length } (F/N') \lneq \text{length } (F/N)$$

Proof. Writing $N=P^*$, $F \subset P \subset \bar{F}$, (a) is clear by reflexivity. To prove (b) use (a) to get N' with $C.F \subset N' \subset F$ and extend the isomorphism $i : N' \cong N$ to an isomorphism $N' \otimes K \cong N \otimes K = K^n$ so $i \in Gl_n(K)$. As $i(C\cdot F) \subset N \subset F$ all the entries of i are in $\tilde{O}$ so $i \in M_n(\tilde{O})$. It is easy to verify

$$\text{length } (\tilde{F}/i(\tilde{F})) = \text{length } (\tilde{O}/\det(i)).$$

It follows that the length $(F/N') = \text{length } (F/N) - \text{length } (\tilde{O}/\det(i))$. Suppose det (i) is a unit in O so $i \in Gl_n(\tilde{O})$. Then as $C\cdot F \subset N'$, $C\cdot F \subset i(C\cdot F) \subset i(N') = N$ which contradicts our assumption. So i is not in $Gl_n(O)$ and hence length $(O/\det(i)) > 0$. This proves the proposition.

Remark 3.2 : Given B a complete Gorenstein domain and M a torsion free Cohen Macaulay B module of rank n we can construct a family of rings and modules by choosing a subring $k[[t_1,\dots t_d]] \subset B$ where $d = \dim(B) - 1$. Of course M can be embedded in $(B)^n$ but using D'Souza's Deformation Theorem we can find an embedding that lifts any embedding of $(M/t.M)$ in $(B/t.B)^n$. If $B/t.B$ is sufficiently simple then this gives a strong normalization of M as an embedded module.

Lemma 3.3 : Let X be a Gorenstein curve. Then $Quot^m(n,X)$ is irreducible for every m $\Leftrightarrow$ every torsion free O_X module N can be deformed to a locally free module.

Proof : Let N be an O_X module of rank n, $N \subset O_X^n$ with finite cokernel of length m. Suppose $Quot^m(n,X)$ is irreducible so N can be deformed to $N(t) \subset O_X^n \otimes k[[t]]$ with quotient supported at m distinct k[[t]] rational primes and none of them singular. Then clearly $N(t) \otimes k((t))$ is locally free on $X \times \text{Spec } k((t))$.

Conversely let every N be deformable to a locally free module and suppose $Quot^m(n,X)$ is irreducible for $m \leq q-1$. Let $N \subset O_X^r$ define a point in $Quot^q$. Recall that as X is irreducible the quotients O_X^n/N supported at q distinct smooth point form an irreducible open subset

$U = U^q(n,X)$. Also if O_X^n/N is supported at smooth points of X it lies in the closure of U as may be verified by treating O_X^n/N as a sheaf on the normalization of X. Suppose O_X^n/N is supported at several points a,b,.. h. Then for each point w we have s=length $(O_{X,w}^n/N_w) < q$. Since each quotient $O_{X,w}^n/N_w$ defines a point in $Quot^s(n,X)$ which is in the closure of $U^s(n,X)$ there exists a deformation of O_X^n/N generically having support at $q = \sum s$ distinct smooth points. We may therefore assume that O_X^n/N and all its small deformations are supported at one point $x \in X$. This means that the Punctual Quot scheme $Quot_x^q(n,X)$ contains a component W of $Quot^q(n,X)$ and the map $Quot_x^q(n,X) \hookrightarrow Quot^q(n,X)$ is bijective in a neighbourhood of O_X^n/N. By assumption N can be deformed to a locally free O_X module module. Let $N[t]$ be an $O_X \otimes k[[t]]$ module defining this deformation. If we localize around x then we can use D'Souza's Theorem to lift the given imbedding $N_x \subset O_{X,x}^n$ to $N_x[t] \hookrightarrow O_{X,x}^n \otimes k[[t]]$ with cokernel a free k[[t]] module of rank q. Now the imbedding $N_x[t] \hookrightarrow O_{X,x}^n \otimes k[[t]]$ is the restriction of an inclusion $N'[t] \hookrightarrow O_X^n \otimes k[[t]]$ with the same cokernel and N'[t] is generically a vector bundle specializing to N on X. As $Quot_x^q(n,X)$ is bijective with $Quot^q(n,X)$ in a neighbourhood of O_X^n/N, $O_X^n \times k[[t]]/N'[t]$ is supported at (x)x Spec k[[t]] and so there are points of W in every neighbourhood of O_x^n/N defined by vector bundles. We will derive a contradiction.

Let $h_1,\dots h_n \in O_{X,x}^n$ define a free $O_{X,x}$ module P with $O_{X,x}^n/P$ having length q. Then the deformation $(h_i + t)$ is a flat deformation that is not supported at x. But then $Quot^q(n,X)$ cannot be bijective with $Quot_x^q(n,X)$ in a neighbourhood of O_X^n/P and the proposition is proved.

Remark 3.4 : The proof of the above proposition yields the fact that if $N \subset O_X^n$ is deformable over k[[t]] to a vector bundle then the given injection lifts to a (possibly different) deformation that is supported generically at m distinct points when m = length (O_X^n/N). We will use this remark later.

Corollary 3.4.1 : If every N can be deformed to a locally free module then

$$\dim Quot_x^m(n,X) \leqslant n.m - 1$$

for all $x \in X$ and $m \geqslant 1$.

Proof : Since $Quot_x^m$ is a proper closed subset of Quot and $U^m(n,X)$ has dimension n.m the result follows.

Suppose now that X is a curve in a smooth surface Z.

Lemma 3.5 : If v is the multiplicity of O_X and $\mathfrak{m}$ the maximal ideal then

$$C \subset \mathfrak{m}^{v-1}, C \not\subset \mathfrak{m}^v$$

Proof : See R[2].

Remark 3.6 In characteristic zero if f=o defines X in Z then if u, w are local coordinates at $x \in X$ we have the polar f_u which is in C but not in $(u,w)^v$. In general we call such a g a polar.

4. Main Theorem :

THEOREM A' : Let X be an integral curve on a smooth surface Z. Then $Quot^m(n,X)$ is irreducible for all m,n.

Proof : Since the problem is local around the singular points we use induction on the multiplicity of one singular point, $x \in X$. Assume the result true for a curve with multiplicity less than $v = multi(O_{X,x})$. By Lemma 3.5 there is a $g \in O_Z$ with g of order (v-1) and g defining an element of C. We have

$$Quot^m(n,O_X/C) \subset Quot_x^m(n, O_Z/(g)).$$

By adding a general element of O_Z of high order to g we can assume that g=0 defines (locally) a reduced curve in Z irreducible in a neighbourhood of x. Now induction and Cor3.4.1 gives

$$\dim Quot^m(n,O_X/C) \leqslant \dim Quot_x^m(n,O_Z/(g)) \leqslant n.m-1$$

If a,b...h are the singular points of X and N a torsion free rank n O_X module then to show that N is deformable to a vector bundle it suffices to know this for all stalks N_e. We may therefore assume that X has one singular point, x.

Assume all O_X modules N which have an embedding $N \subset O_X^n$ with length $(O_X^n/N) < q$ can be deformed to locally free modules. By Remark (3.4) this is equivalent to the assumption that $U^m(n,X)$ is dense in $Quot^m(n,X)$ for $m < q$ and in particular $Quot^m(n,X)$ is irreducible for $m < q$. We want to show that $Quot^q(n,X)$ is irreducible.

Suppose that $N \subset O_X^n$ is supported at several points a,b,...,h. Then at each point e the length $(O_{X,e}^n/N_e) < q$. Since each quotient defines a point in $Quot^s(n,X)$ (where $s < q$) which is in the closure of $U^s(n,X)$ there exists a deformation of O_X^n/N generically having support at q distinct smooth points. Hence the quotient is in the closure of $U^q(n,X)$. Therefore suppose that the quotient is supported at one point. At smooth points there is no problem so we may take this point to be x.

We can thus restrict ourselves to quotients in $Quot_x^q(n,X)$. The above discussion says that the open set $U = Quot^q(n,X) - Quot_x^q$ has $U^q(n,X)$ as a dense (irreducible) subset. If $Quot^q(n,X)$ is reducible then $Quot_x^q(n,X)$ must contain a component W of $Quot^q(n,X)$. We will prove the theorem by deriving a contradiction.

Let $P \in Quot_x^q(n,X) \subset Quot^q(X)$ be a general point of W so that if O_X^n/N represents P every small deformation of O_X^n/N is supported at x. By Remark (3.4) N cannot be deformed to a locally free module. If $C \cdot O_X^n \not\subset N$ then by Proposition (3.1) there is an N' with $N'_x \cong N_x$ and with $C. O_X^n \subset N' \subset O_X^n$. Further

$$\text{length}(O^n_X/N') \quad < \quad \text{length}(O^n_X/N) = q$$

and so by assumption N' is deformable to a vector bundle. But for $y \neq x$, $N'_y = O^n_{X,y}$ and $N'_x = N_x$ so N is also deformable to a vector bundle. This is a contradiction and hence we may assume $C \cdot O^n_X \subset N$. As O^n_X/N is a general point of W the natural map $\text{Quot}^q(n, O_X/C) \hookrightarrow \text{Quot}^q(n,X)$ is a bijection in a neighbourhood of P. Of course, as W is a component, $\text{Quot}^q(n,X)$ is bijective with $\text{Quot}^q_x(n,X)$ in a neighbourhood of P.

If X' is an affine open subset of X then $\text{Quot}^m(n,X')$ is an open subset of $\text{Quot}^m(n,X)$ and if $x \in X'$ then $\text{Quot}^m_x(n,X') = \text{Quot}^m_x(n,X)$. As we will encounter only quotients supported at x and their small deformations we can restrict ourselves to an open neighbourhood of x. Let $Z' \subset Z$ be an affine open set with $X' = X \cap Z'$ satisfying $x \in X'$ and X' defined by one equation, $\{f=0 \mid f \in O_Z\}$. Let $g \in O_Z$ define a polar of X' at x. If g is chosen sufficiently general and Z' small enough we can arrange so that $(f=0) \cap (g=0) = z \in Z$ and $(f+tg=0) \subset Z' \times \text{Spec}\, k[[t]]$ is smooth outside $(z) \times \text{Spec}\, k[[t]]$, where of course, $O_{Z',z}/(f) = O_{X,x}$. Write $S = \text{Spec}\, k[[t]]$ and $(f+tg=0) = X'_S$ and $\varphi : X'_S \longrightarrow S$ the restriction of the projection map. The family φ is smooth outside $(z) \times S$ and the singular point of the generic fibre has multiplicity $v-1$ = (order g at z), by the definition of a polar. As in the proper case we have a corresponding family $p' : \text{Quot}^q(n, X'_S|S) \rightarrow S$ where the fibres of p' are open subsets of the Quot schemes of the fibres of φ. Since $f+tg \in C$, O^n_X/N or rather $O^n_{X \times S}/N \otimes O_S$, defines a section σ of p'. By induction on the multiplicity of the singular point the generic fibre of p' is irreducible of dimension $n \cdot q$. As the section $\sigma(S)$ passes through $P \in W' = W \cap \text{Quot}^q(n,X')$, semi continuity of dimension gives $\dim_P \text{Quot}^q(n,X) \geqslant n \cdot q$ But we have seen that

$$\text{Quot}^q(n, O_X/C) \hookrightarrow \text{Quot}^q_x(n,X) \hookrightarrow \text{Quot}^q(n,X)$$

are bijections around P so we have

$$(4.1) \quad \dim_P \text{Quot}^q_x(n, O_X/C) \geqslant n \cdot q$$

However g defines an element of C and hence O_X/C is a quotient of $O_Z/(g)$. This means $\text{Quot}^q(n, O_X/C) \subset \text{Quot}^q_y(n, O_Z/(g))$ where $y \in \text{Spec}\, O_Z/(g)$ maps to $z \in Z$. As $\text{Quot}^q(n, O_Z/(g))$ is irreducible of dimension $n \cdot q$ and $\text{Quot}^q_y(n, O_Z/(g))$ is a proper closed subset of $\text{Quot}^q(n, O_Z/(g))$ we have

$$\dim_P \text{Quot}^q_x(n, O_X/C) < n \cdot q$$

This contradicts (4.1) and proves the theorem.

NOTE: A complete and systematic treatment of the semi-continuity question (prop. 2.2) appears in this volume [K]. For finer results than Proposition 2.1 one should use the methods of [G-K].

REFERENCES

[B] H. BASS : On the Ubiquity of Gorenstein Rings, Math. Zeit., Vol. 82 1963, p. 8-28.

[G-K] G.-M. GREUEL, H. KNÖRRER: Einfache Kurvensingularitäten und torsionsfreie Moduln, Math. Ann. 270 (1985), pp. 417-425.

[K] H. KNÖRRER: Torsionsfreie Moduln bei Deformation von Kurvensingularitäten (This volume).

[O-S] ODA, C.S. SESHADRI: Compactifications of the generalized Jacobian. T.A.M.S., Vol. 253, 1979.

[R1] C.J. REGO: The compactified Jacobian. Ann. Scient. E.N.S., Tome 13, 1980

[R2] C.J. REGO: Compactification of the Space of Vector Bundles on a Singular Curve, Comm. Hevetici, Vol. 57, 1982

C.J. REGO
19, 'Prasanna'
Nesbit Road,
Bombay 400 010
INDIA

A CHARACTERISATION OF STRICTLY UNIMODULAR PLANE CURVE SINGULARITIES

Albert Schappert
Fachbereich Mathematik
der Universität Kaiserslautern
D-6750 Kaiserslautern

ABSTRACT: Let R be the local ring of a reduced plane curve singularity. It is shown that R is the ring of a strictly unimodular singularity if and only if it admits finitely many at most 1-parameter families of isomorphism classes of torsion free rank 1 modules.

0. Introduction

G.M. Greuel and H. Knörrer have found a characterisation of simple curve singularities by maximal Cohen Macaulay modules (see [G-K]). We refer to the survey article of F.O. Schreyer ([S]) in this volume for details and for further results (see also [B-G-S] and [Kn1]). A generalisation of these methods to more complicated singularities leads to the following theorem.

THEOREM: *Let R be the local ring of a reduced plane curve singularity. Then the following statements are equivalent:*

(1) R is the local ring of a strictly unimodular singularity.

(2) There are only finitely many at most 1-parameter families of isomorphism classes of torsion free rank 1 modules over R.

(1) $\Longrightarrow$ (2) is proved by an explicit determination of all isomorphism classes of torsion free rank 1 modules over strictly unimodular curve singularities. The complete lists are too long to be presented here, but we describe our method in section 3 and illustrate it by an example. The other implication is proved in section 2 by using Knörrer's semicontinuity theorem (see [Kn2]). In section 1 we give the parametrisations of all unimodular plane curve singularities.

This result is the first classification theorem for non-simple singularities by module properties. However, only rank 1 modules are considered, because the calculations for higher ranks become extremely voluminous (even for rank 2). The complete module category is described up to now only for some examples of unimodular singularites, namely for simple elliptic singularities of type $T_{*,*}$ (see [D], [Ka]), but this is done using different methods.

By the above theorem the singularities that admit at most one parameter families of torsionfree rank 1 modules are not the unimodular singularities (neither in the sense of right-equivalence nor in the sense of contact-equivalence). However, they are precisely the strictly unimodular singularities (compare definition below). Up to now, the semicontinuity theorem for curves (theorem 2) is the only hint for this connection. We do not know if there is a deeper relation between deformation properties and module properties.

Notation

Let R be the local ring of a reduced plane curve singularity, m its maximal ideal. We identify R with its parametrisation, i.e. its image in the normalisation $\bar{R} = \oplus_{i=1}^{r} \mathbb{C}\{t\}$, r the number of irreducible components.

The valuation on $\bar{R}$ given by the valuation on the components is denoted by

$$v = (v_1,\dots,v_r):\ \bar{R} \longrightarrow (\mathbb{N} \cup \{\infty\})^r.$$

$\bar{R}$ is a principal ideal ring, thus every $\bar{R}$-ideal i is given by an element $(l) = (l_1,\dots,l_r) \in (\mathbb{N} \cup \{\infty\})^r$, because

$$J_{(l)} := J_{(l_1,\dots,l_r)} := (t^{i_1},\dots,t^{i_r})\bar{R} = i \quad \text{where } t^{\infty} = 0.$$

In the following we consider only modules M such that $R \subset M \subset \bar{R}$ (i.e. modules of rank 1). We denote by $c(M)$ the *conductor* of an R-module M, i.e. the maximal $\bar{R}$-ideal contained in M.

1. Classification of Plane Curve Singularitites.

Let $C = (C,0)$ be the germ of a reduced plane curve singularity given as the zero set of a holomorphic function germ $f:(\mathbb{C}^2,0) \longrightarrow (\mathbb{C},0)$. Two singularities are called ***equivalent***, if the germs are isomorphic. A singularity C is called ***unimodular*** (i.e. contact unimodular in the sense of Wall, c.f. [W]), if a sufficiently small deformation leads to finitely many at most one parameter families of distinct equivalence classes and if C is not simple. Of special interest to us are singularities with the following deformation property. Let $F:X \longrightarrow S$ be a miniversal deformation of a unimodular plane curve singularity C. C is called ***strictly unimodular*** (c.f. [W], p. 626), if every nearby fibre $F^{-1}(s)$ contains at most one unimodular singularity.

THEOREM 1 [W]: *The unimodular reduced plane curve singularities are given by the following list:*

Type	Equation	Parametrisation	
$T_{3,6}$	$x^3 + x^2y^2 + \underline{a}y^6$	$\mathbb{C}\{(t,t,\underline{a}t),(0,t^2,t^2)\}$	
$T_{4,4}$	$x^4 + \underline{a}x^2y^2 + y^4$	$\mathbb{C}\{(t,0,t,\underline{a}t),(0,t,t,t)\}$	
$T_{p,q}$	$x^p + x^2y^2 + y^q$	$\mathbb{C}\{(t^2,t^{p-2}),(t^{q-2},t^2)\}$	p,q odd
	$1/p+1/q < 1/2$	$\mathbb{C}\{(t,t,t^{p-2}),(0,t^{q/2-1},t^2)\}$	q even, p odd
		$\mathbb{C}\{(t,0,t,t^{p/2-1}),(0,t,t^{q/2-1},t)\}$	p, q even
W_{12}	$x^4 + y^5$	$\mathbb{C}\{t^4,t^5\}$	
W_{13}	$x^4 + xy^4$	$\mathbb{C}\{(t,t^3),(0,t^4)\}$	
W_{17}	$x^4 + xy^5$	$\mathbb{C}\{(t,t^3),(0,t^5)\}$	
W_{18}	$x^4 + y^7$	$\mathbb{C}\{t^4,t^7\}$	
$W_{1,0}$	$x^4 + x^2y^3 + \underline{a}y^6$	$\mathbb{C}\{(t^2,t^2),(t^3,\underline{a}t^3)\}$	
$W_{1,p}$	$x^4 + x^2y^3 + y^{6+p}$	$\mathbb{C}\{(t^2,t^2),(t^3,t^{3+p})\}$	p even
		$\mathbb{C}\{(t,t,t^2),(0,t^{(p+3)/2},t^3)\}$	p odd
$W^{\#}_{1,p}$	$(x^2+y^3)^2+\begin{cases} x^2y^{p/2+3} \\ xy^{(p+9)/2} \end{cases}$	$\mathbb{C}\{(t^2,t^2),(t^3,t^3+t^{p/2+3})\}$	p even
		$\mathbb{C}\{t^4,t^6+t^{p+6}\}$	p odd
E_{6m+6}	$x^3 + y^{3m+4}$	$\mathbb{C}\{t^3,t^{3m+4}\}$	
E_{6m+7}	$x^3 + xy^{2m+3}$	$\mathbb{C}\{(t,t^2),(0,t^{2m+3})\}$	
E_{6m+8}	$x^3 + y^{3m+5}$	$\mathbb{C}\{t^3,t^{3m+5}\}$	
$E_{m+1,0}$	$x^3+x^2y^{m+1}+\underline{a}y^{3(m+1)}$	$\mathbb{C}\{(t,t,t),(0,t^{m+1},\underline{a}t^{m+1})\}$	
$E_{m+1,p}$	$x^3+x^2y^{m+1}+y^{3(m+1)+p}$	$\mathbb{C}\{(t,t,t),(0,t^{m+1},t^{m+1+p/2})\}$	p even
		$\mathbb{C}\{(t,t^2),(t^{m+1},t^{2(m+1)+p})\}$	p odd
Z_{6m+5}	$y(x^3 + y^{3m+1})$	$\mathbb{C}\{(t,t^{3m+1}),(0,t^3)\}$	
Z_{6m+6}	$y(x^3 + xy^{2m+1})$	$\mathbb{C}\{(t,0,t^2),(0,t,t^{2m+1})\}$	
Z_{6m+7}	$y(x^3 + y^{3m+2})$	$\mathbb{C}\{(t,t^{3m+2}),(0,t^3)\}$	
$Z_{m-1,0}$	$y(x^3+ x^2y^m+ \underline{a}y^{3m})$	$\mathbb{C}\{(t,0,t,t),(0,t,t^m,\underline{a}t^m)\}$	
$Z_{m-1,p}$	$y(x^3+x^2y^m+y^{3m+p})$	$\mathbb{C}\{(t,0,t,t),(0,t,t^m,t^{m+p/2})\}$	p even
		$\mathbb{C}\{(t,t^m,t^{2m+p}),(0,t,t^2)\}$	p odd

These singularities are strictly unimodular if the index m is not greater than 2.

Remark: a) The parameter $\underline{a}$ ranges over the complex numbers except some values of degeneracy.
b)Besides the quasihomogenous equivalence classes of the above list there exists always a second equivalence class given by a semiquasihomogenous representative.
c) Under the classification according to right-equivalence (compare [Arn]) the above singularities are 1-modular or (if there is an index m) m-modular. Hence the strictly unimodular singularities are the right-1- and right-2-modular singularities.

2. Proof of (2) $\Longrightarrow$ (1)

The main ingredient for the proof is the semi-continuity theorem of H.Knörrer. The number of parameters of torsion free rank n modules over a local ring R of a reduced curve singularity is denoted by $\mathrm{par}_n(R)$ (see [Kn2] for a precise definition). Let $F: X \longrightarrow S$, S smooth, be a deformation of a reduced curve singularity.

Theorem 2 [*Kn2*]: *The mapping*

$$Z: S \longrightarrow \mathbb{N}$$
$$s \longrightarrow Z(s) := \sum_{x \in F^{-1}(s)} \mathrm{par}_n(\mathcal{O}_{F^{-1}(s),x})$$

is upper semicontinous (for $n \in \mathbb{N}$ *fixed).*

Remark 1: Let R be the local ring of a unimodular plane curve singularity which is not strictly unimodular. Then there exists at least a two-parameter family of modules over R.
This follows from theorem 2 and the characterisation of the simple singularities (see also [G-K]). □

Remark 2: Every germ $f : (\mathbb{C}^2,0) \longrightarrow (\mathbb{C},0)$ of right-modality greater than 2 deforms into one of the singularities $J_{4,0}$, $X_{2,0}$, $Z^1_{2,0}$ and N (we follow the notation of [Arn]).
This follows from adjancency-diagrams published by V.I. Arnold ([Arn] pp. 30-35): □

Finally we need a generalisation of a lemma of G.-M. Greuel and H. Knörrer. For a reduced local ring R we define the embedding dimension $\mathrm{edim}\, R := \dim_{\mathbb{C}} \mathfrak{m}/\mathfrak{m}^2$ and the multiplicity $\mathrm{mult}\, R := \dim_{\mathbb{C}} \bar{R}/\mathfrak{m}\bar{R}$.

Lemma 3: a) If there is an $l = (l_1,\ldots,l_r) \in \mathbb{N}^r$ *such that*

$$\dim_{\mathbb{C}} J_{(l)}/J_{2(l)} + \mathfrak{m}J_{(l)} + \mathfrak{m} \cap J_{(l)} \geq m + 2$$

then for every n ∈ ℕ *there exist* (mn + 1)*-parameter families of isomorphism classes of indecomposable torsion free rank* n *modules over* R.
b) *If* mult(R) ≥ edim R + (m+2), *then for every* n ∈ ℕ *there exist* (mn + 1)*-parameter families of isomorphism classes of indecomposable torsion free rank* n *modules over* R.

Proof a) We set

$$M_{(\lambda)} := R^n + \left(f_1 \mathrm{id} + f_2 \begin{bmatrix} 1.\lambda_o & & 0 \\ & \ddots & \lambda_o \\ 0 & & 1^0 \end{bmatrix} R + \sum_{3 \leq j \leq m+2} f_j \begin{bmatrix} \lambda_{j1} & & 0 \\ & \ddots & \\ 0 & & \lambda_{j1} \end{bmatrix}\right) R^n$$

where f_j, $1 \leq j \leq m+2$, are linearly independent elements. Now the proof proceeds exactly like the proof of lemma 4 in [G-K].
b) This follows in the same way from a) as lemma 5 from lemma 4 in [G-K]. □

Now we can prove the implication (2) ⟹ (1): Let C be a singularity that is not strictly unimodular, i.e. neither 1-modular nor 2-modular under right equivalence.
If C is simple, then the singularity admits only finitely many modules by the result of [G-K].
If the right-modality of C is greater than 2, it follows from remark 2, that f deforms into one of the 4 singularities $J_{4,0}$, $X_{2,0}$, $Z^1_{2,0}$ or N. By semicontinuity it is sufficent to show, that these singularities admit 2-parameter families of modules.
a) consider $J_{4,0}$: $J_{4,0} = E_{4,0}$ is unimodular but not strictly unimodular (see theorem 1). By remark 1 $J_{4,0}$ admits 2-parameter families of modules.
b) consider $Z^1_{2,0}$: the same argument applies to $Z^1_{2,0} = Z_{2,0}$.
c) consider $X_{2,0}$: One easily sees (by blowing up) that the singularity $X_{2,0}$ consists of 4 distinct branches having the same tangent. Thus a parametrisation is given by

$$\mathbb{C}\{(t, t+p_1, t+p_2, t+p_3), (0, p_4, p_5, p_6)\},$$

where $p_i \in \mathbb{C}[t]$ and $v(p_i) \geq 2$. Setting $l = (1,1,1,1)$ lemma 3a) gives the required statement.
d) consider N: Since mult(N) = 5 ≥ 2 + (1+2) = edim N + (1+2), lemma 3b) implies, that the singularity N admits 2-parameter-families of rank 1 modules.
This proves (2) ⟹ (1). □

3. Proof of (1) ⟹ (2)

In this section we state the basic results needed to determine the isomorphism classes of torsion free rank 1 modules over the local ring of a reduced curve singularity. Furthermore we present an example to illustrate these methods (developed explicitly in [Sch]).

3.1. Torsion free rank 1 modules

Let M be a finitely generated R-module of rank 1, R the local ring of a reduced plane curve singularity. Then there is an embedding $M \subset \bar{R}$ such that $R \subset M \subset \bar{R} = \oplus_{i=1}^{r} \mathbb{C}((t))$. We identify R and M with their images. $\bar{R}^*$ (R^*) denotes the group of units of $\bar{R}$ (R).

<u>*Lemma 4:*</u> *Let M and N be isomorphic R-modules with $R \subset M, N \subset \bar{R}$. Then the set of all isomorphisms from M to N is given by*

$$\mathrm{Iso}_R(M,N) = \{x \in \bar{R}^* \mid xM = N\}.$$

<u>Proof</u>: Let $0 \neq m \in M$. There is a non-zerodivisor $r \in R$ such that $r\,m \in c(R)$. Let φ be an isomorphism. Then we have

$$r\,\varphi(m) = \varphi(r\,m) = r\,m\,\varphi(1) \Longrightarrow \varphi(m) = m\,\varphi(1) \Longrightarrow \varphi(1) \in \bar{R}^*.$$

The other direction is trivial. □

3.2. Short systems of generators

We define a mapping

$$I : \bar{R}-\{0\} \longrightarrow \mathbb{N}^2$$
$$\underline{a} \longrightarrow (i_0, v_{i_0}(\underline{a}))$$

where i_0 is chosen such that $v_j(\underline{a}) = \infty$ if $j < i_0$ and $v_{i_0}(\underline{a}) < \infty$. This mapping marks the first non-vanishing component of $\underline{a}$ together with its subdegree. The lexicographical ordering of $\mathbb{N}^2$ induces a half-ordering on $\bar{R}-\{0\}$ via I.

Now let M, $R \subset M \subset \bar{R}$, be an R-module given by a minimal set of generators $E = \langle e_1, \dots, e_n \rangle$, where

$$e_i = (e_{i1}, \dots, e_{ir}),\ r \text{ the number of branches}$$

and

$$e_{ij} = \sum_{k \geq k_0} \lambda^i_{j,k}\, t^k \quad \text{with } \lambda^i_{j,k} \in \mathbb{C}.$$

One can achieve that

i) for every $e_i \in E$ the parameters $\lambda^i_{j,k}$ vanish if $(j,k) \in I(me_i + \sum_{s \neq i} Re_s)$

ii) for every $e_i \in E$, $\lambda^i_{I(e_i)} = \lambda^i_{i_0, v_{i_0}(e\ )} = 1$

iii) $e_1 = (1, \dots, 1)$

and

iv) E is ordered via I, i.e. $I(e_1) < I(e_2) < \dots < I(e_n)$.

Definition: *A minimal system of generators satisfying i)-iv) is called short.*

Lemma 5: *Let M, $R \subset M \subset \bar{R}$, be an R-module. Then there exists a uniquely determined short system of generators.*

3.3. Invariants

Let every R-module M, $R \subset M \subset \bar{R}$, be represented by a short system E of generators.

Lemma 6: *I(M) is an invariant of the isomorphism class of M.*

Proof: Let M and N be isomorphic R-modules represented by short systems of generators $E = \{e_1,\dots,e_n\}$ and $F = \{f_1,\dots,f_n\}$ and let $\varphi: M \longrightarrow N$ be an isomorphism. We prove $I(E) = I(F)$ by induction.

Assume $I(e_1,\dots,e_k) = I(f_1,\dots,f_k)$ and $I(e_{k+1}) \neq I(f_{k+1})$. Then we have

$$I(\varphi e_{k+1}) = I(e_{k+1}) \notin I(e_1,\dots,e_k) = I(f_1,\dots,f_k).$$

Since $\varphi(e_{k+1}) \in N$ and $\varphi \in \bar{R}^*$ we have

$$\begin{aligned} I(e_{k+1}) &= I(\sum_{j<j_o} a_j f_j + a_{j_o} f_{j_o} + \sum_{j>j_0} a_j f_j) \text{ with } a_j \in \mathfrak{m} \text{ if } j < j_0 \\ &\qquad\qquad \text{and } a_{j_0} \in R^* \\ &= \min\,(I(f_{j_0}),\ I(\sum_{j\neq j_0} a_j f_j)) \text{ since } I(f_{j_0}) \neq I(\sum_{j\neq j_0} a_j f_j) \text{ by} \\ &\qquad\qquad \text{the shortness of } F. \\ &= I(f_{j_0}) \quad \text{since } I(e_{k+1}) = I(\sum_{j\neq j_0} a_j f_j) = I(\sum_{j<j_o} a_j f_j) \in I(\mathfrak{m}N) \\ &\qquad\qquad = I(\mathfrak{m}M) \text{ would violate the shortness of } E \\ &> I(f_{k+1}) \quad \text{by the induction hypothesis.} \end{aligned}$$

By applying the same arguments to φ^{-1} we get a contradiction. □

3.4 Determination of isomorphism classes

The isomorphism classes are determined by an explicit calculation of the orbits under the action of the group of units of $\bar{R}$.

Lemma 7: *Let M,N be R-modules, $R \subset M$, $N \subset \bar{R}$, and suppose $\varphi \in \bar{R}^*$. Then the following are equivalent*

(1) $\varphi : M \longrightarrow N$ is an isomorphism

(2) $I(E) = I(F)$ and $M/\mathfrak{m}M \subset N/\mathfrak{m}N$

(3) $I(E) = I(F)$ and $M/c(M) \subset N/c(N)$, $c(M) = c(N)$, $\dim M/c(M) = \dim N/c(N)$.

<u>Proof</u>: (1) $\Longrightarrow$ (2),(3) is clear.
(2),(3) $\Longrightarrow$ (1): Since $\varphi \in \bar{R}^*$ we have $c(\varphi M) = \varphi c(M)$ and $\varphi M/\varphi S \simeq M/S$ for every submodule S. Because of $I(M) = I(N)$ M and N have the same number of generators. Let $SM \in \{mM, c(M)\}$ (SN respecitvely), then

$$N/SN = \varphi M/\varphi SM \simeq M/SM.$$

Hence Nakayama's lemma gives the demanded isomorphisms. □

Later we shall work with statenent (3), since it simplifies the calculations.
Now fix the local ring R of a reduced plane curve singularity. If M is an R-module as above given by a short system of generators E of length n we consider $I(M) := I(E)$ as an element of $(\mathbb{N}^2)^n$.
Choose an ordered tuple

$$I = ((i_1,j_1),\dots,(i_n,j_n)) \in (\mathbb{N}^2)^n$$

such that $1 \leq i_s \leq r$ and $0 \leq j_s < c_{i_s}$ (for all s, $1 \leq s \leq n$) where

$$J_{(c_1,\dots,c_r)} = c(R).$$

If there exists a module M with $I(M) = I$, then the short system of generators $E = \langle e_1,\dots,e_n\rangle$ is of the form

$$e_1 = (1,\dots,1)$$

$$e_m = (0,\dots,0,\underset{\substack{\uparrow\\ i_m}}{t^{j_m} + \sum_{j_m<s<c_{i_m}} \lambda^m_{i_m,s} t^s},\dots,\sum_{0\leq s<c_r} \lambda^m_{r,s} t^s)$$

for $2 \leq m \leq n$ with $I(e_m) = (i_m,j_m)$, $\lambda^*_{*,*} \in \mathbb{C}$

Hence the set of modules M, such that $I(M) = I$ can be parametrized by $\mathbb{C}^N$, N suitably large.
Two modules $M_{(\lambda)}, M_{(\mu)}$, $(\lambda),(\mu) \in \mathbb{C}^N$, with $I(M_{(\lambda)}) = I(M_{(\mu)})$ are isomorphic if and only if there is a unit

$$\varphi = (\varphi_{10}+\varphi_{11}t+\dots+\varphi_{1c_1-1}t^{c_1-1},\dots,\varphi_{r0}+\dots+\varphi_{rc_r-1}t^{c_r-1}) \in \bar{R}^*$$

such that

$$(*) \qquad \varphi e_s^{(\lambda)} \in M_{(\mu)} \quad \text{for all } 1 \leq s \leq n,$$

where $\langle e_s^{(\lambda)}\rangle_{1\leq s\leq n}$ is a short system of generators of $M_{(\lambda)}$. This is equivalent to the existence of a solution of the linear system of equations

$$A: \quad (0 = \varphi e_s^{(\lambda)} - \Sigma_{j=1}^n b_{j,1}^{(\lambda),(\mu)} e_j^{(\mu)}),\ 1 \leq s \leq n$$

depending on (λ) and (μ), such that the constant terms $\varphi_{10},\dots,\varphi_{r0}$ of φ are not zero. Hence we have to determine the kernel of a linear mapping $A(\lambda,\mu)$.

<u>Remark:</u> An immediate consequence is that the space of isomorphism classes of torsion free rank 1 R-modules is constructible.

<u>3.5 Example</u>

Let R be the local ring of the singularity E_{12}:

$$R = \mathbb{C}\{t^3, t^7\}\ ,\ c(R) = J_{(12)}.$$

Among all tuples of invariants I that admit a module, consider for example the following:

$$((1,0),(1,4),(1,5)),\ ((1,0),(1,1)) \text{ and } ((1,0),(1,4)).$$

In many cases the invariant determines the representative of its isomorphism class completely. For instance, the tuple $I = ((1,0),(1,4),(1,5))$ leads to the module generated by

$$e_1 = 1,\ e_2 = t^4,\ e_3 = t^5.$$

Now we consider the second tuple: $I = ((1,0),(1,1))$. Simple calculations give $c(M) = J_{(6)}$. Hence we can calculate modulo $(t^6)R$.

Thus $M^{(\lambda)}$ is generated by

$$e_1 = 1 \quad \text{and} \quad e_2 = t + \lambda_2 t^2 + \lambda_3 t^3 + \lambda_4 t^4 + \lambda_5 t^5.$$

The system of linear equations arising from (*) and given by

$$\varphi_0 + \varphi_1 t + \varphi_2 t^2 + \varphi_3 t^3 + \varphi_4 t^4 + \varphi_5 t^5 =$$
$$= \alpha\, 1 + \beta\, t^3 + \gamma\, (t + \lambda_2 t^2 + \lambda_3 t^3 + \lambda_4 t^4 + \lambda_5 t^5) + \delta\, (t^4 + \lambda_2 t^5)$$
$$\varphi_0 t + (\varphi_1 + \varphi_0\mu_2)t^2 + (\varphi_2 + \varphi_1\mu_2 + \varphi_0\mu_3)t^3 + (\varphi_3 + \varphi_2\mu_1 + \varphi_1\mu_2 + \varphi_0\mu_3)t^4$$
$$+ (\varphi_4 + \varphi_3\mu_2 + \varphi_2\mu_3 + \varphi_1\mu_4 + \varphi_0\mu_5)t^5 =$$
$$= \bar{\alpha}\, 1 + \bar{\beta}\, t^3 + \bar{\gamma}\, (t + \lambda_2 t^2 + \lambda_3 t^3 + \lambda_4 t^4 + \lambda_5 t^5) + \bar{\delta}\, (t^4 + \lambda_2 t^5)$$

is solvable independently from λ and μ such that $\varphi_0 \neq 0$. Hence there is only one isomorphism class. It is represented by $e_1 = 1,\ e_2 = t$.

Consider the third invariant $I = ((1,0),(1,4))$. There are two possible values for the conductor, choose $c(M) = J_{(6)}$. Then the module can be represented by generators

$$e_1 = 1,\ e_2 = t^4 + \lambda\, t^5,\ \lambda \in \mathbb{C}-\{0\}$$

($\lambda = 0$ gives the second case: $c(M) = J_{(9)}$).

The matrix of the linear system of equations is

$$A = \begin{bmatrix}
1 & & & & & & -1 & & & & & \\
 & 1 & & & & & & & & & & \\
 & & 1 & & & & & & & & & \\
 & & & 1 & & & & -1 & & & & \\
 & & & & 1 & & & -a & -1 & & & \\
 & & & & & 1 & & & -\lambda & & & \\
 & & & & & & & & & -1 & & \\
 & & & & & & & & & & -1 & \\
1 & & & & & & & & & & & -1 \\
\mu & 1 & & & & & & & & & & -\lambda
\end{bmatrix}$$

Transformation leads to

$$A' = \begin{bmatrix} 1 & & & -1 \\ & \ddots & & \vdots \\ & & 1 & \vdots \\ 0 & & 0 & (\lambda-\mu) \end{bmatrix} \Longrightarrow \lambda = \mu, \text{ since } \varphi_0 \text{ must not vanish,}$$

and there is a 1-parameter family of isomorphism classes

$e_1 = 1$, $e_2 = t^4 + \lambda t^5$, $\lambda \in \mathbb{C}-\{0\}$.

Finally, over the singularity E_{12} there are the following modules given by the following generators over R:

$\langle 1\rangle$, $\langle 1, t\rangle$, $\langle 1, t^2\rangle$, $\langle 1, t^4+\lambda t^5\rangle$, $\langle 1, t^5\rangle$, $\langle 1, t^8\rangle$, $\langle 1, t^{11}\rangle$,

$\langle 1, t, t^2\rangle$, $\langle 1, t, t^5\rangle$, $\langle 1, t^2, t^4\rangle$, $\langle 1, t^4, t^5\rangle$, $\langle 1, t^4, t^8\rangle$.

There is exactly one 1-parameter-family, namely $\langle 1, t^4 + \lambda t^5\rangle R$, $\lambda \in \mathbb{C}$.

The calculations for the other strictly unimodular singularities given in theorem 1 are similar, but rather tedious and lenghty.

References

[Arn] **V.I. ARNOL'D**: Critical Points of Smooth Functions and their Normal Forms, RMS 30:5 (1975), 1-75

[B-G-S] **R.-O. BUCHWEITZ, G.-M. GREUEL, F.-O. SCHREYER**: Cohen-Macaulay-Modules on Hypersurface Singularities, Part II; Invent. math. 88, pp 165 - 182, 1987

[D] **E. DIETERICH**: Classification of the Indecomposable Representations of the Cyclic Group of Order Three in a Complete Discrete Valuation Ring of Ramification Degree Four. Preprint Universität Bielefeld, 1985.

[G-K] **G.-M. GREUEL, H. KNÖRRER**: Einfache Kurvensingularitäten und torsionsfreie Moduln; Math. Ann. 270 (1985), 417-425

[Ka] **C. KAHN**: Reflexive Moduln auf einfach elliptischen Flächensingularitäten; Preprint Max Planck Institut für Mathematik, Bonn 1987

[Kn1] **H. KNÖRRER**: Cohen-Macaulay-Modules on Hypersurface Singularities, Part I; Invent. math. 88, pp. 153 - 164, 1987

[Kn2] **H. KNÖRRER**: Torsionsfreie Moduln bei Deformation von Kurvensingularitäten; these Proceedings

[S] **F.-O. SCHREYER**: Finite and Countable CM-Representation Type; these Proceedings

[Sch] **A. SCHAPPERT**: Die Klassifikation aller torsionsfreier Moduln vom Rang 1 über den unimodularen Kurvensingularitäten; Diplomarbeit, Kaiserslautern 1985

[W] **C.T.C. WALL**: Classification of Unimodal Isolated Singularities of Complete Intersections; Proceedings of Symp. in Pure Mathematics, 40 (1983), p. 625-640

POLAR CURVES, RESOLUTION OF SINGULARITIES AND THE FILTERED MIXED HODGE STRUCTURE ON THE VANISHING COHOMOLOGY

Joseph Steenbrink*
Mathematical Institute
University of Leiden
Postbus 9512
2300 RA Leiden, The Netherlands.

and

Steven Zucker*
Department of Mathematics
Johns Hopkins University
Baltimore, MD 21218
U.S.A.

ABSTRACT. LÊ D.T., F. MICHEL and C. WEBER have obtained results which connect the polar invariants of a plane curve singularity with the Waldhausen decomposition associated to its Milnor fibration. In this paper we give proofs of most of these results and investigate the relation between the polar filtration of the Milnor fibre and the mixed Hodge structure on its cohomology.

INTRODUCTION

Let $f: (\mathbb{C}^2,0) \to (\mathbb{C},0)$ be a holomorphic function germ, $X = f^{-1}(0)$. Let $\Phi = (\ell,f): (\mathbb{C}^2,0) \to (\mathbb{C}^2,0)$ where ℓ is a general linear form. A <u>polar quotient</u> of f is the first Puiseux exponent of some component of the discriminant of Φ. The polar quotients of f can be obtained from a good resolution of f, and they give rise to a filtration $F_1 \subset F_2 \subset \dots \subset F_g$ of the Milnor fibre of f, the <u>polar filtration.</u> These properties have been studied by LÊ, MICHEL and WEBER [6], and summarized by LÊ in [4] and [5]. We refer to these papers for a discussion of the history of the subject.

In this paper, we describe the Hodge-theoretic properties of the polar filtration.

*Supported by the SLOAN foundation.

The groups $H^i(F_j)$, $H^i(F_j, F_{j-1})$ carry natural mixed Hodge structures; those on $H^i(F_j, F_{j-1})$ are of a particularly simple kind (see our Theorem 2 in §3).
In order to prove our result, we felt obliged to reprove most of the results of [6]. This is done in § 2: we claim no priority there, though our proof is substantially different and gives a slightly more general result. The generalization to functions of more than two variables, predicted by LÊ, is treated in §4.

We thank LÊ for suggesting this problem to us; we also thank the Institute of Advanced Study in Princeton for its hospitality during January 1986, where most of this work was done.

§1. RESOLUTION OF PLANE CURVE SINGULARITIES.

Let $f: (\mathbb{C}^2, 0) \to (\mathbb{C}, 0)$ be a germ of an analytic function, $X = f^{-1}(0)$. We assume that $0 \in X$ is a singular point of the reduced variety X^{red}.

Let $\pi: Z \to \mathbb{C}^2$ give the minimal good embedded resolution of X, in which $0 \in \mathbb{C}^2$ is replaced by a connected union of smooth rational curves, such that for $\tilde{f} = f\pi$, $\tilde{f}^{-1}(0)$ is a divisor on Z with normal crossings. It is constructed by successively blowing up points, creating a tower of modifications

$$(1) \qquad \begin{array}{ccc} Z_{j+1} & \xrightarrow{\pi_{j+1}} & \mathbb{C}^2 \\ \downarrow{\scriptstyle \varepsilon_j} & \nearrow{\scriptstyle \pi_j} & \\ Z_j & & \end{array} \qquad (j \geq 1)$$

If we write $\tilde{f}_j = f\pi_j$, a point of $\tilde{f}_j^{-1}(0) \subset Z_j$ is blown up to produce Z_{j+1} only if $\tilde{f}_j^{-1}(0)$ fails to have normal crossings at that point. Thus, the process can be described by saying that in Z_{j+1}, there is one additional exceptional curve E_{j+1} beyond what one has in Z_j. In case f has only one distinct irreducible factor (up to units, of course), Z_{j+1} is always obtained by blowing up a point of E_j. The process terminates after a finite number of steps, with Z being the final Z_j.

The graph of the resolution -the following description makes sense for any sequence of blow-ups -consists of a vertex for each exceptional curve, and an edge between two vertices if the corresponding curves intersect. The graph is a tree, with one distinguished vertex α^* corresponding to the first exceptional curve, E_1.As such, the set V of vertices forms a partially ordered set, with order defined by "distance" to the distinguished vertex: $\beta < \alpha$ if and only if the chain from β to α^* passes through α. In the case where f has one irreducible factor, the graph can be layed down as a "mobile" ([1] p. 698), e.g.

(2)

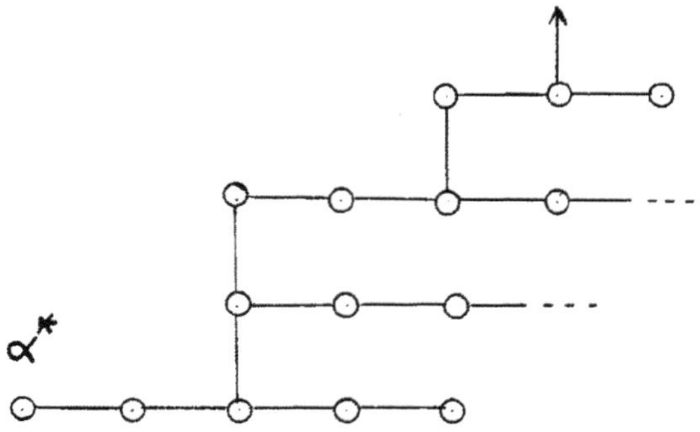

We have started above employing the notation that an arrow denotes the presence of a component of the proper transform of X.

The picture (2) arises as follows. Recall that blowing up a point is the process that distinguishes the tangent directions. One begins by blowing up points until the proper transform $\tilde{X}$ of X becomes tangent to an exceptional component. This produces, in decreasing order, a chain in the graph. At the next stage, the proper transform passes through the intersection of two exceptional components. The succeeding blow-up then inserts in the chain a component between these two. This process repeats until the proper transform no longer passes through a crossing point. At the next stage, one moves upward in the mobile.

REMARK 1. In the minimal resolution one cannot have

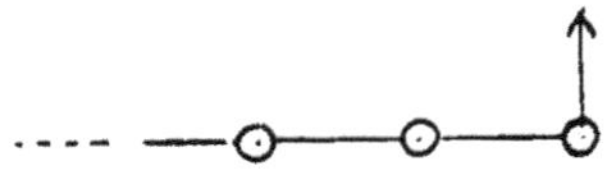

at the top of (2), for then the last step in the resolution process would have been superfluous.

In the reducible case, one must perform the blowings up while watching all of the irreducible components. This gives in the initial chain several arrows in the picture, some perhaps meeting the same component in the same or different points, e.g.

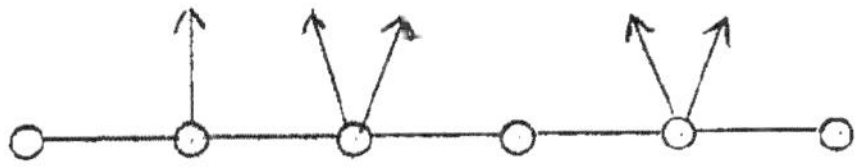

One proceeds as before, going "up" from each of the points of intersection with the proper transform.

REMARK 2. In the minimal resolution, for any point $\alpha \in V$ which is connected to at least 3 other points, their exists $\beta < \alpha$ such that $\tilde{X}$ intersects E_β. This is easily deduced from the construction described above.

§2. THE RELATIVE POLAR CURVES OF f.

Let

$$(3) \qquad \prod_j f_j^{\nu_j}$$

be the factorization of f. Note that any directional derivative of f is divisible by $f_j^{\nu_j - 1}$ for each j, and by no higher power of f_j. For generic directions u, the topological type of the curve defined by the vanishing of

$$(4) \qquad D_u f / \prod_j f_j^{\nu_j - 1}$$

is constant: it is called a <u>relative polar curve</u> associated to f.

Let ℓ be a linear form on $\mathbb{C}^2$. One defines the mapping

$$(5) \qquad \Phi: (\mathbb{C}^2, 0) \to (\mathbb{C}^2, 0)$$

by $\Phi = (\ell, f)$. For generic choice of ℓ , the union of the components of the critical locus of Φ , other than those of X , is a polar curve Γ. We make the elementary, but useful, observation:

PROPOSITION 1: <u>Outside</u> $f^{-1}(0)$, <u>the polar curve is the set of points where the restriction of</u> f <u>to the level curve of</u> ℓ <u>has a critical point. Moreover,</u> $\Gamma \cap f^{-1}(0) = \{(0,0)\}$.

One puts $\Delta = \Phi(\Gamma)$. If (z,w) denote the coordinates on $\mathbb{C}^2$, then each component Δ_i of Δ is tangent to the z-axis ([11], Prop. (1.2)). As such, it has a Puiseux series

$$z = a_i w^{r_i} + \text{ higher order terms,}$$

with $r_i < 1$. If Γ_i is a component of Γ mapping onto Δ_i, we can see that

$$r_i = \Gamma_i \cdot L / \Gamma_i \cdot X, \tag{6}$$

where L is the line defined by $\ell = 0$, and $A \cdot B$ denotes intersection multiplicity of A and B. One calls (6) a <u>polar ratio</u> of f.

Consider next an arbitrary composite of blow-ups, $\pi: Z \to \mathbb{C}^2$, and let $\tilde{X}, \tilde{L}$ and $\tilde{\Gamma}_i$ denote the proper transforms of X, L and Γ_i respectively. Let $\tilde{\ell} = \ell \circ \pi$. We have on Z:

$$(7) \qquad \begin{aligned} (\tilde{\ell}) &= \tilde{L} + \sum m_\alpha E_\alpha \\ (\tilde{f}) &= \tilde{X} + \sum d_\alpha E_\alpha \end{aligned}$$

PROPOSITION 2: <u>Assume that</u>

i) $\tilde{\Gamma}_i \cap \tilde{X} = \emptyset$

ii) $\tilde{\Gamma}_i$ <u>meets only one exceptional component</u> $E_{\alpha(i)}$.

<u>Then</u> $r_i = m_{\alpha(i)}/d_{\alpha(i)}$.

PROOF. We have by the projection formula (see [3]. Prop. 2 on p. 182)

$$\Gamma_i \cdot L = \tilde{\Gamma}_i \cdot (\tilde{L} + \sum_\alpha m_\alpha E_\alpha) = m_{\alpha(i)} \tilde{\Gamma}_i \cdot E_{\alpha(i)}$$

(and Γ_i and L have different tangents at 0 in case $\alpha(i) = \alpha^*$). Likewise,

$$\Gamma_i \cdot X = d_{\alpha(i)}\ \tilde{\Gamma}_i \cdot E_{\alpha(i)}$$

and the desired result follows.

We are now almost ready to state a result from [4]. To the graph of the minimal resolution, we add an arrow connected to α^*, representing the presence of $\tilde{L}$. One then makes the following:

DEFINITION: <u>A rupture point</u> of V is a vertex to which three or more lines (edges or arrows) connect.

THEOREM 1. <u>The collection of numbers r_i coincides with the set of values of m_α/d_α at the rupture points of the graph of the minimal resolution.</u>

There are two main steps in proving the theorem:

i) one studies the properties of the function $r: V \to \mathbb{Q}$ given by $r(\alpha) = m_\alpha/d_\alpha$;

ii) one locates the components $\tilde{\Gamma}_i$ of the proper transform of the polar curve of f.

We fix the following notation. Recall that α^* is the unique maximal element of V. For $\alpha \in V$ we let $c_\alpha = \tilde{X} \cdot E_\alpha$, $b_\alpha = -E_\alpha \cdot E_\alpha$. Note that $b_\alpha > 0$. If $\alpha \neq \alpha^*$, the successor of α is the unique $\gamma \in V$ with $\gamma > \alpha$ and $E_\gamma \cdot E_\alpha = 1$.

PROPOSITION 3. Let $\alpha \in V$, $\alpha \neq \alpha^*$, and let γ be its successor. Then $r(\gamma) \geq r(\alpha)$, and $r(\gamma) = r(\alpha)$ if and only if $c_\alpha = 0$ and $r(\beta) = r(\alpha)$ for all $\beta < \alpha$.

PROOF. As $\alpha \neq \alpha^*$, $E_\alpha \cdot \tilde{L} = 0$. Let $B_\alpha = \{\beta \in V: \beta < \alpha$ and β is connected to $\alpha\}$. From $E_\alpha \cdot (\tilde{l}) = E_\alpha \cdot (\tilde{f}) = 0$, we get

$$(8)\qquad \begin{aligned} 0 &= m_\gamma - m_\alpha b_\alpha + \textstyle\sum_{\beta\in B_\alpha} m_\beta, \\ 0 &= c_\alpha + d_\gamma - d_\alpha b_\alpha + \textstyle\sum_{\beta\in B_\alpha} d_\beta . \end{aligned}$$

Put $r = r(\alpha)$. By (8) the claim is true for minimal α.
We assume by induction

(9) $\quad m_\beta < rd_\beta \quad$ if $\quad \beta \in B_\alpha$.

Inserting this into (8), we get

$$m_\gamma > rd_\alpha b_\alpha - \textstyle\sum_{\beta\in B_\alpha} rd_\beta,$$

$$d_\gamma = d_\alpha b_\alpha - c_\alpha - \textstyle\sum_{\beta\in B_\alpha} d_\beta .$$

Dividing, we obtain

$$(10)\qquad r(\gamma) > \left[\frac{d_\alpha b_\alpha - \sum d_\beta}{d_\alpha b_\alpha - c_\alpha - \sum d_\beta}\right] r.$$

The coëfficient of r in (10) is at least 1, and equals 1 precisely when $c_\alpha = 0$. Also, the inequality in (10) is strict unless all inequalities in (9) are actually equalities.

COROLLARY. In the graph of the minimal resolution, let $I \subset V$ be a maximal connected subset on which r is constant, $\#I > 2$. Then I is a chain leading from a minimal element of V to the first rupture point it meets on its way to α^*.

PROOF. Use Remark 2 of §1.

We next show:

PROPOSITION 4: i) Suppose that a component of $\tilde{X}$ passes through a point of E_α in the minimal resolution. Then no component $\tilde{\Gamma}_i$ of the proper transform of the polar curve passes through that point.
ii) If a polar component $\tilde{\Gamma}_i$ passes through $E_\alpha \cap E_\beta$, then $r(\alpha) = r(\beta)$.

PROOF: Choose local coordinates (s,t) centered at the point in question, so that in these coordinates, $s = 0$ defines E_α, $t = 0$ defines $\tilde{X}$, and

$$\tilde{f} = (s^m t)^\nu .$$

In terms of linear coordinates (x,y) for $\mathbb{C}^2$ with $\ell = y$,

$$x = s^d u_1 \quad \text{and} \quad y = s^d u_2 ,$$

where u_1 and u_2 are units. By the chain rule,

$$\frac{\partial f}{\partial s} = \frac{\partial f}{\partial x}\frac{\partial x}{\partial s} + \frac{\partial f}{\partial y}\frac{\partial y}{\partial s},$$

$$\frac{\partial f}{\partial t} = \frac{\partial f}{\partial x}\frac{\partial x}{\partial t} + \frac{\partial f}{\partial y}\frac{\partial y}{\partial t}.$$

Solving for $\frac{\partial f}{\partial x}$, we obtain

(11) $$\frac{\partial f}{\partial x} = \left(\frac{\partial f}{\partial s}\frac{\partial y}{\partial t} - \frac{\partial f}{\partial t}\frac{\partial y}{\partial s}\right)\Big/\frac{\partial(x,y)}{\partial(s,t)} .$$

Since $\frac{\partial f}{\partial x}$ gets expressed as a power series in s and t, it suffices to check that s and t are the only factors, up to units, of the numerator. One computes the latter to be

$$\nu s^{m\nu+d-1} t^{\nu-1}\left[mt(\partial u_2/\partial t) - (du_2+s(\partial u_2/\partial s))\right].$$

Since u_2 is a unit, so is the expression in brackets; this gives (i).

For (ii), we consider the local situation at the intersection point of $E_\alpha \cap E_\beta$, where in suitable coordinates

$$\tilde{f} = s^{d_\alpha} t^{d_\beta}, \quad x = s^{m_\alpha} t^{m_\beta} u_1, \quad y = s^{m_\alpha} t^{m_\beta} u_2 .$$

This time, one computes the numerator in (11) to be

$$s^{d_\alpha+m_\alpha-1} t^{d_\beta+m_\beta-1}\left[(d_\alpha m_\beta - d_\beta m_\alpha)u_2 + t\frac{\partial u_2}{\partial t} - s\frac{\partial u_2}{\partial s}\right].$$

It contains a non-unit factor other than s and t only if

$$d_\alpha m_\beta - d_\beta m_\alpha = 0,$$

i.e., $\quad r(\alpha) = r(\beta)$.

COROLLARY: <u>If a component</u> $\tilde{\Gamma}_i$ <u>in the minimal resolution passes through</u> E_α, <u>then the associated Puiseux exponent</u> r_i <u>is given by</u> $r(\alpha)$.

PROOF: One may continue to blow up the minimal resolution Z until assumption (ii) of Prop. 2 is satisfied. From Props. 3 and 4, it follows that the new exceptional components E_β will have

$$r(\beta) = r(\alpha).$$

Now apply Prop. 2. (Alternatively, one may mimic the proof of Prop. 2 for the case $\tilde{\Gamma}_i \cap E_\alpha \cap E_\beta \neq 0$.)

The last task is to show that the values $r(\alpha)$ for exceptional curves meeting the polar components are precisely those at the rupture points of the graph of the minimal resolution. To this end, let $I \subset V$ denote a maximal connected set on which r is constant. By the Corollary to Prop 3. we know that I is a chain. Let U_α $(\alpha \in V)$ be a small tubular neighbourhood of E_α in Z as in [2], Theorem 4.1. Let

(12) $$U_I = \bigcup_{\alpha \in I} U_\alpha,$$

a regular neighbourhood of $E_I = \bigcup_{\alpha \in I} E_\alpha$ in Z, and

(13) $$S_I = U_I \cap \{z \in Z : \tilde{\ell}(z) = \varepsilon\},$$

for ε sufficiently close to 0. By Prop. 1, the presence of polar components intersecting E_I will manifest itself in the form of critical points for $\tilde{f}|S_I$. We relate this to topological properties of S_I.

Let α be the maximal element of I and let

(14) $$\tilde{B}_\alpha = B_\alpha \cup \{\gamma\}$$

(notation as in the proof of Prop. 3). Let

(15) $$\begin{aligned} g_I &= \gcd\{m_\beta : \beta \in I \cup \tilde{B}_\alpha\}, \\ g_\beta &= \gcd\{m_\alpha, m_\beta\} \qquad (\beta \in \tilde{B}_\alpha) \end{aligned}$$

LEMMA A: *Assume that* $\alpha^* \notin I$. *Then* S_I *is a genus zero Riemann surface with boundary. It has* g_I *connected components, and*

$$\sum_{\beta\in\tilde{B}_\alpha} g_\beta$$

simple closed curves as boundary. It follows that the Euler characteristic of S_I *is*

$$\chi(S_I) = 2g_I - \sum_{\beta\in\tilde{B}_\alpha} g_\beta. \tag{16}$$

PROOF: S_I is of genus 0, as it is a subset of a level set of a linear form. It has g_I components since $\tilde{\ell}$ has a g_I-th root in U, and no higher. Near $E_\alpha \cap E_\beta$, we can write in local coordinates

$$\tilde{\ell} = s^{m_\alpha} t^{m_\beta}.$$

The boundary of S_I can be taken to be defined by $|s| = \delta, \tilde{\ell}=\varepsilon$, and this set clearly has g_β connected components.

We will use the following case of the Poincaré-Hopf theorem (essentially Morse theory):

PROPOSITION 5: *Let* ϕ *be a smooth real-valued function on* S_I *with only non-degenerate critical points, such that at the boundary,* $\nabla\phi$ *is transverse to* ∂S_I. *Then for* ϕ,
$\#(\text{maxima}) + \#(\text{minima}) - \#(\text{saddle points}) = \chi(S_I)$.

We will take ϕ to be of the form $|h|^2$, where h is holomorphic on S_I. The following is verified by direct calculation:

LEMMA B: i) *The critical points of* $|h|^2$ *are the points where either* h *or* h' *vanish;*

ii) *At all critical points of* $|h|^2$ *where* $h \neq 0$, *the index of* ∇h *is negative,*

iii) *The Hessian of* $|h|^2$ *equals*

$$4\left[|h'|^4 - |h|^2 |h''|^2\right].$$

PROPOSITION 6.: *Let* z *be a global coordinate on* S_I. *Then there exists a neighbourhood* W *of* $0 \in \mathbb{C}$ *such that for generic* $\lambda, \mu \in W$, *the function* $h = \tilde{f} - \lambda z - \mu$ *satisfies:*

i) h <u>has no zeros or critical points on</u> ∂S_I,

ii) h <u>has the same number of zeros in</u> S_I <u>as</u> $\tilde{f}$,

iii) $|h|^2$ <u>has only non-degenerate critical points in</u> S_I.

PROOF: Since the local form for $\tilde{f}$ near $E_\alpha \cap E_\beta$ is

$$\tilde{f} = s^{d_\alpha} t^{d_\beta}, \tag{17}$$

we have also

$$\frac{d\tilde{f}}{\tilde{f}} = d_\alpha \frac{ds}{s} + d_\beta \frac{dt}{t}. \tag{18}$$

Neither (17) nor (18) vanishes on ∂S_I. The same holds for h if λ is sufficiently small, giving (i). By applying the argument principle. we also get (ii). Finally, (iii) is a consequence of Lemma B, since we only need h to have simple zeros, and to have its critical points off the finite set of zeros of the function $\tilde{f}''(z)$.

For $|h|^2$ with h as above, we know by the maximum modulus principle that there are no maxima in S_I, and the minima coincide with the zeros of h. Let c'_α denote the number of components of $\tilde{X}$ that meet E_α (so $c'_\alpha = c_\alpha$ if $\tilde{X}$ is reduced); each such component meets S_I in m_α points. Let σ_I denote the number of saddle points for $|h|^2$. From Prop. 5, we obtain

$$0 < \sigma_I = C_I + \sum_{\beta \in \tilde{B}_\alpha} g_\beta - 2g_I;$$

where

$$C_I = \sum_{\alpha \in I} c'_\alpha m_\alpha.$$

In view of Proposition 1, we can state:

PROPOSITION 7: <u>For</u> $I \neq \{\alpha^*\}$, <u>there are polar components meeting</u> E_I <u>if and only if</u>

$$\sigma_I = C_I + \sum_{\beta \in \tilde{B}_\alpha} g_\beta - 2g_I > 0.$$

Before enumerating the cases that can occur, we state a useful fact:

LEMMA C: Let J be a chain that receives no branches of $\tilde{X}$. Then for all pairs β_1, β_2 of adjacent vertices in J, the value of

$$\gcd(m_{\beta_1}, m_{\beta_2})$$

is constant. In case J contains a minimal element ε of V, the common value is m_ε.

PROOF: Everything can be deduced by induction from the formula

$$b_\beta m_\beta = m_{\beta^+} + m_{\beta^-},$$

whenever $\beta_- < \beta < \beta_+$ is a 3-element subchain of C.

We now discuss the possibilities for the numerical invariants of I (when V comes from the minimal resolution):

CASES. 1. $\# \tilde{B}_\alpha > 2$, or $\# \tilde{B}_\alpha = 2$ and $C_I > 0$. Then α is clearly a rupture point, with $\sigma_I > 0$, as each g_β is always a multiple of g_I.

2. $I = \{\alpha\}$, $\alpha \neq \alpha^*$, $\# \tilde{B}_\alpha = 2$, $C_I = 0$. Such α is a typical non-rupture point. By Lemma C, $g_\beta = g_I$ when $\beta \in \tilde{B}_\alpha$. Thus $\sigma_I = 0$.

3. $\# \tilde{B}_\alpha = 1$, $C_I > 0$, $\alpha \neq \alpha^*$. Then α must be minimal, $I = \{\alpha\}$ and $c'_\alpha > 2$ by Remark 1 of §1. So $\tilde{B}_\alpha = \{\gamma\}$, $g_I = g_\gamma = m_\alpha$ and $\sigma_I = C_I - m_\alpha = (c'_\alpha - 1)\, m_\alpha > 0$, and α is a rupture point.

4. $I = \{\alpha^*\}$. Here, α^* is a rupture point if and only if in the (first) blow-up of the origin in $\mathbb{C}^2$, E_{α^*} has components of the (then) proper transform of X at two or more points, i.e. if X has at least two distinct tangents. (Note that $r(\alpha^*) = 1/m$, where m is the multiplicity of f at 0.) If the latter holds, it follows from the factorization that the polar curve has a tangent different from those of X, hence a polar component intersects E_{α^*}.

REMARK 3. In [5], LÊ also considers the case of function germs on isolated surface singularities. Our proof (and theorem) works with minor modifications for the case $f: (M,0) \to (\mathbb{C},0)$ where M is an isolated surface singularity for which the local level curves of sufficiently general linear forms near 0 have genus 0.

In particular, this is true when $(M,0)$ is a rational singularity with reduced fundamental cycle. In the more general case, where V is just supposed to be a tree, one can generalize Lemma A by computing the genus of the surface S_I. The corresponding statement of Theorem 1 becomes much more complicated, however.

§3. POLAR FILTRATION OF THE MILNOR FIBRE AND MIXED HODGE STRUCTURE.

We keep the notation of the preceding sections. Let $\{r_1,\dots,r_g\}$ be the set of all distinct polar exponents of f at 0, and assume that $r_1 > r_2 > \dots > r_g$. Choose $B_i \in \mathbb{R}$ sufficiently large (see [5], (2.5)) and $0 < \delta \ll \varepsilon \ll 1$, and let

$$F_i = \{z \in \mathbb{C}^2 \mid |z| < \varepsilon,\ |\ell(z)| < B_i\delta^{r_i} \text{ and } f(z) = \delta\}.$$

Then F_g is diffeomorphic to the Milnor fibre of f and the filtration $F_1 \subset \dots \subset F_g$ is called a polar filtration of f. The diffeomorphism class of the g-tuple $(F_1,\dots,F_g)$ is an analytic invariant of the germ of f at 0 (loc. cit. Théorème (2.7)).

An alternative description of the F_i can be given in terms of the minimal resolution of f, using the results of §2. We define an increasing chain of subsets $V(i)$ of V, $i = 1,\dots,g$, by

$$(19) \qquad V(i) = \{\alpha \in V \mid r(\alpha) > r_i\}.$$

The $V(i)$ are connected subgraphs of V. To these, there correspond curves

$$(20) \qquad E(i) = \bigcup_{\alpha \in V(i)} E_\alpha$$

with neighbourhoods

$$(21) \qquad U(i) = \bigcup_{\alpha \in V(i)} U_\alpha$$

in Z (cf.(12)).

PROPOSITION 8. With notation as above, for $\delta > 0$ sufficiently small, the filtration

$$\overline{U(1)} \cap f^{-1}(\delta) \subset \ldots \subset \overline{U(g)} \cap f^{-1}(\delta)$$

<u>is diffeomorphic to a polar filtration of</u> f.

PROOF. For $\alpha \in V$ let $\overset{\circ}{U}_\alpha = U_\alpha \setminus \bigcup_{\beta \neq \alpha} U_\beta$. Let $h_i : Z \setminus f^{-1}(0) \to \mathbb{R}$ be given by $h_i = |\ell| \cdot |f|^{-r_i}$. If $\alpha \in V$ with $r(\alpha) < r_i$, then $\lambda_i(\delta) := \inf\{h_i(z) \mid z \in \overset{\circ}{U}_\alpha \cap f^{-1}(\delta)\}$ tends to infinity as $\delta \to 0$; if $r(\alpha) > r_i$, then $\lambda_i(\delta)$ stays bounded as $\delta \to 0$. We take B_i, δ such that B_i is a regular value of $h_i | f^{-1}(\delta)$, $r(\alpha) < r_i \Rightarrow \lambda_i(\delta) > B_i$ and $r(\alpha) > r_i \Rightarrow \lambda_i(\delta) < B_i$. It follows, that the boundary of $F_i = \{z \in f^{-1}(\delta) \mid h_i(z) < B_i\}$ is contained in $\bigcup\{U_\alpha \cap U_\beta \cap f^{-1}(\delta) \mid r(\alpha) < r_i < r(\beta)\}$. We will construct our diffeomorphism in such a way that it will differ from the identity only on these open sets. In local coordinates (s,t) near $E_\alpha \cap E_\beta$ on Z, we will have

$$U_\alpha : |s| < 1, \; U_\beta : |t| < 1,$$

$$\tilde{f}(s,t) = s^{d_\alpha} t^{d_\beta}, \; \tilde{\ell}(s,t) = u\, s^{m_\alpha} t^{m_\beta}$$

with u a unit. Then

$$h_i(s,t) = |u| \cdot |s|^{-a} |t|^{b} \text{ with } a > 0,\; b > 0.$$

We claim that there exists a diffeomorphism ϕ of $\bar{U}_\alpha \cap \bar{U}_\beta \cap f^{-1}(\delta)$ which maps $\bar{U}_\beta \cap f^{-1}(\delta)$ to $\bar{U}_\alpha \cap \bar{U}_\beta \cap f^{-1}(\delta) \cap h^{-1}([0,B_i])$. To do this, we may take slightly bigger U'_α with $\bar{U}_\alpha \subset U'_\alpha$, for which the above still holds, and work on $\bar{U}'_\alpha \cap \bar{U}'_\beta \cap f^{-1}(\delta)$. Each connected component of this is biholomorphic to an annulus $G_{c,d} = \{z \in \mathbb{C} \mid c < |z| < d\} \subset \mathbb{C}$. The proposition thus follows from the following lemma, whose proof is left to the reader:

LEMMA D. <u>Let</u> $0 < a_1 < a_2 < c < b_2 < b_1 < \infty$. <u>Let</u> $g: G_{a_1,b_1} \to \mathbb{C}$ <u>be a holomorphic function without zeros. Suppose</u> B <u>is a regular value of</u> $|g|$ <u>such that</u> $|g(z)| < B$ <u>on</u> G_{a_1,a_2} <u>and</u> $|g(z)| > B$ <u>on</u> G_{b_2,b_1}. <u>Let</u> $F = \{z \in G_{a_1,b_1} \mid |g(z)| < B\}$.

<u>Then there exists a diffeomorphism</u>

ϕ <u>of</u> G_{a_1,b_1} <u>such that</u> $\phi(z)=z$ <u>for</u> $z \in G_{a_1,a_2} \cup G_{b_2,b_1}$ <u>and</u> $\phi(F) = G_{a_1,c}$. □

Let us consider the cohomology groups of the pieces F_i of a polar filtration of the Milnor fibre. We will show that the groups

$$H^j(F_i),\ H^j(F_i,F_{i-1})$$

carry a natural mixed Hodge structure (MHS) and determine the simple nature of the MHS on $H^1(F_i,F_{i-1})$ for $i = 1,\dots,g$. Our description is a slight variation on the construction in [7] and [8].In view of Prop. 8, we call (20) the <u>polar filtration</u> of E.

Extend the graph V to a graph V^* by adding a vertex β for each component $\tilde{X}_\beta$ of $\tilde{X}$, and for such β let d_β be its multiplicity in $\tilde{f}^{-1}(0)$. Let d be a positive integer divisible by all d_β for $\beta \in V^*$. We consider the normalization $\tilde{Z}$ of the fibre product

$$Z \times_{\mathbb{C}} \tilde{\mathbb{C}}$$

with $\tilde{\mathbb{C}}$ a copy of $\mathbb{C}$ and $\tilde{\mathbb{C}} \to \mathbb{C}$ given by $t \to t^d$. Then $\tilde{Z}$ is a surface with only cyclic quotient singularities; let $p\colon \tilde{Z} \to Z$ be the corresponding d-fold cyclic cover; $\tilde{E}= p^{-1}(E)$ is a reduced divisor with normal crossings on $\tilde{Z}$. It has a graph $\tilde{V}$ with a natural finite-to-one map $\tilde{V} \to V$: for a vertex $\alpha \in V$, its inverse image in $\tilde{V}$ consists of $\gcd\{d_\beta \mid \beta \in V^*$ linked to $\alpha\}$ vertices, and above an edge $\alpha\beta$ there lie $\gcd\{d_\alpha, d_\beta\}$ edges of $\tilde{V}$. The map $\tilde{E}_\alpha = p^{-1}(E_\alpha) \to E_\alpha$ has degree d_α. We lift the partial ordering to $\tilde{V}$ as follows:

$\tilde{\alpha} < \tilde{\beta}$ iff there exists a chain in $\tilde{V}$ from $\tilde{\alpha}$ to $\tilde{\beta}$ mapping to an increasing chain in V.

Define $\tilde{Y} = p^{-1}(\tilde{X})$. We also define $\tilde{r}\colon \tilde{V} \to \mathbb{Q}$ by composing $\tilde{V} \to V$ with r. For $I \subset \tilde{V}$ let $\tilde{E}_I = \bigcup_{\alpha \in I} \tilde{E}_\alpha$ and $I' = \tilde{V} \setminus I$. We define a complex of sheaves, supported on $\tilde{E}_I$, by

$$K^\bullet(I)=\left[\mathcal{O}_{\tilde{E}_I} \xrightarrow{d} \omega_{\tilde{E}_I}(\delta_I)\right]$$

where $\omega_{\tilde{E}_I}$ is the dualizing sheaf of $\tilde{E}_I$ and δ_I is the divisor of points where $\tilde{E}_I$ and $\tilde{E}_{I'} \cup \tilde{Y}$ intersect (compare [8] p. 523). Then

$$\mathbb{H}^*(K^\bullet(\tilde{V}(i))) \cong H^*(F_i,\mathbb{C}).$$

This provides the latter with a MHS. For $i = g$ we recover the usual MHS on the vanishing cohomology. For $J \subset I$ we have

$$\omega_{\tilde{E}_J}(\delta_J) \cong \omega_{\tilde{E}_I}(\delta_I) \otimes \mathcal{O}_{\tilde{E}_J} \cong \omega_{\tilde{Z}}(\tilde{E}+Y) \otimes \mathcal{O}_{\tilde{E}_J}$$

so we obtain a surjective morphism

$$K^\bullet(I) \longrightarrow K^\bullet(J).$$

We let $K^\bullet(I,J)$ denote its kernel, i.e. the complex

$$(22) \qquad \mathcal{O}_{\tilde{E}_{I-J}}(-\tilde{E}_J) \longrightarrow \omega_{\tilde{E}_{I-J}}(\delta_I|\tilde{E}_{I-J}).$$

This becomes a mixed Hodge complex (via a cone construction, compare (27)) and leads to a MHS on the relative groups. In particular

$$\mathbb{H}^*(K^\bullet(\tilde{V}(i),\tilde{V}(i-1)) \cong H^*(F_i,F_{i-1}) .$$

In order to describe $H^*(F_i,F_{i-1})$, we let $C_i = \bigcup\{E_\alpha | \alpha$ a rupture pt of V with $r(\alpha) = r_i\}$. Then $p^{-1}(C_i)$ is a disjoint union of components of $\tilde{E}$. From $p^{-1}(C_i) = C_i'$ omit the set Σ_i' of the intersection points with those $\tilde{E}_\beta$ for which $r(\beta) < r_i$, to obtain a smooth curve $\tilde{C}_i$. Let

$$\Sigma_i = \bigcup\{\tilde{C}_i \cap \tilde{E}_\beta | \tilde{r}(\beta) > r_i\}.$$

THEOREM 2. <u>There is a natural isomorphism of mixed Hodge structures</u>

$$H^*(F_i,F_{i-1}) \xrightarrow{\sim} H^*(\tilde{C}_i,\Sigma_i),$$

transferring the monodromy action on $H^*(F_i,F_{i-1})$ to the action of a covering transformation of $\tilde{C}_i \longrightarrow C_i$ on $H^*(\tilde{C}_i,\Sigma_i)$.

We first prove:

LEMMA E. Let $\alpha\in I\subset\tilde{V}$, $J = I\setminus\{\alpha\}$, $I = \tilde{V}\setminus I$. Then $\mathbb{H}^*(K(I,J)) \cong H^*(\tilde{E}_\alpha\setminus\Sigma',\Sigma)$ as MHS, where $\Sigma' = \tilde{E}_\alpha \cap (\tilde{E}_I \cup \tilde{Y})$ and $\Sigma = \tilde{E}_\alpha \cap \tilde{E}_J$.

PROOF. We have by (22):

$$K^\bullet(I,J) = \left[\mathcal{O}_{\tilde{E}_\alpha}(-\Sigma) \xrightarrow{d} \Omega^1_{\tilde{E}_\alpha}(\log\Sigma')\right]$$

which is an incarnation of the standard mixed Hodge complex for $(\tilde{E}_\alpha\setminus\Sigma',\Sigma)$. □

PROPOSITION 9. $H^*(F_i,F_{i-1}) \cong \mathbb{H}^*(K^\bullet(\tilde{V}(i),\tilde{V}(<i))$ where $\tilde{V}(<i) = \{\alpha\in\tilde{V} \mid \tilde{r}(\alpha)<r_i\}$.

PROOF. From Lemma E we conclude that $\mathbb{H}^*(K(I,J)) = 0$ in case $\tilde{E}_\alpha$ is a rational curve with $\#\Sigma = \#\Sigma' = 1$. By repeated application we get $\mathbb{H}^*(\tilde{V}(<i),\tilde{V}(i-1)) = 0$ because $\tilde{V}(i-1)$ can be obtained from $\tilde{V}(<i)$ by successive deletion of points α such that $\tilde{E}_\alpha$ has that property; $\tilde{V}(<i)\setminus\tilde{V}(i-1)$ consists of linear chains connecting two consecutive rupture points. □

To finish the proof of Theorem 2, we let $M_i = \bigcup\{\tilde{E}_\alpha \mid \alpha\in\tilde{V}$ with $\tilde{r}(\alpha) = r_i$, α not a rupture point$\}$ and consider the morphism

$$\phi\colon C_i \cup M_i \longrightarrow C_i$$

which is the identity on C_i and that maps each connected component of M_i to its intersection point with C_i. Recall from (22) that

$$K^\bullet(\tilde{V}(i),\tilde{V}(<i)) = \left[\mathcal{O}_{C_i^! \cup M_i}(-\Sigma_i) \longrightarrow \omega_{C_i^! \cup M_i}(\Sigma_i^!)\right].$$

To identify the hypercohomology of this complex with that of the complex

$$L^\bullet = \left[\mathcal{O}_{C_i^!}(-\Sigma_i) \longrightarrow \Omega^1_{C_i^!}(\Sigma_i^!)\right]$$

we use the Leray spectral sequence of ϕ. One has by direct verification:

$$\phi_* \mathcal{O}_{C_i' \cup M_i}(-\Sigma_i) = \mathcal{O}_{C_i'}(-\Sigma_i),$$

$$\phi_* \omega_{C_i' \cup M_i}(-\Sigma_i) = \Omega^1_{C_i'}(\Sigma_i')$$

and both $R^1\phi_*$-sheaves are zero. Hence

$$\begin{array}{ccc} \mathbb{H}^*(K^\bullet(\tilde{V}(i),\tilde{V}(<i))) & \cong & \mathbb{H}^*(L^\bullet). \\ \wr\| & & \qquad \searrow\!\wr \\ H^*(F_i,F_{i-1}) & & \qquad\qquad H^*(\tilde{C},\Sigma_i). \end{array}$$

The identification of monodromy with the action of a covering transformation follows from [7] proof of Thm. (2.13)(3) and the fact that the monodromy has finite order on $H^*(F_i,F_{i-1})$, (see [12]).

EXAMPLE. Let $f(x,y) = x^5 + x^2y^2 + y^5$. We have drawn $V,\tilde{V}$ and the values of d,m and r on V below (● denotes a rupture point of V).

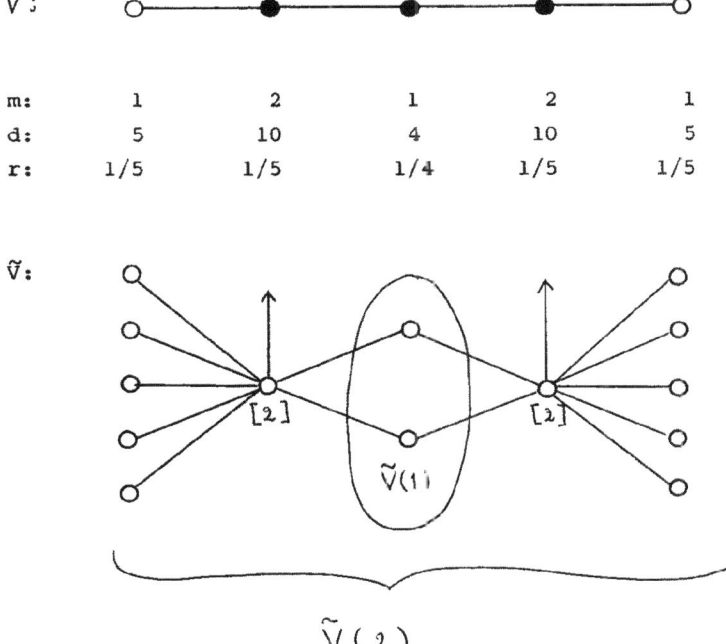

V:

m:	1	2	1	2	1
d:	5	10	4	10	5
r:	1/5	1/5	1/4	1/5	1/5

$\tilde{V}$:

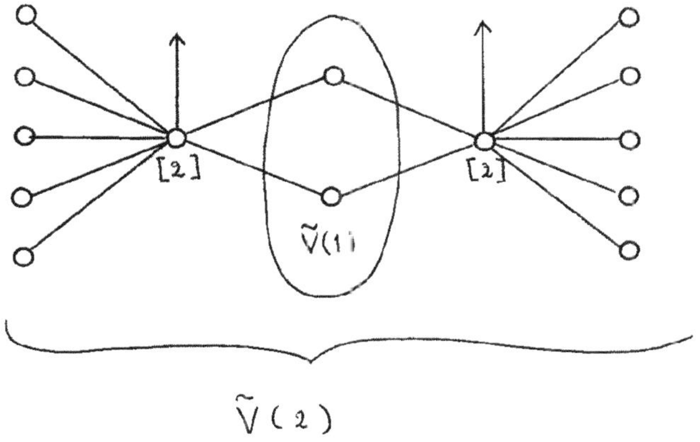

The figure [2] means that the corresponding component of $\tilde{E}$ has genus 2; the other components have genus 0.
The curve $\tilde{C}_1$ is a union of 2 copies of $\mathbb{C}^*$, whereas $\tilde{C}_2$ is 2 punctured curves of genus 2; moreover $\Sigma_1 = \emptyset$ and Σ_2 has 2 points on each component of $\tilde{C}_2$.

§4. THE FILTERED MIXED HODGE STRUCTURE FOR HIGHER DIMENSIONAL SINGULARITIES.

Let D be a divisor with normal crossings on the complex manifold Z. For any sub-divisor E of D, we put for $p,q > 0$

$$(23) \qquad K_Z^{p,q}(D,E) = \Omega_Z^{p+q+1}(\log D)/W(E)_q\Omega_Z^{p+q+1}(\log D),$$

where $W(E)$ denotes the weight filtration with respect to the components of E only. If

$$\theta \in H^0(Z,\Omega_Z^1(\log D))$$

is a closed form, we obtain a double complex, whose associated single complex we denote by $K^\bullet$, by defining

$$d': K^{p,q} \longrightarrow K^{p+1,q}$$

to be induced by exterior differentation, and

$$d'': K^{p,q} \longrightarrow K^{p,q+1}$$

induced from $\omega \to \theta \wedge \omega$. Examples of this construction already occur in the literature.

PROPOSITION 10. Let $g: Z \longrightarrow \Delta$ be a holomorphic mapping, smooth over Δ^*, such that $E = g^{-1}(0)$ is a divisor with normal crossings. Take $D = E$, $\theta = g^*(dt/t)$. If g has unipotent monodromy, then $K^\bullet$ is the complex $A^\bullet$ (see [8] (2,7)) - quasi-isomorphic to $R\Psi_g\,\mathbb{C}$, the sheaf of vanishing cycles along E - that determines the limit mixed Hodge structures of the degeneration when g is proper and Kähler.

REMARK 4. In the above, take instead $D = E + T$, where T is a divisor on Z with relative normal crossings over Δ^* (such that D is also a divisor with normal crossings). Then $K^\bullet$ is the complex $A^\bullet$ of [9] (5.5) that determines the filtered limit mixed Hodge structure of the degeneration $g|_{Z-T}$.

More to the point, we have also:

PROPOSITION 11. _Let_ $f: (Y,0) \longrightarrow (\mathbb{C},0)$ _be a germ of a holomorphic mapping, such that_ Y _and_ f _are smooth except perhaps at_ $\Sigma = \mathrm{Sing}(X)$ _where_ $X = f^{-1}(0)$. _Let_ $p: Z \longrightarrow Y$ _be a resolution of singularities such that for_ $h = fp$, $D = h^{-1}(0)$ _is a reduced divisor with normal crossings. We write_ $D = \tilde{X} + E$, _where_ $\tilde{X}$ _is the proper transform of_ X, _and_ E _is the exceptional divisor. Take_ $\theta = h^*(dt/t)$. _Then_ $K^\bullet$ _is the complex that determines the mixed Hodge structure of the cohomology of the Milnor fiber of_ f.

PROOF. By passing to a suitable representative of the germ, we may assume Y and X to be contractible Stein spaces. The description of $K^\bullet$ here coincides with that of [8] via the equality

$$W(E)_q\Omega_Z^{p+q+1}(\log D) = W_q\Omega_Z^{p+q}(\log E) \wedge \Omega_Z^1(\log \tilde{X}) .$$

The key points in the proof are that $K_Z^\bullet(D,E)$ is supported on E, and moreover,

$$S^\bullet = \ker\{A^\bullet = K_Z^\bullet(D,D) \rightarrow K_Z^\bullet(D,E)\}$$

has trivial hypercohomology. The terms of $S^\bullet$ are

$$\begin{aligned} S^{p,q} &= (W(E)_q/W_q)\Omega_Z^{p+q+1}(\log D) \\ &= \mathrm{Gr}_1^{W(X)}\mathrm{Gr}_q^{W(E)}\Omega_Z^{p+q+1}(\log D) \\ &= \mathrm{Gr}_q^{W(E)}\Omega_{\tilde{X}}^{p+q}(\log E\cap\tilde{X}) , \end{aligned}$$

hence

$$S^\bullet[-1] = \mathrm{Cone}\ (\mathbb{C}_{\tilde{X}} \longrightarrow \mathbb{C}_{E\cap\tilde{X}}) ,$$

which gives as cohomology

$$H^\bullet(\tilde{X}, E\cap\tilde{X}) = H^\bullet(X,\Sigma) = 0$$

because $\Sigma(E\cap\tilde{X})$ is a deformation retract of X (resp. $\tilde{X}$) . □

A reasoning as in the preceding proof has been used by V. Navarro Aznar to put a mixed Hodge structure on the vanishing cohomology of any holomorphic map germ (as opposed to the case of a smoothing above).

Let $\{E(i)\}$ be, for the moment, any increasing filtration of E, in the situation of Prop. 11. Then $\{K^\bullet_Z(D,E(i))\}$ defines an increasing
cofiltration of $K^\bullet_Z(D,E)$, i.e. a sequence of quotients. One has a canonical morphism

$$(24) \qquad K^\bullet_Z(D,E(i)) \longrightarrow (R\Gamma_h\ \mathbb{C})_{E(i)-E'(i)}$$

where $E'(i)$ is the divisor complementary to $E(i)$, and (24) induces an isomorphism on hypercohomology.

Equally interesting is the corresponding decreasing filtration S, defined by

$$(25) \qquad S^iK^\bullet = \ker\{K^\bullet_Z(D,E) \longrightarrow K^\bullet_Z(D,E(i-1))\} .$$

Note that

$$(26) \qquad Gr^i_S K^{p,q} = W(E(i-1))_q/W(E(i))_q)\Omega^{p+q+1}_Z(\log D) .$$

Because of the way in which the filtrations are defined, one sees that the cohomological mixed Hodge complex $K^\bullet_Z(D.E)$ is filtered by S (in the technical sense; see [9] §6). It follows that the filtration induced by S on $\mathbb{H}^\bullet(K^\bullet)$ is a filtration by mixed Hodge substructures, and moreover that the isomorphism

$$Gr^i_S\ \mathbb{H}^k(K^\bullet) \longrightarrow {}_SE_2^{i,k-i}(R\Gamma K^\bullet)$$

is one of mixed Hodge structures (see [10] § 3).

Assume that $\dim Z = 2$. In this case, the complex $Gr^i_S K^\bullet$ is just

$$(27)\qquad \begin{array}{ccc} (W(E(i-1))_1/W(E(i))_1)\Omega^2_Z(\log D) & & \\ \uparrow \wedge h^*(dt/t) & & \\ (W(E(i-1))_0/W(E(i))_0)\Omega^1_Z(\log D) & \xrightarrow{d} & (W(E(i-1))_0/W(E(i))_0)\Omega^2_Z(\log D) \end{array}$$

By taking residues along $E(i)$, we see that (27) is isomorphic to

$$(28)\qquad \begin{array}{c} \mathcal{O}_P \\ \uparrow r \\ \bigoplus_C \mathcal{O}_C \to \omega_{E(i,i-1)}(\log E'(i)), \end{array}$$

where C runs over the irreducible components of $E(i,i-1) = E(i) \cap E'(i-1)$, P is the set of double points of $E(i)$ that are not internal to $E(i-1)$, and r is, up to signs, the restriction mapping. By replacing the left hand column of (28) by the kernel of r, we recover (22).

As described in [7] (3.11), nilpotent monodromy logarithms N are induced by the nilpotent endomorphism ν of $K^\bullet$, defined, again up to signs, by the canonical surjections (recall (23))

$$\nu: K^{p,q} \to K^{p-1,q+1}.$$

We look at this for $G^\bullet = G(i)^\bullet := Gr^i_S K^\bullet$. There is a long exact sequence

$$0 \longrightarrow \mathbb{H}^0(G^\bullet) \longrightarrow H^0(G^0) \longrightarrow H^0(G^1) \longrightarrow \mathbb{H}^1(G^\bullet) \longrightarrow H^1(G^0) \longrightarrow \cdots$$

which in our case gives

$$\mathbb{C}^{E(i,i-1)} \xrightarrow{(0,r)} H^0(\omega\ldots) \oplus \mathbb{C}^P \longrightarrow \mathbb{H}^1(G^\bullet) \longrightarrow 0.$$

(with ν acting on $H^0(\omega\ldots) \oplus \mathbb{C}^P$ and N acting on $\mathbb{H}^1(G^\bullet)$)

From this, we see:

PROPOSITION 12. <u>The following conditions are equivalent:</u>

(a) $N = 0$ <u>on</u> $\mathbb{H}^1(G^\bullet(i))$;

(b) $\mathrm{Im}(\nu) \subset \mathrm{Im}(r)$;

(c) r <u>is surjective</u> ;

(d) <u>the connected components of</u> $E(i,i-1)$ <u>are trees, each meeting</u> $E(i-1)$ <u>in at most one point.</u>

PROOF. The implications (c) $\Rightarrow$ (b) $\Leftrightarrow$ (a) are trivial.
(c) $\Leftrightarrow$ (d): Let U be a connected component of E(i,i-1). Consider the map

$$r_U : \bigoplus_C H^0(\mathcal{O}_C) \longrightarrow \mathbb{C}^{P(U)}$$

induced by r, where C now runs over the irreducible components of U and $P(U) = P \cap U$. Let a_0 be the number of C's and $a_1 = \# P(U)$. Let $D(U) = E(i-1) \cap U \subset P(U)$ and $a_1' = \# D(U)$. Then

$$\operatorname{Ker} r_U = H^0(O_U(-D(U))) = \begin{cases} 0 & \text{if } D(U) \neq \emptyset, \\ \mathbb{C} & \text{if } D(U) = \emptyset. \end{cases}$$

So if $D(U) = \emptyset$, then r_U is surjective $\Leftrightarrow a_1 = a_0 - 1 \Leftrightarrow$ the graph of U is a tree. If $D(U) \neq \emptyset$, then r_U is surjective $\Leftrightarrow a_1 = a_0 \Leftrightarrow$ the graph of U is a tree and $a_1' = 1$ (recall that U is connected so $a_1 - a_1' \geq a_0$).
(b) $\Rightarrow$ (d): If the graph of U is not a tree or $a_1' \geq 2$, it is easy to find a global section of $\omega_U(D(U))$ whose image under ν does not lie in Im(r).

In the case of a hypersurface singularity $f: (\mathbb{C}^{n+1},0) \to (\mathbb{C},0)$ we take a suitable $d \in \mathbb{N}$ such that all eigenvalues of the monodromy of f are d-th roots of unity and let $Y = \mathbb{C}^{n+1} \times_{\mathbb{C}} \tilde{\mathbb{C}}$ where $\tilde{\mathbb{C}} \to \mathbb{C}$ is given by $t \longmapsto t^d$. We slightly change notation and obtain the following diagram ("semi-stable reduction"):

$$\begin{array}{ccc} \tilde{E} \subset \tilde{Z} & \longrightarrow & Z \supset E \\ \downarrow & & \downarrow \\ Y & \longrightarrow & \mathbb{C}^{n+1}. \end{array}$$

In case n = 1 we have from § 3 the polar filtration of E, and hence also of $\tilde{E}$ on the semistable model $\tilde{Z}$. We have

$$\mathbb{H}^*(\mathrm{Gr}^i K^\bullet_{\tilde{Z}}(\tilde{D},\tilde{E})) \cong H^*(F_i, F_{i-1}) .$$

Since the conditions of Prop. 12 are satisfied (as remarked in the proof of Prop. 10), we recover a result from [6] (see [4] (3.4)):

COROLLARY. <u>The monodromy of</u> $H^*(F_i, F_{i-1})$ <u>is of finite order.</u>

Finally, we turn to the case of arbitrary n. We define a complex $C^\bullet$ by

(29) $$C^\bullet = \text{Cone}\ (\ K^\bullet_{\tilde{Z}}(\tilde{D},\tilde{E}) \longrightarrow K^\bullet_{\tilde{L}}(\tilde{D}\cap\tilde{L},\tilde{E}\cap\tilde{L}))\ ,$$

where L denotes a general hyperplane through 0, $\tilde{L}$ its proper transform in $\tilde{Z}$. Then Prop. 11 gives (when $n > 1$)

(30) $$\mathbb{H}^*(C^\bullet) \simeq H^*(F, F\cap L;\ \mathbb{C})\ ,$$

where F denotes the Milnor fiber of f; and $F\cap L$ is the Milnor fiber of $f_{|L}$. By [11] (3.6), one knows that (30) vanishes except possibly in degree n.

The notion of a polar curve is defined exactly as in § 2, with

$$\Phi\ :\ (\mathbb{C}^{n+1},0) \quad\to (\mathbb{C}^2,0)$$

in (5) (see [5] (1.1)). Moreover, (6) and Prop. 2 remain valid. Thus, the polar filtration (20) of E is still defined, and this in turn defines a filtration S of (29) (cf. (25)). Then:

THEOREM 3.

(i) <u>The complex</u> $C^\bullet$, <u>together with its additional filtration</u> S, <u>determines a filtered mixed Hodge structure on</u> $H^n(F, F\cap L)$. <u>The filtration induced here by</u> S <u>is the polar filtration.</u>

(ii) <u>The sequence</u> <u>(use that</u> $H^n(F\cap L) = 0$)

$$0 \longrightarrow H^{n-1}(F) \longrightarrow H^{n-1}(F\cap L) \longrightarrow H^n(F, F\cap L) \longrightarrow H^n(F) \longrightarrow 0$$

<u>is an exact sequence of filtered mixed Hodge structures.</u>

REMARK 6. It follows from (i) that the polar filtration of $H^n(F, F\cap L)$ can be constructed from <u>any</u> resolution of singularities.

CONJECTURE. The mixed Hodge structure on $Gr^i_S H^n(F, F\cap L)$ can be described in an absolute, geometric manner (as opposed to being given in terms of limit mixed Hodge structures; cf. Thm. 2).

One knows that the monodromy of $H^n(F_i,F_{i-1})$ is again of finite order [13]. This should manifest itself in terms of combinatorial data drawn from the resolution (e.g., Prop. 12). It was for this reason, that Lê asked for a mixed Hodge theoretic reformulation.

References

[1] E. BRIESKORN, H. KNÖRRER: Ebene algebraische Kurven. Birkhäuser, Basel etc. 1981.

[2] C.H. CLEMENS, jr.: Picard-Lefschetz theorem for families of nonsingular algebraic varieties acquiring ordinary singularities. Transactions A.M.S. 136, 93-108(1969).

[3] W. FULTON: Algebraic curves. Benjamin, Reading, Mass. 1969.

[4] LÊ D.T.: Courbes polaires et résolution des courbes planes. Paris, Ecole Polytechnique, Feb. 1985.

[5] LÊ D.T.: Exposants polaires et résolution des surfaces. Communication au 3e congrès des mathématiciens Vietnamiens. Paris, Ecole Polytechnique Aug. 1985.

[6] LÊ D.T., F. MICHEL, C. WEBER: Topologie des courbes planes et courbes polaires. In preparation.

[7] J.H.M. STEENBRINK: Mixed Hodge structure on the vanishing cohomology. In: Real and complex singularities, Oslo 1976. Sijthoff & Noordhoff, Alphen a/d Rijn 1977, 525-563.

[8] J.H.M. STEENBRINK: Mixed Hodge structures associated with isolated singularities. Proc. Symp. Pure Math. A.M.S. vol 40(1983) Part 2, 513-536.

[9] J.H.M. STEENBRINK, S. ZUCKER: Variation of mixed Hodge structure, I. Inventiones Math. 80, 489-542 (1985).

[10] S. ZUCKER: Degeneration of mixed Hodge structures. To appear in Proc. A.M.S. Summer Institute in Algebraic Geometry (Bowdoin 1985).

[11] LÊ D.T.: Calcul du nombre de cycles évanouissants d'une hypersurface complexe. Ann.Inst.Fourier Grenoble 23 (4) (1973), 261-270.

[12] LÊ D.T.: Some remarks on relative monodromy. In: Real and Complex Singularities, Oslo 1976. Sijthoff-Noordhoff, Alphen a/d Rijn 1977, 397-403.

[13] LÊ D.T.: The geometry of the monodromy theorem. In: C.P. Ramanujam, a tribute, ed. K.G. Ramanathan. Tata Institute Studies in Math. 8, Berlin etc.: Springer 1978.

ON THE BETTI NUMBERS OF THE MILNOR FIBRE OF A CERTAIN CLASS OF HYPERSURFACE SINGULARITIES

D. van Straten
Mathematisch Instituut
Rijksuniversiteit Leiden
Niels Bohrweg 1
2333 AC Leiden, The Netherlands.

Introduction. For an isolated hypersurface singularity $f: (\mathbb{C}^{n+1},0) \to (\mathbb{C},0)$ the following celebrated formula is valid (see [Mi], p.59):

$$\mu = \dim_{\mathbb{C}} \mathbb{C}\{x_0,\ldots,x_n\}/(\partial_0 f,\ldots,\partial_n f) \ .$$

It relates the topological invariant μ, the Milnor number to a readily computable algebraic invariant.

For a general hypersurface singularity it is improbable that there exist formulae of comparable simplicity for all Betti numbers of the Milnor fibre. However, for a more restrictive class of functions with non isolated singularities this seems to be possible. Siersma [S] studied hypersurfaces with one dimensional complete intersection singular locus along which f has (away from 0) transversally an A_1-singularity, from a topological point of view. In this paper we show that for this class of singularities the relative de Rham cohomology is torsionfree. This fact implies that for these singularities there are simple algebraic formulae for the Betti numbers of the Milnor fibre.

The proof goes as follows. In §1, we prove the coherence of the relative de Rham cohomology for so-called "concentrated singularities". In §2, we consider the spectral sequence for the Gauss-Manin system coming from the "Hodge filtration". When this spectral sequence degenerates at the E_2-level, one gets torsion freeness of the relative de Rham cohomology in the same way as Malgrange's proof of the corresponding result for isolated hypersurface singularities. In §3, finally we check by explicit calculation the degeneration of the spectral sequence for our special class of functions, using a result of Pellikaan [Pe].

§1. Coherence of Relative de Rham cohomology

In the case that $f: (\mathbb{C}^{n+1},0) \to (\mathbb{C},0)$ defines an isolated singularity, Brieskorn [B], by using a projective compactification and Grauert's direct image theorem, proves that the relative hypercohomology groups $\mathbb{R}^i f_*(\Omega^{\cdot}_{X/S}) \simeq H^i(f_*\Omega^{\cdot}_{X/S})$ are coherent $\mathcal{O}_S$-modules. Here $X \xrightarrow{f} S$ is a Milnor representative of f ; i.e. $X = B_\varepsilon \cap f^{-1}(D_\eta)$ $0 < \eta << \varepsilon$ etc.

In [B-G] Buchweitz & Greuel prove a general coherence theorem for certain complexes $K^{\cdot}$ on an analytic space X with a flat map to a curve S, but still with the condition that the fibres have isolated singularities. They use a result of Kiehl and Verdier (see for example [D]).

In [H] Hamm proves the coherence of $\mathbb{R}^i f_*(\Omega^{\cdot}_{X/S})$/torsion in a quite general setting.

Here we give a coherence theorem general enough to be applied in §2 and §3. In absence of an appropriate reference, we include a proof, which is based on [B] and [B-G]. We consider map germs $(X,x) \xrightarrow{f} (S,s)$ with X an an analytic space and S a smooth curve.

<u>Definition 1.</u> A *standard representative* of the map germ $(X,x) \xrightarrow{f} (S,s)$ is a representative $X \xrightarrow{f} S$ of the form

$$X = X_{\varepsilon,\eta} := (B_\varepsilon \cap Y) \cap f^{-1}(D_\eta)$$

$$S = S_\eta := D_\eta$$

with B_ε an open ε-ball in $\mathbb{C}^N$ and D_η an open η-disc in $\mathbb{C}$. For the intersection we use a fixed embedded representative $Y \subset \mathbb{C}^N$, $T \subset \mathbb{C}$ for the germ $(X,x) \xrightarrow{f} (S,s)$.

We put $\partial X := \partial B_\varepsilon \cap Y \cap f^{-1}(D_\eta)$ and $\bar{X} = X \cup \partial X$ (relative boundary and relative closure).

Note that for ε, η small enough $X_{\varepsilon,\eta}$ will be a contractible Stein space.

<u>Definition 2.</u> Let $X \xrightarrow{f} S$ be a standard representative for a germ $(X,x) \to (S,s)$ and $\mathbb{L}$ a sheaf of $\mathbb{C}$-vectorspaces on X.
$\mathbb{L}$ is called *transversally constant* (with respect to U and θ) if there exists an open neighbourhood U of $\overline{\partial X}$ in $\mathbb{C}^N$ and a C^∞-vectorfield θ on U with the following properties:

1) θ is transversal to ∂B_ε.
2) the local θ-flow in U leaves X and the fibres of f in X invariant.
3) the restriction of $\mathbb{L}$ to the local integral curves of θ is a constant sheaf.

<u>Theorem 1.</u> Let $X \xrightarrow{f} S$ be a standard representative of the germ $(X,x) \xrightarrow{f} (S,s)$.
Let $(K^\cdot, d)$ be a finite complex of sheaves on X. Assume:

1) the sheaves K^p are $\mathcal{O}_X$-coherent modules.
2) the differentials are $f^{-1}(\mathcal{O}_S)$-linear.
3) the cohomology sheaves $H^i(K^\cdot)$ are transversally constant (with respect to a single U and θ).

Then $\mathbb{R}^i f_*(K^\cdot)$ is an $\mathcal{O}_S$-coherent module.

<u>Sketch of proof</u>: Let $X = X_{\varepsilon,\eta}$. Now choose an U and θ exhibiting the $H^i(K^\cdot)$ as transversally constant sheaves. By compactness of $\overline{\partial X}$ and transversality of θ we can find $\varepsilon_2 < \varepsilon$ such that $\partial X_{\alpha,\eta} \subset U$ and $\theta \pitchfork \partial X_{\alpha,\eta}$ for all $\alpha \in [\varepsilon_2, \varepsilon]$. Choose $\varepsilon_1 \in (\varepsilon_2, \varepsilon)$. Because θ respects the f-fibres and leaves X invariant we have a commutative diagram

$$\begin{array}{ccc} \bar{X}_1 - \bar{X}_2 & \longleftrightarrow & X_1 - \bar{X}_2 \\ & {\scriptstyle p} \searrow \quad \swarrow {\scriptstyle q} & \\ & \partial X_1 \xrightarrow{f} S & \end{array} \qquad \text{with} \quad X_i = X_{\varepsilon_i,\eta}, \quad \bar{X}_i = \bar{X}_{\varepsilon_i,\eta}$$

Here p and q are the quotient maps induced by the local θ-flow. If $\mathbb{L}$ is a transversally constant sheaf on X (w.r.t. U and θ) then $R^i p_* \mathbb{L} \,|\bar{X}_1 - \bar{X}_2 \xrightarrow{\sim} R^i q_* \mathbb{L}|\, X_1 - \bar{X}_2$ (in fact $= 0$ for $i > 0$). By the spectral sequence for the composition of two maps we get $R^i f_* \mathbb{L}\,|\bar{X}_1 - \bar{X}_2 \xrightarrow{\sim} R^i f_* \mathbb{L}|\, X_1 - \bar{X}_2$. By Mayer-Vietoris we then get $R^i f_* \mathbb{L}\,|\bar{X}_1 \xrightarrow{\sim} R^i f_* \mathbb{L}\,|X_1$. The same argument for $X - X_1 \hookleftarrow \partial X_1$ gives $R^i f_* \mathbb{L}\,|X \xrightarrow{\sim} R^i f_* \mathbb{L}\,|\bar{X}_1 \xrightarrow{\sim} R^i f_* \mathbb{L}\,|X_1$. Apply this to $\mathbb{L} = H^i(K^\cdot)$. This gives an isomorphism of spectral sequences

$$\begin{array}{ccc} R^p f_*(H^q(K^\cdot)|X) & \xrightarrow{\sim} & R^p f_*(H^q(K^\cdot)|X_1) \\ \Downarrow & & \Downarrow \\ \mathbb{R}^{p+q} f_*(K^\cdot|X) & \xrightarrow{\sim} & \mathbb{R}^{p+q} f_*(K^\cdot|X_1) \end{array}$$

showing that shrinking of X does not change the hypercohomology. This fact implies the coherence of $\mathbb{R}^i f_*(K^\cdot)$ as $\mathcal{O}_S$-module, in exactly the same way as in ([B-G],p.250) by applying the main theorem of Kiehl & Verdier. □

<u>Definition 3.</u> Let $X \xrightarrow{f} S$ be a standard representative of $(X,x) \to (S,s)$. A complex of sheaves $(K^\cdot,d)$ on $X = X_{\varepsilon,\eta}$ is called *concentrated* if for all $\varepsilon' \in (0,\varepsilon]$ there exists $\eta' \in (0,\eta]$ such that the restriction of $K^\cdot$ to $X_{\varepsilon',\eta'}$ full-fills conditions 1), 2) and 3) of Theorem 1. A germ $(X,x) \to (S,s)$ is called concentrated if the relative de Rham complex $\Omega^\cdot_{X/S}$ is concentrated for some standard representative of the germ.

<u>Examples.</u>

1) A deformation $(X,x) \xrightarrow{f} (S,s)$ of an isolated singularity $(X_s = f^{-1}(s),x)$ is concentrated (see [B-G],p.248).
2) A hypersurface germ $f\colon (\mathbb{C}^{n+1},0) \to (\mathbb{C},0)$ with a good $\mathbb{C}^*$-action (i.e. all weights >0) is concentrated.
3) A hypersurface germ $f\colon (\mathbb{C}^{n+1},0) \to (\mathbb{C},0)$ such that for a certain representative $X \xrightarrow{f} S$ there are only a finite number of isomorphism classes of germs $(X,x) \to (S,S)$ with $x \in X$, $s = f(x)$, is concentrated.
4) The function $f = y^4 + xy^2z^2 + z^4$ does not define a concentrated germ at 0. The relative de Rham cohomology is not coherent.

We omit the proofs of these facts.

The idea is that for a concentrated complex the things really only happen in one point.

<u>Proposition 1.</u> Let $X \xrightarrow{f} S$ be a contractible Stein standard representative of a germ $(X,x) \to (S,s)$ and let $(K^\cdot,d)$ be a concentrated complex on X. Then:

$$H^i(f_*K^\cdot)_s \xrightarrow{\sim} \mathbb{R}^i f_*(K^\cdot)_s \xrightarrow{\sim} (f_*H^i(K^\cdot))_s \simeq H^i(K^\cdot)_x \simeq H^i(K_x^\cdot)\ .$$

<u>Proof:</u> The first isomorphism follows from the spectral sequence $H^p(R^q f_*K^\cdot) \Rightarrow \mathbb{R}^{p+q} f_*(K^\cdot)$ and the fact that the K^i are coherent and

X is Stein, so $R^q f_*(K^\cdot) = 0 \quad q > 0$. For the second isomorphism we use the other spectral sequence $R^p f_*(\mathcal{H}^q(K^\cdot)) \Rightarrow \mathbb{R}^{p+q} f_*(K^\cdot)$. By concentradness we may replace X by $\bar{X}$ and then apply ([G],II 4.11.1) to obtain $R^p f_*(\mathcal{H}^q(K^\cdot))_s = H^p(f^{-1}(s), \mathcal{H}^q | f^{-1}(s))$. By concentratedness again we may assume there is a contraction of $f^{-1}(s)$ to x such that the restriction of $\mathcal{H}^q$ to the fibres of the contraciton is constant. The proposition then follows from

Lemma 1. Let $\phi: X \times [0,1] \to X$ be a contraction of X to $p \in X$ by homeomorphisms (i.e.: $\phi(x,0) = x$, $\phi(x,1) = p$, $\phi(p,t) = p \quad \forall t \in [0,1]$ and $\phi(-,t): X \tilde{\to} X_t := \phi(x,t)$ homeomorphism $\forall t \in [0,1)$) . Let $\gamma_x: I \to X \; ; \; t \to \phi(x,t)$. Let F be a sheaf on X with $F|\gamma_x([0,1))$ a constant sheaf.

Then $H^i(X,F) = 0 \quad \forall i > 0$.

Proof. Let $U = X-\{p\}$, $U_t = X_t-\{p\}$ and $j: U \to X$ the inclusion map. First we prove the lemma for $F = j_*G$ with G a sheaf on U . We have a spectral sequence $H^p(X, R^q j_* G) \Rightarrow H^{p+q}(U,G)$. But $H^p(X, R^q j_* G) = 0$ $p,q > 0$ because the higher direct images are concentrated at p . By constancy of G along the contraction fibres $H^{p+q}(U,G) \tilde{\to}$ $\tilde{\to} \lim\limits_{t\to 1} H^{p+q}(U_t,G) = H^0(X, R^{p+q} j_* G)$ so we must have $H^p(X, j_*G) = 0$ for $p > 0$. Using

$$0 \to H^0_{\{p\}}(F) \to F \to \bar{F} \to 0$$

$$0 \to \bar{F} \to j_*j^*F \to H^1_{\{p\}}(F) \to 0$$

and the fact that $H^0(X, j_*j^*F) \twoheadrightarrow H^0(X, H^1_{\{p\}}(F))$ the general case follows from the special case. □

For the relative de Rham complex one has of course a link with the topology of the situation:

Proposition 2. Let $X \xrightarrow{f} S$ be a contractible Stein standard representative of a germ $(X,x) \to (S,s)$. Assume that $\Omega^\cdot_{X/S}$ is a concentrated complex and that $f|X-f^{-1}(s): X-f^{-1}(S) \to S-\{s\}$ is a submersion. Then there is a short exact sequence of $\mathcal{O}_S$-modules

$$0 \to (\mathbb{R}^i f_* \mathbb{C}_X) \otimes \mathcal{O}_S \to \mathcal{H}^i(f_*\Omega^\cdot_{X/S}) \to f_*\mathcal{H}^i(\Omega^\cdot_{X/S}) \to 0 \; .$$

Proof. Look at the spectral sequence $R^p f_*(\mathcal{H}^q(\Omega^\cdot_{X/S})) \Rightarrow \mathbb{R}^{p+q} f_*(\Omega^\cdot_{X/S})$ and remark that $\mathcal{H}^0(\Omega^\cdot_{X/S}) = f^{-1}\mathcal{O}_S$ and that $\mathcal{H}^q(\Omega^\cdot_{X/S})$ is concentrated on $f^{-1}(s)$. Use that $R^i f_* f^{-1} \mathcal{O}_S = R^i f_* \mathbb{C}_X \otimes \mathcal{O}_S$ (by an easy adaptation of [L] , p. 138) □

§2. The Gauss-Manin system

Let $X \xrightarrow{f} S$ be a standard representative of a hypersurface germ $f: (\mathbb{C}^{n+1},0) \to (\mathbb{C},0)$. The Gauss-Manin system $\mathcal{H}_X$ is a certain (complex of) $\mathcal{D}_S$-module(s), describing the behaviour of period integrals over cycles in the f-fibres (see [Ph],[S-S]).
In formula ([S-S],p.646):

$$\mathcal{H}_X = \int^{\cdot} \mathcal{O}_X = \mathbb{R} f_*(\Omega_X^{\cdot}[D]) \quad .$$

Here $\Omega_X^{\cdot}[D]$ is a complex of sheaves on X with differential $\underline{d}$

$$\underline{d}(\omega \cdot D^k) = d\omega \cdot D^k - df \wedge \omega \cdot D^{k+1}.$$

On this complex there is an action of t and ∂_t :

$$t \cdot (\omega \cdot D^k) = f \cdot \omega \cdot D^k - k \cdot \omega D^{k-1}$$

$$\partial_t(\omega D^k) = \omega \cdot D^{k+1} \quad .$$

One should think of the symbol $\omega \cdot D^k$ as representing the differential form

$$\operatorname{Res}_{X_t}\left(\frac{k!\,\omega}{(f-t)^{k+1}}\right)$$

on the Milnor fibre X_t . One can consider the complex $(\Omega_X^{\cdot}[D],\underline{d})$ as the associated single complex of the double complex $(K^{\cdot\cdot};d,-df\wedge)$ with $K^{pq} = \Omega_X^{p+q}$ for $q \geq 0$, $K^{pq} = 0$ for $q < 0$. This complex carries a so called "Hodge filtration", obtained by cutting off vertically.
In formula:

$$F^p \Omega_X^k[D] := \bigoplus_{k-(p+1) \geq \ell} \Omega_X^k \cdot D^\ell \quad .$$

This filtration gives rise to a spectral sequence.

<u>Question.</u> Under what conditions does this spectral sequence degenerate at E_2 ? (i.e. $d_i = 0 \quad i \geq 2$).

Is this true for concentrated singularities in the sense of §1?

Remark. For $f = y^4+xy^2z^2 + z^4$ it does not degenerate at E_2 .

We introduce some notation: Put $\Omega^{\bullet} = \Omega^{\bullet}_X$.

$$S^{\bullet} := \ker(df\wedge : \Omega^{\bullet} \to \Omega^{\bullet+1})$$

$$C^{\bullet} := df \wedge \Omega^{\bullet-1}$$

$H^{\bullet} := S^{\bullet}/C^{\bullet}$ (the Koszul cohomology)

$\Omega^{\bullet}_f := \Omega^{\bullet}/C^{\bullet}$ (the relative de Rham complex).

The relations between these complexes, which carry all a differential induced and denoted by d , are summarized in the following diagram with exact rows and columns.

$$\begin{array}{ccccccccc}
 & & 0 & & 0 & & & & \\
 & & \downarrow & & \downarrow & & & & \\
 & & C^{\bullet} & \cong & C^{\bullet} & & & & \\
 & & \downarrow & & \downarrow & & & & \\
0 & \to & S^{\bullet} & \to & \Omega^{\bullet} & \xrightarrow{df\wedge} & C^{\bullet}[1] & \to & 0 \\
 & & \downarrow & & \downarrow & & \downarrow & & \\
0 & \to & H^{\bullet} & \to & \Omega^{\bullet}_f & \xrightarrow{df\wedge} & C^{\bullet}[1] & \to & 0 \\
 & & \downarrow & & \downarrow & & & & \\
 & & 0 & & 0 & & & &
\end{array}$$

Now the E_2-term of the spectral sequence of the Hodge filtration on $(K^{\bullet\bullet};d,-df\wedge)$ can be written as:

$$E_2^{pq} = \begin{cases} 0 & \text{if } q < 0 \\ H^p(S^{\bullet}) & \text{if } q = 0 \\ H^{p+q}(H^{\bullet}) & \text{if } q > 0 . \end{cases}$$

(Here we abbreviate $H^p(f_*S^{\bullet})$ to $H^p(S^{\bullet})$ etc.)
Thus we get a collection of maps $d_2 \colon H^p(H^{\bullet}) \to H^{p+1}(S^{\bullet})$ $p=0,\dots,n+1$.

Due to the peculiar shape of the complex $(K^{\bullet\bullet};d,-df\wedge)$ we have

Lemma 2. If $d_2 \colon H^p(H^{\bullet}) \to H^{p+1}(S^{\bullet})$ $p=1,\dots,n$ is the zero map, then the spectral sequence degenerates, i.e. $E_2 = E_\infty$.

Proof. A form $w \in \Omega^p$ represents a class in $H^p(H^\cdot)$ iff $df \wedge w = 0$ and $dw = df \wedge w_1$ for a certain $w_1 \in \Omega^p$. Then $d_2[w]$ is represented by dw_1, considered as an element in $H^{p+1}(S^\cdot)$. This element represents zero iff $dw_1 = d\eta$ with $df \wedge \eta = 0$ for a certain $\eta \in \Omega^p$. This means that we can change w_1 to $\tilde{w}_1 = w_1 - \eta$, which is closed. So we have: $d_2[w] = 0$ means: If $df \wedge w = 0$ and $dw = df \wedge w_1$, then we can choose w_1 closed.
Now suppose we have a form w representing a cycle for the differential d_r. This means that we can find $w_1,\dots,w_r$ such that $df \wedge w = 0$ and $dw = df \wedge w_1$, $dw_k = df \wedge w_{k+1}$ $k=1,\dots,r-1$ but already $dw = df \wedge w_1$ implies that we can choose w_1 closed, so we can take $w_k = 0$ $k=2,\dots r$. Hence $d_{r+1}[w] = [dw_r] = 0$. □

Remark. $H^0(H^\cdot) = H^{n+2}(S^\cdot) = 0$, so the map is only interesting for $p = 1,\dots,n$.

We will now give an alternative description of the d_2-map. Look at the long exact cohomology sequences

$$\cdots \to H^p(C^\cdot) \to H^p(S^\cdot) \to H^p(H) \to \cdots\cdots$$

$$\cdots \to H^p(S^\cdot) \to H^p(\Omega^\cdot) \to H^{p+1}(C^\cdot) \to \cdots\cdots$$

coming from the diagram. If $p \geq 1$, then $H^p(\Omega^\cdot) = 0$, so we get an isomorphism $H^p(C^\cdot) \xrightarrow{\sim} H^p(S^\cdot)$ $(p > 2)$. We call this isomorphism ∂_t. If an element of $H^p(C^\cdot)$ is reppesented by $df \wedge \eta$, $\eta \in \Omega^{p+1}$ then $\partial_t([df\wedge\eta]) = [d\eta]$.
We can eliminate $H^p(C^\cdot)$ from the first long exact sequence using this isomorphism. So we get:

$$\begin{array}{ccccccc} \cdots & \to H^p(H^\cdot) & \to & H^{p+1}(C^\cdot) & \xrightarrow{j} & H^{p+1}(S^\cdot) & \to \cdots\cdots \\ & & \searrow^{\alpha} & \downarrow \partial_t & \nearrow_{\partial_t^{-1}} & & \\ & & & H^{p+1}(S^\cdot) & & & \end{array} \qquad (p \geq 1)$$

Claim. $\alpha = d_2$.

Proof. The map $H^p(H^\cdot) \alpha H^{p+1}(C^\cdot)$ can be described as follows: If w represents a class in $H^p(H^\cdot)$ then $df \wedge w = 0$ and there is an w_1 such that $dw = df \wedge w_1$. The image in $H^{p+1}(C^\cdot)$ is then just $[dw] = [df\wedge w_1]$. Applying ∂_t to this element gives $[dw_1]$, so $\alpha([w]) = d_2([w])$. □

The map j above is induced by the inclusion $C^{\cdot} \subset S^{\cdot}$ and although the induced map $H^{p+1}(S^{\cdot}) \to H^{p-1}(S^{\cdot})$ is not really the inverse of ∂_t, we denote it by ∂_t^{-1}. One has $\partial_t^{-1} \circ \partial_t = j$. Observe that j is $\mathcal{O}_S$-linear whereas ∂_t is a derivation over j.
Similarly we have an exact sequence and isomorphism involving $H^p(\Omega_f^{\cdot})$:

$$\begin{array}{ccccccc} \cdots & \to H^p(H^{\cdot}) & \to H^p(\Omega_{=}^{\cdot}) & & \to H^{p+1}(C^{\cdot}) & & \to H^{p+1}(H^{\cdot}) \\ & & & \partial_t^{-1} \searrow & \uparrow \partial_t \wr & \nearrow & \\ & & & & H^p(\Omega_f^{\cdot}) & & \end{array}$$

In this diagram ∂_t is represented as follows: A class in $H^p(\Omega_f^{\cdot})$ is represented by $w \in \Omega^p$ such that $dw = df \wedge \eta$. Then $\partial_t([w]) = [df \wedge \eta]$.
As we have isomorphisms of the maps

$$\begin{array}{ccccc} H^p(\Omega_f^{\cdot}) & \xrightarrow{\approx} & H^{p+1}(C^{\cdot}) & \xrightarrow{\sim} & H^{p+1}(S^{\cdot}) \\ \downarrow \partial_t^{-1} & & \downarrow \partial_t^{-1} & & \downarrow \partial_t^{-1} \\ H^p(\Omega_f^{\cdot}) & \xrightarrow{\sim} & H^{p+1}(C^{\cdot}) & \xrightarrow{\sim} & H^{p+1}(S^{\cdot}) \end{array}$$

(where the horizontal maps are all called ∂_t) we get:

<u>Corollary</u>. Equivalent are

1) $d_2: H^p(H) \to H^{p+1}(S^{\cdot})$ is the zero map
2) $\partial_t^{-1}: H^p(\Omega_f^{\cdot}) \circlearrowleft$, $H^{p+1}(C^{\cdot}) \circlearrowleft$ or $H^{p+1}(S^{\cdot}) \circlearrowleft$ is injective
3) $H^p(\Omega_f^{\cdot}) \xrightarrow{j} H^{p+1}(C^{\cdot})$ or $H^{p+1}(C^{\cdot}) \xrightarrow{j} H^{p+1}(S^{\cdot})$ is injective.

Now, philosophically at least, the operator ∂_t^{-1} should be similar to multiplication by t. Injectivity of ∂_t^{-1} should learn about injectivity of t, i.e. torsion freeness of $H^p(\Omega_f^{\cdot})$ as an $\mathcal{O}_S$-module. The modules $H^p(\Omega_f^{\cdot})$, $H^{p+1}(C^{\cdot})$ and $H^{p+1}(S^{\cdot})$ are analoguous to the modules of Brieskorn [B] H, H' and H'' respectively: on $S-\{s\}$ they are locally free of rank $b_p(F)$, the p-th Betti number of the Milnor fibre $F = f^{-1}(t)$, $t \neq s$. The isomorphism on $S-\{s\}$ is given by the map $j|S-\{s\}$, so $\ker j$ and $\operatorname{cok} j$ are both modules supported on the point $\{s\}$. Further we have isomorphisms $H^p(\Omega_f^{\cdot}) \xrightarrow{\partial_t} H^{p+1}(S^{\cdot})$ and $H^{p+1}(C^{\cdot}) \xrightarrow{\partial_t} H^{p+1}(S^{\cdot})$. The relation $\partial_t \cdot t - t\partial_t = j$ is easily seen to hold. We repeat Malgrange's proof of the Sebastiani theorem (see [Ma], p.416): the torsion freeness of the Brieskorn module $H'' = H^{n+1}(S^{\cdot})$ in the case of an isolated singularity.

<u>Theorem 2.</u> Assume that $H^p(\Omega_f^\bullet)$, $H^{p+1}(C^\bullet)$ and $H^{p+1}(S^\bullet)$ are coherent $\mathcal{O}_S$-modules. If d_2: $H^p(H^\bullet) \to H^{p+1}(S^\bullet)$ is the zero map, then $H^p(\Omega_f^\bullet)$, $H^{p+1}(C^\bullet)$ and $H^{p+1}(S^\bullet)$ are torsion free.

<u>Proof.</u> Put $E = H^{p+1}(C^\bullet)$, $F = H^{p+1}(S^\bullet)$. We have an isomorphism $E \overset{\partial_t}{\to} F$ and if $d_2 = 0$ an $\mathcal{O}_S$-linear injection $E \overset{j}{\to} F$ with $F/j(E)$ $\mathcal{O}_S$-torsion, i.e. we have an (E,F)-connection in the sense of Malgrange.
We derive a contradiction by assuming Torsion (F) $\neq 0$. So let $t\cdot\omega = 0$, $0 \neq \omega \in F$. By $E \underset{\sim}{\overset{\partial_t}{\to}} F$ we find an $\eta \in E$ such that $\partial_t\eta = \omega$. Now $t^k\eta \neq 0$ $\forall k$, because if $t^k\eta = 0$, with k smallest as possible, then $0 = \partial_t\ t^k\eta = k\cdot t^{k-1}\cdot j\cdot\eta + t^k\partial_t\eta = k\cdot t^{k-1}\cdot j\cdot\eta$. By injectivity of j it follows that $t^{k-1}\eta = 0$, so contradiction. By coherence of E as $\mathcal{O}_S$-module it follows that $\eta|S-\{s\} \neq 0$, but $\partial_t\eta|S-\{s\} = 0$. But now we use the link with the topology, by integrating η over a horizontal family of vanishing cycles $\gamma(t)$, $t \in [0,1]$. One has

$$0 = \int_{\gamma(t)} \partial_t\eta = \frac{d}{dt}\int_{\gamma(t)} \eta$$

so the period $t \to \int_{\gamma(t)}\eta$ is constant. Because η is holomorphic on the whole of X , and has closed restriction to the f-fibres, we know however that this integral has to go to zero. (Here one has to use an extension of Lemma 4.5 of [Ma] to the case of p-forms, which can be proved quite in the same way). Hence $\int_{\gamma(t)}\eta = 0$ $t \in [0,1]$. As this is true for every horizontal family of cycles we conclude that η represents the zero form. Contradiction, hence torsion (F) = 0 .
The rest of the proof is obtained by remarking that via the $\mathcal{O}_S$-linear map j $H^p(\Omega_f^\bullet)$ and $H^{p+1}(C^\bullet)$ are submodules of $H^{p+1}(S)$. □

<u>Remark.</u> The proof of the theorem shows that one really needs coherence modulo torsion of the module $H^{p+1}(C^\bullet)$, which follows from the results of Hamm [H]. In order to keep this paper as selfcontained as possible, we prefer to use the direct coherence theorem of §1 for the singularities we are interested in.

There is an obvious kind of converse to Theorem 2.

Proposition 3. Assume $H^p(H^\cdot)$ coherent. Then if $H^{p+1}(S^\cdot)$ is torsion free, then $d_2: H^p(H^\cdot) \to H^{p+1}(S^\cdot)$ is the zero map.

Proof. $H^p(H^\cdot)$ is an $\mathcal{O}_S$-module concentrated at s. By coherence, it is torsion. Hence the $\mathcal{O}_S$-linear map d_2 has to be zero. □

Of course, if one knows that $H^p(\Omega_f^\cdot)$ is a torsion free $\mathcal{O}_S$-module, then one gets relatively nice formulae for the Betti numbers of the Milnor fibre.

For a concentrated singularity one has the exact sequence of Proposition 2, §1:

$$0 \to R^i f_* \mathbb{C}_X \otimes \mathcal{O}_S \to H^i(f_*\Omega_f^\cdot) \to f_* H^i(\Omega_f^\cdot) \to 0 .$$

The first sheaf has stalk 0 at s and $\mathbb{C}^{b_i} \otimes \mathcal{O}_{S,t}$ at $t \neq s$ where $b_i = b_i(F)$ is the i-th Betti number of the Milnor fibre. The second sheaf is $\mathcal{O}_S$-coherent with stalk $f_* H^i(\Omega_f^\cdot) = H^i(\Omega_{f,x}^\cdot)$ at s. If we know that t acts injectively one thus finds.

$$b_i(F) = \dim_{\mathbb{C}} H^i(\Omega_{f,x}^\cdot)/t \cdot H^i(\Omega_{f,x}^\cdot) .$$

By Malgranges index theorem ([Ma],p.408) this number is also equal to $\dim_{\mathbb{C}} H^i(\Omega_{f,x}^\cdot)/\partial_t^{-1} H^i(\Omega_{f,x}^\cdot) = \dim_{\mathbb{C}} H^{i+1}(H_x^\cdot)$.

Conclusion. For a concentrated singularity where

$$\partial_t^{-1}: H^i(\Omega_{f,x}^\cdot) \hookrightarrow H^i(\Omega_{f,x})$$

we have: $b_i(F) = \dim_{\mathbb{C}} H^{i+1}(H_x^\cdot)$ $(i > 0)$.

§3. A special class of singularities

We now specialize our situation to the case of a hypersurface germ $f: (\mathbb{C}^{n+1},0) \to (\mathbb{C},0)$ with a one dimensional singular locus. This is the simplest situation where the map d_2 of §2 can be nontrivial. (In the sequel a fixed appropriate contractible Stein representative $X \overset{f}{\to} S$ is understood).

We will give the singular locus the non reduced structure defined by the jacobi ideal $J_f = (\partial_0 f, \ldots, \partial_n f)$ and denote it by $\tilde{\Sigma}$. So we put $\mathcal{O}_{\tilde{\Sigma}} = \mathcal{O}/J_f$, where $\mathcal{O} = \mathcal{O}_X$. We also will consider the curve Σ, defined by the ideal I, which is obtained from J_f by removing the M-primary component. In other words, Σ is the largest Cohen-Macaulay curve contained in $\widetilde{\Sigma}$. Thus we have an exact sequence:

$$0 \to I/J_f \to \mathcal{O}_{\widetilde{\Sigma}} \to \mathcal{O}_{\Sigma} \to 0$$

where I/J_f is an M-primary $\mathcal{O}$-module. In [Pe] modules like I/J_f have been studied and they are called "jacobi-modules".

In order to study the map d_2 we first need a description of the Koszul cohomology groups H^i. One easily sees (use for instance the "Lemme d'Acyclicité, see [P-S]) that the Koszul complex on the generators $\partial_i f$, $i=0,1,\ldots n$, acting on $\mathcal{O}$, is exact except possibly in degrees 0 and 1 .
One has (where $H_i(\mathcal{O};\partial_0 f,\ldots,\partial_n f)$ denotes Koszul homology)

$$\begin{aligned}
\mathcal{O}/I_f = H_0(\mathcal{O};\partial_0 f,\ldots,\partial_n f) &\simeq H^{n+1} = \Omega^{n+1}/df\wedge\Omega^n \\
H_1(\mathcal{O};\partial_0 f,\ldots,\partial_n f) &\simeq H^n \quad = \ker(df\wedge:\Omega^n \to \Omega^{n+1})/df\wedge\Omega^{n-1} \\
H_i(\mathcal{O};\partial_0 f,\ldots,\partial_n f) &= 0 \qquad i \geq 2 .
\end{aligned}$$

Note that H^n and H^{n+1} are $\mathcal{O}_{\widetilde{\Sigma}}$-modules. The funny thing about H^n is, that although it is defined in terms of the function f , its structure as a module is only dependent on the singular locus Σ . This is always the case with the first non vanishing Koszul cohomology group. It turns out that this cohomology group as a module is always isomorphic to the dualizing module ω_Σ of the singular locus. For our purpose it is important to have an explicit isomorphism between H^n and ω_Σ . The description of this isomorphism is due to R. Pellikaan [Pe], and can be formulated as follows:

We consider the following diagram:

$$\begin{array}{ccccccccccccc}
0 & & & & & & & & & & & & \\
\uparrow & & & & & & & & & & & & \\
\mathcal{O}_\Sigma & \leftarrow & \mathcal{O} & \leftarrow & \mathcal{O}^{d_1} & \leftarrow & \mathcal{O}^{d_2} & \leftarrow\cdots & \mathcal{O}^{d_n} & \leftarrow & 0 & & \\
\uparrow & & & & \uparrow{\scriptstyle\phi_1} & & \uparrow{\scriptstyle\phi_2} & & \uparrow{\scriptstyle\phi_n} & & & & (*) \\
\mathcal{O}_{\widetilde{\Sigma}} & \leftarrow & \mathcal{O} & \leftarrow & \Theta & \leftarrow & \Lambda^2\Theta & \leftarrow\cdots & \Lambda^n\Theta & \leftarrow & \Lambda^{n+1}\Theta & \leftarrow & 0 \\
\uparrow & & & & & & & & & & & & \\
I/J_f & & & & & & & & & & & & \\
\uparrow & & & & & & & & & & & & \\
0 & & & & & & & & & & & &
\end{array}$$

In the top row we put the minimal resolution of $\mathcal{O}_\Sigma$ as an $\mathcal{O}$-module. The bottom row is a natural incarnation of the Koszul complex on the generators $\partial_i f$ $i=0,\ldots,n$; Θ is the module of tangent vectors and $\Theta \to \mathcal{O}$ is the map $\Sigma a_i \partial_i \to \Sigma a_i \partial_i f$. The vertical maps ϕ_i are induced from ϕ_1 , which expresses the fact that $J_f \subset I$. Dualizing this diagram with respect to $\mathcal{O}$ and taking homology produces a map

$$[\phi_n^T]: \mathrm{Ext}^n_{\mathcal{O}}(\mathcal{O}_\Sigma,\mathcal{O}) \to \mathcal{H}^n .$$

<u>Theorem.</u> (R. Pellikaan [Pe],p.152)

$[\phi_n^T]$ is an isomorphism. □

So the choice of a volume form $\Omega \in \Omega^{n+1}$ will give a natural map $\omega_\Sigma \to \mathcal{H}^n$.

We now restrict to an even more special situation: From now on we assume that Σ is a *reduced complete intersection curve*.

This is precisely the class of singularities studied by Siersma from a topological and by Pellikaan from an algebraic point of view.

Reducedness of Σ is equivalent to the condition that the function f defines a singularity which around a point $p \in \Sigma-0$ is right equivalent to $f(x_0,\ldots,x_n) = \Sigma_{i=1}^n x_i^2$ ("generically transversal A_1 ").

If Σ is a complete intersection curve, we can write $I = (g_1,\ldots,g_n)$. From the reducedness it now follows that $f \in I^2$, so we can write $f = \frac{1}{2}\Sigma h_{ij} g_i g_j$. The function $h := \det(h_{ij})$, which is called the transversal Hessian, is non-zero on a generic point of Σ (for these facts, see [Pe]).

As Σ is a complete intersection, defined by $g_1,\ldots,g_n$, we can resolve $\mathcal{O}_\Sigma$ by the Koszul complex. This implies that in diagram (*) we can take $\phi_i = \Lambda^i \phi_1$. Using Pellikaans theorem we can write down a generator for $\mathcal{H}^n$ as $\mathcal{O}_\Sigma$-module as $\omega_1 \wedge \omega_2 \wedge \ldots \wedge \omega_n$, where we put $df = \Sigma \omega_i g_i$ with $\omega_i \in \Omega^1$. So $\mathcal{H}^n = \mathcal{O}_\Sigma \; \omega_1 \wedge \omega_2 \wedge \ldots \wedge \omega_n$. (In concrete terms: Write $\partial_i f = \Sigma A_{ij} g_j$ with A_{ij} a $n \times (n+1)$-matrix. Then $\omega_1 \wedge \ldots \wedge \omega_n = \Sigma \Delta_i d\hat{x}_i$ with $\Delta_i = (-1)^i$. i-th $n \times n$ minor of (A_{ij}) , and $dx_i \wedge d\hat{x}_i = dx_0 \wedge \ldots \wedge dx_n$).

In order to study the map $d: \mathcal{H}^n \to \mathcal{H}^{n+1}$ we first project $\mathcal{H}^{n+1} = \mathcal{O}_{\tilde{\Sigma}} \otimes \Omega^{n+1}$ to $\mathcal{O}_\Sigma \otimes \Omega^{n+1} = \Omega^{n+1}/I\cdot\Omega^{n+1}$ and study the composed map $\underline{d}: \mathcal{H}^n \to \Omega^{n+1}/I\cdot\Omega^{n+1}$. The first step is to compute $\omega_1 \wedge \ldots \wedge \omega_n$ and $d(\omega_1 \wedge \ldots \wedge \omega_n) \bmod I$.

<u>Proposition 4.</u> With the notation as above we have:

a. $\omega_1 \wedge \ldots \wedge \omega_n = h \cdot dg_1 \wedge \ldots \wedge dg_n \mod I \cdot \Omega^n$

b. $d(\omega_1 \wedge \ldots \wedge \omega_n) = \frac{1}{2} dh \wedge dg_1 \wedge \ldots \wedge dg_n \mod I \cdot \Omega^{n+1}$.

<u>Proof.</u> Write $f = \frac{1}{2} \Sigma h_{ij} g_i \cdot g_j$. Then we have

$$df = \sum_{i,j} (h_{ij} dg_j + \tfrac{1}{2} dh_{ij} g_j) g_i$$

so we can take

$$\omega_i = \sum_j (h_{ij} dg_j + \tfrac{1}{2} dh_{ij} \cdot g_j) \ .$$

Hence

$$\omega_i = \sum_j h_{ij} dg_j \mod I \cdot \Omega^1$$

$$d\omega_i = \sum dh_{ij} \wedge dg_j - \tfrac{1}{2} dh_{ij} \wedge dg_j = \tfrac{1}{2} \sum_j dh_{ij} \wedge dg_j \ .$$

So $\omega_1 \wedge \omega_2 \wedge \ldots \wedge \omega_n = \det(h_{ij}) dg_1 \wedge \ldots \wedge dg_n \mod I \cdot \Omega^n$ and

$$d(\omega_1 \wedge \ldots \wedge \omega_n) = \sum_i (-1)^i \omega_1 \wedge \ldots \wedge d\omega_i \wedge \ldots \wedge \omega_n =$$

$$= \sum_i (-1)^i (\Sigma h_{1j} dg_j) \wedge \ldots \wedge (\Sigma \tfrac{1}{2} dh_{ij} \wedge dg_j) \wedge \ldots$$

$$\ldots \wedge (\Sigma h_{nj} dg_j) \mod I \cdot \Omega^{n+1} = \tfrac{1}{2} dh \wedge dg_1 \wedge \ldots \wedge dg_n \ . \qquad \square$$

Using this proposition, we can compute $\underline{d}$:

$$\underline{d}(P\omega_1 \wedge \ldots \wedge \omega_n) = dP \wedge \omega_1 \wedge \ldots \wedge \omega_n + Pd(\omega_1 \wedge \ldots \wedge \omega_n)$$

$$= h dP \wedge dg_1 \wedge \ldots \wedge dg_n + \tfrac{1}{2} P \cdot dh \wedge dg_1 \wedge \ldots \wedge dg_n \ .$$

Introducing the vectorfield θ , dual to $dg_1 \wedge \ldots \wedge dg_n$ (i.e.: $i_\theta (dx_0 \wedge \ldots \wedge dx_n) = dg_1 \wedge \ldots \wedge dg_n$ where i_θ is the contraction operator) we can interpret $\underline{d}$ as a map $D: \mathcal{O}_\Sigma \to \mathcal{O}_\Sigma$; $P \to D(P) = h \cdot \theta(P) + \frac{1}{2}\theta(h) \cdot P$ making the following diagram commutative:

$$\begin{array}{ccc} H^n & \xrightarrow{\underline{d}} & \Omega^{n+1}/I\Omega^{n+1} \\ \uparrow & & \uparrow \\ \mathcal{O}_\Sigma & \xrightarrow{D} & \mathcal{O}_\Sigma \end{array} \ .$$

Here $\mathcal{O}_\Sigma \tilde{\to} H^n$ is given by $P \to P\cdot\omega_1 \wedge \ldots \wedge \omega_n$ and $\mathcal{O}_\Sigma \tilde{\to} \Omega^{n+1}/I\Omega^{n+1}$ by $P \to P\cdot dx_0 \wedge \ldots \wedge dx_n$.
The vectorfield θ is tangent to Σ and non-zero on $\Sigma-\{0\}$.

Now we can prove:

Theorem 3. Let $f: (\mathbb{C}^{n+1},0) \to (\mathbb{C},0)$ define a singularity which has a one dimensional singular locus, which is a reduced complete intersection. Write $f = \frac{1}{2}\Sigma\, h_{ij}g_ig_j$ with $I = (g_1,\ldots,g_n)$ the ideal of Σ and put $h = \det(h_{ij})$. Then

If h is not a unit then $H^n \xrightarrow{d} H^{n+1}$ is injective

If h is a unit then $H^n \to H^{n+1}$ has a one dimensional kernel, which can be represented by a closed form.

Proof. Let $P\cdot\omega_1 \wedge \ldots \wedge \omega_n \in H^n$ be an element in the kernel of the operator d . Then also $\underline{d}(P\omega_1 \wedge \ldots \wedge \omega_n) = 0$ i.e.: $D(P) = 0$. In the ring $\mathcal{O}_\Sigma[h^{\frac{1}{2}}]$ we can write the operator D as follows:

$$D(P) = h\theta(P) + \tfrac{1}{2}\,\theta(h)P = h^{\frac{1}{2}}\cdot\theta(h^{\frac{1}{2}}\cdot P) \ .$$

Because h is a function that is non-zero on $\Sigma-\{0\}$ we conclude $\theta(h^{\frac{1}{2}}\cdot P) = 0$. Because θ is a vectorfield that is tangent to Σ and non-vanishing on $\Sigma-\{0\}$ it follows that $h^{\frac{1}{2}}\cdot P = C \bmod I\cdot\mathcal{O}_\Sigma[h^{\frac{1}{2}}]$, where C is a constant. If this constant is non-zero, then one must have that h is a unit in $\mathcal{O}_{\Sigma,0}$. If this constant is zero it follows that $P \in I$, i.e. $P\,\omega_1 \wedge \ldots \wedge \omega_n$ represents zero hence $d: H^n \to H^{n+1}$ is injective.
If h is a unit, then we can "diagonalize" the matrix h_{ij} by a change of generators for the ideal I from the g_i to $\tilde{g}_i$, achieving the form $f = \frac{1}{2}\,\Sigma\tilde{g}_i^2$ for our function f . (see [S],p.23). But then $df = \Sigma\, d\tilde{g}_i\cdot\tilde{g}_i$, hence the generator of H^n is represented by $d\tilde{g}_1 \wedge \ldots \wedge d\tilde{g}_n$ which is a closed form. It is easy to see that every element in the kernel is a scalar multiple of $d\tilde{g}_1 \wedge \ldots \wedge dg_n$. □

Corollary. Under the hypothesis of theorem 3 and with notations of §2 we have:

1) $H^n(\Omega_f^\cdot)$, $H^{n+1}(C^\cdot)$ and $H^{n-1}(S^\cdot)$ are free $\mathcal{O}_S$-modules of rank $b_n(F)$.
2) $H^{n+1}(\Omega_f^\cdot)$, $H^n(C^\cdot)$ and $H^n(S^\cdot)$ are free $\mathcal{O}_S$-modules of rank $b_{n-1}(F)$.

3) $b_n(F) = \dim_{\mathbb{C}} H^{n+1}(H^\cdot) = \dim_{\mathbb{C}}(\Omega^{n+1}/df \wedge \Omega^n + dH^n)$.

4) $b_{n-1}(F) = 1$ if h is a unit
$= 0$ if h is not a unit.

The corollary follows by remarking that the complexes $\Omega_f^\cdot$, $C^\cdot$ and $S^\cdot$ are concentrated for these singularities, and the fact that the d_2-map is the zeromap, as follows from the fact that the kernel of $d: H^n \to H^{n+1}$ can be represented by a closed form.

It is interesting to note that $\Omega^{n+1}/df \wedge \Omega^n + dH^n$, which is a vector space of dimension $b_n(F)$, does not have a structure of an $\mathcal{O}_X$-module, as in the case of an isolated singularity.
The proof of Theorem 3 shows a bit more: if h is not a unit then $dH^n \cap I\Omega^{n+1} = 0$. This fact gives an exact sequence

$$0 \to I/J_f \otimes \Omega^{n+1} \to \Omega^{n+1}/df \wedge \Omega^n + dH^n \to \Omega^{n+1}/I\Omega^{n+1} + dH^n \to 0$$

leading to the formula

$$b_n(F) = \dim_{\mathbb{C}}(I/J_f) + \dim_{\mathbb{C}}(\mathcal{O}_\Sigma/D(\mathcal{O}_\Sigma)) .$$

The first part, $\dim(I/J_f)$, is called the jacobi number of f . Pellikaan has proved a conjecture of Siersma, stating that this number j_f is equal to $\#A_1$-points $+\#D_\infty$-points in a generic approximation of f , making the singular locus into a smooth curve.
The second part, $\dim(\mathcal{O}_\Sigma/D(\mathcal{O}_\Sigma))$ has to be equal to $\mu(\Sigma) + \#D_\infty - 1$, by comparison with Siersma's formula ([S],p.4). We will give an algebraic proof of this fact.

First note the formula of Buchweitz and Greuel for the Milnor number of a curve: $\mu(\Sigma) = \dim(\omega_\Sigma/d\mathcal{O}_\Sigma)$ (see [B-G],p.244). Secondly, the number of D_∞ points in a deformation can be computed as $\dim(\mathcal{O}_\Sigma/h\cdot\mathcal{O}_\Sigma)$ (see [Pe],p.83).
Now assume that $h^{\frac{1}{2}} \in \mathcal{O}_\Sigma$. Then it is easy to see that we can consider $D: \mathcal{O}_\Sigma \to \mathcal{O}_\Sigma$ as the composition of the following four maps

$$\mathcal{O}_\Sigma \xrightarrow{h^{\frac{1}{2}}} \mathcal{O}_\Sigma \xrightarrow{d} \omega_\Sigma \xrightarrow{h^{\frac{1}{2}}} \omega_\Sigma \approx \mathcal{O}_\Sigma$$

where the first and the third maps are multiplications and the last one is the identification of ω_Σ with $\mathcal{O}_\Sigma$ by the generator $[dx_0 \wedge \ldots \wedge dx_n/dg_1 \wedge \ldots \wedge dg_n]$. By additivity of the index we find:

$$\text{Index}(D) = \text{Index}(h^{\frac{1}{2}}) + \text{Index}(d) + \text{Index}(h^{\frac{1}{2}}) .$$

$$\dim(\mathcal{O}_\Sigma/D(\mathcal{O}_\Sigma)) = \dim(\mathcal{O}_\Sigma/h\cdot\mathcal{O}_\Sigma) + \mu(\Sigma)-1 \ .$$

The proof in the case that $h^{\frac{1}{2}} \notin \mathcal{O}_\Sigma$ is similar.

In the case of a line singularity, i.e. Σ is a smooth curve, one can choose coordinates $(x,y_1,\ldots,y_n)$ such that $I = (y_1,\ldots,y_n)$ and $h = x^\alpha$. As in this case $\theta = \partial_x$ we get a particularly nice form for the operator: $D = x^{\alpha-1}\cdot(x\partial_x + \frac{\alpha}{2})$.

Concluding remarks and questions.

1) There should be some clear 'geometry" in the map $d: H^n \to H^{n+1}$. The expression $D = x^{\alpha-1}(x\partial_x + \frac{\alpha}{2})$ for line singularities suggests that it describes the monodromy of the transversal vanishing cycle by a connection on Σ . However, in general Σ is singular and can have several irreducible components and it is not clear in what sense d is a connection.
2) It is a shame that this theory does not cover the case of $f = x\cdot y\cdot z$; the singular locus is not a complete intersection. Here $b_1(F) = 2$. Is it always true that $b_{n-1}(F) \leq$ Gorenstein type (Σ) when Σ is a reduced curve? Numerous examples confirm this guess.
3) There are many other examples of function for which one can verify the degeneration of the spectral sequence. For example for the singularities studied by T. de Jong in [dJ] one can check this often.
4) The vectorbundle $H^{n+1}(f_*S^\cdot)$ sitting in the Gauss-Manin system does not seem to play the same rôle as in the isolated singularities case in the sense of characteristic exponents. We will study this in in a later paper.

References

[B-G] R. BUCHWEITZ and G-M. GREUEL: The Milnor number and Deformations of Complex Curve Singularities. Inventiones math. 58, 241-281 (1980).

[B] E. BRIESKORN: Die Monodromie der Isolierten Singularitäten von Hyperflächen. Manuscripta Math. 2, 103-161 (1970).

[D] A. DOUADY - J.L. VERDIER: Séminaire de Géométrie Analytique. Asterisque 16 (1974).

[G] R. GODEMENT: Topologie algebrique et théorie des faisceaux. Hermann (1958).

[H] H. HAMM: Habilitationsschrift, Göttingen (1974).

[dJ] Th. DE JONG: Line singularities transversal type A_2,A_3,D_4,E_6,E_7 or E_8 . Preprint Leiden (1986).

[L] E. LOOIJENGA: Isolated Singular Points on Complete Intersections. L.M.S. Lecture notes 77 Cambridge University Press (1984).

[Ma] B. MALGRANGE: Intégral asymptotiques et monodromie. Ann. Sci. Ec. Norm. Super. IV, Sér. 7, 405-430 (1974).

[Mi] J. MILNOR: Singular Points of Complex Hypersurfaces. Ann. of Math. Studies 61, Princeton (1968).

[Pe] G. PELLIKAAN: Hypersurface singularities and Resolutions of Jacobi Modules. Thesis Rijksuniversiteit Utrecht (1985).

[Ph] F. PHAM: Singularités des systèmes différentiels de Gauss-Manin Progress in Math., Vol. 2, Birkhäuser (1979).

[P-S] C. PESKINE and L. SZPIRO: Dimension projective finie et cohomologie locale, Publ. IHES 42 (1973).

[S] D. SIERSMA: Singularities with Critical Locus a 1-dimensional ICIS and Transversal type A_1. Preprint Utrecht (1986).

[S-S] J. SCHERK and J. STEENBRINK: On the Mixed Hodge Structure on the Cohomology of the Milnor Fibre. Math. Ann. 271, 641-665 (1985).

Reflexive modules on cyclic quotient surface singularities

J. Wunram
Mathematisches Seminar
Universität Hamburg
Bundesstraße 55
2000 Hamburg 13

Introduction

The quotient surface singularities are characterized to be those normal surface singularities which have finitely many isomorphism classes of indecomposable reflexive modules (see Herzog [5]). For such a quotient singularity $\mathbb{C}^2/G$ where G is a finite subgroup $G \subset Gl(2,\mathbb{C})$ with no pseudoreflections one obtains a one-to-one correspondence between the isomorphism classes of indecomposable reflexive modules and the isomorphism classes of irreducible representations of G

To give a geometrical explanation of the McKay correspondence for rational double points ($G \subset Sl(2,\mathbb{C})$) (see McKay [7]) G. Gonzalez-Sprinberg and J.L. Verdier (see [3]) associated to each non-trivial indecomposable reflexive module M the first Chern class of its pull-back on the minimal resolution of the singularity. They showed that the first Chern class intersects exactly one irreducible curve of the exceptional set of the minimal resolution and that this Chern class determines the indecomposable reflexive module M .

The main aim of this paper is to characterize geometrically the indecomposable reflexive modules on a cyclic quotient surface singularity by the first Chern class of its pull-back sheaves: From now on let $(X,x) = (\mathbb{C}^2/G,o)$ be the germ of the quotient singularity with quotient map $\mu : \mathbb{C}^2 \to \mathbb{C}^2/G$ where $G = G_{n,q} \subset Gl(2,\mathbb{C})$ is the finite cyclic group generated by $\varphi_{n,q} := \begin{pmatrix} \zeta_n & \\ & \zeta_n^q \end{pmatrix}$, $0<q<n$, $(n,q) = 1$, and ζ_n

is a primitive n-th root of unity. Let $\pi : \tilde{X} \to X$ be the minimal desingularization of X with exceptional system $E = \{E_i\}_{1 \leq i \leq r}$. If $n/q = b_1 - 1 \rfloor\overline{b_2} - 1 \rfloor\overline{\ } \dots - 1 \rfloor\overline{b_r}$, $b_i \in \mathbb{N}$, $b_i \geq 2$, $i = 1, \dots, r$, is the continued fraction expansion of n/q then the resolution graph of π is

$$\underset{E_1}{\overset{-b_1}{\bullet}} \!\!-\!\!-\!\!-\!\! \underset{E_2}{\overset{-b_2}{\bullet}} \quad \cdots \quad \underset{E_{r-1}}{\overset{-b_{r-1}}{\bullet}} \!\!-\!\!-\!\!-\!\! \underset{E_r}{\overset{-b_r}{\bullet}} \qquad \text{(c.f. [11])}$$

For every character χ_α (E_{χ_α} resp.) of G with $\chi_\alpha : \varphi_{n,q} \mapsto \zeta_n^\alpha$, $\alpha = 0, \dots, n-1$, one considers the associated reflexive module $M^{(\alpha)} := (\mu_*(\mathcal{O}_{\mathbb{C}^2} \otimes_{\mathbb{C}} E_{\chi_\alpha}))^G$. This defines the bijection between the set $\{ \chi_\alpha : \alpha = 0, \dots, n-1 \}$ and the set of all isomorphism classes of indecomposable reflexive modules on (X,x). The so-called "full" modules $\mathcal{M}^{(\alpha)} := \pi^* M^{(\alpha)}/\text{torsion}$ are locally free and determined by an effective divisor $D^{(\alpha)}$ on $\tilde{X}$ which intersects the exceptional locus E transversally in generic points and represents the first Chern class: $\mathcal{M}^{(\alpha)} \cong \mathcal{O}_{\tilde{X}}(D^{(\alpha)})$. There is a nice numerical algorithm to compute the intersection numbers $(D^{(\alpha)}.E_1, \dots, D^{(\alpha)}.E_r) \in \mathbb{N}^r$ as function of n, q and α. To derive this one constructs global generating sections of $M^{(\alpha)}$ using the theory of continued fraction expansions and invariant theoretic methods, so that one can restrict to the case where $\alpha = s_i$ with $D^{(s_i)}.E_j = \delta_{ij}$, $1 \leq i,j \leq r$. Now, applying Fujiki's resolution of cyclic quotient surface singularities the proof is an easy induction.

In case of arbitrary quotient surface singularities neither the first Chern class c_1 (see Esnault [2]) nor the Chern character (c_1, rk) (see [11]) of the pull-back module determines the corresponding reflexive module. However, there is a nice connection given by c_1 between the Auslander-Reiten quiver of the singularity (see Auslander [1]) and the resolution graph of the minimal desingularization which generalizes the McKay correspondence for rational double points. This is another part of my thesis which will be published later (see [12]).

Finally, I would like to thank Prof. O. Riemenschneider for suggesting these problems to me and Dr. K. Behnke for many helpful discussions.

A geometric description of the full modules

In this chapter we will prove the following theorem for the cyclic quotient surface singularity (X,x) of type (n,q) :

Theorem. Determine integers $s_0, s_1, \dots, s_r \in \mathbb{N}$ which are defined by the continued fraction expansion of n/q :

$$(1)\quad \begin{cases} s_0 := n \ ; \quad s_1 := q \ ; \\ s_j = b_{j+1}\cdot s_{j+1} - s_{j+2} \ , \quad 0<s_{j+2}<s_{j+1} \ , \qquad 0\leq j\leq r-2 \ ; \\ (s_r = 1) \end{cases}$$

For every $\alpha \in \mathbb{N}$ with $0\leq\alpha\leq r-1$ there are unique non-negative integers $d_1, \dots, d_r$ with:

$$(2)\quad \begin{cases} \alpha = d_1\cdot s_1 + t_1 & t_1 \in \mathbb{N} \quad 0\leq t_1 < s_1 \ ; \\ t_j = d_{j+1}\cdot s_{j+1} + t_{j+1} \ , & t_{j+1} \in \mathbb{N} \ , \ 0\leq t_{j+1}<s_{j+1} \ , \quad 1\leq j\leq r-1 \ ; \\ (t_r = 0) \end{cases}$$

Then the full module $M^{(\alpha)} \simeq \mathcal{O}_{\tilde{X}}(D^{(\alpha)})$ is given by

$$D^{(\alpha)}.E_i = d_i \ , \qquad 1\leq i\leq r \ .$$

In the first step of the proof we determine semi-invariants in $(\mathbb{C}[x,y] \otimes_{\mathbb{C}} E_{\chi_\alpha})^G$ which define global sections generating $M^{(\alpha)}$.
For this let us derive some properties of the continued fraction expansions used here:

The integers s_i defined in the theorem are characterized by the property $s_i/s_{i+1} = b_{i+1} - 1\rfloor\!\!\lceil b_{i+2} - 1\rfloor\!\!\lceil \dots -1\rfloor\!\!\lceil b_r$ and the following lemma gives an alternative interpretation of the tuples $(d_1, \dots, d_r)$ defined in (2).

Lemma 1. The set of $(d_1, \dots, d_r) \in \mathbb{N}^r$ for which there is an α , $0\leq\alpha\leq n-1$, with (2) equals the set of all $(d_1, \dots, d_r) \in \mathbb{N}^r$ with the two properies:

i) $0 \leq d_i \leq b_i - 1$, $1\leq i\leq r$

ii) If $d_i = b_i - 1$ and $d_j = b_j - 1$ with $1\leq i<j\leq r$, then there is an index i_0 with $i<i_0<j$ and $d_{i_0} \leq b_{i_0} - 3$.

Proof. We may assume $b_1 = b_2 = \dots = b_w = 2$, $b_{w+1} > 2$ with $0\leq w\leq r$.
Then formula (1) gives
$(b_{w+1} - 1)\cdot s_{w+1} - s_{w+2} = s_w - s_{w+1} = \dots = s_0 - s_1 = n - q$. Thus we have the continued fraction expansion

$$(n-q)/s_{w+1} = (b_{w+1} - 1) - 1\rfloor\!\!\lceil b_{w+2} - 1\rfloor\!\!\lceil \dots - 1\rfloor\!\!\lceil b_r \ .$$

Now, the lemma can be proved by induction: For $\underline{n = 1}$ there is nothing to prove. Assume $\underline{n>1}$: By induction hypothesis on q the corresponding tuple $(d_1, \ldots, d_r)$, $d_1 = 0$, in case $\alpha = 0, \ldots, q-1$ has the desired properties i) and ii). On the other hand every $(d_1, \ldots, d_r)$ with $d_1 = 0$, i) and ii) corresponds in this way to an $\alpha = 0, \ldots, q-1$. For $\alpha = q, q+1, \ldots, n-1$ the algorithm (2) shows $(d_1, \ldots, d_r) = (1,0, \ldots, 0, d_{w+1}, \ldots, d_r)$ where $d_{w+1} \leqslant b_{w+1} - 2$. Otherwise we would have $\alpha - q \geqslant (b_{w+1} - 1)\cdot s_{w+1} \geqslant n-q$ or $\alpha - q \geqslant s_i \geqslant n - q$, $1 \leqslant i \leqslant w$, respectively which is in contradiction to $\alpha \leqslant n-1$. Thus one gets the desired result by induction hypothesis for $n-q$. □

Let q' be the unique integer with $0<q'<n$ and $q\cdot q' \equiv 1 \pmod n$. By [10], p. 61, we have $n/q' = b_r - 1 \rfloor\!\overline{} b_{r-1} - 1 \rfloor\!\overline{} \ldots - 1 \rfloor\!\overline{} b_1$.
As in (1) we construct non-negative integers $\tilde{s}_{r+1}, \tilde{s}_r, \ldots, \tilde{s}_1$ by

$$(1)^\sim \begin{cases} \tilde{s}_{r+1} := n \ ; \quad \tilde{s}_r := q' \ ; \\ \tilde{s}_j \quad = b_{j-1}\cdot \tilde{s}_{j-1} - \tilde{s}_{j-2} \ , \quad 0<\tilde{s}_{j-2}<\tilde{s}_{j-1} \ , \qquad r+1 \geqslant j \geqslant 3 \\ (\tilde{s}_1 \quad = 1) \end{cases}.$$

Then the connection between s_i and $\tilde{s}_i$ is given by

Lemma 2. Let $0 \leqslant \alpha \leqslant n-1$ and $(d_1, \ldots, d_r) \in \mathbb{N}^r$ the corresponding numbers. Then the unique integer $\tilde{\alpha}$ with $0 \leqslant \tilde{\alpha} \leqslant n-1$ and $q\cdot\tilde{\alpha} \equiv \alpha \pmod n$ is defined by $\tilde{\alpha} = d_1\cdot\tilde{s}_1 + \ldots + d_r\cdot\tilde{s}_r$.

Proof. For $\alpha = s_i$, $1 \leqslant i \leqslant r$, we prove the lemma by induction:
"i=1" : $q\cdot\tilde{s}_1 = q\cdot 1 = s_1$
"i→i+1" : By induction hypothesis one has: $q\cdot\tilde{s}_{i+1} = q\cdot b_i\cdot\tilde{s}_i - q\cdot\tilde{s}_{i-1} \equiv b_i\cdot s_i - s_{i-1} = s_{i+1} \pmod n$.
Thus for $\alpha = d_1\cdot s_1 + \ldots + d_r\cdot s_r$ and $\tilde{\alpha} = d_1\cdot\tilde{s}_1 + \ldots + d_r\cdot\tilde{s}_r$ the formula $q\cdot\tilde{\alpha} \equiv \alpha \pmod n$ is valid and Lemma 1 applied to n/q and n/q' respectively shows: $0 \leqslant \tilde{\alpha} \leqslant n-1$. □

An important aid for constructing minimal sets of generators of the invariants related to certain characters of the cyclic group G will be

Lemma 3. Let $1 \leqslant \alpha \leqslant n-1$ and $(d_1, \ldots, d_r) = (d_1, \ldots, d_w, 0, \ldots, 0)$, $d_w > 0$, $1 \leqslant w \leqslant r$, the corresponding tuple determined by (2). Let i,j be integers with $0 \leqslant i,j \leqslant n-1$ and $i+q\cdot j \equiv \alpha \pmod n$. Then the inequality $i < s_w$ implies $j \geqslant \tilde{s}_w$.

Proof. If $i=0$ the integer j is equal to $\tilde{\alpha}$ and Lemma 2 implies the statement. So, assume $0<i<s_w$ and determine $\hat{d}_{w+1}, \ldots, \hat{d}_r \in \mathbb{N}$ by

$$\begin{cases} s_w - i = \hat{d}_{w+1}\cdot s_{w+1} + t_{w+1} , & t_{w+1} \in \mathbb{N} , \ 0 \leq t_{w+1} < s_{w+1} ; \\ t_j = \hat{d}_{j+1}\cdot s_{j+1} + t_{j+1} , & t_{j+1} \in \mathbb{N} , \ 0 \leq t_{j+1} < s_{j+1} , \ w+1 \leq j \leq r-1 ; \\ (t_r = 0) & \end{cases}$$

Formula (2) shows

$$\alpha - i = d_1\cdot s_1 + \ldots + d_{w-1}\cdot s_{w-1} + (d_w - 1)\cdot s_w + \hat{d}_{w+1}\cdot s_{w+1} + \ldots + \hat{d}_r\cdot s_r$$

and by Lemma 2 we have

$$j := (\alpha - i)^{\sim} = d_1\cdot \tilde{s}_1 + \ldots + d_{w-1}\cdot \tilde{s}_{w-1} + (d_w - 1)\cdot \tilde{s}_w + \underbrace{\hat{d}_{w+1}\cdot \tilde{s}_{w+1} + \ldots + \hat{d}_r\cdot \tilde{s}_r}_{> \tilde{s}_w} .$$

The integer j has the property $0 \leq j < n$, $j > \tilde{s}_w$ and $i + q\cdot j \equiv \alpha \pmod n$. □

Now, we can easily determine suitable invariants:

<u>Lemma 4.</u> Let α , $1 \leq \alpha \leq n-1$, be an integer and χ_α ($E_{\chi_\alpha} \cong \mathbb{C}$ resp.) the corresponding character of G . Then a set of generators for the $\mathbb{C}[x,y]^G$-module $(\mathbb{C}[x,y] \otimes_{\mathbb{C}} E_{\chi_\alpha})^G$ is given by the monomials $x^i\cdot y^j$, $0 \leq i,j \leq n-1$ and $i + q\cdot j \equiv \alpha \pmod n$.

<u>Proof.</u> By E. Noether (see [8]) a desired set of generators is $\{ \mu_\alpha(x^i\cdot y^j) : 0 \leq i,j < n \}$ where μ_α is the canonical map

$$\mu_\alpha : \begin{cases} \mathbb{C}[x,y] & \to (\mathbb{C}[x,y] \otimes_{\mathbb{C}} E_{\chi_\alpha})^G \\ f & \to 1/n \cdot \sum_{g \in G} \chi_\alpha(g)\cdot f^g \end{cases} .$$

Thus the following computation proves the lemma:

$$\mu_\alpha(x^i\cdot y^j) = (1/n \cdot \sum_{\nu=0}^{n-1} (\zeta_n^{\alpha-i-qj})^\nu)\cdot x^i\cdot y^j =$$

$$= \begin{cases} 1/n \dfrac{1-(\zeta_n^{\alpha-i-q\cdot j})^n}{1-\zeta_n^{\alpha-i-q\cdot j}} x^i\cdot y^j = 0 , & \text{if } \zeta_n^{\alpha-i-q\cdot j} \neq 1 \\ x^i\cdot y^j , & \text{if } \zeta_n^{\alpha-i-q\cdot j} = 1 . \end{cases}$$

□

<u>Lemma 5.</u> i) If $\alpha = s_i$, $1 \leq i \leq r$, then the $\mathbb{C}[x,y]^G$-module $(\mathbb{C}[x,y] \otimes_{\mathbb{C}} E_{\chi_\alpha})^G$ is generated by x^α and $y^{\tilde{\alpha}}$.

ii) For $\alpha = d_1\cdot s_1 + \ldots + d_r\cdot s_r$ with (2) the monomial $x^i\cdot y^j \in$ $\in (\mathbb{C}[x,y] \otimes_{\mathbb{C}} E_{\chi_\alpha})^G$ (i.e. $i + q\cdot j \equiv \alpha \pmod n$) can be decomposed in a product

$$x^i\cdot y^j = f_0\cdot f_1^{(1)}\cdot \ldots \cdot f_1^{(d_1)}\cdot \ldots \cdot \ldots \cdot f_r^{(1)}\cdot \ldots \cdot f_r^{(d_r)}$$

where $f_\ell^{(k)} \in (\mathbb{C}[x,y] \otimes_{\mathbb{C}} E_{X_{s_\ell}})^G$, $1 \leqslant \ell \leqslant r$, is a monomial as above in i) and $f_0 \in \mathbb{C}[x,y]^G$ is an absolut invariant of G .

Proof. Write $\alpha = d_1 \cdot s_1 + \ldots + d_w \cdot s_w$ as in Lemma 3 and consider $x^i . y^j$ with $i+q \cdot j \equiv \alpha \pmod n$. Lemma 3 implies

$$x^i . y^j = \begin{cases} x^{s_w} \cdot (x^{i-s_w} \cdot y^j) , & \text{if } i \geqslant s_w \\ y^{\tilde{s}_w} \cdot (x^i \cdot y^{j-\tilde{s}_w}) , & \text{if } i < s_w \end{cases}$$

where $x^{i-s_w} \cdot y^j$ and $x^i \cdot y^{j-\tilde{s}_w} \in (\mathbb{C}[x,y] \otimes_{\mathbb{C}} E_{X_{\alpha - s_w}})^G$. Inductively we get the desired result. □

A consequence of Lemma 5 for the reflexive modules $M^{(\alpha)}$ is

Lemma 6. Let $\alpha = d_1 \cdot s_1 + \ldots + d_r \cdot s_r$ with (2). Then there is an isomorphism

$$M^{(\alpha)} \simeq (M^{(s_1) \otimes d_1} \otimes_{\mathcal{O}_X} \ldots \otimes_{\mathcal{O}_X} M^{(s_r) \otimes d_r}) \ / \ \text{torsion}$$

in a neighbourhood of the singular point $x \in X$.

Proof. By Lemma 5, ii) the canonical multiplication map

$$M^{(s_1) \otimes d_1} \otimes_{\mathcal{O}_X} \ldots \otimes_{\mathcal{O}_X} M^{(s_r) \otimes d_r} \to M^{(\alpha)}$$

is an epimorphism in a neighbourhood of x . The reflexive sheaf $M^{(\alpha)}$ is torsion-free and the kernel of the morphism is concentrated in $\{x\}$. □

Now, to prove the theorem we <u>always</u> may assume <u>$\alpha = s_i$</u> (see [4]):

$$\begin{aligned} M^{(\alpha)} &\simeq \pi^*((M^{(s_1) \otimes d_1} \otimes \ldots \otimes M^{(s_r) \otimes d_r})/\text{torsion}) \ / \ \text{torsion} \\ &\simeq \pi^*(M^{(s_1) \otimes d_1} \otimes \ldots \otimes M^{(s_r) \otimes d_r}) \ / \ \text{torsion} \\ &\simeq M^{(s_1) \otimes d_1} \otimes \ldots \otimes M^{(s_r) \otimes d_r} \end{aligned}$$

In the following we consider for $\alpha = s_i$ and for generic $[\lambda_1 : \lambda_2] \in \mathbb{P}^1$ the strict transform of $\{ \lambda_1 \cdot x^\alpha + \lambda_2 \cdot y^{\tilde{\alpha}} = 0 \} \subset X$ under $\pi : \tilde{X} \to X$. For this we sketch Fujiki's construction to resolve the singularity (X,x) (see [9]) : Let $V(x,y) = \mathbb{C}^2$ be the space on which the cyclic group of order n , G_n , operates like $G = G_{n,q}$. The covering map

$$\begin{cases} Y(y_1,y_2) = \mathbb{C}^2 & \to \quad V \\ (y_1,y_2) & \mapsto \quad (y_1,y_2{}^q) \end{cases}$$

is the quotient map $Y \to Y/G_q$ where a generator $\varkappa$ of G_q operates by $(y_1,y_2) \mapsto (y_1,\zeta_q y_2)$. If $\sigma : \tilde{Y} \to Y$ denotes the blow up of Y at the origin and $\tilde{Y}$ is divided by G_n and G_q then the quotient has exactly one cyclic singularity of typ (s_1,s_2) :

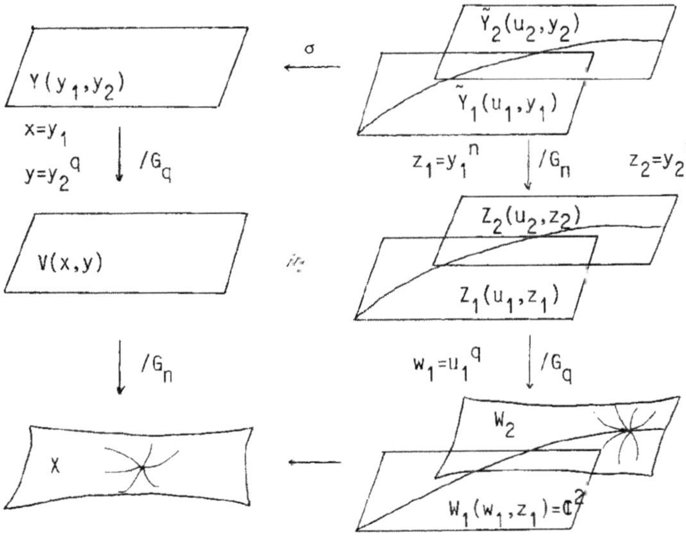

The generators η of G_n and $\varkappa$ of G_q act by:

operation of G_n on V :	(x,y)	$\mapsto$	$(\zeta_n \cdot x\ ,\ \zeta_n^q \cdot y)$
operation of G_n on Y :	(y_1,y_2)	$\mapsto$	$(\zeta_n \cdot y_1, \zeta_n \cdot y_2)$
operation of G_q on Y :	(y_1,y_2)	$\mapsto$	$(y_1\ ,\ \zeta_q \cdot y_2)$
operation of G_n on $\tilde{Y}_1$:	(u_1,y_1)	$\mapsto$	$(u_1\ ,\ \zeta_n \cdot y_1)$
operation of G_n on $\tilde{Y}_2$:	(u_2,y_2)	$\mapsto$	$(u_2\ ,\ \zeta_n \cdot y_2)$
operation of G_q on $\tilde{Y}_1$:	(u_1,y_1)	$\mapsto$	$(\zeta_q \cdot u_1,\ y_1)$
operation of G_q on $\tilde{Y}_2$:	(u_2,y_2)	$\mapsto$	$(\zeta_q^{-1} u_2, \zeta_q \cdot y_2)$
operation of G_q on Z_1 :	(u_1,z_1)	$\mapsto$	$(\zeta_q \cdot u_1,\ z_1)$
operation of G_q on Z_2 :	(u_2,z_2)	$\mapsto$	$(\zeta_q^{-1} u_2, \zeta_q^n \cdot z_2)$

The blow up σ is given by

$$(y_1,y_2) = (y_1,u_1 \cdot y_1) \quad , \quad (u_1,y_1) \in \tilde{Y}_1$$
$$(y_1,y_2) = (u_2 \cdot y_2,y_2) \quad , \quad (u_2,y_2) \in \tilde{Y}_2 \quad .$$

By this construction the singularity (X,x) can be resolved inductively and we obtain the following lemma as a consequence of it.

Lemma 7. For $\alpha = s_i$ and generic $[\lambda_1:\lambda_2] \in \mathbb{P}^1$ the strict transform $D^{(\alpha)}$ of $\{\lambda_1 \cdot x^\alpha + \lambda_2 \cdot y^{\tilde{\alpha}} = 0\}$ under π meets the exceptional divisor E transversally in one point:

$$D^{(\alpha)}.E_j = \delta_{ij}$$

To prove this lemma we need

claim. For $\alpha = s_i$ we have

i) $0 \leqslant \frac{q \cdot \tilde{\alpha} - \alpha}{n} < q$

ii) $s_2 \cdot (\frac{q \cdot \tilde{\alpha} - \alpha}{n}) \equiv \alpha \pmod q$

iii) $q \cdot \tilde{\alpha} - \alpha = 0 \iff \alpha = s_1 = q$

Proof of claim. to i): There are implications

$q \cdot \tilde{\alpha} \equiv \alpha \pmod n \implies q \cdot \tilde{\alpha} \geqslant \alpha \implies \frac{q \cdot \tilde{\alpha} - \alpha}{n} \geqslant 0$

and

$0 \leqslant \tilde{\alpha} < n \implies q \cdot \tilde{\alpha} < q \cdot n \implies \frac{q \cdot \tilde{\alpha} - \alpha}{n} < q$.

to ii): We have

$s_2 \cdot (\frac{q \cdot \tilde{\alpha} - \alpha}{n}) = (b_1 \cdot s_1 - n) \cdot (\frac{q \cdot \tilde{\alpha} - \alpha}{n}) \equiv -n \cdot (\frac{q \cdot \tilde{\alpha} - \alpha}{n}) \equiv \alpha \pmod q$.

to iii): Since $\alpha = s_i$ and $s_i < q$ for $i \geqslant 2$ this statement is trivial. □

Proof of Lemma 7. The strict transform of $\{\lambda_1 \cdot x^\alpha + \lambda_2 \cdot y^{\tilde{\alpha}} = 0\} \subset X$ in W_1 and Z_2 respectively is defined by:

$$W_1 : \quad \{\lambda_1 + \lambda_2 \cdot z_1^{(\frac{q \cdot \tilde{\alpha} - \alpha}{n})} \cdot w_1^{\tilde{\alpha}} = 0\}$$

$$Z_2 : \quad \{\lambda_1 \cdot u_2^{\alpha} + \lambda_2 \cdot z_2^{(\frac{q \cdot \tilde{\alpha} - \alpha}{n})} = 0\} .$$

case $r = q = 1$: In this case we have $W_2 = Z_2$ with coordinate transforms in $W_1 \cap W_2$:

$$w_1 = 1/u_2 \ , \quad z_1 = u_2^n \cdot z_2 \ .$$

The strict transform intersects the exceptional locus for generic $[\lambda_1:\lambda_2] \in \mathbb{P}^1$ transversally in exactly one generic point ($\alpha = \tilde{\alpha} = q = 1$!):

$$\{w_1 = -\lambda_1/\lambda_2\} = \{u_2 = -\lambda_2/\lambda_1\} .$$

$r \to r+1$: For generic $[\lambda_1:\lambda_2] \in \mathbb{P}^1$ and $\alpha \neq q$ there is no point of intersection of the strict transform with the exceptional locus in W_1 (see claim iii)) . Because of claim i), ii) the induction hypothesis implies the statement.

On the other hand for $\alpha = s_1 = q$ the strict transform meets the exceptional locus in exactly one point of W_1, of course transversally, and there are no other points of intersection. □

The following lemma finally proves the theorem.

<u>Lemma 8.</u> For $\alpha = s_i$ we have

$$M^{(\alpha)} \simeq \mathcal{O}_{\tilde{X}}(D^{(\alpha)}) \quad ,$$

where $D^{(\alpha)}$ is the divisor defined in Lemma 7.

<u>Proof.</u> (see [6], § 5) For generic $[\lambda_1 : \lambda_2] \in \mathbb{P}^1$ the divisor of the global section $\lambda_1 \cdot \pi^* x^\alpha + \lambda_2 \cdot \pi^* y^{\tilde{\alpha}} \in \Gamma(\tilde{X}, M^{(\alpha)})$ is equal to $D^{(\alpha)}$ plus eventually one divisor supported on E. But $\lambda_1 \cdot \pi^* x^\alpha + \lambda_2 \cdot \pi^* y^{\tilde{\alpha}}$ generates $M^{(\alpha)}$ on $\tilde{X} - D^{(\alpha)}$:

By Lemma 5 the global sections $\pi^* x^\alpha$, $\pi^* y^{\tilde{\alpha}} \in \Gamma(\tilde{X}, M^{(\alpha)})$ generate $M^{(\alpha)}$ and one has

$$\pi^* x^\alpha = \frac{1}{\lambda_1 + \lambda_2 \cdot \pi^*(y^{\tilde{\alpha}}/x^\alpha)} \cdot (\lambda_1 \cdot \pi^* x^\alpha + \lambda_2 \cdot \pi^* y^{\tilde{\alpha}})$$

and

$$\pi^* y^{\tilde{\alpha}} = \frac{1}{\lambda_1 \cdot \pi^*(x^\alpha / y^{\tilde{\alpha}}) + \lambda_2} \cdot (\lambda_1 \cdot \pi^* x^\alpha + \lambda_2 \cdot \pi^* y^{\tilde{\alpha}})$$

respectively. The coefficients are global meromorphic functions which do not have any poles along one component E_j of E . Otherwise we would have $\pi^*(y^{\tilde{\alpha}}/x^\alpha) \equiv -\lambda_1/\lambda_2$ or $\pi^*(x^\alpha/y^{\tilde{\alpha}}) \equiv -\lambda_2/\lambda_1$ along E_j which is in contradiction to the generic choice of $[\lambda_1 : \lambda_2]$ □

<u>Example.</u> In case $(n,q) = (7,3)$ the resolution graph of π is

$$\underset{-3}{\bullet} \!\!-\!\!-\!\!-\!\! \underset{-2}{\bullet} \!\!-\!\!-\!\!-\!\! \underset{-2}{\bullet} \quad ,$$

that is: $b_1 = 3$, $b_2 = 2$, $b_3 = 2$. The theorem states $s_0 = 7$, $s_1 = 3$, $s_2 = 2$, $s_3 = 1$ and

	$(D^{(\alpha)}.E_1, D^{(\alpha)}.E_2, D^{(\alpha)}.E_3)$
$\alpha = 1$	(0,0,1)
$\alpha = 2$	(0,1,0)
$\alpha = 3$	(1,0,0)
$\alpha = 4$	(1,0,1)
$\alpha = 5$	(1,1,0)
$\alpha = 6$	(2,0,0)

□

Remark. The minimal number of generators $ck(M^{(\alpha)})$ of the $\mathcal{O}_X$-module $M^{(\alpha)}$ is given by

$$ck(M^{(\alpha)}) = \sum_{i=1}^{r} d_i + 1$$

This is a special case of a more general formula for the minimal number of generators of a reflexive module M on a rational surface singularity (see [12]).

References.

[1] *Auslander, M.*: Rational singularities and almost split sequences. Trans. AMS **293**, 511 - 531 (1986).

[2] *Esnault, H.*: Reflexive modules on quotient surface singularities. J. Reine Angew. Math. **362**, 63 - 71 (1985).

[3] *Gonzalez-Sprinberg, G., Verdier, J.-L.*: Construction géométrique de la correspondance de McKay. Ann. Sci. Ecole Norm. Sup. **16**, 409 - 449 (1983).

[4] *Grauert, H., Riemenschneider, O.*: Verschwindungssätze für analytische Kohomologiegruppen auf komplexen Räumen. Invent. Math. **11**, 263 - 292 (1970).

[5] *Herzog, J.*: Ringe mit nur endlich vielen Isomorphieklassen von maximalen unzerlegbaren Cohen-Macaulay-Moduln. Math. Ann. **233**, 21 - 34 (1978).

[6] *Knörrer, H.*: Group representations and resolution of rational double points. Contemp. Math. **45**, 175 - 222 (1985).

[7] *McKay, J.*: Graphs, singularities and finite groups. In: Finite Groups, Santa Cruz 1979. Proc. Sympos. Pure Math. **37**, 183 - 186 (1980).

[8] *Noether, E.*: Der Endlichkeitssatz der Invarianten endlicher Gruppen. Math. Ann. **77**, 89 - 92 (1916).

[9] *Pinkham, H.*: Singularité de Klein - I,II. In: Séminaire sur les Singularités des Surfaces, Palaiseau 1976 - 1977. Lecture Notes in Mathematics 777, pp. 1 - 20. Berlin - Heidelberg - New York: Springer 1980.

[10] *Randow, R.v.*: Zur Topologie von dreidimensionalen Baummannigfaltigkeiten. Bonner Math. Schriften 14 (1962).

[11] ***Riemenschneider, O***: Deformation von Quotientensingularitäten (nach zyklischen Gruppen). Math. Ann. **209**, 211-248 (1974).

[12] *Wunram*, J.: Reflexive Moduln auf zweidimensionalen Quotientensingularitäten. Dissertation Fachber. Math. Univ. Hamburg (1986).

ALMOST SPLIT SEQUENCES FOR $\mathbb{Z}$-GRADED RINGS

Maurice Auslander[1] and Idun Reiten

Our main purpose in this paper is to prove the following. Let $R = k[X_1,\ldots,X_d]$ be a polynomial ring in indeterminates $X_1,\ldots,X_d$ over a field k together with a $\mathbb{Z}$-grading such that the degrees of the X_i are at least one and the constants have degree zero. Suppose Λ is a $\mathbb{Z}$-graded R-algebra (see §1 for definition) which is a Cohen-Macaulay ring and let $CM(\mathrm{gr}\ \Lambda)_0$ denote the category of finitely generated graded (maximal) Cohen-Macaulay Λ-modules with degree zero morphisms. Then an indecomposable nonprojective C in $CM(\mathrm{gr}\ \Lambda)_0$ has an almost split sequence $0 \to A \to B \to C \to 0$ in $CM(\mathrm{gr}\ \Lambda)_0$ if and only if $C_{\underline{p}}$ is $\Lambda_{\underline{p}}$-projective for all nonmaximal prime ideals $\underline{p}$ of R. The method of proof is similar to that used in [1] and [8] to prove the analogous nongraded theorem where R is a complete Gorenstein ring. When $d = 2$, this theorem was proven in [5] in the case Λ is a polynomial ring with the usual $\mathbb{Z}$-grading modulo a homogeneous ideal, using very different methods.

As an application of this result, we show that the $\mathbb{Z}$-graded ring $k[X_1,\ldots,X_5]/(X_1X_3-X_2^2, X_1X_5-X_2X_4, X_2X_5-X_3X_4)$ where $k[X_1,\ldots,X_5]$ has the usual $\mathbb{Z}$-grading has only a finite number of nonisomorphic indecomposable graded Cohen-Macaulay modules, up to shift. To do this we first establish a criterion for proving finite Cohen-Macaulay type, and here we would like to thank Sverre Smalø for helpful conversations. The above ring is a graded scroll. The corresponding complete ring was shown to be of finite Cohen-Macaulay

---

[1] Written with partial support of the National Science Foundation contract # MCS 83-03348

type in [6], using analogous methods. It would be interesting to be able to deduce one result directly from the other, rather than by imitating the method. As in [6] we get a description of which graded scrolls are of finite graded Cohen-Macaulay type.

We also give a translation of our results for almost split sequences to certain subcategories of locally free coherent sheaves on a nonsingular projective variety. Except for curves, these subcategories depend on the embedding of the projective variety in projective space, and there can even be a finite number of indecomposables (up to shift) in one embedding, and an infinite number in another.

We refer the reader to [12] for general facts about graded modules over $\mathbb{Z}$-graded rings, and to [13] for basic facts about almost split sequences.

1. Existence of almost split sequences for $\mathbb{Z}$-graded rings.

Before we can state our main theorem, we need to introduce some notation and terminology.

Let k be a field and R the polynomial ring $k[X_1,\ldots,X_d]$. We consider R as a $\mathbb{Z}$-graded ring, where X_i has a fixed degree $n_i \geq 1$.

Assume that Λ is a $\mathbb{Z}$-graded R-algebra, that is, we have a ring homomorphism $i\colon R \to \Lambda$ of degree zero, with $i(R)$ contained in the center of Λ, and assume that Λ is a finitely generated projective R-module. The only modules we consider over R and Λ are graded by $\mathbb{Z}$. A morphism $f\colon A \to B$ between graded Λ-modules is homogeneous of degree t if $f(A_i) \subset B_{i+t}$ for all i. If A is finitely generated, every element in $\operatorname{Hom}_\Lambda(A,B)$ is a sum of homogeneous maps. For a graded Λ-module A, $A(t)$ denotes the graded Λ-module with $A(t)_i = A_{t+i}$.

We denote by $\operatorname{Mod}(\operatorname{gr}\Lambda)$ the graded Λ-modules, by $\operatorname{mod}(\operatorname{gr}\Lambda)$ the finitely generated graded (left) Λ-modules, by $CM(\mathrm{gr}\ \Lambda)$ the subcategory whose objects are projective R-modules. Further $L_P(\mathrm{gr}\ \Lambda)$ denotes the full subcategory of $CM(\mathrm{gr}\ \Lambda)$ whose objects C are such that $C_{\underline{p}}$ is $\Lambda_{\underline{p}}$-projective for each nonmaximal prime in R, and $L_I(\mathrm{gr}\ \Lambda)$ denotes the full subcategory of $CM(\mathrm{gr}\ \Lambda)$ whose objects A are such that $A_{\underline{p}}$ is an injective object in $CM(\mathrm{gr}\ \Lambda_{\underline{p}})$ for each nonmaximal prime ideal $\underline{p}$ in R. For each of the categories a subscript 0 will indicate that we consider the degree zero morphisms. $\underline{L_P(\mathrm{gr}\ \Lambda)_0}$ denotes $L_P(\mathrm{gr}\ \Lambda)_0$ modulo projectives, and $\overline{L_I(\mathrm{gr}\ \Lambda)_0}$ denotes $L_I(\mathrm{gr}\ \Lambda)_0$ modulo injectives in $CM(\mathrm{gr}\ \Lambda)$.

Since the Krull-Schmidt theorem holds in $\mathrm{mod}(\mathrm{gr}\ \Lambda)_0$, the objects have minimal projective resolutions. For X in $\mathrm{mod}(\mathrm{gr}\ \Lambda)$, let $P_i \xrightarrow{f} P_{i-1} \to \cdots \to P_1 \to P_0 \to X \to 0$ be a minimal projective resolution in $\mathrm{mod}(\mathrm{gr}\ \Lambda)_0$. Then $\Omega^i X =$ Ker f, and $\mathrm{Tr}\ X$ is given by the exact sequence $\mathrm{Hom}_\Lambda(P_0,\Lambda) \to \mathrm{Hom}_\Lambda(P_1,\Lambda) \to \mathrm{Tr}\ X \to 0$ in $\mathrm{mod}(\mathrm{gr}\ \Lambda^{op})_0$. We define $\mathrm{Tr}_L X = \Omega^d\, \mathrm{Tr}\, X$, where $d = \dim R$, and for X in $\mathrm{mod}(\mathrm{gr}\ \Lambda)$ we define $D(X) = \mathrm{Hom}_R(X,R)$ in $\mathrm{mod}(\mathrm{gr}\ \Lambda^{op})$.

It should be noted that $\mathrm{Tr}_L X$ and therefore $D\,\mathrm{Tr}_L X$ are in $CM(\Lambda^{op})$ and $CM(\Lambda)$ respectively. Also if X in $CM(\Lambda)$ is not projective and $X_{\underline{p}}$ is $\Lambda_{\underline{p}}$ projective for all nonmaximal prime ideal $\underline{p}$ of R, then $D\,\mathrm{Tr}_L X$ is indecomposable. This follows easily from the fact that $\mathrm{Tr}\ X$ is indecomposable and $\mathrm{Ext}^i_\Lambda(\mathrm{Tr}\ X,\Lambda) = 0$ for $i = 1,\ldots,d$ (see [1, 7.5]).

We have the following existence theorem for almost split sequences.

<u>Theorem 1.1</u>. Let $R = k[X_1,\ldots,X_d]$ be as before, and let Λ be a Z-graded R-algebra which is a projective graded R-module. Let $\tilde{d} = n_1 + \cdots + n_d$, where n_i is the degree of X_i.

(a) Let C be indecomposable nonprojective in $CM(\mathrm{gr}\ \Lambda)$ (A an indecomposable noninjective object in $CM(\mathrm{gr}\ \Lambda)$). Then

C is in $L_P(\mathrm{gr}\ \Lambda)$ (A is in $L_I(\mathrm{gr}\ \Lambda)$) if and only if there is an almost split sequence $0 \to A \to B \to C \to 0$ in $CM(\mathrm{gr}\ \Lambda)_0$. When there is such an almost split sequence, then $A = D\,\mathrm{Tr}_L\, C(-\tilde{d})$.

(b) Let C be indecomposable in $CM(\mathrm{gr}\ \Lambda)$ (A indecomposable in $CM(\mathrm{gr}\ \Lambda)$). Then C is in $L_P(\mathrm{gr}\ \Lambda)$ (A is in $L_I(\mathrm{gr}\ \Lambda)$) if and only if there is a minimal right almost split map $B \to C$ (a minimal left almost split map $A \to B$) in $CM(\mathrm{gr}\ \Lambda)_0$.

Proof. The proof is modeled on the complete case [8], and we just give a brief outline, stressing the points which are particular to the graded case. Usually it is the question of checking that the morphisms involved are of degree zero. As in [8] an essential part of our proof is to establish the following formula, which is proved by checking that the morphisms involved are of degree zero.

Proposition 1.2. Let $R = k[X_1,\dots,X_d]$ be graded by an abelian group G, and let Λ be a G-graded R-algebra. For each $i \in G$ we have a natural degree zero isomorphism $\alpha\colon \mathrm{Ext}^1_\Lambda(C, D\,\mathrm{Tr}_L\, X(i)) \to \mathrm{Hom}_R(\underline{\mathrm{Hom}}_\Lambda(X,C), I_d(i))$, for C in $CM(\mathrm{gr}\ \Lambda)$ and X in $L_P(\mathrm{gr}\ \Lambda)$, which is functorial in C.

Here $\underline{\mathrm{Hom}}$ denotes the ordinary maps modulo those factoring through projectives, and I_d denotes the last term in the minimal injective resolution $0 \to R \to I_0 \to \dots \to I_d \to 0$ in $\mathrm{Mod}(\mathrm{gr}\ \Lambda)_0$.

Assume again that we have our usual Z-grading. Then we have by Proposition 1.2 for each C in $L_P(\mathrm{gr}\ \Lambda)$ a natural degree zero isomorphism $\beta\colon \mathrm{Ext}^1_\Lambda(C, D\,\mathrm{Tr}_L\,(C(-\tilde{d})) \xrightarrow{\sim} \mathrm{Hom}_R(\underline{\mathrm{Hom}}_\Lambda(C,C), I_d(-\tilde{d}))$. Since C is in $L_P(\mathrm{gr}\ \Lambda)$, $\Gamma = \underline{\mathrm{Hom}}_\Lambda(C,C)$ is a finite dimensional Z-graded algebra over k. Then the subgroup $\underline{r}$ which is the direct sum of all Γ_i for $i \neq 0$ and $\mathrm{rad}\,\Gamma_0$ is an ideal in Γ. We here use that $\mathbf{Z}$ is a torsionfree group (see [5,II 1.7]). Hence there is a nonzero degree zero map from Γ to $\Gamma_0/\mathrm{rad}\,\Gamma_0$ and consequently a nonzero degree zero R-map f from Γ to the simple graded R-module $R/\underline{m} = k$, where $\underline{m} = (X_1,\dots,X_d)$, with the property that $f(\text{rad } \Gamma_0) = 0$. We next want to show that $\text{Hom}(R/\underline{m}, I_d(-\tilde{d}))_0 \neq 0$. The Koszul complex gives a minimal projective resolution $0 \to R(-\tilde{d}) \to \dots \to R \to R/\underline{m} \to 0$ in $\text{mod}(\text{gr } R)_0$ (see [10;6.13]). This shows $\text{Ext}^d_R(R/\underline{m}, R(-\tilde{d}))_0 \neq 0$. Considering the injective resolution in $\text{Mod}(\text{gr } \Lambda)_0$ $0 \to R(-\tilde{d}) \to \dots \to I_d(-\tilde{d}) \to 0$ and the induced complex $0 \to \text{Hom}_R(R/\underline{m}, R(-\tilde{d})) \to \dots \to \text{Hom}_R(R/\underline{m}, I_d(-\tilde{d})) \to 0$, we see that $\text{Hom}_R(R/\underline{m}, I_d(-\tilde{d}))_0 \neq 0$. We then have a nonzero degree zero map $\bar{f}\colon \underline{\text{Hom}}(C,C) \to I_d(-\tilde{d})$ such that $\bar{f}(\text{rad } \underline{\text{Hom}}(C,C)_0) = 0$. Since the above isomorphism β is of degree zero, $\beta^{-1}(\bar{f})$ gives a nonsplit exact sequence $0 \to D\,\text{Tr}\, C(-\tilde{d}) \to E \to C \to 0$ in $\text{mod}(\text{gr } \Lambda)_0$. Since $D\,\text{Tr}\, C(-\tilde{d})$ is indecomposable, standard arguments show that this sequence is almost split in $CM(\text{gr } \Lambda)_0$ (see [3,Prop. 4.1]).

It follows as in the nongraded case in [2, Prop. 6.6] that $CM(\text{gr } \Lambda)_0$ is contravariantly finite in $\text{mod}(\text{gr } \Lambda)_0$ (see [7, Prop. 2.3]). This means that given any T in $\text{mod}(\text{gr } \Lambda)$, there is a degree zero map $g\colon B \to T$ with B in $CM(\text{gr } \Lambda)$, such that for any degree zero map $h\colon X \to T$ with X in $CM(\text{gr } \Lambda)$, there is some $t\colon X \to B$ with $gt = h$. Applying this to $T = \text{rad } C$ when C is indecomposable projective in $\text{mod}(\text{gr } \Lambda)_0$, we get a minimal right almost split map $g\colon B \to C$ in $CM(\text{gr } \Lambda)_0$. It follows by duality that if A is an indecomposable injective object in $CM(\text{gr } \Lambda)_0$, we have a minimal left almost split map $h\colon A \to B$.

That it is necessary that C is in $L_P(\text{gr } \Lambda)$ and A is in $L_I(\text{gr } \Lambda)$ to have an almost split sequence $0 \to A \to B \to C \to 0$ follows as in [2, Prop. 3.1, Lemma 3.2] using that $\mathbb{Z}$ is a torsionfree group.

We note that most of the proof of the above theorem is valid when R is graded by a torsionfree group. To get a nice formula for the shift it is however convenient to have a $\mathbb{Z}$-grading of the type we assumed.

It follows from Theorem 1.1 that if $\text{Hom}_R(\Lambda, R)$ is in $L_P(\text{gr } \Lambda)$ or equivalently $L_P(\text{gr } \Lambda) = L_I(\text{gr } \Lambda) = CM(\text{gr } \Lambda)$, we have the existence of almost split sequences and minimal left and right almost split maps in $CM(\mathrm{gr}\ \Lambda)_0$. We then also have the standard connection with irreducible maps (see [4, section 2]).

2. Application to proving finite graded Cohen-Macaulay type.

Using the results in section one, we prove a criterion for finite graded Cohen-Macaulay type, and then apply this criterion to a concrete example. To avoid introducing extra terminology, we do not state the most general result. However it is worth noting that the criterion holds with the same proof in other graded contexts where almost split sequences exist, like in the situation discussed in [7].

Theorem 2.1. Let $R = k[X_1,\dots,X_d]$ be a $\mathbb{Z}$-graded ring as before, and let Λ be an indecomposable $\mathbb{Z}$-graded R-algebra which is a finitely generated free R-module. Assume that $CM(\mathrm{gr}\ \Lambda) = L_p(\mathrm{gr}\ \Lambda)$. Let $\mathcal{D}$ be a nonempty subset of indecomposable objects in $CM(\mathrm{gr}\ \Lambda)$, having only a finite number of indecomposables up to shift. Then each of the following conditions is sufficient for $\mathcal{D}$ to consist of all indecomposables in $CM(\mathrm{gr}\ \Lambda)$.

(a) $\mathcal{D}$ is closed under irreducible maps.

(b) No nonzero projective in $CM(\mathrm{gr}\ \Lambda)_0$ is injective in $CM(\mathrm{gr}\ \Lambda)_0$, $\mathcal{D}$ contains all projectives and injectives in $CM(\mathrm{gr}\ \Lambda)_0$, and $\mathcal{D}$ is closed under almost split sequences, that is, given C in $\mathcal{D}$ which is indecomposable nonprojective in $CM(\mathrm{gr}\ \Lambda)_0$ (or A in $\mathcal{D}$ which is indecomposable noninjective in $CM(\mathrm{gr}\ \Lambda)_0$), then all indecomposable summands of the terms in the almost split sequence $0 \to A \to B \to C \to 0$ are in $\mathcal{D}$.

Proof. (a) Assume to the contrary that there are indecomposable objects in $CM(\mathrm{gr}\ \Lambda)$ which are not in $\mathcal{D}$. We want to get a contradiction to Λ being an indecomposable $\mathbb{Z}$-graded algebra by showing that the category $CM(\mathrm{gr}\ \Lambda)_0$ is not indecomposable. Namely, we want to show that if Y is in $\mathcal{D}$

and X is an indecomposable in $CM(\mathrm{gr}\,\Lambda)$ which is not in $\mathcal{D}$, then $\mathrm{Hom}_{\mathrm{gr}\,\Lambda}(Y,X)_0 = 0 = \mathrm{Hom}_{\mathrm{gr}\,\Lambda}(X,Y)_0$.

Assume now that we have a nonzero degree zero map $g: Y \to X$. By considering the minimal left almost split map $h: Y \to E$ in $CM(\mathrm{gr}\,\Lambda)_0$ we can find an irreducible map $f_1: Y = Y_0 \to Y_1$ and a map $g_1: Y_1 \to X$ in $CM(\mathrm{gr}\,\Lambda)_0$ such that $g_1 f_1$ is not zero. Continuing this standard argument from the theory of artin algebras, we get for any i a chain of maps $Y = Y_0 \xrightarrow{f_1} Y_1 \xrightarrow{f_2} Y_2 \longrightarrow \cdots \xrightarrow{f_i} Y_i \xrightarrow{g_i} X$, where the f_j are irreducible and the composition is nonzero. Since $\mathrm{End}(Y_j)_0$ is an artin ring, its radical is nilpotent. Hence there is a bound on the number of times a given Y_j can occur in such a chain. Since we have a nonzero degree zero map $Y \to X$, there is for each Y_j only a fixed finite number of shifts which can occur. Since by assumption all Y_j are in $\mathcal{D}$, and $\mathcal{D}$ has only a finite number of indecomposables up to shifts, we now have a contradiction to $\mathrm{Hom}_{\mathrm{gr}\,\Lambda}(Y,X)_0 \neq 0$. By considering minimal right almost split maps we show similarly that $\mathrm{Hom}_{\mathrm{gr}\,\Lambda}(X,Y)_0 = 0$, and we are done with (a).

(b) We want to show that $\mathcal{D}$ is closed under irreducible maps. If A is in $\mathcal{D}$ and is not injective in $CM(\mathrm{gr}\,\Lambda)$ and $f: A \to X$ is irreducible with X indecomposable, then X is in $\mathcal{D}$, since in the almost split sequence $0 \to A \to B \to C \to 0$, X is a summand of B. Similarly, if $g: Y \to A$ is irreducible and A is not projective and in $\mathcal{D}$, then Y is in $\mathcal{D}$. If A is injective and $f: A \to X$ is irreducible, we can assume that X is not projective, so that we have an almost split sequence $0 \to Y \to A \amalg A' \to X \to 0$. Since by assumption A is not projective, we have Y in $\mathcal{D}$, and consequently X is in $\mathcal{D}$. Similarly if $g: Y \to A$ is irreducible and A is projective, then Y is in $\mathcal{D}$.

We now show how Theorem 2.1 can be applied to proving finite graded Cohen-Macaulay type.

<u>Theorem 2.2</u>. Let k be a field and X_0, X_1, X_2, Y_0, Y_1 indeterminates. Then the ring

$S = k[X_0,X_1,X_2,Y_0,Y_1]/(X_0X_2 - X_1^2, X_0Y_1 - X_1Y_0, X_1Y_1 - X_2Y_0)$ with the natural $\mathbb{Z}$-grading has only a finite number of indecomposable graded Cohen-Macaulay modules up to shift.

Proof. We write x_i and y_i for the images of X_i and Y_i in S. The polynomial ring $R = k[X_0,Y_1,X_2-Y_0]$ is a graded subring of S such that S is a finitely generated graded projective R-module. Then S is known to be an isolated singularity, so that $CM(\text{gr } S) = L_P(\text{gr } S)$, and Theorems 1.1 and 2.1 apply.

We showed in [6] that the ring $\hat{S}$ obtained by completion of S at $\underline{m} = (X_0,X_1,X_2,Y_0,Y_1)$ has only five indecomposable Cohen-Macaulay modules, and these are completions of the graded Cohen-Macaulay S-modules $\{S, A = (x_0,x_1),$ $B = (x_0,x_1,y_0)$, $C = (x_0,x_1,x_2)$, $K = \Omega^1 B\}$. We imitate the method from [6] to show that $\{S(i),A(i),B(i),C(i),K(i); i \in Z\}$ are exactly the indecomposable graded Cohen-Macaulay S-modules.

Since $\dim R = 3$, we know by Theorem 1.1 that if X is indecomposable nonprojective in $CM(\text{gr } \Lambda)$, then $(D\,\mathrm{Tr}_L X)(-3)$ gives the left hand term for the almost split sequence with X on the right, where $\mathrm{Tr}_L X = \Omega^3 \mathrm{Tr}\, X = \Omega^1 X^*$ and $X^* = \mathrm{Hom}_\Lambda(X,\Lambda)$. Modifying the computations in [6], it is easily seen that $D(S(i)) = A(2-i)$, $D(A(i)) = S(2-i)$, $D(B(i)) = C(2-i)$, $D(C(i)) = B(2-i)$, $D(K(i)) = K(4-i)$. Further, $B(i)^* = A(1-i)$, and the exact sequence $0 \to B(j-1) \to S(j-1) \amalg S(j-1) \to A(j) \to 0$ then gives $\mathrm{Tr}_L B(i) = B(-i)$. Since $A(i)^* = B(1-i)$, the exact sequence $0 \to K(j) \to S(j-1) \amalg S(j-1) \amalg S(j-1) \to B(j) \to 0$ gives $\mathrm{Tr}_L A(i) = K(1-i)$. $C(i)^* = X(-i)$, where $X = (x_0, y_0^2)$, and the exact sequence $0 \to C(i-1) \to S(i) \amalg S(i-1) \to X(i) \to 0$ shows that $\mathrm{Tr}_L C(i) = C(-i-1)$.

Keeping track of the shifts, we get from [6] the nonsplit exact sequences $0 \to S(i-2) \to A(i-1) \amalg A(i-1) \amalg B(i-1) \to K(i) \to 0$, and by dualizing, $0 \longrightarrow K(i) \longrightarrow S(i-1) \amalg S(i-1) \amalg C(i-1) \longrightarrow A(i) \longrightarrow 0$. It is easy to see that $\mathrm{End}(A(i))_0 = k$, and since $D\,\mathrm{Tr}_L(A(i))(-3) = DK(1-i)(-3) = K(i)$, we know as in [3, Prop. 5.1] that the second set of sequences are almost split, and hence by duality the first set. Since $D\,Tr_L\,B(i)(-3) = C(i-1)$ and $D\,Tr_L\,C(i)(-3) = B(i)$, we can deduce (as in [6]) that $0 \to C(i-1) \to A(i) \amalg S(i-1) \to B(i) \to 0$ and $0 \to B(i) \to K(i+1) \to C(i) \to 0$ are almost split. Since the projectives $S(i)$ are not injective in $CM(gr\ \Lambda)_0$, we are done by Theorem 2.1(b).

S is a graded scroll of type $(n_1,\dots,n_t)$ over a field k if S is the polynomial ring over k in the indeterminates $X_j^{(i)}$, $1 \leq i \leq t$, $0 \leq j \leq n_i$, with the natural Z-grading, modulo the ideal generated by the determinants of 2×2 minors of the matrix

$$\begin{pmatrix} X_0^{(1)} \cdots X_{n_1-1}^{(1)} & \Big| & \cdots & \Big| & X_0^{(t)} \cdots X_{n_t-1}^{(t)} \\ X_1^{(1)} \cdots X_{n_1}^{(1)} & \Big| & \cdots & \Big| & X_1^{(t)} \cdots X_{n_t}^{(t)} \end{pmatrix}.$$

For general information on scrolls we refer to [9].

Combining Theorem 2.2 with earlier results we know exactly when a graded scroll has a finite number of indecomposable Cohen-Macaulay graded modules up to shift, and the result is analogous to the corresponding result for $\hat{S}$.

Theorem 2.3. A graded scroll S over an infinite field k has only a finite number of indecomposable Cohen-Macaulay graded modules up to shift, if and only if $\hat{S}$ is of type (n), (1,1) or (2,1).

Proof. If S is not of type (n), (1,1) or (2,1), infinite graded Cohen-Macaulay type was proved in [6]. The case (2,1) is treated in Theorem 2.2. Theorem 2.1 can also be used to show that S of type (1,1) is of finite graded Cohen-Macaulay, as has been done by Ø. Solberg. Note that if char k ≠ 2, it follows from the work of Knörrer [11]. The case (n) is well known.

§3. Vector Bundles.

Let k be an infinite field. Let X be a nonsingular projective variety of dimension d over k. In this section we briefly discuss the question of the existence of almost split sequences in the category of locally free coherent sheaves over X. If $d = 1$, then we know that the category of locally free coherent sheaves on X has almost split sequences ([5], [14]). If $d > 1$, then it is not difficult to see that the category of locally free coherent sheaves on X does not have almost split sequences. It is our purpose in this section to apply the results of §1 to show that it is sometimes possible to obtain subcategories of the category of locally free coherent sheaves on X which have almost split sequences.

We assume now that we are given an embedding g of X in $\mathbb{P}^n(k)$ whose homogeneous coordinate ring S is Cohen-Macaulay. Since X nonsingular, we know that $S_{\underline{p}}$ is regular for all prime ideals $\underline{p}$ other than the unique maximal homogeneous ideal of S. Thus S is an integrally closed domain of dimension $d+1$. Therefore the sheafification functor $\sim: \operatorname{mod} \operatorname{gr} S \to \operatorname{Coh}(X)$, where $\operatorname{Coh}(X)$ is the category of coherent sheaves on X, induces an equivalence of categories $\sim: \operatorname{Ref} \operatorname{mod} \operatorname{gr} S \to \operatorname{Ref} \operatorname{Coh}(X)$ where $\operatorname{Ref} \operatorname{mod} \operatorname{gr} S$ is the category of reflexive modules in $\operatorname{mod} \operatorname{gr} S$ and $\operatorname{Ref} \operatorname{Coh}(X)$ is the category of reflexive coherent sheaves on X. Thus sheafification induces an equivalence of categories $\sim: CM(\operatorname{gr} S) \to \widetilde{CM(\operatorname{gr} S)}$, where $\widetilde{CM(\operatorname{gr} S)}$ consists of all F in $\operatorname{Coh}(X)$ such that $F \cong \tilde{M}$ for some M in $CM(\operatorname{gr} S)$. Our aim now is to show that $\widetilde{CM(\operatorname{gr} S)}$ has almost split sequences.

Since k is infinite and $S = \coprod_{i \geq 0} S_i$ is Cohen-Macaulay, we can find $X_0, \ldots, X_d$ in S_1 which is a regular S-sequence. From this it follows that the $X_0, \ldots, X_d$ are algebraically independent over k and S is a graded algebra over R, the polynomial subring $k[X_0, \ldots, X_d]$ of S. Moreover S is a free R-module, so we can apply the results of §1 to $CM(\operatorname{gr} S)$.

Since S_p is regular for all prime ideals of S different from the unique maximal homogeneous ideal $\underline{m}$ of S, we have $CM\,gr(S) = L_p(gr\ S) = L_I(gr\ S)$. Therefore we have that $CM(gr\ S)$ has almost split sequences by Theorem 1.1. Hence the subcategory $\widetilde{CM(gr\ S)}$ of locally free coherent sheaves on X has almost split sequence since $CM(gr\ S)$ and $\widetilde{CM(gr\ S)}$ are equivalent categories. More specifically, given an indecomposable H in $\widetilde{CM(gr\ S)}$ not isomorphic to $\mathcal{O}_X(n)$ for any n, there is an almost split sequence $0 \to F \to G \to H \to 0$ in $\widetilde{CM(gr\ S)}$ where $F \cong D\,\widetilde{Tr_L\,\Gamma_*}(H)(-d-1)$ and $\Gamma_*(H)$ is the unique M in $CM(gr\ S)$ such that $\tilde{M} = \Gamma_*(H)$.

Also for S in $CM(gr\ S)$ we know that there is a minimal right almost split morphism $B \to S$ in $CM(gr\ S)$ by Theorem 1.1. Hence $\tilde{B} \to \tilde{S} = \mathcal{O}_X$ is a minimal right almost split map in $\widetilde{CM(gr\ S)}$. Thus $\tilde{B}(n) \to \mathcal{O}_X(n)$ for each n in $\mathbf{Z}$ is a minimal right almost split morphism in $\widetilde{CM(gr\ S)}$.

It should be observed that for $d > 1$, $\widetilde{CM(gr\ S)}$ is a proper subcategory of the locally free coherent sheaves on X, depending on the particular embedding of X in projective space given by the homogeneous coordinate ring S. For example, the graded scrolls S_1 and S_2 of type (2,1) and (3,2) respectively are the coordinate rings of the same nonsingular projective variety X given by different embeddings of X in projective spaces. But the subcategories $\widetilde{CM(gr\ S_1)}$ and $\widetilde{CM(gr\ S_2)}$ of the locally free coherent sheaves on X are radically different since the first has only a finite number of nonisomorphic indecomposable objects up to shift while the second has an infinite number.

References

1. M. Auslander, Functors and morphisms determined by objects, Representation theory of algebras. Lecture Notes in pure and applied math., Vol. 37, Marcel Dekker, New York and Basel, 1-244.

2. M. Auslander, Isolated singularities and existence of almost split sequences, Proc. ICRA IV, Lectures Notes in Math. 1178 (1986), 194-241.

3. M. Auslander, I. Reiten, Representation theory of artin algebras III, Comm. Algebra 3 (1975), 239-294.

4. M. Auslander, I. Reiten, Representation theory of artin algebras IV, Comm. Algebra 5 (1977), 443-518.

5. M. Auslander, I. Reiten, Almost split sequences in dimension two, Adv. in Math. (to appear).

6. M. Auslander, I. Reiten, The Cohen-Macaulay type of Cohen-Macaulay rings.

7. M. Auslander, I. Reiten, Almost split sequences for abelian group graded rings.

8. M. Auslander, I. Reiten, Almost split sequences for Cohen-Macaulay modules.

9. D. Eisenbud, J. Harris, On varieties of minimal degree, Symp. Pure Math. AMS, ed. by S. Block (1987).

10. N. Jacobson, Basic Algebra II, W.H. Freeman and Company (1980).

11. H. Knörrer, Cohen-Macaulay modules on hypersurface singularities I.

12. C. Nastasescu, F. Van Oystayen, Graded and filtered rings and modules, Lecture Notes in Math. 758, Springer-Verlag (1979).

13. I. Reiten, Finite dimensional algebras and singularities, this volume.

14. A. Schofield, unpublished.

THE AUSLANDER-REITEN QUIVER OF AN ISOLATED SINGULARITY

Ernst Dieterich
Brandeis University
Department of Mathematics
Waltham, Mass. 02254/USA

The first part of this note (sections 1-4) is devoted mainly to the non-specialist reader. It contains a brief introduction to the concept of the Auslander-Reiten quiver of an isolated singularity, leading up to a basic structure theorem for stable Auslander-Reiten components which contain a periodic point. This is illustrated in the second part (section 5) in the case of isolated hypersurface singularities. The third part (sections 6-8) outlines how various consequences can be derived from the basic structure theorem, when studying isolated singularities with a reduction ideal. This last part follows rather closely the talk which I gave on the Lambrecht conference, except that it contains some additional results which have been obtained in the meanwhile (Theorems 13, 16 and 17).

0. Preliminaries

Throughout this note, modules are understood to be left modules. For any ring A, we write Mod A for the category of all left A-modules, and mod A for the category of all finitely generated left A-modules. If V is a vectorspace over a skewfield $\mathfrak{f}$, then we write $[V:\mathfrak{f}]$ for the dimension of V over $\mathfrak{f}$. For any object C in any category $\mathcal{C}$ we denote by $[C]$ the isomorphism class of C, and by $[\mathcal{C}]$ the set of all isomorphism classes of $\mathcal{C}$. We agree that $\mathbb{N} = \{1,2,3,\ldots\}$, whereas $\mathbb{N}_0 = \{0,1,2,\ldots\}$.

We write $\mathbb{C}[[X_1,\ldots,X_n]]$ for the ring of formal power series in n variables $X_1,\ldots,X_n$ over $\mathbb{C}$. Cohen-Macaulay modules over a commutative noetherian local ring are always understood to be maximal Cohen-Macaulay modules, i.e. the depth of the module equals the Krull dimension of the ring. For any commutative ring S, we write Spec(S) for the spectrum of S, Reg(S) for the regular locus of S and Sing(S) for the singular locus of S. The dimension of S is understood to be the Krull dimension of S, and denoted by dim S.

Throughout, R denotes a commutative noetherian complete local ring, and Λ denotes an R-algebra which is finitely generated as R-module. Let $\mathcal{L}$ be the class of all R-algebras which arise in this way. Usually we shall consider algebras from subclasses of $\mathcal{L}$ by assuming in addition, for example, that R is Cohen-Macaulay or even regular, or else that Λ is finitely generated free as R-module or even a commutative local Cohen-Macaulay ring. But, unless otherwise stated, Λ is not assumed to be commutative. We write $\mathfrak{m}$ for the unique maximal ideal of R, d for the dimension of R, and K for the field of fractions of R. In case Λ is commutative local, we write M for the unique maximal ideal of Λ. For any algebra $\Lambda \in L$ and $M,N \in \mathrm{mod}\,\Lambda$ we set $\underline{\mathrm{Hom}}_\Lambda(M,N) = \mathrm{Hom}_\Lambda(M,N)/\mathrm{Hom}'_\Lambda(M,N)$ where $\mathrm{Hom}'_\Lambda(M,N)$ consists of all homomorphisms in $\mathrm{Hom}_\Lambda(M,N)$ which factor through a projective Λ-module.

Given an R-algebra Λ in L, we denote by $\mathrm{mod}_R\Lambda$ the full subcategory of $\mathrm{mod}\,\Lambda$ consisting of all objects which are projective as R-modules, and we write $\mathrm{ind}_R\Lambda$ for the full subcategory of $\mathrm{mod}_R\Lambda$ consisting of all indecomposable objects of $\mathrm{mod}_R\Lambda$. (For any twosided ideal I contained in the radical of Λ, Λ is complete with respect to the I-adic topology and idempotents can be lifted from Λ/I to Λ. Consequently every indecomposable object in $\mathrm{mod}\,\Lambda$ has local endomorphism ring, and therefore Krull-Schmidt's Theorem holds in $\mathrm{mod}\,\Lambda$ as well as in $\mathrm{mod}_R\Lambda$.) For any $M \in \mathrm{ind}_R\Lambda$ we set $\delta(M) = \mathrm{End}_\Lambda M/\mathrm{rad}(\mathrm{End}_\Lambda M)$. A short exact sequence in $\mathrm{mod}_R\Lambda$ is understood to be a short exact sequence $0 \to A \to B \to C \to 0$ in $\mathrm{mod}\,\Lambda$, such that A,B,C are in $\mathrm{mod}_R\Lambda$. An object $P \in \mathrm{mod}_R\Lambda$ is said to be a projective object in $\mathrm{mod}_R\Lambda$ if every short exact sequence $0 \to A \to B \to P \to 0$ in $\mathrm{mod}_R\Lambda$ splits. Dually, an object $I \in \mathrm{mod}_R\Lambda$ is said to be an injective object in $\mathrm{mod}_R\Lambda$ if every short exact sequence $0 \to I \to B \to C \to 0$ in $\mathrm{mod}_R\Lambda$ splits. (In particular, if Λ is finitely generated free as R-module, then the projective objects in $\mathrm{mod}_R\Lambda$ and the finitely generated projective Λ-modules coincide.)

For any $M \in \mathrm{mod}_R\Lambda$ we set $\rho(M) = [K \otimes_R M : K]$, and we call $\rho(M)$ the R-rank of M. Usually ρ will be considered as a function on $[\mathrm{ind}_R\Lambda]$. An R-algebra Λ in L is said to be of <u>finite type</u> in case $[\mathrm{ind}_R\Lambda]$ is a finite set, respectively of <u>infinite type</u> in case $[\mathrm{ind}_R\Lambda]$ is an infinite set.

A special but important subclass of L arises in the following way. Suppose we are given a commutative noetherian complete local ring S. Then there exists a commutative noetherian complete regular local subring $R \subset S$ such that S is finitely generated as R-module. R is called a <u>Noether normalization</u> of S. In this situation, the category $\mathrm{mod}_R S$ coincides with the category of Cohen-Macaulay S-modules. In particular, S is finitely generated free as R-module if and only if S is a Cohen-Macaulay ring.

1. Auslander's characterization of isolated singularities

Assume that R is a commutative noetherian complete regular local ring, and that Λ is an R-algebra which is finitely generated free as R-module. We do not assume that Λ is commutative.

We define the <u>regular locus of</u> Λ <u>in</u> R to be $\mathrm{Reg}_R(\Lambda) = \{p \in \mathrm{Spec}(R) \mid \mathrm{gldim}\,\Lambda_p = \dim R_p\}$, and the <u>singular locus of</u> Λ <u>in</u> R to be $\mathrm{Sing}_R(\Lambda) = \{p \in \mathrm{Spec}(R) \mid \mathrm{gldim}\,\Lambda_p \neq \dim R_p\}$. Following Auslander [Au 84] we say that Λ is <u>nonsingular</u> if $\mathrm{Sing}_R(\Lambda) = \emptyset$, respectively that

Λ is an <u>isolated singularity</u> if $\mathrm{Sing}_R(\Lambda) = \{m\}$, respectively that Λ is a <u>nonisolated singularity</u> if $\mathrm{Sing}_R(\Lambda) \not\subset \{m\}$.

These notions are compatible with the corresponding notions from commutative algebra. Namely, in case Λ is commutative, let $\phi\colon \mathrm{Spec}(\Lambda) \to \mathrm{Spec}(R)$ be the continuous surjective map of spectra given by $\phi(P) = P \cap R$. Then, for each $p \in \mathrm{Spec}(R)$, the following statements are equivalent.

(i) $p \in \mathrm{Reg}_R(\Lambda)$.

(ii) Λ_p is a regular ring.

(iii) $P \in \mathrm{Reg}(\Lambda)$, for all $P \in \phi^{-1}(p)$.

It follows that $\mathrm{Sing}_R(\Lambda) = \phi(\mathrm{Sing}(\Lambda))$. In particular, if Λ is a commutative noetherian complete local Cohen-Macaulay ring, then the notions of nonsingularity, isolated singularity and nonisolated singularity as defined above in terms of $\mathrm{Sing}_R(\Lambda)$ and m, coincide with the corresponding notions defined in terms of $\mathrm{Sing}(\Lambda)$ and M.

We now return to the general situation, where Λ is not assumed to be commutative. A short exact sequence $0 \to A \xrightarrow{\Phi} B \xrightarrow{\Psi} C \to 0$ in $\mathrm{mod}_R\Lambda$ is said to be an <u>Auslander-Reiten sequence</u> (in $\mathrm{mod}_R\Lambda$) if it is nonsplit, if A and C are indecomposable, and if it has the factorization property that for any $Z \in \mathrm{mod}_R\Lambda$ and any $\zeta \in \mathrm{Hom}_\Lambda(Z,C)$ which is not a splittable epimorphism, there exists $\zeta' \in \mathrm{Hom}_\Lambda(Z,B)$ such that $\zeta = \psi\zeta'$, and dually, for any $X \in \mathrm{mod}_R\Lambda$ and any $\xi \in \mathrm{Hom}_\Lambda(A,X)$ which is not a splittable monomorphism, there exists $\xi' \in \mathrm{Hom}\,(B,X)$ such that $\xi = \xi'\phi$. We say that the category $\mathrm{mod}_R\Lambda$ <u>has Auslander-Reiten sequences</u>, in case for any indecomposable object C in $\mathrm{mod}_R\Lambda$ which is not projective in $\mathrm{mod}_R\Lambda$ there exists an Auslander-Reiten sequence ending in C, and for any indecomposable object A in $\mathrm{mod}_R\Lambda$ which is not injective in $\mathrm{mod}_R\Lambda$ there exists an Auslander-Reiten sequence starting in A.

Much of the present note is based on M. Auslander's characterization of isolated singularities, which we now quote. We refer to [Au 84] for proof and further details.

<u>Theorem 1</u> (Auslander). Let R be a commutative noetherian complete regular local ring, and let Λ be an R-algebra which is finitely generated free as R-module. Then the following statements are equivalent.

(i) The R-algebra Λ is an isolated singularity or nonsingular.

(ii) The category $\mathrm{mod}_R\Lambda$ has Auslander-Reiten sequences.

(iii) For all $M,N \in \mathrm{mod}_R\Lambda$, the R-module $\underline{\mathrm{Hom}}_\Lambda(M,N)$ has finite length.

(iv) For all $M \in \mathrm{mod}_R\Lambda$ and all $p \in \mathrm{Spec}(R)\setminus\{m\}$, the Λ_p-module M_p is projective.

Moreover, Λ is nonsingular if and only if all objects in $\mathrm{mod}_R\Lambda$ are projective.

Let us illustrate Theorem 1 in some cases of particular interest.

1) Suppose $d = 0$. Then Λ is a finite-dimensional algebra over

the field R, $\mathrm{mod}_R\Lambda = \mathrm{mod}\,\Lambda$ and Λ is always an isolated singularity or nonsingular. It is nonsingular if and only if it is semisimple. We obtain that, for any finite-dimensional algebra Λ over a field, the category of finitely generated Λ-modules has Auslander-Reiten sequences. This result was originally proved in [Au/Re 75].

2) Suppose $d = 1$. Then Λ is an order over the complete discrete valuation ring R, contained in the finite-dimensional K-algebra $K \otimes_R \Lambda$, and $\mathrm{mod}_R\Lambda$ is the category of Λ-lattices. Here, Λ is nonsingular or an isolated singularity if and only if $K \otimes_R \Lambda$ is semisimple. It is nonsingular if and only if it is hereditary. We obtain that, for any order Λ over a complete discrete valuation ring R, the category of Λ-lattices has Auslander-Reiten sequences if and only if $K \otimes_R \Lambda$ is semisimple. This result goes back to [Au 75] and [Ro/Schm 76].

3) Let Λ be any commutative noetherian complete local Cohen-Macaulay ring. Choose a Noether normalization $R \subset \Lambda$ such that Λ becomes an R-algebra which is finitely generated free as R-module. Combining Theorem 1 with the remark preceding it, we obtain that the category of Cohen-Macaulay Λ-modules has Auslander-Reiten sequences if and only if $\mathrm{Sing}(\Lambda) \subset \{M\}$.

4) In particular, let $\Lambda = \hat{\mathcal{O}}_{H,0} = \mathbb{C}[[X_0,\dots,X_d]]/(f(X))$ be the complete local ring of a hypersurface H at 0, where H is given by a polynomial $f(X) = f(X_0,\dots,X_d)$ in $\mathbb{C}[X_0,\dots,X_d]$. Let $J_f = (\frac{\partial f}{\partial X_0},\dots,\frac{\partial f}{\partial X_d})$ be the Jacobi ideal of $f(X)$ in $\mathbb{C}[[X_0,\dots,X_d]]$, and consider its image $\mathcal{J}_f = (J_f + (f(X)))/(f(X))$ in Λ. Applying Jacobi's Criterion we obtain that, for any polynomial $f(X) \in \mathbb{C}[X_0,\dots,X_d]$ with $f(0) = 0$, the category of Cohen-Macaulay modules over its complete local ring at 0 has Auslander-Reiten sequences if and only if $\mathcal{J}_f$ is either equal to Λ or M-primary.

2. Language of quivers

Some consequences of Theorem 1 can be formulated conveniently in the language of quivers, which we proceed to recall.

A quiver $Q = (Q_0,Q_1)$ is given by a set of points Q_0 and a set of arrows Q_1, linked by a map $(s,e)\colon Q_1 \to Q_0 \times Q_0$ which assigns to each arrow α an ordered pair of points $(s(\alpha),e(\alpha))$, its starting point $s(\alpha)$ and its end point $e(\alpha)$. We write $x \overset{\alpha}{\to} y$ in order to indicate that $(s(\alpha),e(\alpha)) = (x,y)$. With any quiver Q there is associated its underlying grapy $\bar{Q}$, obtained from Q by forgetting the orientation of the arrows. A quiver Q is said to have no multiple arrows if the map $(s,e)\colon Q_1 \to Q_0 \times Q_0$ is injective. An arrow α of a quiver Q is called a loop if $s(\alpha) = e(\alpha)$. A quiver Q is said to be finite if both sets Q_0 and Q_1 are finite. Given a quiver Q and any point $x \in Q_0$, we denote by x^- the set of all points $y \in Q_0$ such that there exists an arrow $y \to x$ in Q_1, and dually we denote by x^+ the set of all points

$y \in Q_0$ such that there exists an arrow $x \to y$ in Q_1. We call x^- the set of direct predecessors of x and we call x^+ the set of direct successors of x. A quiver Q is said to be locally finite if for each point $x \in Q_0$ the set $x^- \cup x^+$ is finite. For any quiver $Q = (Q_0, Q_1)$ and any subset $Q_0' \subset Q_0$, the full subquiver Q' of Q defined by Q_0' is given by $Q' = (Q_0', Q_1')$, where Q_1' is the set of all arrows in Q_1 which have both starting point and end point in Q_0'.

A translation quiver $Q = (Q_0, Q_1, \tau)$ is given by a quiver $Q = (Q_0, Q_1)$ together with two distinguished subsets Q_0', Q_0'' of Q_0 and a bijective mapping $\tau\colon Q_0' \to Q_0''$, such that Q is locally finite and has no multiple arrows, and moreover, $x^- = (\tau x)^+$ for all $x \in Q_0'$. (Note that a translation quiver in our sense may have loops, in this respect differing from the notion "Darstellungsköcher", as introduced by Riedtmann in [Rie 80].) Given a translation quiver Q and an arrow $x \overset{\alpha}{\to} y$ such that $y \in Q_0'$, we denote by $\sigma\alpha$ the uniquely determined arrow $\tau y \overset{\sigma\alpha}{\to} x$. Given a translation quiver $Q = (Q_0, Q_1, \tau)$, we call τ its translation map, the points in $Q_0 \setminus Q_0'$ its projective points, and the points in $Q_0 \setminus Q_0''$ its injective points. Moreover, a point x of a translation quiver Q is said to be stable if $\tau^n(x)$ is defined for all $n \in \mathbb{Z}$. It is said to be periodic of period n, $n \in \mathbb{N}$, if $\tau^n(x) = x$.

With any translation quiver Q we associate its stable subquiver Q_s which is the full subquiver of Q defined by the set of all stable points of Q. With the restriction of τ as a translation map, the stable subquiver is again a translation quiver, and the restriction of τ induces an automorphism of Q_s. A translation quiver is said to be stable, if $Q = Q_s$.

A valued quiver $Q = (Q_0, Q_1, (a,a'))$ is given by a quiver $Q = (Q_0, Q_1)$, together with a map $(a,a')\colon Q_1 \to \mathbb{N} \times \mathbb{N}$. We call (a,a') the valuation of the valued quiver. For any arrow $x \overset{\alpha}{\to} y$ in a valued quiver $Q = (Q_0, Q_1, (a,a'))$ we set $(a_\alpha, a_\alpha') = (a_{xy}, a_{xy}') = (a,a')(\alpha)$, and we write $x \xrightarrow{(n,n')} y$ in order to indicate that $(a_{xy}, a_{xy}') = (n,n')$. Moreover, we follow the usual convention that $x \to y$ has the same meaning as $x \xrightarrow{(1,1)} y$.

A valued translation quiver $Q = (Q_0, Q_1, \tau, (a,a'))$ is given by a translation quiver $Q = (Q_0, Q_1, \tau)$ together with a valuation $(a,a')\colon Q_1 \to \mathbb{N} \times \mathbb{N}$, such that $(a_{\sigma\alpha}, a_{\sigma\alpha}') = (a_\alpha', a_\alpha)$. for all $\alpha \in Q_1$ with $e(\alpha) \in Q_0'$.

A subadditive function ϕ for a valued translation quiver $Q = (Q_0, Q_1, \tau, (a,a'))$ is given by a function $\phi\colon Q_0 \to \mathbb{N}$, satisfying the inequality

$$\phi(x) + \phi(\tau x) \geq \sum_{y \in x^-} \phi(y) a_{yx}',$$

for all $x \in Q_0'$. A subadditive function is called additive, if this inequality is an equality for all $x \in Q_0'$.

3. Definition of the Auslander-Reiten quiver

Again assume that R is a commutative noetherian complete regular local ring, and that Λ is an R-algebra which is finitely generated free as R-module. In this section we define the Auslander-Reiten quiver of Λ. It is a valued quiver attached to the category $\mathrm{mod}_R\Lambda$. It contains full information about the set of isomorphism classes of indecomposable objects of $\mathrm{mod}_R\Lambda$, and it contains some information about morphisms in the category. We shall see that the Auslander-Reiten quiver of Λ is a valued translation quiver in case Λ is an isolated singularity or nonsingular.

First we have to recall the notion of the radical of the category $\mathrm{mod}_R\Lambda$ and the related notion of an irreducible morphism in $\mathrm{mod}_R\Lambda$. If A and B are indecomposable objects in $\mathrm{mod}_R\Lambda$, then $\mathrm{rad}\,(A,B)$ is the set of all nonisomorphisms in $\mathrm{Hom}_\Lambda(A,B)$. If A and B are arbitrary objects in $\mathrm{mod}_R\Lambda$ and if $A = \bigoplus_{j=1}^{n} A_j$, $B = \bigoplus_{i=1}^{m} B_i$ are decompositions of A and B into indecomposable direct summands, then $\mathrm{rad}_\Lambda(A,B)$ is the set of all morphisms $\phi = (\phi_{ij})_{\substack{i=1,\dots,m\\ j=1,\dots,n}}$ in $\mathrm{Hom}_\Lambda(A,B)$ such that $\phi_{ij} \in \mathrm{rad}_\Lambda\,(A_j,B_i)$, for all $j = 1,\dots,n$ and $i = 1,\dots,m$. Observe that $\mathrm{rad}_\Lambda(A,A)$ coincides with the Jacobson radical of $\mathrm{End}_\Lambda A$. Moreover, $\mathrm{rad}^2_\Lambda(A,B)$ is the set of all morphisms $\phi \in \mathrm{Hom}_\Lambda(A,B)$ such that there exist $X \in \mathrm{mod}_R\Lambda$, $\alpha \in \mathrm{rad}_\Lambda(A,X)$, $\beta \in \mathrm{rad}_\Lambda(X,B)$ subject to $\phi = \beta\alpha$. In this manner we associate with each ordered pair (A,B) of objects in $\mathrm{mod}_R\Lambda$ a descending chain of $(\mathrm{End}\,B,\mathrm{End}\,A)$-bimodules $\mathrm{Hom}_\Lambda(A,B) \supset \mathrm{rad}_\Lambda(A,B) \supset \mathrm{rad}^2_\Lambda(A,B)$. A morphism $\phi\colon A \to B$ in $\mathrm{mod}_R\Lambda$ is said to be <u>irreducible</u> if it is neither a splittable monomorphism nor a splittable epimorphism and if, in addition, every factorization $\phi = \phi''\phi'$ of ϕ in $\mathrm{mod}_R\Lambda$ has the property that either ϕ' is a splittable monomorphism or ϕ'' is a splittable epimorphism. Now suppose that both A and B are indecomposable. Then a morphism $\phi\colon A \to B$ is irreducible if and only if it is in $\mathrm{rad}_\Lambda(A,B)\setminus\mathrm{rad}^2_\Lambda(A,B)$. The factorbimodule $\mathrm{irr}_\Lambda(A,B) = \mathrm{rad}_\Lambda(A,B)/\mathrm{rad}^2_\Lambda(A,B)$ is an $(\mathfrak{f}(B),\mathfrak{f}(A))$-bimodule which measures the multitude of irreducible morphisms from A to B. Therefore it is called the <u>bimodule of irreducible morhpisms</u> (from A to B).

<u>Definition</u>. The <u>Auslander-Reiten quiver</u> of Λ is the valued quiver $A(\Lambda) = (A_0,A_1,(a,a'))$, associated with the category $\mathrm{mod}_R\Lambda$ as follows. The set of points A_0 is given by the set of isomorphism classes $[A]$ of indecomposable objects in $\mathrm{mod}_R\Lambda$; the set of arrows A_1 is given by the set of symbols $[A] \to [B]$, where $[A]$ and $[B]$ are elements in A_0 such that $\mathrm{irr}_\Lambda(A,B) \neq 0$; the valuation $(a,a')\colon A_1 \to \mathbb{N}\times\mathbb{N}$ is given by $(a_{AB},a'_{AB}) = ([\mathrm{irr}_\Lambda(A,B):\mathfrak{f}(B)],[\mathrm{irr}_\Lambda(A,B):\mathfrak{f}(A)])$, for each arrow $[A] \to [B]$ in A_1. (Recall our notation $\mathfrak{f}(X) = \mathrm{End}_\Lambda X/\mathrm{rad}(\mathrm{End}_\Lambda X)$, for any $X \in \mathrm{ind}_R\Lambda$.)

If $0 \to A \to B \to C \to 0$ and $0 \to A' \to B' \to C' \to 0$ are two Auslander-Reiten sequences in $\mathrm{mod}_R \Lambda$, then it is easily seen that $A \cong A'$ if and only if $C \cong C'$. Therefore, setting $A_0' = \{[C] \in A_0 \mid$ there exists an Auslander-Reiten sequence ending in $[C]\}$ and $A_0'' = \{[A] \in A_0 \mid$ there exists an Auslander-Reiten sequence starting in $A\}$, Auslander-Reiten sequences in $\mathrm{mod}_R \Lambda$ define a bijective mapping $\tau: A_0' \to A_0''$, given by $\tau([C]) = [A]$ whenever A is starting term of an Auslander-Reiten sequence ending in C. This mapping τ is called the Auslander-Reiten translation (on A_0).

We also have to recall the notions of sink morphism and source morphism in $\mathrm{mod}_R \Lambda$ (minimal right almost split morphism and minimal left almost split morphism in the language of [Au/Re 75]). Let C be an indecomposable object in $\mathrm{mod}_R \Lambda$. A morphism $\xi: X \to C$ in $\mathrm{mod}_R \Lambda$ is called a sink morphism for C if it has the following properties.

(a) It is not a splittable epimorphism.

(b) For any morphism $\xi': X' \to C$ in $\mathrm{mod}_R \Lambda$ which is not a splittable epimorphism there exists $\psi: X' \to X$ such that $\xi' = \xi\psi$.

(c) For all $\varepsilon \in \mathrm{End}_\Lambda X$, if $\xi\varepsilon = \xi$ then ε is an automorphism of X.

Let A be an indecomposable object in $\mathrm{mod}_R \Lambda$. A morphism $\eta: A \to Y$ in $\mathrm{mod}_R \Lambda$ is called a source morphism for A if it has the following properties.

(a') It is not a splittable monomorphism.

(b') For any morphism $\eta': A \to Y'$ in $\mathrm{mod}_R \Lambda$ which is not a splittable monomorphism there exists $\zeta: Y \to Y'$ such that $\zeta\eta = \eta'$.

(c') For all $\varepsilon \in \mathrm{End}_\Lambda Y$, if $\varepsilon\eta = \eta$ then ε is an automorphism of Y.

If a sink morphism for C exists, then it is uniquely determined up to isomorphism, and dually, if a source morphism for A exists, then it is uniquely determined up to isomorphism. For our context, the following properties of sink and source morphisms are of importance.

(1) If Λ is an isolated singularity or nonsingular, then every indecomposable object C in $\mathrm{mod}_R \Lambda$ has a sink morphism, and every indecomposable object A in $\mathrm{mod}_R \Lambda$ has a source morphism.

(2) Let A be an indecomposable object in $\mathrm{mod}_R \Lambda$ and let $\phi: A \to B$ be a source morphism for A. Let $B = \bigoplus_{i=1}^{n} B_i^{b_i}$ be a decomposition of B into indecomposable and pairwise nonisomorphic direct summands $B_1, \ldots, B_n$. Let $\phi = (\phi_{ij})_{\substack{i=1,\ldots,n \\ j=1,\ldots,b_i}}$ be the corresponding decomposition of ϕ and denote by $\bar{\phi}_{ij}$ the residue class of ϕ_{ij} in $\mathrm{irr}_\Lambda(A, B_i)$. Then the following holds.

(i) The set $\{\bar{\phi}_{i1}, \ldots, \bar{\phi}_{ib_i}\}$ is a basis for the left $\mathfrak{k}(B_i)$-vectorspace $\mathrm{irr}_\Lambda(A, B_i)$, for all $i = 1, \ldots, n$.

(ii) If X is an indecomposable object in $\mathrm{mod}_R\Lambda$ such that $\mathrm{irr}_\Lambda(A,X) \neq 0$, then $X \cong B_i$ for some $i = 1,\ldots,n$.

(2') Let C be an indecomposable object in $\mathrm{mod}_R\Lambda$ and let $\psi: B \to C$ be a sink morphism for C. Then the dual statements (i') and (ii') hold, describing the connection between the right $\mathfrak{f}(B_i)$-vectorspace $\mathrm{irr}_\Lambda(B_i,C)$ and the components $(\psi_{ij})_{j=1,\ldots,b_i}$ of ψ.

Proof of (1): If C is not projective in $\mathrm{mod}_R\Lambda$, then the epimorphism of the Auslander-Reiten sequence ending in C is a sink morphism for C. If A is not injective in $\mathrm{mod}_R\Lambda$, then the monomorphism of the Auslander-Reiten sequence starting in A is a source morphism for A. If C is projective in $\mathrm{mod}_R\Lambda$, then there exists a morphism $\xi: X \to C$ in $\mathrm{mod}_R\Lambda$ which has properties (a) and (b), since $\mathrm{mod}_R\Lambda$ is functorially finite in $\mathrm{mod}\,\Lambda$ [Au 84]. Out of the set of all such morphisms $\xi: X \to C$ choose $\xi_0: X_0 \to C$ to be minimal with respect to restriction of ξ to direct summands of X. Then ξ_0 is a sink morphism for C. If A is injective in $\mathrm{mod}_R\Lambda$, then the existence of a source morphism for A follows from the last remark by means of the duality $\mathrm{Hom}_R(\ ,R)$: $\mathrm{mod}_R\Lambda^{op} \to \mathrm{mod}_R\Lambda$.

For the proof of (2) and (2') the reader is referred to [Rin 84, § 2.2 Lemma 3]. The proof given there in the case of a Krull-Schmidt category over a field literally carries over to our situation.

Proposition 2. If Λ is an isolated singularity or nonsingular, then its Auslander-Reiten quiver $A(\Lambda) = (A_0, A_1, (a,a'))$ together with the Auslander-Reiten translation $\tau: A_0' \to A_0''$ is a valued translation quiver.

Proof. By definition, $A(\Lambda)$ has no multiple arrows and $\tau: A_0' \to A_0''$ is a bijective mapping. In view of statements (1), (2) and (2') above it is clear that $A(\Lambda)$ is locally finite and that $[C]^- = \tau([C])^+$ for all $[C] \in A_0'$. Finally, let $[E] \overset{\alpha}{\to} [C]$ be any arrow in A_1 such that $e(\alpha) \in A_0'$, and consider the arrow $[A] \overset{\sigma\alpha}{\to} [E]$. Then we have the dimension equalities

$$[\mathrm{irr}_\Lambda(E,C): k] = a_\alpha[\mathfrak{f}(C): k] = a_\alpha'[\mathfrak{f}(E): k] ,$$

$$[\mathrm{irr}_\Lambda(A,E): k] = a_{\sigma\alpha}[\mathfrak{f}(E): k] = a_{\sigma\alpha}'[\mathfrak{f}(A): k] ,$$

where k is the residue class field of R. We know from (2) and (2') that $a_{\sigma\alpha} = a_\alpha'$. It is easily seen that $[A] = \tau([C])$ implies $\mathfrak{f}(A) \cong \mathfrak{f}(C)$. It follows that $a_{\sigma\alpha}' = a_\alpha$. q.e.d.

If Λ is an isolated singularity or nonsingular, then we define its stable Auslander-Reiten quiver $A_s(\Lambda)$ to be the stable subquiver of $A(\Lambda)$. Of course, $A_s(\Lambda)$ is a stable valued translation quiver. Observe that if Λ is nonsingular, then $A_s(\Lambda) = \emptyset$.

4. Structure of stable Auslander-Reiten components containing a periodic point

There is a canonical way of constructing stable valued translation quivers. Starting with any valued oriented tree $T = (T_0, T_1, (a,a'))$ (i.e. T is a locally finite valued quiver such that its underlying graph $\bar{T}$ is a tree) we define the stable valued translation quiver $\mathbb{Z}T = ((\mathbb{Z}T)_0, (\mathbb{Z}T)_1, \tau, (a,a'))$ as follows: $(\mathbb{Z}T)_0 = \mathbb{Z} \times T_0$; $(\mathbb{Z}T)_1 = (\mathbb{Z} \times T_1) \dot{\cup} \sigma(\mathbb{Z} \times T_1)$, where $(n,\alpha) = \{(n,x) \to (n,y)\}$ and $\sigma(n,\alpha) = \{(n-1,y) \to (n,x)\}$, for all $n \in \mathbb{Z}$ and $\{x \overset{\alpha}{\to} y\} \in T_1$; $\tau(n,x) = (n-1,x)$, for all $(n,x) \in (\mathbb{Z}T)_0$, $(a,a')(n,\alpha) = (a,a')(\alpha)$, for all $(n,\alpha) \in \mathbb{Z} \times T_1$, and $(a,a')\sigma(n,\alpha) = (a',a)(\alpha)$, for all $\sigma(n,\alpha) \in \sigma(\mathbb{Z} \times T_1)$. Moreover, if G is a group of automorphisms of $\mathbb{Z}T$, then the orbit quiver $\mathbb{Z}T/G$ is again a stable valued translation quiver.

We say that a group of automorphisms of $\mathbb{Z}T$ is <u>weakly admissible</u> if $(n,x)^+ \cap g((n,x)^+) = \emptyset$, for all $(n,x) \in (\mathbb{Z}T)_0$ and all $g \in G\setminus\{1\}$. Then we recall the following facts.

(i) If T and T' are valued oriented trees, then $\mathbb{Z}T \cong \mathbb{Z}T'$ (as stable valued translation quivers) if and only if $\bar{T} \cong \bar{T}'$ (as valued graphs).

(ii) Let Q be any connected stable valued translation quiver. Then there exists a pair (T,G), where T is a connected valued oriented tree and G is a weakly admissible group of automorphisms of $\mathbb{Z}T$, such that $Q \cong \mathbb{Z}T/G$.

(iii) If (T,G) and (T',G') are two pairs as in (ii), such that $\mathbb{Z}T/G \cong \mathbb{Z}T'/G'$, then $\bar{T} \cong \bar{T}'$ and $G = G'$ (after suitable identification of $\mathbb{Z}T$ and $\mathbb{Z}T'$).

(For proofs we refer the reader to [Rie 80], [Hap/Pr/Rin 79], [Hap/Pr/Rin 80]. Riedtmann deals with quivers without valuations and without loops. Note that in this case Riedtmann's notion of an admissible group replaces our notion of a weakly admissible group. [Hap/Pr/Rin 79] deals with valued quivers without loops, and [Hap/Pr/Rin 80] covers our situation.)

Let (Γ,G) be a pair consisting of a connected valued tree Γ (locally finite) and a group G of automorphisms of $\mathbb{Z}T$, where T is any valued oriented tree such that $\bar{T} = \Gamma$. Then, in view of (i), the assignment $(\Gamma,G) \mapsto \mathbb{Z}\Gamma/G := \mathbb{Z}T/G$ is well-defined up to isomorphism. In view of (ii) and (iii) it induces a bijection between the isomorphism classes of all such pairs and the isomorphism classes of connected stable valued translation quivers. Following [Hap/Pr/Rin 79] we call Γ the <u>Cartan class</u> of $\mathbb{Z}\Gamma/G$.

We are now ready to state the basic structure theorem for stable Auslander-Reiten components which contain a periodic point.

Theorem 3. Let Λ be an isolated singularity. Let $C = (C_0, C_1, (a,a'))$ be a connected component of the stable Auslander-Reiten quiver $A_s(\Lambda)$, and denote by $\rho\colon C_0 \to \mathbb{N}$ the rank function. Assume that C_0 contains a periodic point. Then the following statements hold.

(i) The Cartan class of C is either a Dynkin diagram, or a Euclidean diagram, or one of the infinite diagrams $\mathbb{A}_\infty$, $\mathbb{B}_\infty$, $\mathbb{C}_\infty$, $\mathbb{D}_\infty$, $\mathbb{A}_\infty^\infty$.

(ii) If ρ is not additive on C, then the Cartan class of C is either a Dynkin diagram or $\mathbb{A}_\infty$.

(iii) If ρ is unbounded on C_0, then the Cartan class of C is A_∞.

Since, by Proposition 2, C is a connected stable valued translation quiver on which ρ is a subadditive function, Theorem 3 follows immediately from [Hap/Pr/Rin 79] in case C contains no loop, and from [Hap/Pr/Rin 80] in the general case.

For convenience of the reader we include the list of Dynkin diagrams, Euclidean diagrams and infinite diagrams which in Theorem 3 occur as Cartan class of C. Recall the convention that o——o has the same meaning as o$\overset{(1,1)}{\text{——}}$o.

Dynkin diagrams.

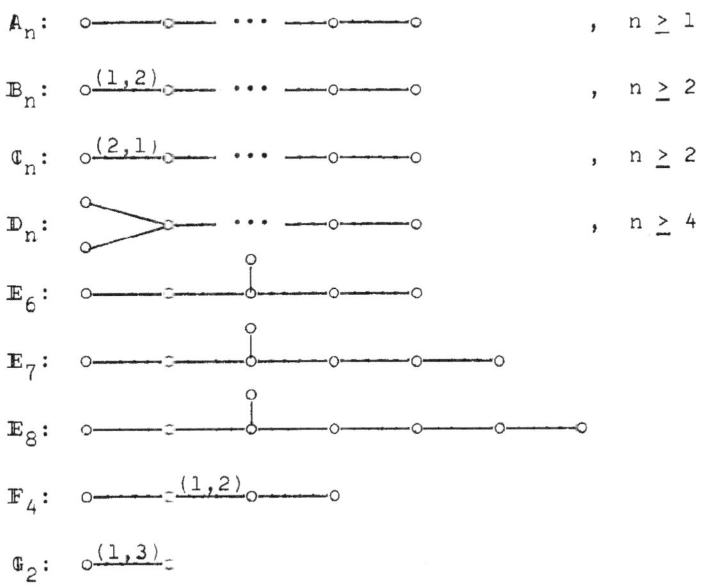

Euclidean diagrams.

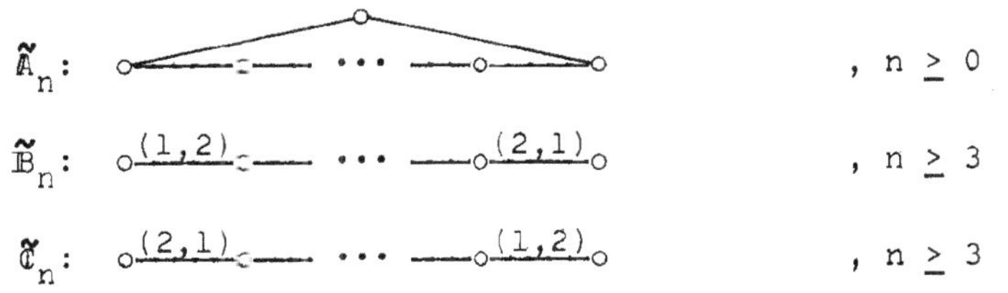

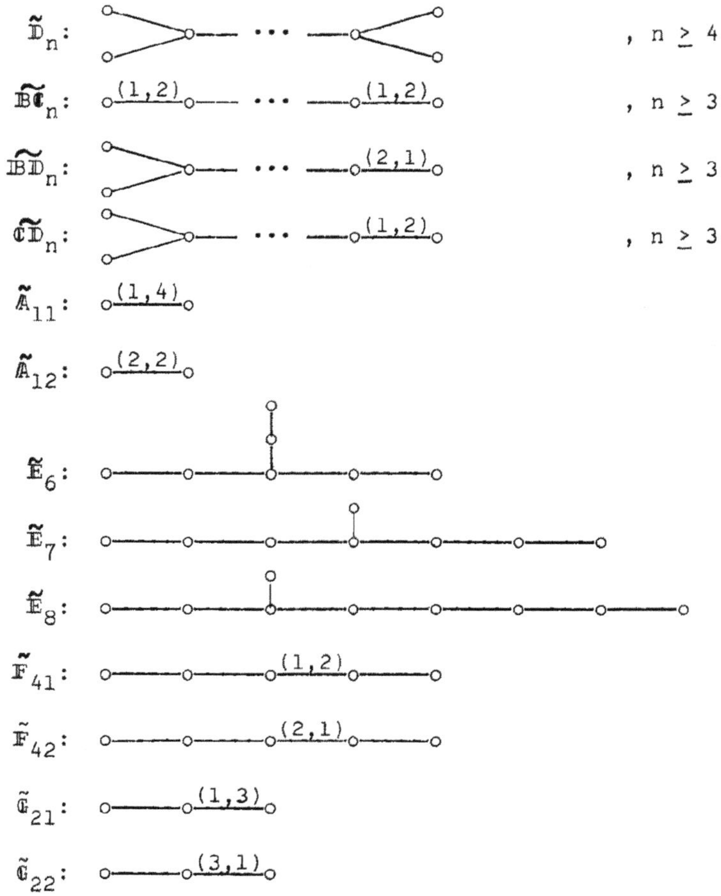

Infinite diagrams.

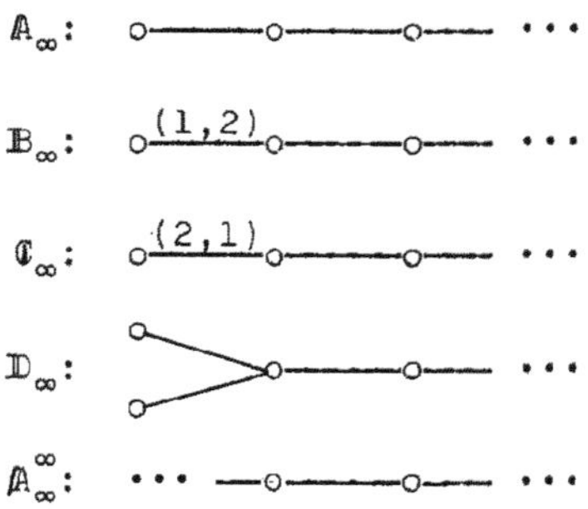

5. Examples (isolated hypersurface singularities)

Assume that $f(X) = f(X_0,\ldots,X_d)$ is a polynomial in $\mathbb{C}[X_0,\ldots,X_d]$, such that the hypersurface $H \subset \mathbb{A}^{d+1}(\mathbb{C})$ given by $f(X)$ has an isolated singularity at 0. Let $\Lambda = \hat{\mathcal{O}}_{H,0} = \mathbb{C}[[X_0,\ldots,X_d]]/(f(X))$ be the complete local ring of H at 0, and let $R \subset \Lambda$ be a chosen Noether normalization. We call such an R-algebra Λ an isolated hypersurface singularity. It is an isolated singularity in the sense of section 1. This section is devoted to illustrate Theorem 3 in the case of isolated hypersurface singularities.

So assume that Λ is an isolated hypersurface singularity, and let M be any indecomposable nonprojective object in $\mathrm{mod}_R\Lambda$. Since Λ is a Gorenstein R-order in the sense of [Au 78], we have that $\tau([M]) = [\Omega^{2-d}(M)]$, by [Au 78, III, Proposition 1.8]. On the other hand, because Λ is a hypersurface, we also have that $[\Omega^2(M)] = [M]$, by [Ei 80, Theorem 6.1]. Therefore $\tau^2([M]) = [M]$, and if d is even then $\tau([M]) = [M]$. In particular, $[M]$ is a periodic point of $A(\Lambda)$, and $[\Lambda]$ is the unique nonstable point, the unique projective point and the unique injective point of $A(\Lambda)$. The stable Auslander-Reiten quiver $A_s(\Lambda)$ is obtained from $A(\Lambda)$ by removing the single point $[\Lambda]$. In other words, $A_s(\Lambda)$ is as close to $A(\Lambda)$ as it can be.

5.1 Isolated hypersurface singularities of finite type

It is known that an isolated hypersurface singularity $\Lambda = \hat{\mathcal{O}}_{H,0}$ is of finite type if and only if $(H,0)$ is a simple singularity (i.e. there exist only finitely many isomorphism classes of singularities in the semiuniversal deformation of $(H,0)$) [Kn 86], [Bu/Gr/Schr 86]. The simple hypersurface singularities are classified [Ar 74] and they are known to correspond to the (trivially valued) Dynkin diagrams via the intersection graph of the universal resolution of $(H,0)$ in dimension 2 [Va 34]. The simple hypersurface singularity of Dynkin type Δ and of dimension d $(d \in \mathbb{N})$ is given by the polynomial $f_\Delta(X) = f_\Delta(X_0,\dots,X_d)$ as follows.

$$\mathbb{A}_n:\quad X_0^{n+1} + X_1^2 + X_2^2 + \cdots + X_d^2, \qquad n \geq 1$$

$$\mathbb{D}_n:\quad X_0^{n-1} + X_0X_1^2 + X_2^2 + \cdots + X_d^2, \qquad n \geq 4$$

$$\mathbb{E}_6:\quad X_0^4 + X_1^3 + X_2^2 + \cdots + X_d^2$$

$$\mathbb{E}_7:\quad X_0^3X_1 + X_1^3 + X_2^2 + \cdots + X_d^2$$

$$\mathbb{E}_8:\quad X_0^5 + X_1^3 + X_2^2 + \cdots + X_d^2$$

Let $\Lambda = \Lambda_{\Delta,d} = \mathbb{C}[[X_0,\dots,X_d]]/(f_\Delta(X))$ be the simple isolated hypersurface singularity of Dynkin type Δ and of dimension d. The Auslander-Reiten quiver of $\Lambda = \Lambda_{\Delta,d}$ has been determined. (See [Di/Wi 86] for $d = 1$, [Au 86] for $d = 2$, and [Kn 86] for $d \geq 3$.) It turns out that $A_s(\Lambda)$ is connected (except when $\Delta = \mathbb{A}_1$ and d is odd) and therefore, by Theorem 3, it must be of the form $A_s(\Lambda) \cong \mathbb{Z}\Gamma/G$, where Γ is the Cartan class of $A_s(\Lambda)$ and G is a weakly admissible group of automorphisms of $\mathbb{Z}\Gamma$. We proceed to describe $A_s(\Lambda)$ in terms of the pair (Γ,G).

In case $d = 1$, the pair (Γ, G) associated with the Dynkin diagram Δ is given as follows. (For the last column, we define, for some Cartan classes Γ, a reflection on Γ by depicting the axis of reflection as an interrupted line.)

Δ		Γ	G	σ induced by reflection in Γ
$\mathbb{A}_1$		$\mathbb{A}_1 \dot{\cup} \mathbb{A}_1$	$<\tau\sigma>$	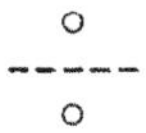
$\mathbb{A}_3$		$\mathbb{A}_3$	$<\tau\sigma>$	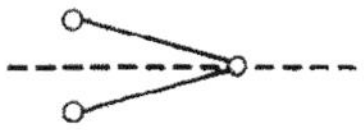
$\mathbb{A}_{n+1}$	$(n = 2m,\ m \in \mathbb{N}\setminus\{1\})$	$\mathbb{D}_{m+2}$	$<\tau\sigma>$	
$\mathbb{A}_n$	$(n = 2m,\ m \in \mathbb{N})$	$\mathbb{A}_n$	$<\sqrt{\tau}\sigma>$	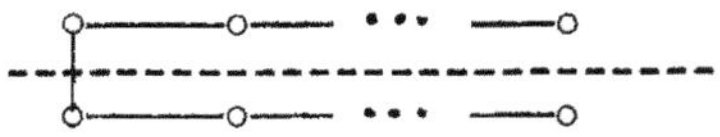
$\mathbb{D}_{n+1}$	$(n = 2m,\ m \in \mathbb{N}\setminus\{1\})$	$\mathbb{A}_{2n-1}$	$<\tau\sigma>$	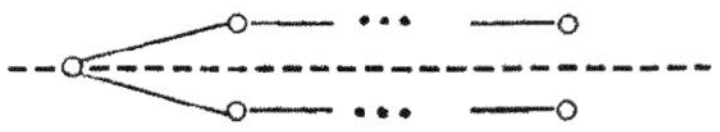
$\mathbb{D}_n$	$(n = 2m,\ m \in \mathbb{N}\setminus\{1\})$	$\mathbb{D}_n$	$<\tau^2>$	
$\mathbb{E}_6$		$\mathbb{E}_6$	$<\tau\sigma>$	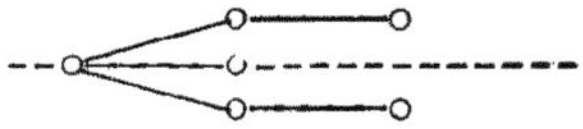
$\mathbb{E}_7$		$\mathbb{E}_7$	$<\tau^2>$	
$\mathbb{E}_8$		$\mathbb{E}_8$	$<\tau^2>$	

In case $d = 2$ we have $A_s(\Lambda_{\Delta,2}) \cong \mathbb{Z}\Delta/<\tau>$, in other words the pair (Γ, G) associated with Δ is given by $(\Delta, <\tau>)$, for all Dynkin diagrams Δ. Finally, the case $d \geq 3$ is described by the fact that $A_s(\Lambda_{\Delta,d}) \cong A_s(\Lambda_{\Delta,d+2})$, for all Dynkin diagrams Δ and all $d \in \mathbb{N}$.

In order to obtain a full description of $A(\Lambda)$ we only have to know how to join $[\Lambda]$ to $A_s(\Lambda)$. In this respect we refer to [Di/Wi 86], [Au 86], [Kn 86].

5.2 Isolated hypersurface singularities of infinite type

If $\Lambda = \mathbb{C}[[X_0, \ldots, X_d]]/(f(X))$ is an isolated hypersurface singularity of infinite type, then $A(\Lambda) \cong C \dot{\cup} (\dot{\bigcup}_{i \in I} \mathbb{Z}\mathbb{A}_\infty / <\tau^{n(i)}>)$, where C is the unique connected component of $A(\Lambda)$ which contains $[\Lambda]$, I is an index set, and $n: I \to \{1,2\}$ is a function. Moreover, $n \equiv 1$ if d is even, and the Cartan class of each connected component of C_s is either a Dynkin diagram or $\mathbb{A}_\infty$. This result is proven in detail in [Di 86], and we shall give an outline of proof in the following sections.

Taking this result for granted, it is natural to ask for the index set I, the function $n: I \to \{1,2\}$, and the connected component C associated with an isolated hypersurface singularity of infinite type. Knowledge of these data solves the classification problems for $\text{mod}_R\Lambda$. So far there is just one case in which a complete answer to this question is known, namely the case of the simple elliptic curve singularity of type $\tilde{\mathbb{E}}_8$. To be precise, let $\Lambda = \mathbb{C}[[X,Y]]/(f(X,Y))$, when $f(X,Y) = Y(Y-X^2)(Y-aX^2)$, with $a \in \mathbb{C}\setminus\{0,1\}$. Then $A_s(\Lambda) \cong \dot{\bigcup}_{i\in\hat{I}} \mathbb{Z}\mathbb{A}_\infty/<\tau^{n(i)}>$, $\hat{I} = \mathbb{P}^1(\mathbb{Q})\times\mathbb{P}^1(\mathbb{C})$, and there are four "exceptional parameters" $\varepsilon_1,\ldots,\varepsilon_4$ in $\mathbb{P}^1(\mathbb{C})$ such that $n(i) = 1$ for all $i \in \mathbb{P}^1(\mathbb{Q})\times(\mathbb{P}^1(\mathbb{C})\setminus\{\varepsilon_1,\ldots,\varepsilon_4\})$ and $n(i) = 2$ for all $i \in \mathbb{P}^1(\mathbb{C})\times\{\varepsilon_1,\ldots,\varepsilon_4\}$. (In this case we actually have $\hat{I} = I \,\dot{\cup}\, \{\iota\}$, and $C_s \cong \mathbb{Z}\mathbb{A}_\infty/<\tau^{n(\iota)}>$, with $\iota = (0,\varepsilon_1)$.)

This result is proven in detail in [Di 85]. (In fact, it is proven in case $\Lambda = RC_3$, where C_3 is the cyclic group of order 3 and R is a complete discrete valuation ring of ramification degree 4. However, the classification problem for the category of lattices over this group ring is analogous to the classification problem for the category of Cohen-Macaulay modules over the simple elliptic curve singularity of type $\tilde{\mathbb{E}}_8$.) The proof consists in a long and technically involved reduction in which matrix calculations play a key role. This eventually reduces the classification problem for $\text{mod}_R\Lambda$ to the classification problem for $\text{mod}\, A$, where A is a finite-dimensional $\mathbb{C}$-algebra. It turns out that A is a noncommutative algebra which is nondomestic tame of tubular type $\tilde{\mathbb{D}}_4$, in the language of [Rin 84]. Application of the theory developed in [Rin 84] yields the solution described above.

6. Isolated singularities with reduction ideal

Let Λ be an R-algebra which is an isolated singularity, and assume the hypothesis of Theorem 3. If $d = \dim R = 0$, then a much stronger statement than the assertion of Theorem 3 is known to be true, namely in that case the Cartan class of C is always either a Dynkin diagram or $\mathbb{A}_\infty$ [Hap/Pr/Rin 79]. Moreover we have observed in the last section (by reference to literature, respectively by anticipation of a result to be outlined in this note), that this stronger statement is also valid for isolated hypersurface singularities. Therefore it seems natural to try to generalize that strengthened version of Theorem 3 to a class of isolated singularities which sould contain the finite-dimensional algebras over a field as well as the isolated hypersurface singularities. Since the proof given in [Hap/Pr/Rin 79] for the case $d = 0$ depends on a Lemma of Harada and Sai which fails to be true for $d > 0$, it is clear that there is no direct way of generalizing this proof to higher dimensional situations. However, it turns out that we can overcome this difficulty if we impose an additional condition on the isolated singularity, namely the existence of a reduction ideal. In that case, some of the ideas which have been developed in representation theory of artin algebras can be transferred to higher dimensional situations and can be utilized there.

The objective of the following sections is to outline various consequences which can be drawn from Theorem 3 in case the isolated singularity has a reduction ideal, and to discuss the question of existence of a reduction ideal. We shall be precise with definitions and statements, but with proofs we shall be content to describe their main ideas. For detailed proofs we refer to the forthcoming paper [Di 86].

Now assume that R is a commutative noetherian complete local ring, and that Λ is an R-algebra which is finitely generated as R-module.

Definition. A twosided ideal I of Λ is called a reduction ideal of Λ if it has the following properties.

(a) $I \subset m\Lambda$

(b) Λ/I is artinian.

(c) The functor $F_I = \Lambda/I \otimes_\Lambda : \mathrm{mod}_R\Lambda \to \mathrm{mod}(\Lambda/I)$ preserves indecomposability and separates isomorphism classes.

If there exists a reduction ideal I, then we call F_I its reduction functor. It reduces the dimension of the ground ring from d to 0, it maps nonisomorphisms to nonisomorphisms, and it induces an inclusion mapping between the sets of isomorphism classes of indecomposable objects, $\bar{F}_I : [\mathrm{ind}_R\Lambda] \hookrightarrow [\mathrm{ind}(\Lambda/I)]$.

A classical example for a reduction ideal is the following. Let R be a complete discrete valuation ring and let Λ be an R-order, such that $K \otimes_R \Lambda$ is a separable K-algebra. Denote by c the conductor in R of a maximal order $\Lambda' \supset \Lambda$ into Λ. Then $mc\Lambda$ is a reduction ideal of Λ. (See [Cu/Re 81] or [Ro/Hu 70] as a general reference for the case $d = 1$.)

Lemma 4. Let Λ be an R-algebra as above, given by a ring homomorphism $\varphi: R \to \Lambda$, and assume that Λ has a reduction ideal I. Set $i = \varphi^{-1}(I)$ and $\ell = \mathrm{length}_R(R/i)$. If $M_1 \xrightarrow{\psi_1} M_2 \xrightarrow{\psi_2} \cdots \xrightarrow{\psi_{2^{\ell b}-1}} M_{2^{\ell b}}$ is a chain of $2^{\ell b}-1$ nonisomorphisms in $\mathrm{mod}_R\Lambda$, such that all modules $M_1, \ldots, M_{2^{\ell b}}$ are indecomposable and of R-rank $\leq b$, then the image of the composed morphism $\psi_{2^{\ell b}-1} \cdot \ldots \cdot \psi_1$ is contained in $M_{2^{\ell b}}$.

If R is a field and $I = 0$, then we literally recover the Lemma of Harada and Sai [Har/Sa 70], [Rin 79]. The proof of Lemma 4 is straightforward, using the reduction functor F_I together with the validity of the Lemma of Harada and Sai in $\mathrm{mod}(\Lambda/I)$.

<u>Proposition 5</u>. Let Λ be an isolated singularity or nonsingular. Assume that Λ is connected (i.e. indecomposable as a twosided ideal) and that Λ contains a reduction ideal I. Let $C = (C_0, C_1)$ be a connected component of the Auslander-Reiten quiver $A(\Lambda)$, on which the rank function $\rho\colon C_0 \to \mathbb{N}$ is bounded. Then $A(\Lambda) = C$, and $A(\Lambda)$ is finite.

If R is a field, this result is due to Auslander [Au 74]. Using the existence of source and sink morphisms and their close connection with irreducible morphisms in $\mathrm{mod}_R\Lambda$ (see (1), (2) and (2') in section 3), together with Lemma 4, it is not difficult to adapt Auslander's original proof to our situation.

As easy consequences of Proposition 5, respectively of Theorem 3 together with Proposition 5, we obtain the following two Corollaries.

<u>Corollary 6</u>. Let Λ be an isolated singularity with reduction ideal, and assume that Λ is of infinite type. Then there exists an infinite sequence $(M_i)_{i\in\mathbb{N}}$ of indecomposable objects in $\mathrm{mod}_R\Lambda$ such that $\rho(M_i) < \rho(M_{i+1})$, for all $i \in \mathbb{N}$.

<u>Corollary 7</u>. Let Λ be an isolated singularity with reduction ideal. Let $C = (C_0, C_1)$ be a connected component of the stable Auslander-Reiten quiver $A_s(\Lambda)$ and assume that C_0 contains a periodic point. Then the Cartan class of C is either a Dynkin diagram or $\mathbb{A}_\infty$.

7. <u>Existence of reduction ideals</u>

Again assume that R is a commutative noetherian complete local ring, and that Λ is an R-algebra which is finitely generated as R-module.

<u>Definition</u>. The <u>Ext-annihilating ideal</u> of Λ is, by definition, the annihilator ideal (in R) of the bifunctor $\mathrm{Ext}^1_\Lambda(\ ,\ )\colon \mathrm{mod}_R\Lambda \times \mathrm{mod}\,\Lambda \to \mathrm{mod}\,R$. We denote the Ext-annihilating ideal of Λ by $\mathfrak{a}$.

Turning to the question of existence of a reduction ideal of Λ, our first step is to prove the following criterion: If R is Cohen-Macaulay and the Ext-annihilating ideal $\mathfrak{a}$ of Λ is $\mathfrak{m}$-primary, then there exists a reduction ideal of Λ. The proof of this criterion is based on the following Lemma.

<u>Lemma 8</u> (Maranda). Let S be a full subcategory of $\mathrm{mod}_R\Lambda$. Let a be a nonunit and nonzerodivisor of R such that $a\,\mathrm{Ext}^1_\Lambda(M,N) = 0$, for all

$M,N \in S$. Then the functor $F_{a^2} = \Lambda/a^2\Lambda \otimes_\Lambda : S \to \mathrm{mod}_{R/a^2R}(\Lambda/a^2\Lambda)$ preserves indecomposability and separates isomorphism classes.

If R is a complete discrete valuation ring, Λ an R-order, and $S = \mathrm{mod}_R\Lambda$, then Lemma 9 is known as "Maranda's Theorem". Its proof, as presented for example in [Cu/Re 81], carries over to our more general situation without essential change.

Now suppose that R is Cohen-Macaulay and that $\mathfrak{a}$ is $\mathfrak{m}$-primary. Then there exists a maximal R-regular sequence $a_1,\ldots,a_d$ in $\mathfrak{a}$. Set $\mathfrak{a}_0 = (0)$ and $\mathfrak{a}_i = (a_1^2,\ldots,a_i^2)$, for all $i = 1,\ldots,d$, and let S_i be the full subcategory of $\mathrm{mod}_{R/\mathfrak{a}_i}(\Lambda/\mathfrak{a}_i\Lambda)$ given by the class of objects $\{M/\mathfrak{a}_iM \mid M \in \mathrm{mod}_R\Lambda\}$, for all $i = 0,\ldots,d$. Then, applying Lemma 8 successively to the functors $F_{a_i^2} = \Lambda/a_i^2\Lambda \otimes_\Lambda : S_{i-1} \to S_i$, we obtain the announced existence criterion. We reformulate it in the following more constructive way.

<u>Theorem 9</u>. Let R be a commutative noetherian complete local Cohen-Macaulay ring, and let Λ be an R-algebra which is finitely generated as R-module. Assume that the Ext-annihilating ideal $\mathfrak{a}$ of Λ is $\mathfrak{m}$-primary. Then for every maximal R-regular sequence $a_1,\ldots,a_d$ contained in $\mathfrak{a}$, $(a_1^2,\ldots,a_d^2)\Lambda$ is a reduction ideal of Λ.

Now let Λ be an isolated singularity. The problem of existence of a reduction ideal of Λ is brought down, by Theorem 9, to the question whether the Ext-annihilating ideal $\mathfrak{a}$ of Λ is $\mathfrak{m}$-primary. Apart from the case $d \leq 1$, there are three classes of isolated singularities for which we can give an affirmative answer to this latter question. The case of isolated hypersurface singularities (Proposition 11) has been pointed out to me by G.-M. Greuel and F.-O. Schreyer, and the more general case of affine algebraic isolated Cohen-Macaulay singularities (Proposition 12) has been outlined by M. Auslander immediately after my talk on the Lambrecht conference.

<u>Proposition 10</u>. If Λ is an isolated singularity of finite type, then the Ext-annihilating ideal $\mathfrak{a}$ of Λ is $\mathfrak{m}$-primary.

<u>Sketch of proof</u>. Let $M_1,\ldots,M_n$ be a set of representatives for the isomorphism classes of indecomposable objects in $\mathrm{mod}_R\Lambda$, set $M = \bigoplus_{i=1}^{n} M_i$, and consider $\mathfrak{a}_M = \mathrm{ann}_R(\underline{\mathrm{End}}_\Lambda M)$. Then $\mathfrak{a}_M$ is $\mathfrak{m}$-primary, by Theorem 1, and it is easy to see that $\mathfrak{a}_M \subset \mathfrak{a}$.

Proposition 11 (Greuel, Schreyer). Let $\Lambda = \mathbb{C}[[X_0,\ldots,X_d]]/(f(X))$ be an isolated hypersurface singularity, with chosen Noether normalization $R \subset \Lambda$. Then the Ext-annihilating ideal $\mathfrak{a}$ of Λ is $\mathfrak{m}$-primary.

Sketch of proof. Denote by M the maximal ideal of Λ. Let $J_f = (\frac{\partial f}{\partial X_0},\ldots,\frac{\partial f}{\partial X_d})$ be the Jacobi ideal of $f(X)$ in $\mathbb{C}[[X_0,\ldots,X_d]]$, and let $\mathcal{J}_f = (J_f + (f(X)))/(f(X))$ be its image in Λ. Then $\mathcal{J}_f$ is M-primary, by Jacobi's Criterion. Moreover, calculating $\mathrm{Ext}^1_\Lambda(C,Y)$ via the Λ-projective resolution of C which is given by the matrix factorization of $f(X)$ corresponding to C [Ei 80, §6], it is easy to prove that $\mathcal{J}_f \,\mathrm{Ext}^1_\Lambda(C,Y) = 0$ for all $C \in \mathrm{mod}_R\Lambda$ and $Y \in \mathrm{mod}\,\Lambda$. It follows that $\mathcal{J}_f \cap R$ is $\mathfrak{m}$-primary and is contained in $\mathfrak{a}$.

Proposition 12 (Auslander). Let $\Lambda = \hat{\mathcal{O}}_{V,0}$ be the complete local ring of an affine algebraic variety $V \subset \mathbb{A}^n(\mathbb{C})$ at an isolated Cohen-Macaulay singularity $0 \in V$, and let $R \subset \Lambda$ be a Noether normalization. Then the Ext-annihilating ideal $\mathfrak{a}$ of Λ is $\mathfrak{m}$-primary.

Sketch of proof. Let $d = \dim V$ and let $S = \mathcal{O}_{V,0}$ be the local ring of V at 0, with maximal ideal M. Then $\mathrm{pd}_{S_p \otimes_{\mathbb{C}} S_p} S_p \le d$, for all $p \in \mathrm{Reg}(S)$. Consider the augmentation sequence $0 \to J \to S \otimes_{\mathbb{C}} S \to S \to 0$, let $M = \Omega^d(S)$ (the d-th syzygy module of S in $\mathrm{mod}(S \otimes_{\mathbb{C}} S)$), and set $A = \mathrm{ann}_S(\underline{\mathrm{End}}_{S\otimes S}M/J\underline{\mathrm{End}}_{S\otimes S}M)$. Then $M^n \subset A$ for some $n \in \mathbb{N}$, and $A\,\mathrm{Ext}^{d+1}_S(X,Y) = 0$ for all $X,Y \in \mathrm{Mod}\,S$. Analyzing the passage to completion one finds that $\hat{A}$ is $\hat{M}$-primary, and $\hat{A}\,\mathrm{Ext}^1_\Lambda(X,Y) = 0$ for all $X \in \mathrm{mod}_R\Lambda$ and $Y \in \mathrm{mod}\,\Lambda$. It follows that $\hat{A} \cap R$ is $\mathfrak{m}$-primary, and is contained in $\mathfrak{a}$.

8. Main results

Summing up the results of sections 6 and 7, we arrive at the following conclusions.

Theorem 13. Let R be a commutative noetherian complete regular local ring, and let Λ be an R-algebra which is finitely generated free as R-module. Assume that Λ is of finite type. Then each connected component C of the stable Auslander-Reiten quiver of Λ is of the form $C \cong \mathbb{Z}\Delta/G$, where Δ is a Dynkin diagram and G is a group of automorphisms of $\mathbb{Z}\Delta$.

Recall from [Au 84] that, since Λ is of finite type, Λ has to be an isolated singularity or nonsingular. Then Theorem 13 follows immediately from Corollary 7, Theorem 9 and Proposition 10. If R is an algebraically closed field, then we recover Riedtmann's well-known Theorem [Rie 80]. The simple isolated hypersurface singularities provide another special case for the class of algebras described in Theorem 13. But much more generally, Theorem 13 states that Dynkin diagrams always appear in connection with R-algebras Λ of finite type, as soon as the Auslander Reiten quiver contains a stable point.

Theorem 14. Let $\Lambda = \mathbb{C}[[X_0,\ldots,X_d]]/(f(X))$ be an isolated hypersurface singularity of infinite type. Then the following statements hold.

(i) The Auslander-Reiten quiver of Λ is of the form $A(\Lambda) \cong C \,\dot{\cup}\, (\dot{\bigcup}_{i\in I} \mathbb{Z}\mathbb{A}_\infty/\langle\tau^{n(i)}\rangle)$, where C is the connected component of $A(\Lambda)$ which contains $[\Lambda]$, I is an index set, and $n(i) \in \{1,2\}$ for all $i \in I$. Moreover, if d is even then $n(i) = 1$ for all $i \in I$.

(ii) The full subquiver of C which consists of all points different from $[\Lambda]$ is of the form $C_s \cong \dot{\bigcup}_{j\in J} \mathbb{Z}\Delta_j/G_j$, where J is a finite index set and for all $j \in J$, Δ_j is either a Dynkin diagram or $\mathbb{A}_\infty$ and G_j is a group of automorphisms of $\mathbb{Z}\Delta_j$.

(iii) If there is only one direct predecessor of $[\Lambda]$ in $A(\Lambda)$, then the stable Auslander-Reiten quiver of Λ is of the form $A_s(\Lambda) \cong \dot{\bigcup}_{i\in\hat{I}} \mathbb{Z}\mathbb{A}_\infty/\langle\tau^{n(i)}\rangle$, where $\hat{I} = I \,\dot{\cup}\, \{\iota\}$, and $n(i) \in \{1,2\}$ for all $i \in \hat{I}$. Moreover, if d is even then $n(i) = 1$ for all $i \in \hat{I}$.

This is the result we anticipated in section 5.2. It follows from Corollary 7, Theorem 9, Proposition 11 and Proposition 5, taking into account that $\tau^2([M]) = [M]$ for all points $[M] \neq [\Lambda]$ in $A(\Lambda)$ (see section 5).

Theorem 15. Let $\Lambda = \mathbb{C}[[X_0,\ldots,X_d]]/(f(X))$ be an isolated hypersurface singularity of infinite type. Then there exists an arithmetic sequence of natural numbers $r,2r,3r,..$ such that for each mr ($m \in \mathbb{N}$) there exists an infinite sequence $(M_{m,n})_{n\in\mathbb{N}}$ of indecomposable pairwise non-isomorphic Cohen-Macaulay Λ-modules, all of which have rank mr.

In other words, the second Brauer-Thrall conjecture is true for isolated hypersurface singularities. The proof of Theorem 15 is based on the existence of an infinite sequence $(M'_n)_{n\in\mathbb{N}}$ of pairwise nonisomorphic indecomposable Cohen-Macaulay Λ-modules of constant rank r', which is proven in [Bu/Gr/Schr 86]. From this, making use of the "tubular structure" of $A(\Lambda)\setminus C$, as described in Theorem 14. it is not difficult to exhibit a double sequence $(M_{m,n})_{m,n\in\mathbb{N}}$ of indecomposable Cohen-Macaulay Λ-modules which has all the properties described in the statement of Theorem 15.

The following two Theorems are immediate consequences of Theorem 9, Proposition 12, and Corollary 7 respectively Proposition 5.

Theorem 16. Let $\Lambda = \hat{\mathcal{O}}_{V,0}$ be the complete local ring of an affine algebraic variety $V \subset \mathbb{A}^n(\mathbb{C})$ at an isolated Cohen-Macaulay singularity $0 \in V$. Let C be a connected component of the stable Auslander-Reiten quiver of Λ such that C contains a periodic point. Then $C \cong \mathbb{Z}\Delta/G$, when Δ is either a Dynkin diagram or $\mathbb{A}_\infty$, and G is a group of automorphisms of $\mathbb{Z}\Delta$.

Theorem 17. Let $\Lambda = \hat{\mathcal{O}}_{V,0}$ be as in Theorem 16. If Λ is of infinite type, then there exists an infinite sequence $(M_i)_{i\in\mathbb{N}}$ of indecomposable Cohen-Macaulay Λ-modules such that rank $M_i <$ rank M_{i+1}, for all $i \in \mathbb{N}$.

In other words, the first Brauer-Thrall conjecture is true for affine algebraic isolated Cohen-Macaulay singularites. Using different methods, Herzog and Kühl recently have obtained results which are related to Theorem 17 [He/Sa 85].

References

[Ar 74] V.I. Arnol'd: Critical points of smooth functions. Proc. Int. Cong. Math. Vancouver , Vol. I (1974), 19-39.

[Au 74] M. Auslander: Representation theory of artin algebras II. Comm. Alg. 1 (1974), 269-310.

[Au 75] M. Auslander: Existence theorems for almost split sequences. Ring Theory II, Proc. of the second Oklahoma Conf. (Marcel Dekker, New York, 1975), 1-44.

[Au 78] M. Auslander: Functors and morphisms determined by objects. Proc. Conf. on Repr. Theory, Philadelphia (1976). Marcel Dekker (1978), 1-245.

[Au 84] M. Auslander: Isolated singularities and existence of almost split sequences. Notes by L. Unger. Proc. of the fourth international conference on representations of algebras, Carleton University, Ottawa 1984, Springer Lecture Notes 1178, 194-242.

[Au 86] M. Auslander: Rational singularities and almost split sequences. Transactions of the American Math. Soc., Vol. 293, No. 2 (1986), 511-531.

[Au/Re 75] M. Auslander and I. Reiten: Representation theory of artin algebras III, Comm. Alg. 3 (1975), 239-294; IV, ibid., 5 (1977), 443-518.

[Bu/Gr/Schr 86] R. Buchweitz and G.-M. Greuel and F.-O. Schreyer: Cohen-Macaulay modules on hypersurface singularities II. To appear.

[Cu/Re 81] C.W. Curtis and I. Reiner: Methods of Representation Theory I, Pure and applied mathematics (Wiley, London 1981).

[Di 85] E. Dieterich: Classification of the indecomposable representations of the cyclic group of order three in a complete discrete valuation ring of ramification degree four. Preprint, Universität Bielefeld, 1985.

[Di 86] E. Dieterich: Reduction of isolated singularities. Preprint, Brandeis University, 1986.

[Di/Wi 86] E. Dieterich and A. Wiedemann: The Auslander-Reiten quiver of a simple curve singularity. Transactions of the American Math. Soc., Vol. 294, No. 2 (1986), 455-475.

[Ei 80] D. Eisenbud: Homological algebra on a complete intersection, with an application to group representations. Transactions of the American Math. Soc., Vol. 260, No. 1 (1980), 35-64.

[Hap/Pr/Rin 79] D. Happel and U. Preiser and C.M. Ringel: Vinberg's characterization of Dynkin diagrams using subadditive functions with application to DTr-periodic modules. Proc. of the third international conf. on repr. of alg., Carleton Univ., Ottawa 1979. Springer Lecture Notes 832, 280-294.

[Hap/Pr/Rin 80] D. Happel and U. Preiser and C.M. Ringel: Binary polyhedral groups and Euclidean diagrams. Manuscripta Mathematica 31 (1980), 317-329.

[Har/Sa 70] M. Harada and Y. Sai: On categories of indecomposable modules I. Osaka Journal of Mathematics 7 (1970), 323-344.

[He/Sa 85] J. Herzog and H. Sanders: Indecomposable syzygy modules of high rank over hypersurface rings. To appear in Journal of pure and applied algebra.

[Kn 86] H. Knörrer: Cohen-Macaulay modules on hypersurface singularities I. To appear.

[Rie 80] Chr. Riedtmann: Algebren, Darstellungsköcher, Überlagerungen, und zurück. Commentarii Mathematici Helvetici 55 (1980), 199-224.

[Rin 79] C.M. Ringel: Report on the Brauer-Thrall conjectures. Proceedings of the third international conference on representations of algebras, Carleton University, Ottawa 1979. Springer Lecture Notes 831, 104-136.

[Rin 84] C.M. Ringel: Tame algebras and integral quadratic forms. Springer Lecture Notes 1099.

[Ro/Hu 70] K.W. Roggenkamp and V. Huber-Dyson: Lattices over orders I, Springer Lecture Notes 115; II, ibid., 142.

[Ro/Schm 76] K.W. Roggenkamp and J.W. Schmidt: Almost split sequences for integral group rings and orders. Communications in Algebra 4 (1976), 893-917.

[Va 34] P. Du Val: On isolated singularities of surfaces that do not affect the conditions of adjunction I, II, III. Proc. Cambridge Phil. Soc. 30 (1934), 483-491.

A class of weighted projective curves arising in representation theory of finite dimensional algebras

Werner Geigle and Helmut Lenzing

Fachbereich Mathematik-Informatik
Universität-GH Paderborn
Warburger Str. 100
D-4790 Paderborn
West-Germany

Introduction

By means of a suitably graded sheaf theory we introduce a new class of curves, called *weighted projective lines*, having an interpretation as lines in an appropriate *weighted projective space* $\mathbf{P}_n(p)$, with respect to a weight sequence $p = (p_0, \ldots, p_n)$ of integers. We note that our approach to weighted projective spaces is similar to the treatment by Delorme [8], Dolgachev [10] and Beltrametti-Robbiano [6] but differs sensibly in spirit and content. Section 1 summarizes those results of joint investigation with D. Baer and P. Dowbor which are needed to put weighted projective lines into proper perspective; a complete account is under preparation. The main advantage of our approach is that Serre's theorem (1.7) holds true, which removes all the pathologies ([6], Section 3) encountered in the former treatment of these spaces.

As becomes clear from the results of Sections 2 and 5, a weighted projective line C behaves like a smooth projective curve with respect to coherent sheaves and vector bundles on C. This allows us to use all the methods familiar in this latter situation, see [24, 32, 1]. So the category of coherent sheaves coh(C) has Serre-duality (2.2), consequently almost-split sequences (2.3). Each coherent sheaf splits into a direct sum of a vector bundle and a torsion sheaf (2.4). By means of a Riemann-Roch theorem (2.9) we attach a (virtual) genus to C, which is characteristic for the complexity of the classification problem for coh(C) (5.4).

Our motivation to investigate weighted projective lines originates from the representation theory of finite dimensional algebras in an attempt to give a geometric treatment similar to [26] for the so-called canonical algebras, introduced and studied by C. M. Ringel [30]. Actually there is a bijective correspondence between (isomorphism classes of) weighted projective lines and canonical algebras, respectively.

The reader will observe that our present treatment differs sensibly from the previous approach, using a variant of Beilinson's theorem [5] as the basis for the comparison between coh(C) and mod(Λ): By means of a tilting sheaf (3.1) with endomorphism algebra Λ, we establish an equivalence $D^b(\mathrm{coh}(C)) = D^b(\mathrm{mod}(\Lambda^{op}))$ of the derived categories of coh(C) and the category mod(Λ^{op}) of finite dimensional Λ-modules, respectively. A comparison theorem (3.3) deduces the consequences of $D^b(\mathrm{coh}(C)) = D^b(\mathrm{mod}(\Lambda^{op}))$ in the spirit of tilting theory [20,7,21]. As a result, the classification problems for coh(C) and mod(Λ), if Λ denotes the canonical algebra attached to C, are basically equivalent. We note that the subdivision of indecomposable coherent sheaves on C into the two classes of indecomposable vector bundles and torsion sheaves translates by means of tilting into a subdivision of indecomposable Λ-modules into now three classes (cf. (4.3) and [30]).

In Section 5 we give a brief account on the classification of indecomposable bundles on C if C has (virtual) genus one. Not unexpectedly, Atiyah's approach to classify vector bundles on smooth elliptic projective curves [1] also works in this context. Our exposition also relies on the work of Narasimhan and Seshadri (see [32]). Thus the comparison $D^b(\mathrm{coh}(C)) = D^b(\mathrm{mod}(\Lambda^{op}))$ establishes a link between Atiyah's classification of vector bundles on elliptic curves and Ringel's classification for modules over canonical algebras of tubular type [30]. From a geometric point of view this interrelation is explained by Example 5.8.

Theorem 5.1 relates the classification of vector bundles on C to the study of graded Cohen-Macaulay modules; a detailed account will be given elsewhere.

The authors acknowledge the support of the Deutsche Forschungsgemeinschaft (SPP Darstellungstheorie). The second-named author wishes to thank C. S. Seshadri for helpful discussions pointing out the interrelations of weighted projective lines to the concept of curves with parabolic structure [32].

1. Weighted projective spaces and weighted projective lines

Throughout k denotes an algebraically closed field of arbitrary characteristic.

1.1. Let $p = (p_0, \ldots, p_n)$ be an $(n+1)$-tuple of integers $p_i \geq 1$, called the *weight sequence.* The *affine algebraic group*

$$G(p) = \{(t_0, t_1, \ldots, t_n) \in (k^*)^{n+1} \mid t_0^{p_0} = t_1^{p_1} = \cdots = t_n^{p_n}\} \tag{1.1.1}$$

acts on affine $(n+1)$-space $\mathbf{A}_{n+1} = k^{n+1}$ by multiplication

$$(t_0, t_1, \ldots, t_n)(x_0, x_1, \ldots, x_n) = (t_0x_0, \ldots, t_nx_n).$$

If $\mathbf{L}(p)$ denotes the *rank one abelian group* on generators $\vec{x}_0, \vec{x}_1, \ldots, \vec{x}_n$ with relations $p_0\vec{x}_0 = p_1\vec{x}_1 = \cdots = p_n\vec{x}_n$, clearly the group Hopf algebra $k[\mathbf{L}(p)]$ of $\mathbf{L}(p)$ represents the affine algebraic group $G(p)$. Hence the above $G(p)$-action on $\mathbf{A}_{n+1}$ corresponds to a graduation of the polynomial algebra $S = k[X_0, X_1, \ldots, X_n]$ with grading group $\mathbf{L}(p)$; the graduation being specified by defining X_i to be homogeneous of degree $\vec{x}_i$. (Notation: $\deg(X_i) = \vec{x}_i$). Thus S carries a decomposition $S = \bigoplus_{\vec{l} \in \mathbf{L}(p)} S_{\vec{l}}$ into k-subspaces satisfying $S_{\vec{l}} S_{\vec{m}} \subset S_{\vec{l}+\vec{m}}$ for all $\vec{l}, \vec{m} \in \mathbf{L}(p)$; moreover $X_i \in S_{\vec{x}_i}$ for $i = 0, \ldots, n$. We use the notation $S(p)$ for the $\mathbf{L}(p)$-*graded algebra* thus defined.

To each sequence $\lambda = (\lambda_0, \ldots, \lambda_n)$ of pairwise distinct elements of $\mathbf{P}_1(k)$, normalized such that $\lambda_0 = \infty$, $\lambda_1 = 0$, $\lambda_2 = 1$, we attach the two-dimensional subvariety $F(p\ , \lambda)$ of $\mathbf{A}_{n+1}$, given by the equations

$$X_i^{p_i} = X_1^{p_1} - \lambda_i X_0^{p_0}\ , \quad i = 2, \ldots, n. \tag{1.1.2}$$

$F(p\ , \lambda)$ is stable under the $G(p)$-action just described. Accordingly, the elements $f_i = X_i^{p_i} - X_1^{p_1} + \lambda_i X_0^{p_0}$, $(i = 2, \ldots, n)$ generate a homogeneous ideal $I(p\ , \lambda)$ of $S(p)$. Hence

$$S(p\ , \lambda) = k[X_0\ , X_1, \ldots, X_n]/I(p\ , \lambda) = k[x_0\ , x_1, \ldots, x_n] \tag{1.1.3}$$

is again $\mathbf{L}(p)$-graded with $\deg(x_i) = \vec{x}_i$.

We are now going to endow the (set-theoretic) quotients $\mathbf{P}_n(p) = k^{n+1} - \{0\}/G(p)$ and $C(p\ , \lambda) = F(p\ , \lambda)/G(p)$ with an $\mathbf{L}(p)$-graded sheaf theory, defining on $\mathbf{P}_n(p)$ and $C(p\ , \lambda)$ the geometric structure of a weighted projective space, a weighted projective line, respectively.

1.2. Call $\vec{c} = p_0\vec{x}_0 = \cdots = p_n\vec{x}_n$ the *canonical element* of $\mathbf{L}(p)$. For reasons, which will become clear later, $\vec{\omega} = (n-1)\vec{c} - \sum_{i=0}^{n} \vec{x}_i$ is called the *dualizing element*. $\mathbf{L}(p)$ is an *ordered group* with $L_+ = \sum_{i=0}^{n} \mathbf{N}\,\vec{x}_i$ as its set of positive elements. Since $\mathbf{L}(p)/\mathbf{Z}\,\vec{c} \cong \prod_{i=0}^{n} \mathbf{Z}/p_i\mathbf{Z}$ canonically, each $\vec{l} \in \mathbf{L}(p)$ can be uniquely written in *normal form*

$$\vec{l} = \sum_{i=0}^{n} l_i\vec{x}_i + l\vec{c} \ \text{ with } \ 0 \le l_i < p_i \ \text{ and } \ l \in \mathbf{Z}. \tag{1.2.1}$$

If $\vec{l}, \vec{m}$ are both in normal form, $\vec{l} \le \vec{m}$ if and only if $l_i \le m_i$ for $i = 0, \ldots, n$ and $l \le m$.

Since $\vec{c} + \vec{\omega} = \sum_{i=0}^{n} (p_i - 1)\vec{x}_i - \vec{c}$ is in normal form, we see that *each* $\vec{l} \in \mathbf{L}(p)$ *satisfies exactly one of the two possibilities:* $0 \le \vec{l}$ or $\vec{l} \le \vec{c} + \vec{\omega}$.

Let $p = \text{l.c.m.}\,(p_0, \ldots, p_n)$. We define the *degree map* $\delta\colon \mathbf{L}(p) \longrightarrow \mathbf{Z}$ on generators by $\delta(\vec{x}_i) = \frac{p}{p_i}$. δ is an epimorphism of ordered groups, its kernel being the torsion group of $\mathbf{L}(p)$. Note that $\delta(\vec{l}) = 0$ implies $p\vec{l} = 0$.

For an $\mathbf{L}(p)$-graded algebra $\mathrm{Mod}^{\mathbf{L}(p)}(S)$, $\mathrm{mod}^{\mathbf{L}(p)}(S)$ and $\mathrm{mod}_0{}^{\mathbf{L}(p)}(S)$ denote the categories of all (all finitely generated, all finite length) $\mathbf{L}(p)$-graded S-modules, respectively, with morphisms the S-linear homogeneous maps of degree 0. $\mathbf{L}(p)$ acts on each of this categories by twist $(M , \vec{x}) \longrightarrow M(\vec{x})$, where $M(\vec{x})_{\vec{y}} = M_{\vec{x}+\vec{y}}$. Similar notations are used for ungraded, also for $\mathbf{L}_+$-graded modules.

1.3. For the applications we have in mind, the following proposition serves as the basic tool:

Proposition. *$S(p)$ and $S(p , \lambda)$ are $\mathbf{L}(p)$-graded factorial domains, i. e. up to scalars each homogeneous element is a product of homogeneous prime elements. Up to scalars a complete system of homogeneous prime elements for $S(p)$ or $S(p , \lambda)$ is given by*

(i) the elements $X_0 , X_1, \ldots ,X_n \in S(p)$, $(x_0 , x_1, \ldots ,x_n \in S(p , \lambda))$, called the exceptional prime elements of $S(p)$, $S(p , \lambda)$, respectively.

(ii) the elements $f(X_0^{p_0}, \ldots ,X_n^{p_n})$, $(f(x_0^{p_0} , x_1^{p_1}))$, called the ordinary prime elements, where f is an irreducible homogeneous element of the polynomial algebra $k[T_0 , T_1, \ldots ,T_n]$, $(k[T_0 , T_1])$, not associated to any $T_0 , T_1, \ldots ,T_n$, (T_0 , T_1), respectively. Here, as usual, both polynomial algebras are $\mathbf{Z}$-graded by total degree.

Moreover, $x_0 , x_1, \ldots ,x_n$, (x_0 , x_1) is an $\mathbf{L}(p)$-homogeneous regular sequence for $S(p)$ or $S(p , \lambda)$, respectively. In particular, $S(p)$, $S(p , \lambda)$ has graded and ungraded Krull dimension $n+1$ or 2, respectively. Also $x_0 , x_1 , f_2, \ldots f_n$ is a regular sequence for $S(p)$.

Since we assume that k is algebraically closed, the elements

$$x_1^{p_1} - \lambda x_0^{p_0} \ , \ \lambda \in k-\{\lambda_0, \ldots ,\lambda_n\} \tag{1.3.1}$$

constitute a complete set of ordinary primes for $S(p , \lambda)$.

Proof. Let first S denote the algebra $S(p)$. We may view the *restriction*

$$R(p) = \bigoplus_{l=0}^{\infty} R_l \ , \quad R_l = S_{l\vec{c}}$$

of S to the subgroup $\mathbf{Z}\,\vec{c}$ of $\mathbf{L}(p)$ as a $\mathbf{Z}$-graded algebra, called the *core* of S. Clearly, $R(p) = k[X_0^{p_0}, \ldots ,X_n^{p_n}]$ is the $\mathbf{Z}$-graded polynomial algebra over k in the indeterminates $X_0^{p_0}, \ldots ,X_n^{p_n}$, and R_l consists of all homogeneous polynomials of total degree l in $X_0^{p_0}, \ldots ,X_n^{p_n}$. Moreover if $\vec{l} \in \mathbf{L}(p)$ is written in normal form (1.2.1) we have

$$S(p)_{\vec{l}} = X_0^{l_0} \cdots X_n^{l_n} R(p)_l \ , \quad \text{where } 0 \le l_i < p_i \ , \ l \in \mathbf{Z}. \tag{1.3.2}$$

Similarly, the core of $S(p , \lambda)$ is the polynomial algebra $R(p , \lambda) = k[x_0^{p_0} , x_1^{p_1}]$ and

$$S(p , \lambda)_{\vec{l}} = x_0^{l_0} \cdots x_n^{l_n} R(p , \lambda)_l \ , \quad \text{where } 0 \le l_i < p_i \ , \ l \in \mathbf{Z}. \tag{1.3.3}$$

The assertions now follow from (1.3.2) and (1.3.3). □

As is clear from (1.3.2) and (1.3.3) we have $S_{\vec{l}} \neq 0$ if and only if $\vec{l} \geq 0$.

The role of the algebras $S(p\ ,\ \lambda)$, with $p_0, \ldots, p_n$ pairwise coprime, is explained by a theorem of Mori [27], stating that for an algebraically closed base field k these are just the **Z**-graded affine k-algebras S with $S_0 = k$, which are graded factorial of Krull dimension two.

1.4. Let S be either $S(p)$ or $S(p\ ,\ \lambda)$ and R be the core of S. If T is a multiplicative subset of R, consisting of homogeneous elements, the quotient ring $T^{-1}S$ is again $\mathbf{L}(p)$-graded: $(T^{-1}S)_{\vec{l}}$ consists of all homogeneous quotients $\frac{s}{t}$, where $s \in S, t \in T$ both are homogeneous and $\deg(s) - \deg(t) = \vec{l}$. Replacing T by $T^p = \{t^p \mid t \in T\}$, if necessary, we may always assume that T is contained in R. In this case, for $\vec{l} \in \mathbf{L}(p)$ written in normal form (1.2.1) we get

$$(T^{-1}S)_{\vec{l}} = X_0^{l_0} \cdots X_n^{l_n}(T^{-1}R)_l. \qquad (1.4.1)$$

If T is the set of all homogeneous non-zero elements of R, $Q = T^{-1}S$ is the *total ring of homogeneous quotients*. Its zero-component Q_0 is the field of rational functions $k(\frac{X_1^{p_1}}{X_0^{p_0}}, \ldots, \frac{X_n^{p_n}}{X_0^{p_0}})$ or $k(\frac{x_1^{p_1}}{x_0^{p_0}})$, according as $S = S(p)$ or $S = S(p\ ,\ \lambda)$. Moreover, passage to the zero-component

$$\mathrm{mod}^{\mathbf{L}(p)}(Q) \longrightarrow \mathrm{mod}\,(Q_0),\ M \longrightarrow M_0$$

defines a category equivalence. We express this fact, stating that the $\mathbf{L}(p)$-graded algebra Q is *Morita-equivalent* to the algebra Q_0.

1.5. Let $\mathbf{X}$ be either $\mathbf{P}_n(p)$ or its subset $C(p\ ,\ \lambda)$, accordingly S stands either for $S(p)$ or $S(p\ ,\ \lambda)$. For $\vec{l} = l_0\vec{x}_0 + \cdots + l_n\vec{x}_n$ and $t = (t_0\ , t_1, \ldots, t_n) \in G(p)$ we write

$$\vec{l}(t) = t_0^{p_0} \cdots t_n^{p_n}, \qquad (1.5.1)$$

which identifies $\vec{l}$ with a character of $G(p)$, actually $\mathbf{L}(p)$ with $\mathrm{Hom}(G(p)\ ,\ k^*)$. We have

$$f(tx) = \vec{l}(t)\, f(x) \ \text{ for } f \in S_{\vec{l}},\ x \in k^{n+1}\ ,\ t \in G(p). \qquad (1.5.2)$$

This allows to form the sets

$$\mathrm{D}(f) = \{[x] \in \mathbf{X} \mid f(x) = 0\}\ ,\ f \in S \text{ homogeneous}\ ,$$

which form a basis for the *Zariski topology* on $\mathbf{X}$. As usual $\mathbf{V}(f)$ denotes the complement of $\mathrm{D}(f)$ in $\mathbf{X}$. The *structure sheaf* $\mathcal{O}_{\mathbf{X}}$ is the sheaf of $\mathbf{L}(p)$-graded k-algebras attached to the presheaf $\mathrm{D}(f) \rightarrow S_f$, $f \in S$ homogeneous. S_f denotes the $\mathbf{L}(p)$-graded quotient ring $T^{-1}S$, where T is the multiplicative set generated by f. As usual ([17]), we have

$$\Gamma\,(\mathrm{D}(f)\,,\,O_X) = S_f \tag{1.5.4}$$

if f is a homogeneous element of $S_+ = \bigoplus_{\vec{l}>0} S_{\vec{l}}$. By graded normality of S (cf. Proposition 1.3), we also get

$$\Gamma\,(X\,,\,O_X(\vec{x})) = S(\vec{x}) \text{ for each } \vec{x} \in \mathbf{L}(p). \tag{1.5.5}$$

As a result, O_X is an $\mathbf{L}(p)$-graded sheaf of algebras

$$O_X = \bigoplus_{\vec{l} \in \mathbf{L}(p)} (O_X)_{\vec{l}}. \tag{1.5.6}$$

We note that our approach differs sensibly from the traditional treatment of weighted projective spaces [8,10,6] where the 0-component $(O_X)_0$ of $O_{P_n(p)}$ serves as the structure sheaf. (In order to make this comparison possible, we have to assume that $p_0, \ldots, p_n$ are pairwise coprime, so $\mathbf{L}(p) = \mathbf{Z}$.).

The *weight $p(t)$ of a point* $t = [t_0, \ldots, t_n]$ is defined as $\prod\{p_j \mid t_j = 0\}$. A point t is called *ordinary* if $p(t) = 1$, otherwise t is called *exceptional*. It follows from Proposition 1.3 that we may view $C(p\,,\,\lambda)$ as a curve in $\mathbf{P}_n(p)$, which is a complete intersection since the defining equations $f_i = X_i^{p_i} - X_1^{p_1} + \lambda_i X_0^{p_0}$ form a regular sequence of homogeneous elements in $S(p)$. Actually we prefer to call $C(p\,,\,\lambda)$ a *weighted projective line* in $\mathbf{P}_n(p)$: the map

$$\mathbf{P}_n(p) \longrightarrow \mathbf{P}_n(k)\,,\ [x_0\,,x_1, \ldots ,x_n] \longrightarrow [x_0^{p_0}, \ldots ,x_n^{p_n}] \tag{1.5.7}$$

allows to identify both spaces set-theoretically. By means of (1.5.7) the defining equations for $C(p\,,\,\lambda)$ convert into equations defining a line in $P_n(k)$. Accordingly

$$C(p\,,\,\lambda) \longrightarrow P_1(k)\,,\ [x_0\,,x_1, \ldots ,x_n] \longrightarrow [x_0^{p_0}\,,\,x_1^{p_1}] \tag{1.5.8}$$

is a bijection. By means of this correspondence the exceptional points of $C(p\,,\,\lambda)$ are just converted to the system $\lambda_0, \ldots ,\lambda_n$, attaching weight p_i to λ_i.

1.6. The stalk $O_{X\,,\,t}$ of the structure sheaf O_X at $t \in X$ is given by

$$O_{X\,,\,t} = \{\frac{f}{g} \mid f\,,g \in S\,,\ g \text{ homogeneous with } g(t) \neq 0\}. \tag{1.6.1}$$

This ring is always $\mathbf{L}(p)$-graded regular local of dimension n+1 if $X = \mathbf{P}_n(p)$, and an $\mathbf{L}(p)$-graded discrete valuation ring if $X = C(p\,,\,\lambda)$. Notice that R is called a graded local ring if R has a unique homogeneous maximal ideal. Moreover regularity means that R has finite graded global dimension n. If additionally $n = 1$ we deal with a graded discrete valuation ring. For an ordinary point t, i. e. if all homogeneous coordinates t_i are non-zero, $O_{X\,,\,t}$ is Morita-equivalent to its zero-component, which according to (1.4.1) is isomorphic to the (now ungraded) localization of the polynomial algebra $k[T_1, \ldots ,T_n]$ with respect to the maximal ideal $(T_1 - t_1, \ldots ,T_n - t_n)$ or to the localization of $k[T]$ with respect to $(T - t_1)$ according as $X = \mathrm{P}_n(p)$ or $X = C(p\,,\,\lambda)$, respectively. Here, we assume $t_0 = 1$. Thus for an ordinary point, the category $\mathrm{mod}^{\mathbf{L}(p)}(O_{X,t})$ of finitely generated $\mathbf{L}(p)$-graded modules over $O_{X,t}$ has exactly one simple module, up to isomorphism.

By contrast for an *exceptional point* t of X $\mathrm{mod}^{\mathbf{L}(p)}(O_{X,t})$ has exactly $p(t)$ isomorphism classes of simple modules. Accordingly, $\mathrm{mod}^{\mathbf{L}(p)}(O_{X,t})$ has exactly $p(t)$ isomorphism classes of indecomposable projective modules, necessarily of the form $O_{X,t}(\vec{x})$ for some $\vec{x} \in \mathbf{L}(p)$.

For T an indeterminate let A be the discrete valuation ring obtained from $k[T^n]$ by localizing at the maximal ideal (T^n). It is easily checked from (1.4.1) that if $X = C(p, \lambda)$ and t denotes a point of weight n, the graded algebra $O_{X,t}$ is Morita equivalent to the subring of $M_n(k(T))$ given by

$$\begin{bmatrix} A & TA & \cdot & \cdot & \cdot & T^{n-1}A \\ T^{n-1}A & A & & & & \cdot \\ \cdot & \cdot & \cdot & & & \cdot \\ \cdot & & & \cdot & & \cdot \\ \cdot & & & & A & TA \\ TA & & \cdot & \cdot & T^{n-1}A & A \end{bmatrix}.$$

This proves that in the $\mathbf{L}(p)$-graded situation, we basically deal with a non-commutative sheaf theory.

We may attach a *generic point* ξ to X, where the stalk $F_\xi = (\tilde{M})_\xi$ is given by $T^{-1}M$, with T the set of all homogeneous non-zero divisors of S. As was shown in (1.4), O_ξ is Morita-equivalent to the function field K of X. Accordingly, we may view F_ξ as a (finite dimensional) vector space over K, whose dimension is called the *rank* of F.

1.7. Again X stands for either $\mathbf{P}_n(p)$ or $C(p, \lambda)$. The sheaves on X we want to consider are the $\mathbf{L}(p)$-graded sheaves M of O_X-modules. Thus M is an O_X-module, carrying an $\mathbf{L}(p)$-graduation $M = \bigoplus_{\vec{l} \in \mathbf{L}(p)} M_{\vec{l}}$ satisfying $O_{X,\vec{l}} M_{\vec{m}} \subset M_{\vec{l}+\vec{m}}$. Morphisms will be morphisms of graded O_X-modules of degree 0. Let $\mathrm{Mod}^{\mathbf{L}(p)}(O_X)$ be the resulting category of $\mathbf{L}(p)$-graded O_X-modules. The group $\mathbf{L}(p)$ acts on $\mathrm{Mod}^{\mathbf{L}(p)}(O_X)$ by *twist*

$$(\vec{l}, M) \longrightarrow M(\vec{l}),$$

where $M(\vec{l})_{\vec{x}} = M_{\vec{l}+\vec{x}}$. Since $F \longrightarrow \Gamma(X, F)$ commutes with the shift operation, (1.5.5) generalizes to

$$\Gamma(X, O_X(\vec{x})) = S(\vec{x}) \quad \text{for each } \vec{x} \in \mathbf{L}(p). \tag{1.7.1}$$

A *coherent sheaf on* X is by definition an $\mathbf{L}(p)$-graded O_X-module M, where for each $x \in X$ there is an open neighbourhood U of x and an exact sequence

$$\bigoplus_{j=1}^{m} O_X(\vec{l}_j)_{|U} \longrightarrow \bigoplus_{i=1}^{n} O_X(\vec{l}_i)_{|U} \longrightarrow M_{|U} \longrightarrow 0 \tag{1.7.2}$$

of $\mathbf{L}(p)$-graded O_U-modules. coh(X) is the full subcategory of $\mathrm{Mod}^{\mathbf{L}(p)}(O_X)$ consisting of all coherent sheaves on X. Quasicoherent sheaves are similarly defined, allowing infinite direct sums in (1.7.2). Qcoh(X) denotes the category of all quasicoherent sheaves on X. Both coh(X) and Qcoh(X) are stable under the twisting operation of $\mathbf{L}(p)$.

Given M , $N \in \mathrm{Mod}^{\mathbf{L}(p)}(O_X)$, the presheaves of $\mathbf{L}(p)$-graded O_X-modules given by

$$U \longrightarrow M(U)\otimes_{O_X(U)}N(U),$$
$$U \longrightarrow \mathrm{HOM}_{O_X(U)}(M(U)\ ,\ N(U)),$$
$$U \longrightarrow \Lambda^p M(U)\ ,\quad \text{respectively}\ ,$$

allow to define the *tensor sheaf* $M\otimes_{O_X}N$, the *homomorphism sheaf* $Hom_{O_X}(M\ ,\ N)$ and the p-th *exterior power sheaf* $\Lambda^p M$. Note for this purpose that $M(U)\otimes_{O_X(U)}N(U)$ is $\mathbf{L}(p)$-graded, its homogeneous component of degree $\vec{l}$ being spanned by all $m\otimes n$ with $m \in M(U)_{\vec{x}}$, $n \in N(U)_{\vec{y}}$ and $\vec{x}+\vec{y}=\vec{l}$. The $\mathbf{L}(p)$-grading on $\Lambda^p M(U)$ is given by a similar procedure. The homogeneous component of HOM($M(U)$, $N(U)$) of degree $\vec{l}$ consists by definition of all morphisms of $O_X(U)$-modules of degree $\vec{l}$.

Clearly, $M(\vec{l}) = O_X(\vec{l})\otimes_{O_X}M$, moreover coh($X$) is stable under the above operations.

1.8. Following Serre [31] (see also [17]) we define a *sheafification functor* $\tilde{}$: $\mathrm{Mod}^{\mathbf{L}(p)}(S) \longrightarrow \mathrm{Mod}^{\mathbf{L}(p)}(O_X)$, $M \longrightarrow \tilde{M}$, where $\tilde{M}$ is the $\mathbf{L}(p)$-graded O_X-module attached to the presheaf

$$\mathrm{D}(f)\to M_f\ ,\ f\in S \text{ homogeneous}\ , \tag{1.8.1}$$

of $\mathbf{L}(p)$-graded O_X-modules. Hence M_f consists of all fractions $\frac{m}{f^n}$ ($m \in M$, $n \in N$). In contrast to the traditional approach, we do not restrict to the zero-component of M_f. As in (1.5.4) we get

$$\Gamma\,(\mathrm{D}(f)\ ,\ \tilde{M}) = M_f\ ,\quad \text{if } f\in S_+ \text{ is homogeneous.} \tag{1.8.2}$$

Again in contrast to [8, 10, 6], we have Serre's theorem (compare [31, 17]), which allows to remove all pathologies of weighted projective spaces encountered in these papers (compare for instance Section 3 of [6]):

Serre's theorem. *Let X denote either $\mathbf{P}_n(p)$ or $C(p\ ,\ \lambda)$, accordingly S be either the $\mathbf{L}(p)$-graded algebra $S(p)$ or $S(p\ ,\ \lambda)$. Then*

(*i*) *Sheafification* $\tilde{}$: $\mathrm{mod}^{\mathbf{L}_+}(S) \longrightarrow \mathrm{coh}(X)$, $M \longrightarrow \tilde{M}$ *is an exact functor, which admits*

$$\Gamma_+\!: \mathrm{coh}(X) \longrightarrow \mathrm{mod}^{\mathbf{L}_+}(S)\ ,\ M \longrightarrow \bigoplus_{\vec{l}\geq 0} \Gamma\,(X\ ,\ M)_{\vec{l}}$$

as a right adjoint. Γ_+ *is a full embedding satisfying* $\Gamma_+(\tilde{M}) = M$ *for all* $M \in \mathrm{coh}(X)$.

(ii) Sheafification annihilates exactly the $M \in \mathrm{mod}^{\mathbf{L}_+}(S)$, *which have finite length, and induces an equivalence*

$$\mathrm{mod}^{\mathbf{L}_+}(S)/\mathrm{mod}_0^{\mathbf{L}_+}(S) \longrightarrow \mathrm{coh}(X)\ , \quad M \longrightarrow \tilde{M}$$

of abelian categories.

(iii) The full subcategory A *of* $\mathrm{mod}^{\mathbf{L}_+}(S)$ *consisting of all* M, *satisfying*

$$\mathrm{Hom}(E\ , M) = 0 = \mathrm{Ext}^1_S(E\ , M) \ \text{ for all simple objects } E \text{ in } \mathrm{mod}^{\mathbf{L}_+}(S)\ ,$$

is an abelian category. Moreover $\tilde{}$ *and* Γ_+ *induce mutually inverse equivalences* $\tilde{}: A \longrightarrow \mathrm{coh}(X)$ *and* $\Gamma_+: \mathrm{coh}(X) \longrightarrow A$.

(iv) The passage $\mathrm{coh}(X) \longrightarrow \mathrm{mod}^{\mathbf{L}(p)}(O_{X\,,\,\xi})$, $F \to F_\xi$ *is an exact functor and induces an equivalence between* $\mathrm{coh}(X)/\mathrm{Ker}\, T$ *and the category* $\mathrm{mod}(k(X))$ *of finite dimensional vector spaces over the function field of* X. *If* $X = C(p\ , \lambda)$, $\mathrm{Ker}\, T$ *is the subcategory* $\mathrm{coh}_0(X)$ *of all finite length coherent sheaves.*

Proof: Since for a quasicoherent sheaf F, each section $s \in \Gamma\,(\mathrm{D}(f)\ , F)$ extends to a global section of F - up to multiplication with a suitable power of f - each quasicoherent (coherent) sheaf on X has the form $\tilde{M}$ for some $M \in \mathrm{Mod}^{\mathbf{L}(p)}(S)$, $M \in \mathrm{mod}^{\mathbf{L}(p)}(S)$, respectively.

For f a homogeneous element of S_+ , $\mathrm{D}(f)$ is an *affine* open subset of X. By definition this means that the functor

$$\mathrm{Mod}^{\mathbf{L}(p)}(O_{\mathrm{D}(f)}) \longrightarrow \mathrm{Mod}^{\mathbf{L}(p)}(S_f)\ , \quad M \longrightarrow \Gamma\,(\mathrm{D}(f)\ , M)$$

induces category equivalences

$$\Gamma : \mathrm{Qcoh}(\mathrm{D}(f)) \longrightarrow \mathrm{Mod}^{\mathbf{L}(p)}(S_f) \ \text{ and } \ \Gamma : \ \mathrm{coh}(\mathrm{D}(f)) \longrightarrow \mathrm{mod}^{\mathbf{L}}(p)(S_f)\ ,$$

respectively, with inverse given by sheafification $M \longrightarrow \tilde{M}$ similar to the one explained before. Since

$$\Gamma\,(\mathrm{D}(f)\ , \tilde{M}) = M_f\ , \tag{1.8.3}$$

we obtain the formula

$$\tilde{M}_{|\mathrm{D}(f)} = (M_f)\tilde{} \tag{1.8.4}$$

as an equivalent assertion.

If F is quasi-coherent, thus $F = \tilde{M}$, this proves by restriction to the various $\mathrm{D}(f)$ that

$$\beta_F: (\Gamma F)\tilde{} \longrightarrow F\ , \tag{1.8.5}$$

given by the maps

$$\beta_{D(f)}:(\Gamma F)_f \longrightarrow \Gamma\,(D(f)\ , F)\ , \quad \frac{s}{f^n} \longrightarrow \frac{1}{f^n}\, s_{|D(f)}$$

is an isomorphism of sheaves.

Moreover, given $M \in \mathrm{Mod}^{\mathbf{L}(p)}(S)$, $F \in \mathrm{Mod}^{\mathbf{L}(p)}(O_X)$, the map

$$\phi_M\colon \mathrm{Hom}(M\ ,\ \Gamma F) \longrightarrow \mathrm{Hom}(\tilde{M}\ ,\ F)\ ,\ \ u \longrightarrow \beta_F \circ \tilde{u} \tag{1.8.6}$$

is an isomorphism for each $M = S(\vec{l})$. Viewing both expressions as functors in M, this clearly implies that ϕ_M is always an isomorphism. This proves adjointness.

In order to prove assertion (iii), we first show that $\mathrm{mod}_0^{\mathbf{L}_+}(S)$ is localizing in $\mathrm{mod}^{\mathbf{L}_+}(S)$, which means by definition that the natural functor

$$T\colon \mathrm{mod}^{\mathbf{L}_+}(S) \longrightarrow \mathrm{mod}^{\mathbf{L}_+}(S)/\mathrm{mod}_0^{\mathbf{L}_+}(S)\quad ,\quad M \longrightarrow M$$

has a left adjoint. According to ([12], p. 372) this amounts to verify the following property (*) for the full subcategory $\boldsymbol{A}$ of $\mathrm{mod}^{\mathbf{L}_+}(S)$, consisting of all modules M with $\mathrm{Hom}_S(E\ ,\ M) = 0 = \mathrm{Ext}^1_S(E\ ,\ M)$ for all simple E in $\mathrm{mod}^{\mathbf{L}_+}(S)$:

(*) Each M in $\mathrm{mod}^{\mathbf{L}_+}(S)$ has a submodule M', maximal among all submodules of finite length. Moreover if $M' = 0$, there is an exact sequence $0 \longrightarrow M \longrightarrow A \longrightarrow F \longrightarrow 0$ with $A \in \boldsymbol{A}$ and $F \in \mathrm{mod}_0^{\mathbf{L}_+}(S)$.

The first assertion follows since S is noetherian. We note that - if $k = S/S_+$ - the simple S-modules are - up to isomorphism - of the form $k(\vec{l})$, moreover $\mathrm{Ext}^1_S(k(\vec{l})\ ,\ k(\vec{m})) \neq 0$ implies $\vec{l} \leq \vec{m}$. From the graded *Koszul complex*, attached to the regular sequence $X_0\ , X_1, \ldots ,X_n$ for $S = S(\boldsymbol{p})$ and $x_0\ ,\ x_1$ for $S = S(\boldsymbol{p}\ ,\ \boldsymbol{\lambda})$, respectively, we deduce that the set of all $\vec{l} \in \mathbf{L}_+$ with $\mathrm{Ext}^1(k(\vec{l})\ ,\ M) \neq 0$, hence also its closure $E(M)$ with respect to predecessors, is finite. We are now going to prove our second assertion by induction on the cardinality $e(M)$ of $E(M)$. If $e(M) = 0$, we are done. Otherwise, there is a non-split exact sequence $0 \longrightarrow M \longrightarrow \overline{M} \longrightarrow k(\vec{m}) \longrightarrow 0$ with $\vec{m}$ maximal in $E(M)$. We infer from the exactness of

$$0 \longrightarrow \mathrm{Hom}_S(k(\vec{l})\ ,\ k(\vec{m})) \longrightarrow \mathrm{Ext}^1(k(\vec{l})\ ,\ M) \longrightarrow \mathrm{Ext}^1(k(\vec{l})\ ,\ \overline{M}) \longrightarrow \mathrm{Ext}^1(k(\vec{l})\ ,\ k(\vec{m}))$$

that $E(\overline{M}) \subset E(M)$, moreover that $\dim_k \mathrm{Ext}^1(k(\vec{m})\ ,\ \overline{M}) < \dim_k \mathrm{Ext}^1(k(\vec{m})\ ,\ M)$. Repeating this process, we may assume $e(\overline{M}) < e(M)$. Since $\overline{M}$ has no simple submodules, the induction hypothesis applies to $\overline{M}$, so $\overline{M}$, hence M, embeds into some $A \in \boldsymbol{A}$ with a finite length cokernel.

We are now going to prove that $\Gamma_+(F)$ is a finitely generated $\mathbf{L}_+$-graded S-module: By the previous argument, $F = \tilde{M}$ for some $M \in \boldsymbol{A}$. This yields an exact sequence $0 \longrightarrow M \longrightarrow \Gamma_+(\tilde{M}) \longrightarrow H \longrightarrow 0$, where each finitely generated submodule of H has finite length. If $H \neq 0$, we find a simple submodule E of H. Since $M \in \boldsymbol{A}$, E embeds into $\Gamma_+(\tilde{M})$, which is impossible by (1.8.6), since $\tilde{S} = 0$. Note for this purpose that $\tilde{M} = 0$ if and only if $M_f = 0$ for some homogeneous $f \in S_+$. □

1.8.1. Corollary. *All homomorphism spaces of* coh(X) *are finite dimensional over k; in particular* coh(X) *is a Krull-Schmidt category, i.e. each $F \in$* coh(X) *has a decomposition $F = F_1 \oplus \cdots \oplus F_n$ where each F_i is indecomposable with local endomorphism ring. Moreover, we have* Hom$(O(\vec{x}), O(\vec{y})) = S_{\vec{y}-\vec{x}}$ *for all $\vec{x}, \vec{y} \in$* $\mathbf{L}(p)$. □

1.8.2. Corollary. *If K denotes the function field of X, the formula* $\mathrm{r}(F) = \dim_K(F_\xi)_0$ *defines a linear form* $\mathrm{r}: K_0(X) \longrightarrow \mathbf{Z}$, *called the rank function. Moreover if* $X = C(p, \lambda)$, $\mathrm{r}(F) = 0$ *if and only if F has finite length.* □

1.8.3. Corollary. *Each coherent sheaf $F \in$* coh(X) *has an exact resolution*

$$\cdots \longrightarrow L_2 \longrightarrow L_1 \longrightarrow L_0 \longrightarrow F \longrightarrow 0,$$

where each L_i is a finite direct sum of sheaves of the form $O_X(\vec{x})$. Moreover for $X = \mathbf{P}_n(p)$ *we may assume $L_j = 0$ for $j \geq n+2$.*

Proof: Use gl.dim $S(p) = n+1$. □

Actually as may be derived from our comparison theorem in Section 3, a stronger result holds true: We may assume $L_j = 0$ for $j \geq n+1$ or $L_j = 0$ for $j \geq 2$ according as $X = \mathbf{P}_n(p)$ or $X = C(p, \lambda)$.

1.9. A coherent sheaf $F \in$ coh(X) is called a *vector bundle* if F is *locally free*, i. e. X can be covered by open subsets U such that

$$F|_U = \bigoplus_{i=1}^{n} O_X(\vec{l}_i)|_U$$

for suitably chosen $\vec{l}_i$, depending on U. The number n is just the rank of F, defined before.

We denote the full subcategory of coh(X), consisting of all vector bundles on X, by vect(X). Usually we use letters as $F, G, \ldots$ to denote vector bundles. Since all stalks $O_{X,x}$ have graded global dimension n (1, resp.) according as $X = \mathbf{P}_n(p)$ or $X = C(p, \lambda)$, each coherent sheaf F has an (exact) resolution

$$0 \longrightarrow L_n \longrightarrow L_{n-1} \longrightarrow \cdots \longrightarrow L_0 \longrightarrow F \longrightarrow 0 \quad (X = \mathbf{P}_n(p)),$$

$$0 \longrightarrow L_1 \longrightarrow L_0 \longrightarrow F \longrightarrow 0 \quad (X = C(p, \lambda))$$

by vector bundles L_i.

2. Serre duality and Riemann-Roch theorem

Throughout this section, C denotes the curve $C(p, \lambda)$ in weighted projective space $\mathbf{P}_n(p)$.

We recall that the *Picard group* Pic(C) of C consists of all *line bundles* (= rank one vector bundles) on C with multiplication induced by the tensor product.

2.1. Proposition. *The map* $\vec{l} \longrightarrow O_C(\vec{l})$ *allows to identify the graduation group* $\mathbf{L}(p)$ *with the Picard group* $\mathrm{Pic}(C)$.

Proof: $S = S(p\ ,\lambda)$ is $\mathbf{L}(p)$-graded factorial by Proposition 1.3. □

In the following D stands for the formation of the k-dual.

2.2. Serre duality. *For* $F\ , G \in \mathrm{coh}(C)$ *we have an isomorphism*

$$\mathrm{DExt}^1(F\ , G) \longrightarrow \mathrm{Hom}(G\ , F(\vec{\omega}))\ ,$$

which is functorial in F *and* G, *where* $\vec{\omega} = (n-1)\vec{c} - \sum_{i=0}^{n}\vec{x}_i$ *is the dualizing element of* $\mathbf{L}(p)$.

Proof. Proceeding as in [18], we first calculate the $\mathbf{L}(p)$-graded Cech cohomology groups of $X = \mathbf{P}_n(p)$ attached to the affine open covering, consisting of all $\mathrm{D}(X_i)\ ,\ i = 0, \ldots ,n$. By means of the regular sequence $X_0, \ldots ,X_n$ for $S(p)$ it then follows that

$$\mathrm{DH}^i(X\ , O_X) \ = \ \mathrm{H}^{n-i}(X\ , O_X(-\sum_{i=0}^{n}\vec{x}_i)) \tag{2.2.1}$$

for each $i\ =\ 0, \ldots ,n$. Note for this purpose that for each $M \in \mathrm{Mod}^{\mathbf{L}(p)}(S)$, $\mathrm{D}M$ is the $\mathbf{L}(p)$-graded module with components $(\mathrm{D}M)_{\vec{x}} \ = \ \mathrm{D}(M_{-\vec{x}})$. $O(-\sum_{i=0}^{n}\vec{x}_i)$ occurs in (2.1.1) since $S(-\sum_{i=0}^{n}\vec{x}_i)$ is the last non-zero term of the $\mathbf{L}(p)$-graded Koszul complex attached to the regular sequence $X_0, \ldots ,X_n$.

Using ([15], théorème 5.9.2) and ([23], proposition 7.17), it follows that Cech cohomology coincides with sheaf cohomology, calculated by means of injective resolutions either in $\mathrm{Mod}^{\mathbf{L}(p)}(O_X)$ or in $\mathrm{Qcoh}(X)$.

Since the system of defining equations $f_i \ = \ X_i^{p_i} - X_1^{p_1} + \lambda_i X_0^{p_0}\ ,\ (i\ =\ 2, \ldots ,n)$ forms a regular sequence of homogeneous elements of $S(p)$, each of degree $\vec{c}$, it follows by standard arguments (see [6], p.57) the existence of an isomorphism of graded S-modules ($S = S(p\ ,\lambda)$)

$$\mathrm{DH}^1(C\ , O_C) \ = \ \Gamma\ (C\ , O_C(\vec{\omega}))\ , \tag{2.2.2}$$

with $\vec{\omega} \ = \ (n-1)\vec{c} - \sum_{i=0}^{n}\vec{x}_i$. In particular for each $\vec{x} \in \mathbf{L}(p)$ we obtain a uniquely defined $\eta_{\vec{x}} \in \mathrm{DExt}^1(O_C(\vec{x})\ , O_C(\vec{x}+\vec{\omega}))$, corresponding up to twist to the identity in $\mathrm{Hom}(O_C\ , O_C)$ by means of (2.2.2). Yoneda's lemma yields natural transformations

$$\eta_{\vec{x}}\colon\mathrm{Hom}(O_C(\vec{x})\ , -) \longrightarrow \mathrm{DExt}^1(-\ , O_C(\vec{x}+\vec{\omega}))\ , \tag{2.2.3}$$

functorial with respect to morphisms $O_C(\vec{x}) \longrightarrow O_C(\vec{y})$. Actually, we may use the family of all $\eta_{\vec{x}}$ to define natural morphisms

$$\eta_{F,G} : \mathrm{Hom}(F, G) \longrightarrow \mathrm{DExt}^1(G, F(\vec{\omega})) \tag{2.2.4}$$

for each pair F, G of coherent sheaves on C, working with resolutions of F by direct sums of twisted structure sheaves, induced by free resolutions of $\Gamma_+(F)$ in $\mathrm{Mod}^{L_+}(S)$.

Since all stalks of C have graded global dimension one, each quasi-coherent sheaf M has injective dimension at most one (cf. [23], proposition 7.17), so Ext^2 vanishes. By standard arguments (see [24], p. 240) one first proves that $\eta_{F,G}$ is an isomorphism if $F = O_C(\vec{x})$ and G is arbitrary; combining right exactness of Ext^1 with Corollary 1.8.3 now proves the assertion. □

2.3. Corollary. *The category* coh(C) *has almost-split sequences. Moreover, twisting* $F \to F(\vec{\omega})$ *with the dualizing element serves as the Auslander-Reiten translation for* coh(C).

Proof: The assertion means that for each indecomposable coherent sheaf F there exists a non-split exact sequence

$$\eta: 0 \longrightarrow F(\vec{\omega}) \xrightarrow{u} G \xrightarrow{v} F \longrightarrow 0$$

such that for each indecomposable sheaf X, each non-isomorphism $f: X \longrightarrow F$ lifts to G via v.

Since $F(\vec{\omega})$ has a local endomorphism ring, $\mathrm{Hom}(-, F(\vec{\omega}))$ - as an abelian group valued additive functor on $\mathrm{coh}(C)^{op}$ - has a (unique) simple quotient H, necessarily with $H(F(\vec{\omega})) \neq 0$. By Serre duality $\mathrm{D}H$ becomes a simple subfunctor of $\mathrm{Ext}^1(F, -)$ with $\mathrm{D}H(F(\vec{\omega})) \neq 0$. Each non-zero $\eta \in \mathrm{D}H(F(\vec{\omega})) \subset \mathrm{Ext}^1(F, F(\vec{\omega}))$ represents an almost-split sequence. □

For further information on almost-split sequences we refer to [2, 14].

For non-singular projective curves the above proof is due to Schofield. Independently, alternative existence proofs were given by Auslander and Reiten, based on the methods of [3], also by the authors using their comparison theorem (see Section 3). If the weight sequence determines a Dynkin diagram, i.e. for $\sum_{i=0}^{n} \frac{1}{p_i} > n-1$, Propositions 2.2 and 2.3 are covered by [26].

As we may deduce from Serre's theorem, each coherent sheaf F has a greatest subsheaf of finite length tF, called the *torsion sheaf* of F. F/tF has no simple subsheaves, so is *torsion-free*. Note that $F = tF$ if and only if F has finite support $\{x \in C \mid F_x \neq 0\}$.

2.4. Proposition. *Each coherent sheaf F on C has a decomposition $F = tF \oplus F$, where tF is the torsion sheaf of F and $F \cong F/tF$ is a vector bundle. In particular, each subsheaf of a vector bundle is a vector bundle again.*

Proof: We claim that the sequence

$$0 \longrightarrow tF \longrightarrow F \longrightarrow F/tF \longrightarrow 0$$

splits, moreover that F/tF is locally free. Since tF has finite support, both properties are of a local character. Passing to $\mathrm{mod}^{\mathbf{L}(p)}(S_f)$ the proof of the corresponding assertions follows from the fact that S_f is an $\mathrm{L}(p)$-graded Dedekind domain. Note for this purpose that all stalks of O_C are graded discrete valuation domains (cf. (1.6)). □

For $x \in C$ let $\mathrm{coh}^{\mathbf{L}(p)}(C)_x$ denote the full subcategory of coherent sheaves with support in x. Clearly,

$$\mathrm{coh}(C)_x \longrightarrow \mathrm{mod}_0(O_{C\,,\,x})\,, \quad F \rightarrow F_x$$

defines an equivalence; moreover $\mathrm{coh}_o(C)$, the category of all torsion sheaves decomposes into $\coprod_{x \in C} \mathrm{coh}(C)_x$.

2.5. Proposition. *The category $\mathrm{coh}_o(C)$ of torsion sheaves on C decomposes into a coproduct $\coprod_{x \in C} \mathrm{mod}_0{}^{\mathrm{L}(p)}(O_{C\,,\,x})$ of uniserial categories. The number of isomorphism classes of simple modules in $\mathrm{mod}_0(O_{C\,,\,x})$ is given by the weight $p(x)$ of x.*

Proof. Each stalk $O_{C\,,\,x}$ is a graded discrete valuation ring, its number of simple graded modules is given by the weight $p(x)$ of x. □

For later use we give the following explicit description of simple sheaves on C:

If $x = [x_0\,, x_1, \ldots, x_n]$ is an ordinary point and $\lambda = \dfrac{x_1^{p_1}}{x_0^{p_0}}$, multiplication with $u = X_1^{p_1} - \lambda X_0^{p_0}$ leads to an exact sequence

$$0 \longrightarrow O_C \xrightarrow{u} O_C(\vec{c}) \longrightarrow S \longrightarrow 0\ , \tag{2.5.1}$$

where S is the unique simple sheaf concentrated at x.

By contrast, if x is exceptional with $x_i = 0$, multiplication by X_i leads to exact sequences

$$0 \longrightarrow O_C(j\vec{x}_i) \xrightarrow{X_i} O_C((j{+}1)\vec{x}_i) \longrightarrow S_{i\,,\,j} \longrightarrow 0\ , \qquad j \in \mathbf{Z}/p_i\mathbf{Z}\ , \tag{2.5.2}$$

defining p_i mutually non-isomorphic simple sheaves concentrated at x.

As is easily checked,

$$S(\vec{x}) \cong S \quad \text{for all } \vec{x} \in \mathbf{L}(p)\ , \tag{2.5.3}$$

if S is ordinary simple. For the exceptional simple sheaves

$$S_{i,j}(\vec{x}) \cong S_{i,j+l_i}, \quad \text{for } \vec{x} = \sum_{r=0}^{n} l_r \vec{x}_r. \tag{2.5.4}$$

In particular, $S_{i,j}(\vec{\omega}) = S_{i,j-1}$.

So the classification of (indecomposable) coherent sheaves reduces to the classification of vector bundles. Here, the existence of a line bundle filtration serves as the basic tool:

2.6. Proposition. *Each vector bundle F on C has a filtration*

$$0 = F_0 \subset F_1 \subset \cdots \subset F_n = F,$$

whose factors F_i/F_{i-1} are line bundles, hence of the form $O_C(\vec{l}_i)$ for suitably chosen $\vec{l}_i \in \mathbf{L}(p)$.

Proof. We proceed by induction on $r(F)$. As follows from Serre's theorem, we may assume by a suitable twist that $\mathrm{Hom}(O_C, F) \neq 0$, hence O_C may be viewed as a subsheaf of F. If $F_1 \subset F$ is chosen with $F_1/O_C = t(F/O_C)$, F_1 is clearly a line bundle and even a subbundle of F, i. e. F/F_1 is a vector bundle again. The assertion now follows from the induction hypothesis. □

2.7. Corollary. *Let $x \in C$. If F is a vector bundle on C of rank n, there is an exact sequence*

$$0 \longrightarrow F \longrightarrow \bigoplus_{i=1}^{n} L_i \longrightarrow H \longrightarrow 0$$

in coh(C), *where each L_i is a line bundle and H is concentrated at x.*

Proof. We proceed by induction on n. Choose an exact bundle sequence $0 \longrightarrow F_{n-1} \longrightarrow F \longrightarrow L_n \longrightarrow 0$ as in (2.6), where L_n is a line bundle . By assumption F_{n-1} embeds into a direct sum $\bigoplus_{i=1}^{n-1} L_i$ with cokernel concentrated at x. Replacing - if necessary - each L_i by some $L_i(l\,\vec{c})$, by means of Serre duality we may assume that $\mathrm{Ext}^1(L_n, L_j) = 0$ for all $j = 1, \ldots, n$. Passage to the push-out

$$\begin{array}{ccccccccc} 0 & \longrightarrow & F_{n-1} & \longrightarrow & F & \longrightarrow & L_n & \longrightarrow & 0 \\ & & \downarrow & & \downarrow & & \| & & \\ 0 & \longrightarrow & \bigoplus_{j=1}^{n-1} L_j & \longrightarrow & G & \longrightarrow & L_n & \longrightarrow & 0 \end{array}$$

proves that F embeds into $G \cong \bigoplus_{i=1}^{n} F_i$, again with cokernel concentrated at x. □

As follows from the foregoing, the classes $[O_C(\vec{l}_i)]$ form a system of generators for the *Grothendieck group* $K_0(C)$, which is defined as the Grothendieck group of coh(C)

with respect to short exact sequences. The previous discussion shows that equivalently $[\mathcal{O}_C]$ and the classes $[S]$ of simple sheaves generate $K_0(C)$. Actually the classes $[\mathcal{O}_C(\vec{x})]$, for $0 \le \vec{x} \le \vec{c}$, form a **Z**-basis for $K_0(C)$, as we will see later. As usual, we view linear forms on $K_0(C)$ as functions on coh(C), which are additive on short exact sequences.

Since Ext^2 vanishes on coh(C), the *Euler characteristic* given by

$$\chi : K_0(C) \longrightarrow \mathbf{Z} \quad , \quad [F] \rightarrow \sum_{j=0}^{1} (-1)^j \dim_k \mathrm{Ext}^j(\mathcal{O}_C , F) \tag{2.8.1}$$

is a linear form on $K_0(C)$. We recall that $\delta : \mathbf{L}(p) \longrightarrow \mathbf{Z}$ is the group homomorphism, defined on generators by $\delta(\vec{x}_i) = \frac{p}{p_i}$, where $p = l.c.m.(p_0, \ldots, p_n)$.

2.8. Proposition. *There is a linear form* $d : K_0(C) \longrightarrow \mathbf{Z}$, *called the degree, which is uniquely determined by each of the following properties*

(*i*) $d(\mathcal{O}_C(\vec{x})) = \delta(\vec{x})$ *for each* $\vec{x} \in \mathbf{L}(p)$.

(*ii*) $d(\mathcal{O}_C) = 0$, *and* $d(S) = \frac{p}{p(x)}$ *if* S *is a simple sheaf, concentrated at* x.

Proof. Let $\overline{\chi}(F) = \sum_{j=0}^{p-1} \chi(F(-j\vec{\omega}))$. Using Serre duality, formulas (2.5.1) and (2.5.2) yield $\overline{\chi}(S) = p$ if $p(x) = 1$ and $\overline{\chi}(S_{i,j}) = \frac{p}{p_i}$ for $0 \le i \le n$, $1 \le j \le p_i$. Hence $d : K_0(C) \longrightarrow \mathbf{Z}$ given by

$$d(F) = \overline{\chi}(F) - r(F)\overline{\chi}(\mathcal{O}_C) \tag{2.8.2}$$

satisfies condition (*ii*). Condition (*i*) now follows by repeated use of exact sequences of type

$$0 \longrightarrow \mathcal{O}_C(\vec{x}) \longrightarrow \mathcal{O}_C(\vec{x}+\vec{x}_i) \longrightarrow S_{i,0}(\vec{x}) \longrightarrow 0. \quad \square \tag{2.8.3}$$

Actually with formula (2.8.2) we have proved the first assertion of Riemann-Roch's theorem:

2.9. Riemann-Roch theorem. *The averaged Euler characteristic for* C *given by*

$$\overline{\chi}(F) = \sum_{j=0}^{p-1} \chi(F(-j\vec{\omega}))$$

satisfies

$$\overline{\chi}(F) = r(F)\, \overline{\chi}(\mathcal{O}_C) + d(F) \, ,$$

in particular

$$\overline{\chi}(\mathcal{O}_C(\vec{x})) = \overline{\chi}(\mathcal{O}_C) + \delta(\vec{x})$$

holds true for each $\vec{x} \in \mathbf{L}(p)$. *Moreover*

$$\frac{1}{p}\overline{\chi}(O_C) = -\frac{1}{2}\delta(\vec{\omega}) = \frac{p}{2}(\sum_{i=0}^{n}\frac{1}{p_i} - (n-1)).$$

Proof: It remains to prove the last assertion. From Serre duality we get

$$\chi(O_C(j\vec{\omega})) = \dim_k \mathrm{Hom}(O_C\,,\,O_C(j\vec{\omega})) - \dim_k \mathrm{Hom}(O_C(j\vec{\omega})\,,\,O_C(\vec{\omega}))$$

for each integer j. Consequently

$$\overline{\chi}(O_C) + \overline{\chi}(O_C(p\vec{\omega})) = \sum_{j=-(p-1)}^{p} \chi(O_C(j\vec{\omega})) = 0\,,$$

hence $2\,\overline{\chi}(O_C) + p\delta(\vec{\omega}) = 0$. □

We have already seen that the function field $k(C)$ of C is the field $k(t)$ of rational functions in the indeterminate t. Hence it is in accordance with usual terminology to view C as a *curve of genus zero*. However, Riemann-Roch's theorem suggests to consider also the *virtual genus* of C, given by

$$g_v(C) = 1 - \frac{1}{p}\,\overline{\chi}(O_C). \tag{2.9.1}$$

Accordingly,

$$g_v(C) = 1 + \frac{1}{2}\delta(\vec{\omega}). \tag{2.9.2}$$

3. Tilting from sheaves to modules

In this section we investigate the interrelations between coherent sheaves over $C = C(p\,,\,\lambda)$ and representations of a finite dimensional k-algebra Λ, arising as endomorphism algebra of a tilting sheaf on C. In the spirit of Beilinson's Theorem [5] we show that the *derived categories* of coh(C) and mod(Λ^{op}), the category of finite dimensional right Λ-modules, are equivalent. The equivalence $D^b(\mathrm{coh}(C)) \cong D^b(\mathrm{mod}(\Lambda^{op}))$ is made precise by a comparison result (Theorem 3.3) which states together with its consequences that the classification problems for coh(C) and mod(Λ^{op}) are basically equivalent.

For an abelian category $\boldsymbol{A}$, $D^b(\boldsymbol{A})$ denotes the *derived category of bounded complexes* in $\boldsymbol{A}$. We refer to [34] and [23] for the definition and properties of *triangulated* and *derived categories*. We will consider $\boldsymbol{A}$ as a full subcategory of $D^b(\boldsymbol{A})$ viewing $A \in \boldsymbol{A}$ as a complex concentrated at 0. We only note that $D^b(\boldsymbol{A})$ is equipped with a translation functor T given by $(T(X^\bullet))^n = X^{n+1}$ and $(Td_{X^\bullet})^n = -\,d_{X^\bullet}^{n+1}$.

3.1. Definition. *A coherent sheaf T on C is called a tilting sheaf if the following properties hold:*

(1) $\mathrm{Ext}^1(T\,,\,T) = 0$

(2) *T generates* $D^b(\mathrm{coh}(C))$ *as a triangulated category, i.e.* $D^b(\mathrm{coh}(C))$ *is the smallest triangulated subcategory of* $D^b(\mathrm{coh}(C))$ *containing* T.

(3) gl.dim $(\mathrm{End}(T)) < \infty$.

Actually condition (3) is a consequence of (1) and (2), as we will prove elsewhere.

Let T be a tilting sheaf and let $\Lambda = \mathrm{End}(T)$. T induces a functor

$$F = \mathrm{Hom}(T\,,-)\colon \mathrm{Qcoh}(C) \longrightarrow \mathrm{Mod}(\Lambda^{op})\,, \quad F \longrightarrow \mathrm{Hom}(T\,,F).$$

Since $\Lambda = \mathrm{End}(T)$, there is a functor

$$G = -\otimes_\Lambda T\colon \mathrm{Mod}(\Lambda^{op}) \longrightarrow \mathrm{Qcoh}(C).$$

with $G(\Lambda) = T$, which is right exact and commutes with arbitrary direct sums. G is uniquely determined up to isomorphism.

In the language of sheaves this functor is given as follows: For each open subset $U \subset C$, $T(U)$ is a Λ–$\mathcal{O}_C(U)$-bimodule, and $M\otimes_\Lambda T$ is isomorphic to the sheaf associated with the presheaf $U \longrightarrow M\otimes_\Lambda T(U)_{\mathcal{O}_C(U)}$, for each $M \in \mathrm{Mod}(\Lambda^{op})$.

Since Qcoh(C) has enough injectives and finite global dimension, the *right derived functor* of F,

$$R^*F\colon D^b(\mathrm{Qcoh}(C)) \longrightarrow D^b(\mathrm{Mod}(\Lambda^{op}))$$

exists; since Mod(Λ^{op}) has enough projectives and finite global dimension, the *left derived functor* of G,

$$L_*G\colon D^b(\mathrm{Mod}(\Lambda^{op})) \longrightarrow D^b(\mathrm{Qcoh}(C))$$

exists (see [34, 23] for the definition and properties of *derived functors*).

3.2. Theorem. *Let* $T \in \mathrm{coh}(C)$ *be a tilting sheaf. The functors*

$$R^*\mathrm{Hom}(T\,,-)\colon D^b(\mathrm{coh}(C)) \longrightarrow D^b(\mathit{mod}(\Lambda^{op}))$$

and

$$L_*(-\otimes_\Lambda T)\colon D^b(\mathrm{mod}(\Lambda^{op})) \longrightarrow D^b(\mathrm{coh}(C))$$

define equivalences of triangulated categories, mutually inverse to each other.

Proof: $D^b(\mathrm{coh}(C))$ and $D^b(\mathrm{mod}(\Lambda^{op}))$ are full triangulated subcategories of $D^b(\mathrm{Qcoh}(C))$ and $D^b(\mathrm{Mod}(\Lambda^{op}))$, respectively. Note that Λ generates $D^b(\mathrm{mod}(\Lambda^{op}))$ as triangulated subcategory since Λ has finite global dimension. R^*F and L_*G induce equivalences between the full subcategories $\{T\}$ and $\{\Lambda\}$ of Qcoh(C) and Mod(Λ^{op}), respectively. Thus the assertion follows from Beilinson's Lemma [5]. □

The following theorem explains the equivalence $D^b(\mathrm{coh}(C)) \cong D^b(\mathrm{mod}(\Lambda^{op}))$ in the spirit of tilting theory [20, 21, 7].

3.3. Theorem. *Let T be a tilting sheaf. We denote by $\boldsymbol{X}_i \subset \mathrm{coh}(C)$ $(i \geq 0)$ the full subcategory of all F with $\mathrm{Ext}^j(T, F) = 0$ for all $j \neq i$ and by $\boldsymbol{Y}_i \subset \mathrm{mod}(\Lambda^{op})$ $(i \geq 0)$ the full subcategory of all M with $\mathrm{Tor}_j^\Lambda(M, T) = 0$ for all $j \neq i$.*

The functors

$$\mathrm{Ext}^i(T, -) : \boldsymbol{X}_i \longrightarrow \boldsymbol{Y}_i$$

and

$$\mathrm{Tor}_i^\Lambda(-, T) : \boldsymbol{Y}_i \longrightarrow \boldsymbol{X}_i$$

define equivalences, mutually inverse to each other.

Proof: Let $F = \mathrm{Hom}(T, -)$ and $G = - \otimes_\Lambda T$. R^iF and L_iG denote the i-th right derived functor of F and the i-th left derived functor of G, respectively.

For $X_i \in \boldsymbol{X}_i$ we have $H^jR^*F(X_i) \cong R^jF(X_i) = 0$ for $j \neq i$. Thus $R^*F(X_i) \cong T^{-i}R^i(X_i)$ in $D^b(\mathrm{mod}(\Lambda^{op}))$, where T denotes the translation functor. We have $L_jG\, R^iF(X_i) = H^jT^iL_*G\, R^*F(X_i)$, which is isomorphic to X_i for $j = i$ and 0 otherwise, hence $R^iF(X_i) \in \boldsymbol{Y}_i$. Similarly we obtain $L_iG(Y_i) \in \boldsymbol{X}_i$, and $R^iF\, L_iG(Y_i) \cong Y_i$ for $Y_i \in \boldsymbol{Y}_i$. □

We remark, that $\boldsymbol{X}_i = 0$, hence $\boldsymbol{Y}_i = 0$ for all $i > 2$ since $\mathrm{Ext}^2(-, -) = 0$ in $\mathrm{coh}(C)$.

We now investigate, how far the subcategories $(\boldsymbol{X}_0, \boldsymbol{X}_1)$ and $(\boldsymbol{Y}_0, \boldsymbol{Y}_1)$ determine the categories $\mathrm{coh}(C)$ and $\mathrm{mod}(\Lambda^{op})$, respectively.

3.4. Corollary. *$(\boldsymbol{X}_0, \boldsymbol{X}_1)$ is a torsion theory for $\mathrm{coh}(C)$. In particular $\boldsymbol{X}_0$ is the full subcategory of all coherent sheaves generated by T, i.e. having the form T^n/U.*

Proof: Obviously, $\boldsymbol{X}_0 \cap \boldsymbol{X}_1 = 0$, $\boldsymbol{X}_1$ is closed under subobjects, and $\boldsymbol{X}_0$ is closed under homomorphic images. It remains to show that every coherent sheaf F admits an exact sequence

$$0 \longrightarrow F_0 \longrightarrow F \longrightarrow F_1 \longrightarrow 0$$

with $F_0 \in \boldsymbol{X}_0$, $F_1 \in \boldsymbol{X}_1$.

Since F is noetherian, there exist a greatest subsheaf F_0 generated by T. Thus $F_0 \in \boldsymbol{X}_0$ and $F_1 = F/F_0 \in \boldsymbol{X}_1$ follows. Moreover this shows that $F \in \boldsymbol{X}_0$ if and only if there exist an epimorphism $T^n \longrightarrow F$. □

If $F \in \mathrm{coh}(C)$, and $0 \longrightarrow F_0 \longrightarrow F \longrightarrow F_1 \longrightarrow 0$ is exact with $F_0 \in \boldsymbol{X}_0$ and $F_1 \in \boldsymbol{X}_1$, $\mathrm{Hom}(T, F) \cong \mathrm{Hom}(T, F_0)$ and $\mathrm{Ext}^1(T, F) \cong \mathrm{Ext}^1(T, F_1)$. This yields:

3.5. Corollary. *For each coherent sheaf F there is an exact sequence*

$$0 \longrightarrow \mathrm{Hom}(T, F) \otimes_\Lambda T \longrightarrow F \longrightarrow \mathrm{Tor}_1^\Lambda(\mathrm{Ext}^1(T, F), T) \longrightarrow 0. \quad \square$$

3.6. Corollary. $\mathrm{Tor}_j^\Lambda(\mathrm{Ext}^i(T\,,F)\,,T)=0$ *for all* $i\,,j=0\,,1\,,\ i\neq j$ *and all coherent sheaves* F. □

Similar results hold true for $\mathrm{mod}(\Lambda^{op})$:

3.7. Proposition. *For each module* $M\in\mathrm{mod}(\Lambda^{op})$, *there exists an exact sequence*

$$0\longrightarrow \mathrm{Ext}^1(T\,,\mathrm{Tor}_1^\Lambda(M\,,T))\longrightarrow M\longrightarrow \mathrm{Hom}(T\,,M\otimes_\Lambda T)\longrightarrow 0.$$

Proof: Let $P_2\longrightarrow P_1\longrightarrow P_0\longrightarrow M\longrightarrow 0$ be a projective resolution of M. Tensoring with T leads to the complex

$$P_2\otimes_\Lambda T\xrightarrow{u_2}P_1\otimes_\Lambda T\xrightarrow{u_1}P_0\otimes_\Lambda T\xrightarrow{u_0}M\otimes_\Lambda T\longrightarrow 0\,,$$

which is not necessarily exact in $P_1\otimes_\Lambda T$. Let $K_i=\ker u_i$ and $B=\mathrm{im}\,u_2$. By Corollary 3.4 , K_0 and B are contained in X_0. Thus the exact sequence $0\longrightarrow B\longrightarrow K_1\longrightarrow \mathrm{Tor}_1^\Lambda(M\,,T)\longrightarrow 0$ shows that $\mathrm{Ext}^1(T\,,K_1)\cong\mathrm{Ext}^1(T\,,\mathrm{Tor}_1^\Lambda(M\,,T))$. Application of $\mathrm{Hom}(T\,,-)$ to the exact sequences

$$0\longrightarrow K_1\longrightarrow P_1\otimes_\Lambda T\longrightarrow K_0\longrightarrow 0\,,$$
$$0\longrightarrow K_0\longrightarrow P_0\otimes_\Lambda T\longrightarrow M\otimes_\Lambda T\longrightarrow 0$$

leads to the commutative diagram with exact rows

$$\begin{array}{ccccccccc}
0 & \longrightarrow & P_1 & \longrightarrow & P_1 & \longrightarrow & 0 & \longrightarrow & 0\\
 & & \downarrow & & \downarrow & & \downarrow & & \\
0 & \longrightarrow & \mathrm{Hom}(T\,,K_0) & \longrightarrow & P_0 & \longrightarrow & \mathrm{Hom}(T\,,M\otimes_\Lambda T) & \longrightarrow & 0\\
 & & \downarrow & & \downarrow & & \downarrow & & \\
0 & \longrightarrow & \mathrm{Ext}^1(T\,,K_1) & \longrightarrow & M & \longrightarrow & \mathrm{Hom}(T\,,M\otimes_\Lambda T) & \longrightarrow & 0
\end{array}$$

and the assertion follows. □

Using this Lemma, we obtain:

3.8. Corollary. $(Y_1\,,Y_0)$ *is a torsion theory for* $\mathrm{mod}(\Lambda^{op})$. □

3.9. Corollary. $\mathrm{Ext}^i(T\,,\mathrm{Tor}_j^\Lambda(M\,,T))=0$ *for all* $i\,,j=0\,,1\,,\ i\neq j$ *and all modules* $M\in\mathrm{mod}(\Lambda^{op})$. □

From the proof of Theorem 3.3 and the fact that $\mathrm{Ext}^i_A(X\,,Y)=\mathrm{Hom}_{D^b(A)}(X\,,T^iY)$ for all $X\,,Y$ in A where $A=\mathrm{coh}(C)$ or $A=\mathrm{mod}(\Lambda^{op})$, we conclude

$$\mathrm{Ext}^l_\Lambda(\mathrm{Ext}^i(T\,,F_i)\,,\mathrm{Ext}^j(T\,,F_j))\cong\mathrm{Ext}^{l-i+j}(F_i\,,F_j)\qquad(3.10.1)$$

for all $i\,,j\,,l$ and all $F_i\in X_i\,,\ F\in X_j$. This formula yields the following consequences:

3.10. Corollary. *The torsion theory (Y_1, Y_0) is splitting, i.e. each indecomposable module $M \in$ mod(Λ^{op}) is either in Y_0 or in Y_1.* □

3.11. Corollary. *gl.dim* $\Lambda \leq 2$. □

We denote by $K_0(\Lambda^{op})$ the Grothendieck group of mod(Λ^{op}). We get:

3.12. Corollary.

$$f\colon K_0(C) \longrightarrow K_0(\Lambda^{op}), \quad [F] \longrightarrow [\mathrm{Hom}(T, F)] - [\mathrm{Ext}^1(T, F)]$$

is an isomorphism with inverse

$$f^{-1}\colon K_0(\Lambda^{op}) \longrightarrow K_0(C) \quad [M] \longrightarrow [M \otimes_\Lambda T] - [\mathrm{Tor}_1^\Lambda(M, T)] \square$$

4. Sheaves and modules over canonical algebras

In [30] Ringel introduced the class of canonical algebras, which might be defined as follows. Given (p, λ), we consider the quiver

$$\begin{array}{ccccccccccc}
 & & \vec{x}_0 & \xrightarrow{X_0} & 2\vec{x}_0 & \xrightarrow{X_0} & \cdots & \xrightarrow{X_0} & (p_0-1)\vec{x}_0 & & \\
 & \nearrow^{X_0} & & & & & & & & \searrow^{X_0} & \\
0 & \xrightarrow{X_1} & \vec{x}_1 & \xrightarrow{X_1} & 2\vec{x}_1 & \xrightarrow{X_1} & \cdots & \xrightarrow{X_1} & (p_1-1)\vec{x}_1 & \xrightarrow{X_1} & \vec{c} \\
 & \searrow_{X_n} & \vdots & & \vdots & & & & \vdots & \nearrow_{X_n} & \\
 & & \vec{x}_n & \xrightarrow{X_n} & 2\vec{x}_n & \xrightarrow{X_n} & \cdots & \xrightarrow{X_n} & (p_n-1)\vec{x}_n & &
\end{array}$$

with relations given by

$$X_i^{p_i} = X_0^{p_0} - \lambda_i X_1^{p_1} \quad \text{for } i = 2, \ldots, n. \tag{4.1.1}$$

Let $T = \bigoplus_{0 \leq \vec{x} \leq \vec{c}} O_C(\vec{x})$ and $\Lambda = \mathrm{End}(T)$. By construction the full subcategory of coh(C), consisting of all $O_C(\vec{x})$ for $0 \leq \vec{x} \leq \vec{c}$ is equivalent (so Λ is Morita equivalent) to the path algebra of this quiver with respect to the relations (4.1.1). The *canonical configuration* $O_C(\vec{x})$, $0 \leq \vec{x} \leq \vec{c}$ visualizes the quiver in coh(C).

4.1. Proposition. $T = \bigoplus_{0 \leq \vec{x} \leq \vec{c}} O_C(\vec{x})$ *is a tilting sheaf.*

Proof: (1) Let $0 \leq \vec{x}, \vec{y} \leq \vec{c}$. By Serre duality we have

$$\mathrm{DExt}^1(\mathcal{O}_C(\vec{x}), \mathcal{O}_C(\vec{y})) \cong \mathrm{Hom}(\mathcal{O}_C(\vec{y}), \mathcal{O}_C(\vec{x}+\vec{\omega})).$$

Since $\vec{\omega} + \vec{x} - \vec{y} \leq \vec{\omega} + \vec{c}$ is not positive, $\mathrm{Hom}(\mathcal{O}_C(\vec{y}), \mathcal{O}_C(\vec{x}+\vec{\omega})) = 0$ follows. This proves $\mathrm{Ext}^1(T, T) = 0$.

(2) In order to prove that T generates $D^b(\mathrm{coh}(C))$ it is sufficient to show that $\mathrm{coh}(C)$ is the smallest subcategory A of $\mathrm{coh}(C)$, which contains all direct factors of T, and is closed under the formation of kernels of epimorphisms, of cokernels of monomorphisms as well as under extensions. By means of the exact sequence (2.5.2) all exceptional simple sheaves are in A. Since $\mathcal{O}_C$ is in A, we conclude from (2.8.3) that all $\mathcal{O}_C(\vec{x})$ hence all vector bundles and all simple sheaves are in A.

(3) The quiver (4.1.1) has no oriented cycles, thus Λ has finite global dimension. In fact gl.dim $\Lambda \leq 2$. □

The category X_0 consists of all coherent sheaves with $H^1(C, F(\vec{x})) = 0$ for $-\vec{c} \leq \vec{x} \leq 0$; the category X_1 consists of all coherent sheaves with $\Gamma(C, F(\vec{x})) = 0$ for $-\vec{c} \leq \vec{x} \leq 0$. We choose the following notation:

$$\mathrm{coh}^+(C) := X_0 \ , \ \mathrm{coh}^-(C) := X_1. \tag{4.1.2}$$

$$\Gamma_\Lambda(C, -) := \mathrm{Hom}(T, -): \mathrm{coh}^+(C) \longrightarrow Y_0.$$

$$H^1_\Lambda(C, -) := \mathrm{Ext}^1(T, -): \mathrm{coh}^-(C) \longrightarrow Y_1.$$

We note that all sheaves of finite length are contained in $\mathrm{coh}^+(\Lambda)$. Hence all sheaves in $\mathrm{coh}^-(C)$ are locally free. Moreover $\mathcal{O}_C(\vec{x}) \in \mathrm{coh}^+(C)$ for all $\vec{x} \in \mathbf{L}_+$ and $\mathcal{O}_C(\vec{x}) \in \mathrm{coh}^-(C)$ otherwise. This implies that for each F in $\mathrm{coh}(C)$ there exists an $\vec{x} \in \mathbf{L}(p)$ such that $F(\vec{y}) \in \mathrm{coh}^+(C)$ for all $\vec{y} \geq \vec{x}$.

A right Λ-module M may be viewed as a representation

$$(M_{\vec{x}} \ , \ X_i: M_{k\vec{x}_i} \longrightarrow M_{(k-1)\vec{x}_i})$$

of the quiver dual to (4.1.1). We call M *monoform* (*epiform*, respectively) if all the linear maps X_i are monomorphisms (epimorphisms, respectively) but not all of them are isomorphisms.

The *rank* of M is defined by

$$r(M) = \dim_k M_0 - \dim_k M_{\vec{c}}.$$

4.2. Lemma. $r(\Gamma_\Lambda(C, F)) = r(F)$ *for all* $F \in \mathrm{coh}^+(C)$ *and* $r(H^1_\Lambda(C, F)) = -r(F)$ *for all* $F \in \mathrm{coh}^-(C)$. *In particular, if* $F \in \mathrm{coh}^+(C)$ *is locally free,* $\Gamma_\Lambda(C, F))$ *is monoform, and if* $F \in \mathrm{coh}^-(C)$, $H^1_\Lambda(C, F))$ *is epiform.*

Proof: We give the proof only in case $F \in \mathrm{coh}^+(C)$; the case $F \in \mathrm{coh}^-(C)$ is similar. So let $F \in \mathrm{coh}(C)$. From the exact sequence (2.5.2)

$$0 \longrightarrow O_C(k\vec{x}_i) \longrightarrow O_C(k+1)\vec{x}_i) \longrightarrow S_{i,k} \longrightarrow 0$$

we obtain the sequence

$$0 \longrightarrow \mathrm{Hom}(S_{i,k}, F) \longrightarrow \mathrm{Hom}(O_C(k+1)\vec{x}_i, F) \longrightarrow \mathrm{Hom}(O_C(k\vec{x}_i), F) \longrightarrow$$
$$\longrightarrow \mathrm{Ext}^1(S_{i,k}, F) \longrightarrow \mathrm{Ext}^1(O_C((k+1)\vec{x}_i), F) \longrightarrow \mathrm{Ext}^1(O_C(k\vec{x}_i), F) \longrightarrow 0.$$

Since $F \in \mathrm{coh}^+(C)$, $\mathrm{Ext}^1(O_C(k+1)\vec{x}_i, F) \cong \mathrm{Ext}^1(O_C(k\vec{x}_i), F) \cong 0$.

Suppose, $F = O_C(\vec{x})$ is a twisted structure sheaf. Then $\mathrm{Hom}(S_{i,k}, O_C(\vec{x})) = 0$ and $\mathrm{Ext}^1(S_{i,k}, O_C(\vec{x})) \neq 0$ only for one $k \in \{0, \ldots, p_i - 1\}$. In this case $\dim_k \mathrm{Ext}^1(S_{i,k}, O_C(\vec{x})) = 1$ hence $\mathrm{r}(\Gamma_\Lambda(C, O_C(\vec{x})) = 1$.

Now, assume that F is a sheaf of finite length, thus $\mathrm{Hom}(O_C(0), F) \cong \mathrm{Hom}(O_C(\vec{c}), F)$, hence $\mathrm{r}(\Gamma_\Lambda(C, F) = 0$. Finally, suppose that F is locally free of rank n. There is an exact sequence

$$0 \longrightarrow F \longrightarrow \bigoplus_{i=1}^{n} O_C(\vec{y}_i) \longrightarrow L \longrightarrow 0,$$

with L of finite length (2.7). We get $\mathrm{r}(\Gamma_\Lambda(C, F)) = \mathrm{r}(\Gamma_\Lambda(C, \bigoplus_{i=1}^{n} O_C(\vec{y}_i)) = n = \mathrm{r}(F)$. In particular, $\Gamma_\Lambda(C, F)$ is monoform since $\mathrm{Hom}(S_{i,k}, F) = 0$ for all i, k, and there exists some $S_{i,k}$ with $\mathrm{Ext}^1(S_{i,k}, F) \neq 0$. □

Let $\mathrm{mod}^+(\Lambda^{op})$ and $\mathrm{mod}^-(\Lambda^{op})$ be the full subcategories of $\mathrm{mod}(\Lambda^{op})$ consisting of all modules M, such that each indecomposable direct factor is monoform (epiform, respectively). Let $\mathrm{mod}^0(\Lambda^{op})$ be the category of all modules, whose indecomposable direct factors are neither in $\mathrm{mod}^+(\Lambda^{op})$ nor in $\mathrm{mod}^-(\Lambda^{op})$.

From Lemma 4.2 we deduce that $\mathrm{mod}^+(\Lambda^{op})$ is equivalent to the category of locally free sheaves contained in $\mathrm{coh}^+(C)$ by means of functor $\Gamma_\Lambda(C, -)$. Further we have equivalences $\mathrm{H}^1_\Lambda\colon \mathrm{coh}^-(C) \longrightarrow \mathrm{mod}^-(\Lambda^{op})$ and $\Gamma_\Lambda\colon \mathrm{coh}^0(C) \longrightarrow \mathrm{mod}^0(\Lambda^{op})$, thus $\mathrm{mod}^0(\Lambda^{op}) \cong \coprod_{x \in C} \mathrm{mod}^0(\Lambda^{op})_x$.

According to [30], we say that $\mathrm{mod}^0(\Lambda^{op})$ *separates* $\mathrm{mod}^+(\Lambda^{op})$ from $\mathrm{mod}^-(\Lambda^{op})$ if the following two conditions are satisfied:

(1) $\mathrm{Hom}_\Lambda(Z, X) = \mathrm{Hom}_\Lambda(Y, Z) = \mathrm{Hom}_\Lambda(Y, X) = 0$ for all modules $X \in \mathrm{mod}^+(\Lambda^{op})$, $Y \in \mathrm{mod}^0(\Lambda^{op})$, and $Z \in \mathrm{mod}^-(\Lambda^{op})$.

(2) Each morphism $f\colon X \longrightarrow Z$, $X \in \mathrm{mod}^+(\Lambda^{op})$, $Z \in \mathrm{mod}^-(\Lambda^{op})$, admits a factorization $f = [X \longrightarrow Y \longrightarrow Z]$ with $Y \in \mathrm{mod}^0(\Lambda^{op})$. Moreover, Y may be chosen in a prescribed component $\mathrm{mod}^0(\Lambda^{op})_x$.

4.3. Proposition ([30]). *An indecomposable module M is in* $\mathrm{mod}^+(\Lambda^{op})$, $\mathrm{mod}^0(\Lambda^{op})$ *or* $\mathrm{mod}^-(\Lambda^{op})$, *if and only if* $r(M) > 0$, $r(M) = 0$, $r(M) < 0$, *respectively. Moreover,* $\mathrm{mod}^0(\Lambda^{op})$ *separates* $\mathrm{mod}^+(\Lambda^{op})$ *from* $\mathrm{mod}^-(\Lambda^{op})$.

Proof: The first assertion is covered by Lemma 4.2. Let $X \in \mathrm{mod}^+(\Lambda^{op})$, $Y \in \mathrm{mod}^0(\Lambda^{op})$, and $Z \in \mathrm{mod}^-(\Lambda^{op})$. By means of Theorem 3.3 we have $\mathrm{Hom}_\Lambda(Y , X) = 0$, since there are no non-zero morphisms from sheaves of finite length to locally free sheaves and $\mathrm{Hom}_\Lambda(Y , Z) = \mathrm{Hom}_\Lambda(X , Z) = 0$ since $X , Y \in \mathcal{Y}_0$ and $Z \in \mathcal{Y}_1$.

Let $f\colon X \longrightarrow Z$ be a morphism, $F = X \otimes_\Lambda T$ and $G = \mathrm{Tor}_1^\Lambda(Y , T)$. There exists an exact sequence $0 \longrightarrow G \longrightarrow F' \longrightarrow L \longrightarrow 0$, where L belongs to a fixed component of $\mathrm{coh}_0(C)$ and such that $\mathrm{Ext}^1(F , F') = 0$ (Corollary 2.7). From the exactness of

$$0 \longrightarrow \Gamma_\Lambda(C , F') \longrightarrow \Gamma_\Lambda(C , L) \longrightarrow Z \longrightarrow 0$$

we conclude that f can be lifted to $\Gamma_\Lambda(C , L)$ since $\mathrm{Ext}_\Lambda^1(X , \Gamma_\Lambda(C , F')) = 0$. □

In the the language of [30], Proposition 4.3 shows that $\mathrm{ind}^0(\Lambda^{op})$, the category of all indecomposable modules in $\mathrm{mod}^0(\Lambda^{op})$, is a *separating tubular family of type* $(p_0, \ldots ,p_n)$.

The categories $\mathrm{coh}(C)$ and $\mathrm{mod}(\Lambda^{op})$ have almost-split sequences. Typically almost-split sequences in $\mathrm{coh}(C)$ give rise to almost-split sequences in $\mathrm{mod}(\Lambda^{op})$:

4.4. Proposition. *Let* $0 \longrightarrow F \longrightarrow M \longrightarrow G \longrightarrow 0$ *be an almost-split sequence in* $\mathrm{coh}(C)$. *If F and G are in* $\mathrm{coh}^+(C)$,

$$0 \longrightarrow \Gamma_\Lambda(C , F) \longrightarrow \Gamma_\Lambda(C , M) \longrightarrow \Gamma_\Lambda(C , G) \longrightarrow 0$$

is an almost-split sequence in $\mathrm{mod}(\Lambda^{op})$; *if F and G are in* $\mathrm{coh}^-(C)$,

$$0 \longrightarrow \mathrm{H}_\Lambda^1(C , F) \longrightarrow \mathrm{H}_\Lambda^1(C , M) \longrightarrow \mathrm{H}_\Lambda^1(C , G) \longrightarrow 0$$

is an almost-split sequence in $\mathrm{mod}(\Lambda^{op})$. □

We note that the categories $\mathrm{mod}^+(\Lambda^{op})$, $\mathrm{mod}^0(\Lambda^{op})$, and $\mathrm{mod}^-(\Lambda^{op})$ are stable under the Auslander-Reiten translation.

Let F be an indecomposable coherent sheaf not contained in $\mathrm{coh}^+(C)$ and $\mathrm{coh}^-(C)$, and let $0 \longrightarrow F_+ \longrightarrow F \longrightarrow F_- \longrightarrow 0$ be exact with $F_+ \in \mathrm{coh}^+(C)$ and $F_- \in \mathrm{coh}^-(C)$. From formula (3.10.1) we derive $\mathrm{Ext}^1(F_- , F_+) = \mathrm{Ext}_\Lambda^2(\mathrm{H}_\Lambda^1(C , F_-) , \Gamma_\Lambda(C , F_+))$, hence each indecomposable direct factor of $\Gamma_\Lambda(C , F_+)$ has injective dimension two and each indecomposable direct factor of $\mathrm{H}_\Lambda^1(C , F_-)$ has projective dimension two. Conversely, if $G \in \mathrm{coh}^+(C)$ is indecomposable and $\Gamma_\Lambda(C , G)$ has injective dimension two, there exist $F_- \in \mathrm{coh}^-(C)$ and an exact sequence $0 \longrightarrow G \longrightarrow F \longrightarrow F_- \longrightarrow 0$; if $G \in \mathrm{coh}^-(C)$ is indecomposable and $\mathrm{H}_\Lambda^1(C , G)$ has projective dimension two, there exists $F_+ \in \mathrm{coh}^+(C)$ and an exact sequence $0 \longrightarrow F_+ \longrightarrow F \longrightarrow G \longrightarrow 0$.

4.5. Proposition. *(1) Let* $M \in \mathrm{mod}^+(\Lambda^{op})$ *or* $M \in \mathrm{mod}^0(\Lambda^{op})$ *be indecomposable and* $0 \longrightarrow M \longrightarrow E \longrightarrow N \longrightarrow 0$ *be an almost-split sequence in* $\mathrm{mod}(\Lambda^{op})$. *Then* $0 \longrightarrow M\otimes_\Lambda T \longrightarrow E\otimes_\Lambda T \longrightarrow N\otimes_\Lambda T \longrightarrow 0$ *is an almost-split sequence in* $\mathrm{coh}(C)$ *if and only if inj dim* $M \leq 1$.

(2) *Let* $M \in \mathrm{mod}^-(\Lambda^{op})$ *be indecomposable and* $0 \longrightarrow N \longrightarrow E \longrightarrow M \longrightarrow 0$ *be an almost-split sequence in* $\mathrm{mod}(\Lambda^{op})$. *Then* $0 \longrightarrow \mathrm{Tor}_1^\Lambda(N, T) \longrightarrow \mathrm{Tor}_1^\Lambda(E, T) \longrightarrow \mathrm{Tor}_1^\Lambda(M, T) \longrightarrow 0$ *is an almost-split sequence if and only if proj dim* $M \leq 1$.

Proof: We give only a proof for (1), since the proof for (2) is dual. Thus let $M \in \mathrm{mod}^+(\Lambda^{op})$ or $M \in \mathrm{mod}^0(\Lambda^{op})$ be indecomposable. Since M is not injective, an almost-split sequence $0 \longrightarrow M \longrightarrow E \longrightarrow N \longrightarrow 0$ exists. Suppose that the injective dimension of M is one, and $f\colon M\otimes_\Lambda T \longrightarrow F$ is a non-zero morphism where F is indecomposable and f is not an isomorphism. We have to show that f extends to $E\otimes_\Lambda T$. If $F \in \mathrm{coh}^+(C)$ we are done. Otherwise, F is neither in $\mathrm{coh}^+(C)$ nor on $\mathrm{coh}^-(C)$. Thus there exists an exact sequence $0 \longrightarrow F_+ \longrightarrow F \longrightarrow F_- \longrightarrow 0$ with $F_+ \in \mathrm{coh}^+(C)$ and $F_- \in \mathrm{coh}^-(C)$. Since $\mathrm{Hom}(M\otimes_\Lambda T, F_-) = 0$, the morphism f factors through F_+. The morphism $M\otimes_\Lambda T \longrightarrow F_+$ is not a splittable monomorphism, since all direct summands of $\Gamma_\Lambda(C, F_+)$ have injective dimension two. Thus this morphism, hence also f, can be extended to $E\otimes_\Lambda T$.

Now suppose that inj dim $M = 2$ and that $0 \longrightarrow M\otimes_\Lambda T \longrightarrow E\otimes_\Lambda T \longrightarrow N\otimes_\Lambda T \longrightarrow 0$ is an almost-split sequence. Since inj dim $M = 2$, there exist a non-split exact sequence $0 \longrightarrow M\otimes_\Lambda T \longrightarrow F \longrightarrow F_- \longrightarrow 0$ with $F_- \in \mathrm{coh}^-(C)$. The extension property of almost-split sequences leads to a non-zero morphism $N\otimes_\Lambda T \longrightarrow F_-$, a contradiction. □

5. Classification for bundles and modules

Let $S = S(p, \lambda)$, we consider the polynomial algebra $R = k[x_0^{p_0}, x_1^{p_1}]$ now as an $\mathbf{L}(p)$-graded subalgebra. As follows from (1.3.3), the elements

$$x_0^{l_0} \cdots x_n^{l_n} \quad (0 \leq l_i < p_i)$$

form an $\mathbf{L}(p)$-homogeneous basis for S over R, so S is an $\mathbf{L}(p)$-*graded Cohen-Macaulay* algebra. We denote by

$$\mathrm{CM}^{\mathbf{L}(p)}(S)$$

the category of all $M \in \mathrm{mod}^{\mathbf{L}(p)}(S)$, which are finitely generated free as $\mathbf{L}(p)$-graded R-modules. By definition these are the (maximal) $\mathbf{L}(p)$-*graded Cohen-Macaulay modules* over S. Note that all $S(\vec{x})$, $\vec{x} \in \mathbf{L}(p)$ are in $\mathrm{CM}^{\mathbf{L}(p)}(S)$.

We are now going to prove a refinement of Serre's theorem 1.8; again $C = C(p, \lambda)$.

5.1. Theorem. *The* L(**p**)*-graded global sections functor induces an equivalence* Γ: vect(*C*) $\longrightarrow$ $\mathrm{CM}^{\mathbf{L}(p)}(S)$*. Moreover,* $\mathrm{CM}^{\mathbf{L}(p)}(S)$ *consists of all* $M \in \mathrm{mod}^{\mathbf{L}(p)}(S)$*, where* $\mathrm{Hom}(E, M) = 0 = \mathrm{Ext}^1(E, M)$ *holds for each simple* L(**p**)*-graded S-module E.*

We note that (5.1) establishes a link to the study of Cohen-Macaulay modules for the isolated singularity 0 of the surface $F(p, \lambda)$, see for instance [4].

Proof. We denote by $A(S)$ the category consisting of all $M \in \mathrm{mod}^{\mathbf{L}(p)}(S)$, satisfying $\mathrm{Hom}(E, M) = 0 = \mathrm{Ext}^1(E, M)$ for all simple graded modules E. Since S is noetherian, by means of (1.7.1) using a line bundle filtration for F, we conclude that $\Gamma(F)$ is finitely generated over S. Moreover $\Gamma(F) \in A(S)$, as one may deduce from Serre's theorem. Conversely, if $M \in A(S)$, $\tilde{M}$ is a vector bundle over C: First note that $M = (\Gamma(\tilde{M}))$, since $M \in A(S)$, using Serre's theorem for quasicoherent sheaves in combination with ([12], p. 372). If $\tilde{M}$ has a simple subsheaf E, M contains $\Gamma(E) = U$ as a submodule. Since $U_{\vec{x}} \neq 0$ for infinitely many $\vec{x} \leq 0$, U hence M is not finitely generated, a contradiction. We conclude that

$$\Gamma : \mathrm{vect}(C) \longrightarrow \mathrm{CM}^{\mathbf{L}(p)}(S) \text{ and } \tilde{} : \mathrm{CM}^{\mathbf{L}(p)}(S) \longrightarrow \mathrm{vect}(C)$$

define mutually inverse equivalences of categories.

Passing to the special case $C = \mathbf{P}_1(k)$, we see that $\mathrm{vect}(\mathbf{P}_1(k))$ and $A(R)$ are equivalent categories. We infer from (1.7.1) in combination with Grothendieck's theorem [16] that $A(R)$ consists just of all free modules in $\mathrm{mod}^{\mathbf{L}(p)}(R)$.

It thus remains to prove for $M \in \mathrm{mod}^{\mathbf{L}(p)}(S)$ that $M \in A(S)$ if and only $M \in A(R)$. Let

$$0 \longrightarrow S(-2\vec{c}) \xrightarrow{\alpha} S(-\vec{c}) \oplus S(-\vec{c}) \xrightarrow{\beta} S \longrightarrow E \longrightarrow 0 \tag{5.1.1}$$

with $\alpha = (x_0^{p_0}, x_1^{p_1})$ and $\beta = (x_1^{p_1}, -x_0^{p_0})$ be the Koszul complex given by the regular sequence $x_0^{p_0}, x_1^{p_1}$ in S. It is easily checked that $M \in A(S)$ if and only if $\mathrm{Hom}_S(E(\vec{x}), M) = 0 = \mathrm{Ext}^1_S(E(\vec{x}), M)$ for each $\vec{x} \in \mathbf{L}(p)$. An equivalent assertion is the exactness of

$$0 \longrightarrow M_{\vec{x}} \longrightarrow M_{\vec{x}-\vec{c}} \oplus M_{\vec{x}-\vec{c}} \longrightarrow M_{\vec{x}-2\vec{c}} \tag{5.1.2}$$

for each $\vec{x} \in \mathbf{L}(p)$. Let $V \subset \mathbf{L}(p)$ consist of all $\sum_{i=0}^{n} l_i \vec{x}_i$, with $0 \leq l_i < p_i$. If $k = R/R_+$, we have $E = \bigoplus_{\vec{x} \in V} k(\vec{x})$ as graded R-modules. Accordingly $M \in A(R)$ if and only if $\mathrm{Hom}_R(E(\vec{x}), M) = 0 = \mathrm{Ext}^1_R(E(\vec{x}), M)$. By means of (5.1.1) this amounts to exactness of

$$0 \longrightarrow \bigoplus_{\vec{v} \in V} M_{\vec{x}}(\vec{v}) \longrightarrow \bigoplus_{\vec{v} \in V} (M_{\vec{x}-\vec{c}}(\vec{v}) \oplus M_{\vec{x}-\vec{c}}(\vec{v})) \longrightarrow \bigoplus_{\vec{v} \in V} M_{\vec{x}-2\vec{c}}(\vec{v}), \tag{5.3.1}$$

an assertion equivalent to exactness of all sequences (5.1.2). □

For each non-zero vector bundle let $\mu(F) = \frac{d(F)}{r(F)}$, where d and r denote rank and degree, respectively. F is called *semi-stable (stable)* if for each non-zero sub-bundle F' of F we have $\mu(F') \leq \mu(F)$ ($\mu(F') < \mu(F)$, respectively).

As in the case of smooth projective curves we have the following result due to Narasimhan and Seshadri [32]:

5.2. Proposition. *For each $q \in \mathbb{Q}$ let C_q denote the category consisting of the zero bundle and all semi-stable vector bundles F with $\mu(F) = q$. The following properties hold true:*

(i) Each C_q is an exact abelian subcategory of coh(C), *closed under extensions.*

(ii) Each $F \in C_q$ has finite length in C_q. The simple objects in C_q are just the stable bundles; in particular $\mathrm{End}(F) = k$ *if F is stable.*

(iii) If $F \in C_q$, $F' \in C_{q'}$ and $\mathrm{Hom}(F, F') \neq 0$ *then $q \leq q'$.*

Proof: (i) is easily checked. (ii) Assume $0 \neq F' \subset F$ are both in C_q with $r(F') = r(F)$, hence $d(F') = d(F)$. Since $r(F/F') = 0$, the sheaf F/F' has finite length, from $d(F/F') = 0$ we conclude $F/F' = 0$. Thus any chain of subobjects of F within C_q has length $\leq r(F)$. To prove (iii) note that $\mu(F/F') \geq \mu(F)$ for each semi-stable F. □

As in the case of smooth projective curves (cf. [32]) bundles on C have a Harder-Narasimhan filtration as follows from

5.3. Lemma. *Each non-zero bundle F on C has a non-zero sub-bundle F_1 such that each non-zero sub-bundle (sub-sheaf) F' of F satisfies $\mu(F') \leq \mu(F_1)$. F_1 is uniquely determined if we assume additionally that F_1 has maximal rank.*

In particular F_1 is semi-stable, called the *maximal semi-stable* sub-bundle of F.

Proof. If $0 = F_0 \subset F_1 \subset \cdots \subset F_n = F$ is a line bundle filtration for F, each non-zero $F' \subset F$ satisfies

$$\mu(F') \leq \sum_{i=1}^{n} |d(F_i/F_{i-1})|.$$

Now choose $0 \neq F_1 \subset F$ with $\mu(F_1)$ being maximal. □

5.4. Remark. *The complexity of the classification problem for* coh(C), *hence for* vect(C), *depends mainly on the virtual genus of C, equivalently on the degree $\delta(\overline{\omega})$ of the dualizing sheaf $O_C(\overline{\omega})$. We have to distinguish the following cases:*

5.4.1. If $\delta(\vec{\omega}) < 0$ (accordingly $g_v(C) < 1$), we deal with the weight sequence attached to a *Dynkin diagram* $\Delta = \mathbf{A}_{p,q}$ ($p \geq 1$, $q \geq 1$), $\mathbf{D}_n$ ($n \geq 4$), $\mathbf{E}_6$, $\mathbf{E}_7$, $\mathbf{E}_8$, by counting the length of the arms of Δ. If $n = 1$, i.e. $\Delta = \mathbf{A}_{p,q}$, we just deal with the weighted projective line $\mathbf{P}_1(p, q)$; if $n = 2$, i.e. $\Delta = \mathbf{D}_n$ or $\mathbf{E}_6$, $\mathbf{E}_7$, $\mathbf{E}_8$, C is defined by just one equation

$$X_2^{p_2} - X_1^{p_1} + X_0^{p_0} = 0 , \tag{5.4.1}$$

and no parameters λ_i occur.

In this situation the canonical algebra $\Lambda = \Lambda(p_0, p_1, p_2)$ arises as a tilted algebra of a tame hereditary algebra Σ of extended Dynkin type $\tilde{\Delta}$, actually as a so-called concealed quiver algebra. [30]. By means of the comparison theorem we get $D^b(\mathrm{mod}(\Sigma)) = D^b(\mathrm{coh}(C))$, hence the classification for $\mathrm{coh}(C)$ is equivalent to the classification for $\mathrm{mod}(\Sigma)$. We refer to [26] for the details of the transfer of the classification for $\mathrm{mod}(\Sigma)$, due to Nazarova [28] and Donovan-Freislich [11] (see also [9]) to the classification for $\mathrm{coh}(C)$.

Suppose now $k = \mathbf{C}$. Let $G \subset SL(2, \mathbf{C})$ be a binary polyhedral group of Dynkin type $\Delta = (p_0, p_1, p_2)$, i.e. G has generators ξ_0, ξ_1, ξ_2 and relations $\xi_0^{p_0} = \xi_1^{p_1} = \xi_2^{p_2} = \xi_0\xi_1\xi_2$. The *algebra of relative invariants* $A^{G,rel}$ with respect to the natural action of G on $A = k[T_0, T_1]$ is generated by three fundamental relative invariants F_0, F_1, F_2, subject to relation (5.4.1). As was proved already by F. Klein [25]

$$k[X_0, X_1, X_2]/(X_2^{p_2} - X_1^{p_1} + X_0^{p_0}) \cong A^{G,rel}.$$

We refer to the survey of Slodowy [33] for further information.

$A^{G,rel}$ is naturally $G^* \times \mathbf{Z}$-graded, where G^* denotes the character group of G. Moreover $\mathbf{L}(p)$ may be identified with a subgroup of $G^* \times \mathbf{Z}$ attaching to $\vec{x}_i$ the pair (χ, n) where F_i has degree n in A and weight $\chi \in G^*$. So we obtain $S(p, \lambda)$ by restriction of $A^{G,rel}$ to the subgroup $\mathbf{L}(p)$ of $G^* \times \mathbf{Z}$.

5.4.2. $\delta(\vec{\omega}) = 0$, accordingly $g_v(C) = 1$. Here we deal with the cases (2,2,2,2), (3,3,3), (2,4,4), (2,3,6)—called *tubular* by Ringel [30] —corresponding to the extended Dynkin diagrams $\tilde{\mathbf{D}}_4$, $\tilde{\mathbf{E}}_6$, $\tilde{\mathbf{E}}_7$ and $\tilde{\mathbf{E}}_8$, respectively. By means of the comparison $D^b(\mathrm{coh}(C)) = D^b(\mathrm{mod}(\Lambda))$, classification for $\mathrm{coh}(C)$ reduces to the classification for $\mathrm{mod}(\Lambda)$, with Λ a canonical algebra of tubular type [30] and conversely. In the rest of this section we will show that the classification for $\mathrm{coh}(C)$ is possible along the lines of Atiyah's classification for vector bundles ([1] , see also [29]) on smooth elliptic curves. This will relate Atiyah's classification with Ringel's classification [30] for modules over tubular algebras.

5.4.3. $\delta(\vec{\omega}) > 0$, accordingly $g_v(C) > 1$. Here, the corresponding canonical algebras Λ are wild. By comparison $D^b(\text{coh}(C)) = D^b(\text{mod}(\Lambda))$, classification for coh(C) is also a wild problem. The treatment of smooth projective curves by Narasimhan and Seshadri (compare [32]) suggests to develop a classification of stable bundles by means of moduli spaces also for coh(C).

We note that in all three cases it is easy to determine the structure of Auslander-Reiten components using [19].

By means of the Harder-Narasimhan filtration it is easy to deal with the cases $\delta(\vec{\omega}) \leq 0$:

5.5. Proposition. *Let F be an indecomposable vector bundle on $C = C(p\ ,\ \lambda)$:*

(i) If $\delta(\vec{\omega}) < 0$, correspondingly $\Lambda(p\ ,\ \lambda)$ is a concealed tame quiver algebra, F is stable and End(F) $= k$.

(ii) If $\delta(\vec{\omega}) = 0$, accordingly $\Lambda(p\ ,\ \lambda)$ is a tubular algebra, F is semi-stable.

Proof: Let $\delta(\vec{\omega}) \leq 0$. Consider an exact sequence

$$\eta\colon 0 \longrightarrow F_1 \longrightarrow F \longrightarrow F/F_1 \longrightarrow 0\ ,$$

where F_1 is the maximal semi-stable sub-bundle of F. Assume $F/F_1 \neq 0$, so η does not split since F is indecomposable. By Serre duality we get a non-zero $u\colon F_1(-\vec{\omega}) \longrightarrow (F/F_1)$. Since $\delta(\vec{\omega}) \leq 0$, we have $\mu(F_1(-\vec{\omega})) \geq \mu(F_1)$. Hence F/F_1 has a non-zero subsheaf F_2/F_1 with $\mu(F_2/F_1) \geq \mu(F_1)$. We conclude $\mu(F_2) \geq \mu(F_1)$, contradicting the choice of F_1. So $F/F_1 = 0$ and F is semi-stable.

Suppose now $\delta(\vec{\omega}) < 0$. We know from the previous part that $F \in \mathcal{C}_q$ for some $q \in \mathbb{Q}$. We choose an exact sequence in $\mathcal{C}_q$

$$\eta\colon 0 \longrightarrow S \longrightarrow F \longrightarrow G \longrightarrow 0$$

with $S \in \mathcal{C}_q$ simple (= stable). If $G \neq 0$, η does not split, so Hom$(G\ ,\ S(\vec{\omega})) \neq 0$. Hence $q = \mu(G) \leq \mu(S(\vec{\omega})) < \mu(S) = q$, a contradiction. □

5.6. Theorem. *Suppose C has virtual genus one. Then*

(i) Each indecomposable vector bundle F on C is semi-stable.

(ii) Each $\mathcal{C}_q$ is closed under the formation of Auslander-Reiten sequences; in particular $\mathcal{C}_q(\vec{\omega}) = \mathcal{C}_q$.

(iii) Each $\mathcal{C}_q$ is a uniserial category. Accordingly ind$(\mathcal{C}_q)$ *decomposes into Auslander-Reiten components, which all are tubes of finite rank.*

(iv) Hom$(F\ ,\ G) \neq 0$ *for $F \in \mathcal{C}_q\ ,\ G \in \mathcal{C}_{q'}$ implies $q \leq q'$ in* $\mathbb{Q}$.

Actually it is possible to prove a stronger assertion in (iii), namely

$$\mathcal{C}_q = \coprod_{x \in C} \mathcal{C}_{q\,,\,x}$$

with uniserial categories $C_{q,x}$ having $p(x)$ isomorphism classes of simple modules.

Proof: (i) is covered by Proposition 5.5.

(ii) Since $\delta(\vec{\omega}) = 0$, clearly $C_q = C_q(\vec{\omega})$ for each $q \in \mathbf{Q}$, so C_q is closed under Auslander-Reiten translation, hence under the formation of Auslander-Reiten sequences, since C_q is closed under extensions.

(iii) Let S , T be simple objects in C_q. Since C_q is extension- closed in coh(C), calculation of extension of S , T in C_q can be done in coh(C), so by Serre-duality

$$\mathrm{Ext}^1(S\ ,\ T) = \mathrm{Hom}(T\ ,\ S(\vec{\omega})).$$

Hence $\mathrm{Ext}^1(S\ ,\ T) \neq 0$ if and only if $T \cong S(\vec{\omega})$; moreover if $\mathrm{Ext}^1(S\ ,\ T) \neq 0$ it has dimension one over k. By a classical result of representation theory [13], C_q is uniserial hence each $F \in C_q$ is uniquely determined by its simple socle S and its length n. Notation: $F = S^{(n)}$. Since Auslander-Reiten translation $F \longrightarrow F(\vec{\omega})$ is given by an automorphism of finite order (note $p\vec{\omega} = 0$), all Auslander-Reiten components of C_q are actually tubes, whose rank, i.e. the number of isomorphisms classes of simple objects in C_q, is a divisor of p.

(iv) is covered by Proposition 5.2. □

5.7. Remark. Let first C denote a weighted projective line of arbitrary virtual genus. Since Qcoh(C) has global dimension one, each $X \in \mathrm{D}^b(\mathrm{coh}(C))$ decomposes in $\mathrm{D}^b(\mathrm{Qcoh}(C))$, hence in $\mathrm{D}^b(\mathrm{coh}(C))$ into a (finite) direct sum of complexes $T^n A_n$ with $A_n \in \mathrm{coh}(C)$. Hence $\mathrm{ind}(\mathrm{D}^b(\mathrm{coh}(C)))$ can be calculated as the Ext – *category* of ind(coh(C)), whose *objects* are pairs $(F\ ,\ n)$ with $F \in$ ind (coh(C)), $n \in \mathbf{Z}$; *morphisms* are given by $\mathrm{Hom}((F\ ,\ n)\ ,\ (G\ ,\ m)) = \mathrm{Ext}^{m-n}(F\ ,\ G)$, and *composition* is defined by means of the Yoneda composition.

If $g_v(C) = 1$ and Λ is the attached canonical algebra of tubular type. Theorem 5.6 allows easily to calculate the Auslander-Reiten quiver of $\mathrm{ind}(\mathrm{D}^b(\mathrm{mod}\Lambda)) = \mathrm{ind}(\mathrm{D}^b(\mathrm{coh}(C)))$, as done by Happel and Ringel [22] by different methods.

5.8. Example. We are now going to sketch how the curves C of virtual genus one arise as "quotients" T/G of a smooth elliptic curve T by a suitably chosen action of a finite algebraic group G. Here, we restrict to the case $(p\ ,\ \lambda) = (2\ ,\ 2\ ,\ 2\ ,\ 2\ ;\ \lambda)$, $\lambda \neq 0$, 1.

We consider the plane elliptic curve $T \subset P_2(k)$ given by the equation $f = 0$, where

$$f = U_0U_2^2 - U_1(U_1 - U_0)(U_1 - \lambda U_0). \qquad (5.8.1)$$

The projective coordinate algebra

$$A = k[U_0\ ,\ U_1\ ,\ U_2]/(f) \qquad (5.8.2)$$

of T admits a $\mathbf{Z}_2\times\mathbf{Z}$-graduation given by

$$\deg(U_0) = \deg(U_1) = (0\ ,\ 1) \quad , \quad \deg(U_2) = (1\ ,\ 1)\ , \tag{5.8.3}$$

inducing a μ_2-action on T, where μ_2 is the algebraic group of second roots of unity. (On points of T this action is given by $[u_0\ ,\ u_1\ ,\ u_2] \longrightarrow [u_0\ ,\ u_1\ ,\ -u_2]$). If $\psi: T \longrightarrow T/\mu_2 = Y$ denotes the quotient map, the direct image sheaf $\psi_*(O_T) := O_Y$ carries a $\mathbf{Z}_2$-graduation corresponding to the μ_2-action on T. Contrary to the usual approach, where the zero-component of O_Y serves as the structure sheaf for T/μ_2, *we define* coh(Y) *as the category of all coherent* $\mathbf{Z}_2$ *-graded* O_Y*- modules.*

We are now going to prove that $\mathrm{coh}(Y) \cong \mathrm{coh}(C(2\ ,2\ ,2\ ,2\ ;\lambda))$, which justifies our assertion $T/\mu_2 = C$.

First, we note that the map $\phi: \mathbf{Z}_2\times\mathbf{Z} \longrightarrow \mathbf{L}(p)$, given on generators by $\phi(0\ ,\ 1) = 3\vec{x}_0$, $\phi(1\ ,\ 1) = \vec{x}_1 + \vec{x}_2 + \vec{x}_3$ allows to identify $\mathbf{Z}_2\times\mathbf{Z}$ with a subgroup of $\mathbf{L}(p)$. Moreover, with

$$U_0 = x_0^3\ ,\ U_1 = x_0\, x_1^2\ ,\ U_2 = x_1\, x_2\, x_3 \tag{5.8.4}$$

it is easy to see that the restriction $\bigoplus_{(u\ ,\ n)\in\mathbf{Z}_2\times\mathbf{Z}} S_{\phi(u\ ,\ n)}$ of $S = S(p\ ,\ \lambda)$ to $\mathbf{Z}_2\times\mathbf{Z}$ is isomorphic to A as a $\mathbf{Z}_2\times\mathbf{Z}$-graded algebra.

Moreover, one checks that $M \in \mathrm{mod}^{\mathbf{L}(p)}(S)$ has finite length if and only its restriction $\overline{M} = \bigoplus_{(u\ ,\ n)\in\mathbf{Z}_2\times\mathbf{Z}} M_{\phi(u\ ,\ n)}$ has finite length in $\mathrm{mod}^{\mathbf{Z}_2\times\mathbf{Z}}(A)$. Hence restriction $M \longrightarrow \overline{M}$ induces an equivalence

$$\mathrm{mod}^{\mathbf{L}(p)}(S)/\mathrm{mod}_0^{\mathbf{L}}(p)(S) \longrightarrow \mathrm{mod}^{\mathbf{Z}_2\times\mathbf{Z}}(A)/\mathrm{mod}_0^{\mathbf{Z}_2\times\mathbf{Z}}(A) \tag{5.8.5}$$

of quotient categories. The category on the left is just $\mathrm{coh}(C(p\ ,\ \lambda))$, by a variant of Serre's theorem the category on the right is equivalent to coh(Y).

References

1. M. F. Atiyah, "Vector bundles over an elliptic curve," *Proc. London Math. Soc.*, vol. 7, pp. 414-452, 1957.
2. M. Auslander and I. Reiten, "Representation theory of Artin algebras III," *Comm. Algebra* , vol. 3, pp. 239-294, 1975.
3. M. Auslander, "Functors and morphisms determined by objects and applications of morphisms determined by modules," *Proc. conf. representation theory, Philadelphia 1976*, pp. 1-244, Marcel Dekker, New York, 1978.
4. M. Auslander, "Rational singularities and almost split sequences," *Trans. Amer. Math. Soc.*, vol. 293, pp. 511 - 532, 1986.
5. A. A. Beilinson, "Coherent sheaves on P_n and problems of linear algebra," *Funct. Anal. Appl.*, vol. 12, pp. 214-216, 1979.

6. M. Beltrametti and L. Robbiano, *Introduction to the theory of weighted projective spaces*, Preprint MPI Bonn, 1985.
7. K. Bongartz, ''Tilted algebras,'' *Representations of algebras*, pp. 26 - 38, Springer - Verlag, Berlin - Heidelberg - New York, 1981. Lecture Notes in Mathematics 903.
8. C. Delorme, ''Espaces projectifs anisotropes,'' *Bull. Soc. Math. France*, vol. 103, pp. 203-223, 1975.
9. V. Dlab and C. M. Ringel, ''Indecomposable representations of graphs and algebras,'' *Mem. Amer. Math. Soc.*, vol. 173, 1976.
10. I. Dolgachev, ''Weighted projective varieties,'' *Group actions and vector fields*, pp. 34 - 71, Springer - Verlag, Berlin - Heidelberg - New York, 1982. Lecture Notes in Mathematics 956.
11. P. Donovan and M. R. Freislich, *The representation theory of finite graphs and associative algebras*, Ottawa, 1973. Carleton Lecture Notes 5.
12. P. Gabriel, ''Des catégories abéliennes.,'' *Bull. Soc. math. France*, vol. 90, pp. 323-448, 1967.
13. P. Gabriel, ''Indecomposable representations II,'' *Symposia Math. Ist. Naz. Alta Mat.*, vol. 11, pp. 81-107, 1973.
14. P. Gabriel, ''Auslander-Reiten sequences and representation- finite algebras,'' *Proceedings Ottawa 1979*, pp. 1-71, Springer-Verlag, Berlin-Heidelberg-New York, 1980.
15. R. Godement, *Théorie des faisceaux*, Hermann, Paris, 1973.
16. A. Grothendieck, ''Sur la classification des fibrés holomorphes sur la surface de Riemann,'' *Amer. J. Math.*, vol. 79, pp. 121 - 138, 1957.
17. A. Grothendieck, *Eléments de géométrie algébrique II*, Publications Mathématiques 8, Institut des Hautes Etudes Scientifiques, Paris, 1961.
18. A. Grothendieck, *Eléments de géométrie algébrique III*, Publications Mathématiques 11, Institut des Hautes Etudes Scientifiques, Paris, 1961.
19. D. Happel, U. Preiser, and C. M. Ringel, ''Binary polyhedral groups and Euclidean diagrams,'' *Manuscripta math.*, vol. 31, pp. 317-329, 1980.
20. D. Happel and C. M. Ringel, ''Tilted Algebras,'' *Trans. Amer. Math. Soc.*, vol. 274, pp. 399-443, 1982.
21. D. Happel, ''Triangulated categories in representation theory of finite dimensional algebras,'' *Comm. Math. Helv.* , 1986. To appear.
22. D. Happel and C. M. Ringel, ''The derived category of a tubular algebra,'' *Representation theory I, Finite dimensional algebras*, pp. 156 -180, Springer-Verlag, Berlin - Heidelberg - New York, 1986. Lecture Notes in Mathematics 1177.

23. R. Hartshorne, *Residues and duality*, Lecture Notes in Mathematics 20, Springer-Verlag, Berlin-Heidelberg-New York, 1966.

24. R. Hartshorne, *Algebraic geometry*, Graduate Texts in Mathematics 52, Springer-Verlag, Berlin-Heidelberg-New York, 1977.

25. F. Klein, *Vorlesungen über das Ikosaeder und die Auflösung der Gleichungen vom fünften Grade*, Teubner, Leipzig, 1884.

26. H. Lenzing, "Curve singularities arising from the representation theory of tame hereditary Artin algebras," *Representation theory I. Finite dimensional Algebras*, pp. 199-231, Springer-Verlag, Berlin-Heidelberg-New York, 1986. Lecture Notes in Mathematics 1177.

27. S. Mori, "Graded factorial domains," *Japan. J. Math.*, vol. 3, pp. 223-238, 1977.

28. L. A. Nazarova, "Representations of quivers of infinite type," *Izv Akad. Nauk SSR, Ser. Math.*, vol. 37, pp. 752 - 791, 1973.

29. T. Oda, "Vector bundles on an elliptic curve," *Nagoya Math. J.*, vol. 43, pp. 41-72, 1971.

30. C. M. Ringel, *Tame algebras and integral quadratic forms*, Springer, Berlin - Heidelberg - New York, 1984. Lecture Notes in Mathematics 1099.

31. J.-P. Serre, "Faisceaux algébriques cohérents," *Annals of Math.*, vol 61, pp. 197-278, 1955.

32. C. S. Seshadri, "Fibrés vectoriels sur les courbes algébriques," *Astérisque*, vol. 96, pp. 1-209, 1982.

33. P. Slodowy, "Platonic solids, Kleinian singularities and Lie groups," *Algebraic Geometry*, pp. 102-138, Springer-Verlag, Heidelberg - New York, 1983. Lecture Notes in Mathematics 1008.

34. J. L. Verdier, "Catégories dérivés, etat 0," *Séminaire géométrie algébrique, 4 1/2*, pp. 262-311, Springer-Verlag, Berlin - Heidelberg - New York, 1977. Lecture Notes in Mathematics 569.

REPETITIVE CATEGORIES

Dieter Happel

Let A be a finite-dimensional algebra (associative, with 1) over an algebraically closed field k. By mod A we denote the category of finitely generated left A-modules.

As it is well-known we may associate with mod A a triangulated category; namely the derived category of bounded complexes over mod A which will be denoted by $D^b(A)$. (For the basic concepts concerning $D^b(A)$ we refer to the article of Verdier [V]).

Recently we discovered that there is a second possibility [H]. Let T(A) be the trivial extension of A by the minimal injective cogenerator. Then T(A) is a $\mathbb{Z}$-graded algebra and let $\text{mod}^{\mathbb{Z}}T(A)$ be the category of finitely generated $\mathbb{Z}$-graded T(A)-modules with morphisms of degree zero. The associated stable category $\underline{\text{mod}}^{\mathbb{Z}}T(A)$ is a triangulated category (see section 1 for details).

If A has finite global dimension we constructed in [H] a functor $F : D^b(A) \longrightarrow \underline{\text{mod}}^{\mathbb{Z}}T(A)$ and showed that F is a triangle-equivalence.

In this paper we will provide some necessary background for studying $\text{mod}^{\mathbb{Z}}T(A)$ (section 1). Using the above result we focus in section 2 on an example which shows that the work of Bernstein-Gel'fand-Gel'fand [BGG] on vector bundles on the n-dimensional projective space over the complex numbers is equivalent to the work of Beilinson [B].

Finally we indicate in section 3 the construction of a quasi-inverse to F.

Throughout the composition of morphisms $f : X \to Y$ and $g : Y \to Z$ in any category $\mathcal{K}$ is denoted by fg.

For the representation-theoretic terminology as well as for further references we refer to [R]. This also contains a description of the vectorspace category method for solving certain classification problems in the representation theory of finite-dimensional algebras.

1. Repetitive categories

1.1 Let A be a finite-dimensional algebra over an algebraically closed field k. We will assume throughout that A is basic (i.e. $A/\mathrm{rad}\,A = \coprod k$, where $\mathrm{rad}\,A$ is the radical of A). By $\mathrm{mod}\,A$ we denote the category of finitely generated left A-modules. Let $D = \mathrm{Hom}_k(-,k)$ be the standard duality. Then $Q = DA$ is the minimal injective cogenerator which carries a natural A-bimodule structure: given $a', a'' \in A$ and $\varphi \in Q$ then $a'\varphi a''$ is the k-linear map which sends $a \in A$ to $\varphi(a''aa')$.

The trivial extension algebra $T(A)$ of A by Q is the following finite-dimensional k-algebra, whose additive structure is $A \oplus Q$ and the multiplication is defined by

$$(a,\varphi)\cdot(b,\psi) = (ab, a\psi+\varphi b)$$

for $a,b \in A$ and $\varphi,\psi \in Q$.
(A more suggestive interpretation is to consider the elements of $T(A)$ as given by matrices $\begin{pmatrix} a & \varphi \\ 0 & a \end{pmatrix}$, where $a \in A$, $\varphi \in Q$ and the multiplication is the ordinary matrix multiplication.)

The algebra $T(A)$ is a $\mathbb{Z}$-graded algebra, where the elements of $A \oplus 0$ are the elements of degree 0 and those of $0 \oplus Q$ the elements of degree 1. We denote by $\mathrm{mod}^{\mathbb{Z}} T(A)$ the category of finitely generated $\mathbb{Z}$-graded $T(A)$-modules with morphisms of degree zero. We call $\mathrm{mod}^{\mathbb{Z}} T(A)$ the repetitive category associated with A.

In the following we give an alternative description of $\mathrm{mod}^{\mathbb{Z}} T(A)$ which also explains the chosen terminology. Moreover we will outline some properties of this category.

1.2 For this we have to consider (associative) algebras R (defined over k) which are not necessarily finite-dimensional and which are not required to have a unit element. However, we assume that $R^2 = R$, where R^2 denotes the subspace of R generated by all products r_1r_2, with $r_1,r_2 \in R$. Moreover we assume that R has a complete set of pairwise orthogonal primitive idempotents $\{e_i \mid i \in I\}$ such that Re_i and e_iR are finite-dimensional for all $i \in I$. An algebra R satisfying these assumptions is called a locally bounded k-algebra (for a similar usage of this terminology compare [BG]).

Given an algebra R, a (left) R-module is always supposed to satisfy the condition RM = M (as above, RM denotes the subspace of M generated by all elements of the form rm, with $r \in R$ and $m \in M$).

If $\{e_i \mid i \in I\}$ is a complete set of pairwise orthogonal idempotents in R, and M is an R-module, then, as a k-vectorspace, M decomposes in the form $M = \bigoplus_{i \in I} e_i M$. An R-module M is called finitely generated if there are elements $m_1,\dots,m_n \in M$ such that $M = \sum_{i=1}^{n} Rm_i$. We denote by mod R the category of all finitely generated R-modules.

From now on we assume that R is a locally bounded k-algebra, and let $\{e_i \mid i \in I\}$ be a complete set of pairwise orthogonal primitive idempotents. The R-modules we will deal with, always are supposed to be finitely generated. The R-module, Re_i will be denoted by P(i). One obtains in this way all possible (up to isomorphism) indecomposable projective R-modules. Dually, let $Q(i) = \mathrm{Hom}_k(e_iR,k)$. This gives all possible (up to isomorphism) indecomposable injective R-modules.

Note that if R is locally bounded, then any finitely generated R-module is of finite length, hence mod R is an abelian category and the simple R-modules are of the form P(i)/rad P(i) where rad P(i) denotes the radical of P(i). Also in this case, any R-module has both a projective cover and an injective envelope.

1.3 We call an algebra R a Frobenius algebra if it is locally bounded, and the indecomposable projective R-modules coincide with the indecomposable injective R-modules. (Note that we deviate from the usual terminology. But following Heller [He] we are used to call in this case mod R a Frobenius category, so we choose this more consistent terminology).

Given a Frobenius algebra R, we denote by $\underline{\mathrm{mod}}$ R the associated stable category whose objects are the same as those of mod R and given two R-modules M,N, the set of morphisms from M to N is denoted by $\underline{\mathrm{Hom}}(M,N)$, and $\underline{\mathrm{Hom}}(M,N) = \mathrm{Hom}_R(M,N)/\sim$, where $f \sim g$ if and only if $f-g$ factors over a projective

R-module. The residue class of a map $f : M \to N$ in $\underline{\mathrm{mod}}\,R$ is denoted by $\underline{f}$.

1.4 Let R be a Frobenius algebra. The category $\underline{\mathrm{mod}}\,R$ is no longer abelian, but has the structure of a triangulated category in the sense of Verdier [V]. We will briefly indicate the choice of the translation functor and the construction of triangles in $\underline{\mathrm{mod}}\,R$, but refer for the detailed verification of the axioms to section 9 of [H]. Following Heller [He] the suspension functor may be chosen such that it is an automorphism on $\underline{\mathrm{mod}}\,R$. For any object M in mod R choose an injective module $I(M)$ with submodule M, and let $T(M) = I(M)/M$. This will be the translation functor on $\underline{\mathrm{mod}}\,R$.

Let $u : M \to N$ be a map in mod R and consider the induced sequence

$$\begin{array}{ccccccccc} 0 & \longrightarrow & M & \xrightarrow{\mu} & I(M) & \xrightarrow{\pi} & T(M) & \longrightarrow & 0 \\ & & \downarrow{\scriptstyle u} & & \downarrow{\scriptstyle \bar{u}} & & \| & & \\ 0 & \longrightarrow & N & \xrightarrow{v} & C_u & \xrightarrow{w} & T(M) & \longrightarrow & 0 \end{array}$$

where μ denotes the inclusion and π the projection. Then $(M,N,C_u,\underline{u},\underline{v},\underline{w})$ is called a standard triangle, and it turns out that a triangulation of $\underline{\mathrm{mod}}\,R$ is given by the class of sextuples which are isomorphic to standard triangles.

1.5 We now turn to the alternative description of repetitive categories. Given any finite-dimensional k-algebra A, let us construct the *repetitive algebra* $\hat{A}$, as proposed by Hughes and Waschbüsch [HW]. It will be a Frobenius algebra and always infinite-dimensional (except in the trivial case $A = 0$ which we exclude). Recall that $Q = \mathrm{Hom}_k(A,k)$.

The underlying vectorspace of $\hat{A}$ is given by

$$\hat{A} = \Big(\bigoplus_{i\in\mathbb{Z}} A\Big) \oplus \Big(\bigoplus_{i\in\mathbb{Z}} Q\Big),$$

we denote the elements of $\hat{A}$ by $(a_i,\varphi_i)_i$, where $a_i \in A$, $\varphi_i \in Q$, of course with almost all a_i, φ_i being zero. The multiplication is defined by

$$(a_i,\varphi_i)_i \cdot (b_i,\psi_i)_i = (a_ib_i, a_{i+1}\psi_i+\varphi_ib_i)_i .$$

Clearly $\hat{A}$ is a locally bounded k-algebra.

(A more suggestive interpretation is to consider $\hat{A}$ as the doubly infinite matrix algebra, without identity.

$$\hat{A} = \begin{pmatrix} \ddots & & & & 0 \\ \ddots & A_{i-1} & & & \\ & Q_{i-1} & A_i & & \\ & & Q_i & A_{i+1} & \\ 0 & & & \ddots & \ddots \end{pmatrix}$$

in which matrices have only finitely many non-zero entries, $A_i = A$ is placed on the main diagonal, $Q_i = Q$ for all $i \in \mathbb{Z}$, all the remaining entries are zero, and the multiplication is induced from the canonical maps $A \otimes_A Q \longrightarrow Q$, $Q \otimes_A A \longrightarrow Q$ and the zero map $Q \otimes_A Q \longrightarrow 0$.)

It is easily seen that the finitely generated $\hat{A}$-modules can be written in the following way: $M = (M_i, f_i)_i$, where the M_i are A-modules, all but finitely many being zero, the f_i are A-linear maps $f_i : M_i \longrightarrow \mathrm{Hom}_A(Q, M_{i+1})$ such that $f_i \cdot \mathrm{Hom}_A(Q, f_{i+1}) = 0$ for all $i \in \mathbb{Z}$. Instead of $(M_i, f_i)_i$ we also write

$$\dots \; M_{-2} \overset{f_{-2}}{\sim} M_{-1} \overset{f_{-1}}{\sim} M_0 \overset{f_0}{\sim} M_1 \overset{f_1}{\sim} M_2 \; \dots$$

or simply

$$\dots \; M_{-2} \sim M_{-1} \sim M_0 \sim M_1 \sim M_2 \; \dots$$

if we do not want to specify the maps f_i.

Using the description of $\hat{A}$-modules a morphism $h : M = (M_i, f_i) \to N = (N_i, g_i)$ between $\hat{A}$-modules is a sequence $h = (h_i)$ of A-linear maps $h_i : M_i \to N_i$ such that the following diagram commutes for all $i \in \mathbb{Z}$:

$$\begin{array}{ccc} M_i & \xrightarrow{f_i} & \mathrm{Hom}_A(Q, M_{i+1}) \\ \downarrow{\scriptstyle h_i} & & \downarrow{\scriptstyle \mathrm{Hom}_A(Q, h_{i+1})} \\ N_i & \xrightarrow{g_i} & \mathrm{Hom}_A(Q, N_{i+1}) \end{array}$$

It is quite easy to see that $\hat{A}$ is a Frobenius algebra. The indecomposable projective-injective $\hat{A}$-modules are given by

$$\ldots \quad 0 \sim M_i \overset{f_i}{\sim} M_{i+1} \sim 0 \quad \ldots$$

where M_{i+1} is an indecomposable injective A-module, $M_i = \mathrm{Hom}_A(Q, M_{i+1})$ (so M_i is an indecomposable projective A-module) and $f_i = \mathrm{id}_{M_i}$.

Using 1.4 we see that the stable category $\underline{\mathrm{mod}}\ \hat{A}$ is a triangulated category. For more details, such as the existence of a t-structure [BBD] on $\underline{\mathrm{mod}}\ \hat{A}$ we refer to [H].

We have a canonical embedding of mod A into mod $\hat{A}$ which sends $M \in$ mod A onto (M_i, f_i) where $M_0 = M$ and $M_i = 0$ for $i \neq 0$.

Lemma: The composition of this embedding with the canonical functor mod $\hat{A} \to \underline{\mathrm{mod}}\ \hat{A}$ is a full embedding.

Proof: Indeed, given A-modules M,N and a map $f : M \to N$. If $\underline{f} = 0$ in $\underline{\mathrm{mod}}\ \hat{A}$ then f factors through an injective $\hat{A}$-module. In this case, f actually factors through an injective envelope of M in mod $\hat{A}$. In particular we obtain in mod $\hat{A}$

$$\begin{array}{ccccccc}
\ldots & 0 & \sim & M & \sim & 0 & \ldots \\
 & \downarrow & & \downarrow f_1 & & \downarrow & \\
\ldots & \mathrm{Hom}_A(Q,I) & \overset{\mathrm{id}}{\sim} & I & \sim & 0 & \ldots \\
 & \downarrow & & \downarrow f_2 & & \downarrow & \\
\ldots & 0 & \sim & N & \sim & 0 & \ldots
\end{array}$$

with I an A-injective envelope of M and $f = f_1 f_2$. Thus $f_2 = 0$.

We recall from [H, 10.3] a rather useful property of this embedding. Let C be the smallest full triangulated subcategory of $\underline{\mathrm{mod}}\ \hat{A}$ which contains mod A and is closed under isomorphisms. If A has finite global dimension then $C = \underline{\mathrm{mod}}\ \hat{A}$.

Lemma: Let A be a finite-dimensional k-algebra. Then mod $\hat{A}$ and $\mathrm{mod}^{\mathbb{Z}}\, T(A)$ (compare 1.1) are equivalent.

The proof is straightforward from the definitions.

The following easy remark is sometimes useful if one is interested in studying mod T(A). Note that the forgetful functor from $\mathrm{mod}^{\mathbb{Z}}\, T(A)$ to mod T(A) is a Galois covering in the sense of Gabriel [G].

1.6 In this section let A be a finite-dimensional k-algebra of finite global dimension. We denote by ${}_A I$ the full category of mod A given by the injective A-modules. Then it is well-known that the homtopy category of bounded complexes $K^b({}_A I)$ is triangle-equivalent to the derived category of bounded complexes $D^b(A)$.

Denote by $C^b({}_A I)$ the category of bounded complexes over ${}_A I$ and by $C^{\leq o}({}_A I)$ the full subcategory of $C^b({}_A I)$ of bounded complexes which is formed by the complexes vanishing in positive degrees. The translation functor T is defined on $C^{\leq o}({}_A I)$ and the mapping cone $C_{f^\cdot}$ of a morphism $f^\cdot$ in $C^{\leq o}({}_A I)$ is contained in $C^{\leq o}({}_A I)$. For $i \leq o$, denote by $C[-i,o]$ the full subcategory of $C^{\leq o}({}_A I)$ with objectis $X^\cdot = (X^n, d^n)$ such that $X^n = 0$ for $n < -i$. Identify $C[0,0]$ with ${}_A I$.

By induction on i we will construct below functors $F_i : C[-i,o] \to \text{mod } \hat{A}$ such that $F_i|_{C[-i,o]} = F_{i-i}$.

In [H] we have shown that there is the following commutative diagram of categories and functors and that F is a triangle-equivalence.

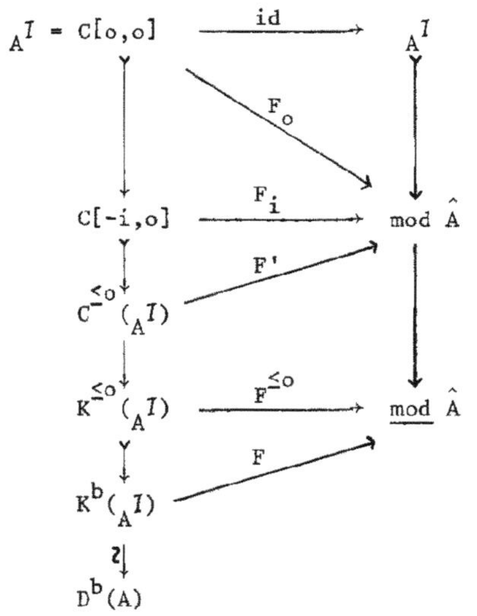

We will now outline the construction of F' and refer for the technicalities involved with the extension of F' to section 10 of [H]. We will recall from [H, 10.4] that there is an exact functor $I : \text{mod } \hat{A} \to \text{mod } \hat{A}$ and a monomorphism

$\mu : id \to I$ such that $I(X)$ is injective for each $X \in \mathrm{mod}\,\hat{A}$. Moreover we set $S(X) = \mathrm{coker}\,\mu(X)$ and denote by $\pi(X) : I(X) \to S(X)$ the canonical projection. This defines an exact functor $S : \mathrm{mod}\,\hat{A} \to \mathrm{mod}\,\hat{A}$.

Using the identification of $C[o,o]$ with ${}_A I$, we define F_o to be the canonical embedding of ${}_A I$ into $\mathrm{mod}\,\hat{A}$. Suppose that $F_{i-1} : C[-i+1,o] \to \mathrm{mod}\,\hat{A}$ is already constructed. Let $X^{\cdot} = (X^n, d^n_X)$ be in $C[-i,o]$. Let $X'^{\cdot} = (X'^n, d^n_{X'})$ be the complex such that $X'^n = o$ for $n \geq o$, $X'^n = X^n$ for $n < o$ and $d^n_{X'} = d^n_X$ for $n < -1$. Then $T^- X'^{\cdot}$ is contained in $C[-i+1,o]$ and d^{-1}_X induces a morphism e_X from $T^- X'^{\cdot}$ to X^o whose mapping cone is $X^{\cdot}$. The functor F_{i-1} is defined on $T^- X'^{\cdot}$, X^o and e_X. Consider the following pushout diagram in $\mathrm{mod}\,\hat{A}$,

$$\begin{array}{ccccccccc}
0 \to & F_{i-1}(T^- X'^{\cdot}) & \xrightarrow{\mu(F_{i-1}(T^- X'^{\cdot}))} & I(F_{i-1}(T^- X'^{\cdot})) & \xrightarrow{\pi(F_{i-1}(T^- X'^{\cdot}))} & S(F_{i-1}(T^- X'^{\cdot})) & \to 0 \\
 & \downarrow{\scriptstyle F_{i-1}(e_X)} & & \downarrow & & \| & \\
0 \to & F_{i-1}(X^o) & \xrightarrow{v_{X^{\cdot}}} & C_{F_{i-1}(e_X)} & \xrightarrow{w_{X^{\cdot}}} & S(F_{i-1}(T^- X'^{\cdot})) & \to 0
\end{array}$$

Then we set $F_i(X^{\cdot}) = C_{F_{i-1}(e_X)}$.

It may be shown by induction on i that $F_i|_{C[-i+i,o]} = F_{i-1}$. Then we define F' to be $\varinjlim_i F_i$.

Note that there exists a canonical isomorphism of functors $F'T \simeq SF'$. Moreover it is easily seen that the composition of F' with the canonical functor from $\mathrm{mod}\,\hat{A}$ to $\underline{\mathrm{mod}}\,\hat{A}$ factors over the homotopy category $K^{-\leq o}({}_A I)$. For this it is clearly enough to show that a complex $X^{\cdot} = (X^n, d^n)$ of $C^{-\leq o}({}_A I)$ with $X^n = 0$ for $n < -1$ and $X^{-1} = X^o$, $d^{-1} = id$ is mapped under F' to a projective-injective $\hat{A}$-module. In particular this yields the functor $F^{-\leq o}$ in the diagram above.

The functor $S : \mathrm{mod}\,\hat{A} \to \mathrm{mod}\,\hat{A}$ defined above maps injectives onto injectives. Thus S induces a functor $\underline{S} : \underline{\mathrm{mod}}\,\hat{A} \to \mathrm{mod}\,\hat{A}$ which turns out to be selfequivalence. The isomorphism $F'T \simeq SF'$ induces an isomorphism of functors $F^{-\leq o} T \simeq \underline{S} F^{-\leq o}$.

It is quite easy to see that a pointwise split exact sequence $0 \to X^{\cdot} \xrightarrow{u} Y \xrightarrow{v} Z^{\cdot} \to 0$ in $C^{-\leq o}({}_A I)$ is transformed under F' to a short exact sequence of $\hat{A}$-modules.

In [H, 10.8] we have shown that there exists a k-linear functor $F : K^b({}_A I) \longrightarrow \underline{\mathrm{mod}}\ \hat{A}$ and an isomorphism of functors $FT \simeq \underline{S}F$ such that $F|_{K^{\leq 0}({}_A I)} = F^{\leq 0}$. This extension uses the trivial fact that for $X^{\cdot} \in K^b({}_A I)$ there exists $t(X) \geq 0$ such that $T^{t(X)}X^{\cdot} \in K^{\leq 0}({}_A I)$. Then let $t(X)$ be minimal with this property. Then we define $F(X^{\cdot}) = S'^{t(X)}F^{\leq 0}(T^{t(X)}X^{\cdot})$ where S' is a quasi-inverse of $\underline{S}$ on $\underline{\mathrm{mod}}\ \hat{A}$. The problem of defining F on morphisms is technically more delicate.

We have defined an automorphism T on $\underline{\mathrm{mod}}\ \hat{A}$ (1.4) which serves us as a translation functor for the triangulated category $\underline{\mathrm{mod}}\ \hat{A}$. Following [He] there exists an isomorphism of functors $\underline{S} \simeq T$. In particular, we obtain an isomorphism of functors $FT \simeq TF$. Moreover, using a previous remark, it is easy to see that F transforms triangles to triangles. Thus F is an exact functor of triangulated categories. Finally one shows, using a simple lemma from [B], that F is an equivalence.

1.7 Let A be a finite-dimensional k-algebra and $\underline{\mathrm{mod}}\ \hat{A}$ the stable category of its repetitive category. There are several selfequivalences (respectively automorphisms) on $\underline{\mathrm{mod}}\ \hat{A}$. First of all we have the translation functor T of the triangulated structure. Next, there is an automorphism η on mod $\hat{A}$ induced from the degree shifting on $\mathrm{mod}^{\mathbb{Z}}T(A)$. This induces an automorphism, again denoted by η on $\underline{\mathrm{mod}}\ \hat{A}$. For the third we have to recall the notion of an Auslander-Reiten triangle as introduced in [H].

Let C be a triangulated category such that $\mathrm{Hom}_C(X,Y)$ is a finite dimensional k-vectorspace for all $X,Y \in C$ and assume that the endomorphism ring of an indecomposable object is local. These assumptions are for example satisfied if $C = \underline{\mathrm{mod}}\ R$ for some Frobenius algebra R. A triangle (X,Y,Z,u,v,w) is called an <u>Auslander-Reiten triangle</u> if the following conditions are satisfied:

(i) X,Z are indecomposable

(ii) $w \neq 0$

(iii) If $f : W \to Z$ is not a retraction, then there exists $f' : W \to Y$ such that $f'v = f$.

In case (X,Y,Z,u,v,w) is an Auslander-Reiten triangle, the objects X and Z determine each other (up to isomorphism), and we write $X = \tau Z$, $Z = \tau^{-}X$, and τ is called the Auslander Reiten translation. We say that $\mathcal{C}$ has Auslander-Reiten triangles, if for any indecomposable object of $\mathcal{C}$ there is an Auslander-Reiten triangle where it occurs in the first position, and one where it occurs in the third position. It is easily seen that $\underline{\text{mod}}$ R has Auslander-Reiten triangles and that τ is a selfequivalence on $\underline{\text{mod}}$ R. Indeed, let X be an indecomposable non-injective R-module and let $o \to X \xrightarrow{u} Y \xrightarrow{v} Z \to o$ be an Auslander-Reiten sequence starting in X. Let $o \to Y \xrightarrow{\mu} I \xrightarrow{\mu'} TY \to o$ be exact, where I is R-injective. Then we obtain the following commutative diagram with exact rows.

$$\begin{array}{ccccccccc} o & \to & X & \xrightarrow{u} & Y & \xrightarrow{v} & Z & \to & o \\ & & \| & & \downarrow{\scriptstyle \mu} & & \downarrow{\scriptstyle w} & & \\ o & \to & X & \xrightarrow{u\mu} & I & \xrightarrow{\pi} & TX & \to & o \end{array}$$

Then it follows from the description of triangles in $\underline{\text{mod}}$ R that $X \xrightarrow{u} Y \xrightarrow{v} Z \xrightarrow{-w} TX$ is a triangle, which by construction is an Auslander-Reiten triangle.

In $\underline{\text{mod}}$ $\hat{A}$ the three functors are related by the classical formula $\eta = T^2\tau$.

2. An example

In this section we will apply the general concepts reviewed in the last section to discuss one example in detail. I am grateful to C.M. Ringel who pointed out this application.

2.1 For this we have to recall the concept of a one-point extension. For more details and the representation theoretic tools available in this context we have to refer to [R]. Let A be a finite-dimensional k-algebra and M a (left) A-module. The <u>one-point extension</u> A[M] of A by M is by definition the finite-dimensional k-algebra:

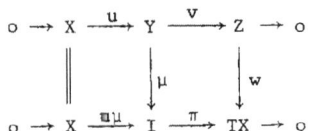

$$A[M] = \left\{ \begin{pmatrix} a & m \\ o & \lambda \end{pmatrix} \;\middle|\; a \in A,\, m \in M,\, \lambda \in k \right\}$$

with multiplication

$$\begin{pmatrix} a & m \\ o & \lambda \end{pmatrix} \cdot \begin{pmatrix} a' & m' \\ o & \lambda' \end{pmatrix} = \begin{pmatrix} aa' & am'+m\lambda' \\ o & \lambda\lambda' \end{pmatrix}.$$

Using vectorspace category methods (see [R]) one can relate mod A[M] with mod A.

For example let A_n be the algebra of $(n\times n)$-upper triangular matrices over k. Then A_n operates on an n-dimensional vectorspace by left multiplication. We denote this A_n-module by M. Then an easy verification shows that $A_n[M] = A_{n+1}$.
From the definition of the one-point extension we see that a necessary condition for an algebra B to be of the form A[M] for some algebra A and an A-module M is that there is a simple injective B-module. This clearly is sufficient. Indeed, if an algebra B admits a simple injective module S. Let P(S) be a projective cover of S and let $e \in B$ be an idempotent such that $P(S) = Be$. Let $A = B/\langle e \rangle$ ($\langle e \rangle$ the two-sided ideal in B generated by e) and $M = \operatorname{rad} P(S)$. Then M is an A-module and it is easily checked that $B = A[M]$.

2.2 We now turn to a more complicated example. Let V be an $(n+1)$-dimensional k-vectorspace and Λ the exterior algebra on V which we consider with its usual grading. We will construct a locally bounded Frobenius algebra $\tilde{\Lambda}$ such that $\operatorname{mod} \tilde{\Lambda} \simeq \operatorname{mod}^{\mathbb{Z}} \Lambda$.
It is common to describe locally bounded algebras by quivers and relations [BG].
Let $\tilde{\Delta}$ be the following quiver

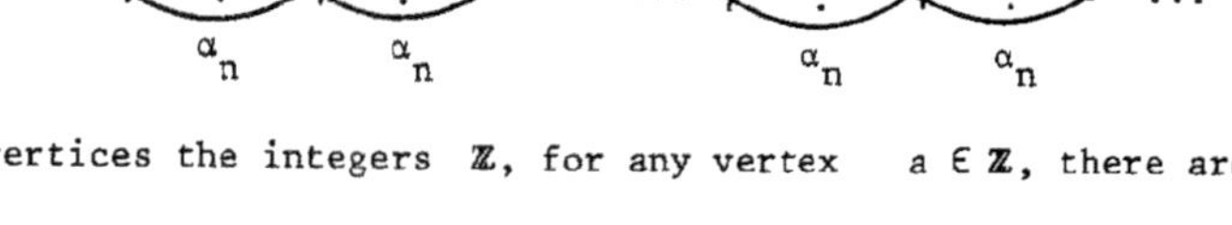

with set of vertices the integers $\mathbb{Z}$, for any vertex $a \in \mathbb{Z}$, there are n+1 arrows

$\alpha_i = \alpha_i^{(a)} : a \to a+1$, $o \leq i \leq n$. Denote by $\tilde{\Lambda}$ the opposite of the path algebra $k\tilde{\Delta}$ over the field k modulo the two-sided ideal generated by $\alpha_i^{(a)}\alpha_i^{(a+1)}$, $\alpha_i^{(a)}\alpha_j^{(a+1)} + \alpha_j^{(a)}\alpha_i^{(a+1)}$ for all i,j with $i < j$ and all $a \in \mathbb{Z}$.
Clearly $\tilde{\Lambda}$ is a locally bounded k-algebra which moreover is a Frobenius algebra. Note that the indecomposable $\tilde{\Lambda}$-projective P with top P the simple corresponding to the vertex $a \in \tilde{\Lambda}$ coincides with the indecomposable $\tilde{\Lambda}$-injective I with soc I the simple corresponding to the vertex $a-n-1$.
The following remark is straightforward.

Remark: $\text{mod}\,\tilde{\Lambda} \simeq \text{mod}^{\mathbb{Z}}\Lambda$.

Given $a \leq b$ in $\mathbb{Z}$, denote by $\tilde{\Lambda}_{a,b}$ the restriction of $\tilde{\Lambda}$ to the full subquiver of $\tilde{\Delta}$ with vertices satisfying $a \leq x \leq b$. We denote by A the finite-dimensional algebra $\tilde{\Lambda}_{o,n}$. Thus A is the opposite of the path algebra $k\Delta$ where Δ is the following quiver

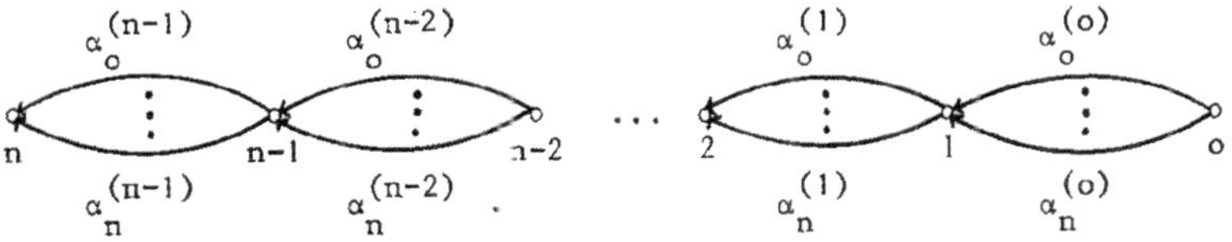

modulo the two-sided ideal generated by

$\alpha_i^{(a)}\alpha_i^{(a+1)}$, $\alpha_i^{(a)}\alpha_j^{(a+1)} + \alpha_j^{(a)}\alpha_i^{(a+1)}$ for all i,j with $i < j$ and $o \leq a < n-1$

Note that A has finite global dimension.

Proposition: With the notation above we have that $D^b(A)$ is triangle-equivalent to $\text{mod}\,\tilde{\Lambda}$.

Proof: By the result mentioned in 1.6 it is clearly enough to show that $\hat{A} \simeq \tilde{\Lambda}$. For this observe that A has a unique simple projective module $P_A(o)$ namely $P_A(o)$ is the simple corresponding to the vertex o. Let $Q_A(o)$ the indecomposable injective A-module with soc $Q_A(o)$ the simple corresponding to the vertex o. Then clearly $A[Q_A(o)] = \tilde{\Lambda}_{o,n+1}$. Inductively, we see that $\tilde{\Lambda}_{o,b}$, for $b > n$, is obtained from $\tilde{\Lambda}_{o,b-1}$ as one-point extension, using the indecomposable injective $\tilde{\Lambda}_{o,b-1}$-module with socle the simple corresponding to the vertex $b-n-1$. Using the dual process of succesive one-point coextensions we finally see that $\hat{A} = \tilde{\Lambda}$.

In the work of Bernstein-Gel'fand-Gel'fand [BGG] the category of graded modules over the exterior algebra was related to the category V_n of vector bundles over the projective space $\mathbb{P}_n(\mathbb{C})$, where $\mathbb{C}$ denotes the field of complex numbers. More precisely they obtained the following result. Let $\mathrm{coh}(\mathbb{P}_n(\mathbb{C}))$ the category of coherent sheaves on $\mathbb{P}_n(\mathbb{C})$. Then $D^b(\mathrm{coh}(\mathbb{P}_n(\mathbb{C}))$ is triangle-equivalent to $\underline{\mathrm{mod}}^{\mathbb{Z}}\Lambda$. In other words using the proposition above and the previous remark we see that $D^b(\mathrm{coh}(\mathbb{P}_n(\mathbb{C})))$ is triangle-equivalent to $D^b(A)$. This is the description of $D^b(\mathrm{coh}(\mathbb{P}_n(\mathbb{C})))$ given by Beilinson [B]. Of course, starting from Beilinson's result we can derive in a similar way the result of Bernstein-Gel'fand-Gel'fand.

For a result describing the derived category of coherent sheaves on Grassmann varieties we refer to an article of Kapranov [K].

3. A quasi-inverse to F

In section 1.6 we indicated the construction of a functor $F : K^b({}_A I) \to \underline{\mathrm{mod}}\ \hat{A}$. Here we will give the basic ingredients for the construction of a quasi-inverse to F. Again let A be a finite-dimensional k-algebra of finite global dimension and denote by $C^b(\mathrm{mod}\ A)$ the category of bounded complexes over mod A and by $C^b({}_A P)$ the category of bounded complexes over the additive category of projective A-modules.

3.1 Lemma: *There exists a functor* $P^\cdot : C^b(\mathrm{mod}\ A) \to C^b({}_A P)$ *and a canonical morphism* $\pi^\cdot : P^\cdot \to \mathrm{id}$ *such that* $\pi^\cdot(X^\cdot)$ *is a quasi-isomorphism for all* $X^\cdot \in C^b(\mathrm{mod}\ A)$.

Proof: Let $t = \mathrm{gl\ dim}\ A$ and let $X \in \mathrm{mod}\ A$. We denote by $P(X) = {}_A A \otimes_k \mathrm{Hom}_A(A,X)$ and by $\pi(X) : P(X) \to X$ the A-linear morphism $(a \otimes \varphi)\pi(X) = a\varphi$. Then $P(X)$ is a projective A-module and $\pi(X)$ is surjective. We consider P as a functor from mod A to ${}_A P$. Indeed, given a map f in mod A we define $P(f) = {}_A A \otimes_k \mathrm{Hom}_A(A,f)$.

Using P we construct a projective resolution $P^\cdot = (P^i, d^i)$ of $X \in \mathrm{mod}\ A$

as follows: $P^0(X) = P(X)$. Denote ι_X the canonical inclusion of $\ker \pi(X)$ into $P(X)$. Then $P^{-1}(X) = P(\ker \pi(X))$ and $d^{-1} = \pi(\ker(\pi(X)))\iota_X$. Inductively we may assume that P^i and d^i are constructed. If $-i < t-1$ then $P^{i-1} = P(\ker d^i)$ and $d^{i-1} = \pi(\ker d^i)\iota_{\ker d^i}$. If $-i = t-1$, then $\ker d^i$ is projective, for $t = \mathrm{gl}\dim A$. Set $P^{-t} = \ker d^i$ and $d^{-t} = \iota_{\ker d^i}$. All the other terms of $P^\cdot$ are equal to zero. It is easily seen that we may consider $P^\cdot$ as a functor from mod A to $C^b({}_A\mathcal{P})$. Denote by $\pi^\cdot(X)$ the canonical morphism from $P^\cdot(X)$ to X. Clearly $\pi^\cdot(X)$ is a quasi-isomorphism. Note that $P^\cdot(X)$ usually will not be a minimal projective resolution.

Finally we extend this construction to $C^b(\text{mod } A)$ and obtain a functor $P^\cdot : C^b(\text{mod } A) \to C^b({}_A\mathcal{P})$ and a canonical morphism $\pi^\cdot(X) : P^\cdot \to X^\cdot$ which is a quasi-isomorphism for all $X^\cdot \in C^b(\text{mod } A)$.

Dually, using $I(X) = \mathrm{Hom}_k(A,X)$ and $\mu(X) : X \to I(X)$ where $x\mu(X) = (a \to ax)$ for $X \in \text{mod } A$ we can construct a functor $I^\cdot : C^b(\text{mod } A) \to C^b({}_A\mathcal{I})$ and a canonical morphism $\mu^\cdot(X^\cdot) : X^\cdot \to I^\cdot(X^\cdot)$ which is a quasi-isomorphism for all $X^\cdot \in C^b(\text{mod } A)$.

3.2 With ${}_A\mathcal{I}$ we have denoted the full subcategory of mod A formed by the injective A-modules. There is a well-known functor $\nu : {}_A\mathcal{P} \to {}_A\mathcal{I}$ called the <u>Nakayama functor</u>. In fact, $\nu = Q \otimes_A -$ is an equivalence with a quasi-inverse being given by $\nu^{-1} = \mathrm{Hom}_A(Q,-)$, where $Q = \mathrm{Hom}_k(A,k)$. This induces an equivalence, again denoted by ν, from $C^b({}_A\mathcal{P})$ to $C^b({}_A\mathcal{I})$.

We denote the composition of $P^\cdot$ and ν by $\bar{\nu}$. Thus $\bar{\nu}$ is a functor from $C^b(\text{mod } A)$ to $C^b({}_A\mathcal{I})$. The restriction of $\bar{\nu}$ to $C^b({}_A\mathcal{I})$ is thus an endo-functor on $C^b({}_A\mathcal{I})$.

3.3 For the construction of a quasi-inverse we need two preliminary considerations.

Let $X,Y \in \text{mod } A$ and let $f : Q \otimes_A X \to Y$ be an A-linear map. Then f induces a morphism $f^\cdot : \bar{\nu}X \to Y$ of complexes such that $f^0 = (Q \otimes \pi(X))\cdot f$ and $f^i = 0$ for $i \neq 0$. We will consider $\bar{\nu}X \xrightarrow{f^\cdot} Y$ as a double complex over mod A.

$$\begin{array}{ccccccccc}
 & \vdots & & \vdots & & \vdots & & \vdots & \\
\cdots & 0 & \longrightarrow & Q \otimes P^{-1}(X) & \xrightarrow{f^{-1}} & 0 & \longrightarrow & 0 & \cdots \\
 & \downarrow & & \downarrow & & \downarrow & & \downarrow & \\
\cdots & 0 & \longrightarrow & Q \otimes P^{o}(X) & \xrightarrow{f^{o}} & Y & \longrightarrow & 0 & \cdots \\
 & \downarrow & & \downarrow & & \downarrow & & \downarrow & \\
\cdots & 0 & \longrightarrow & 0 & \longrightarrow & 0 & \longrightarrow & 0 & \cdots \\
 & \vdots & & \vdots & & \vdots & & \vdots &
\end{array}$$

Using $I^{\cdot}$ (compare 3.1) we obtain the following double complex over ${}_A I$.

$$\begin{array}{ccccccccc}
 & \vdots & & \vdots & & \vdots & & \vdots & \\
\cdots & 0 & \longrightarrow & Q \otimes P^{-1}(X) & \xrightarrow{f^{-1}} & 0 & \longrightarrow & 0 & \cdots \\
 & \downarrow & & \downarrow & & \downarrow & & \downarrow & \\
\cdots & 0 & \longrightarrow & Q \otimes P^{o}(X) & \xrightarrow{f^{o}\mu^{o}} & I^{o}(Y) & \longrightarrow & I^{1}(y) & \cdots \\
 & \downarrow & & \downarrow & & \downarrow & & \downarrow & \\
\cdots & 0 & \longrightarrow & 0 & \longrightarrow & 0 & \longrightarrow & 0 & \cdots \\
 & \vdots & & \vdots & & \vdots & & \vdots &
\end{array}$$

Note that this double complex is bounded.

Now suppose we have A-modules X,Y,Z and A-linear morphisms $f : Q \otimes_A X \to Y$ and $g : Q \otimes_A Y \to Z$ such that $(1 \otimes f) \cdot g = 0$. Above we have defined double complexes $\bar{\nu}X \xrightarrow{f^{\cdot}} Y$ and $\bar{\nu}Y \xrightarrow{g^{\cdot}} Z$. We claim that $\bar{\nu}^{2}X \xrightarrow{\bar{\nu}f^{\cdot}} \bar{\nu}Y \xrightarrow{g^{\cdot}} Z$ is a double complex over mod A, which then using again $I^{\cdot}$ we will consider as double complex over ${}_A I$.

To prove our claim we consider the following commutative diagram:

$$\begin{array}{ccccc}
\downarrow & & \downarrow & & \downarrow \\
Q \otimes P^{-1}(Q \otimes P^{\cdot}(X)) & \xrightarrow{(\bar{\nu}f^{\cdot})^{-1}} & Q \otimes P^{-1}(Y) & \longrightarrow & 0 \\
\downarrow & & \downarrow & & \downarrow \\
Q \otimes P^{o}(Q \otimes P^{\cdot}(X)) & \xrightarrow{(\bar{\nu}f^{\cdot})^{o}} & Q \otimes P^{o}(Y) & \xrightarrow{(Q \otimes \pi(Y)) \cdot g} & Z \\
\downarrow{\scriptstyle Q \otimes \pi^{o}(Q \otimes P^{\cdot}(X))} & & \downarrow{\scriptstyle Q \otimes \pi(Y)} & \nearrow{\scriptstyle g} & \\
Q \otimes Q \otimes P^{o}(X) & \xrightarrow{(Q \otimes Q \otimes \pi(X))(Q \otimes f)} & Q \otimes Y & &
\end{array}$$

We have to show that $(\bar{\nu}f^{\cdot})^{i}g^{i} = 0$ for all $i \in \mathbb{Z}$. This follows from the commutativity of the diagram above and the assumption that $(1 \otimes f)g = 0$.

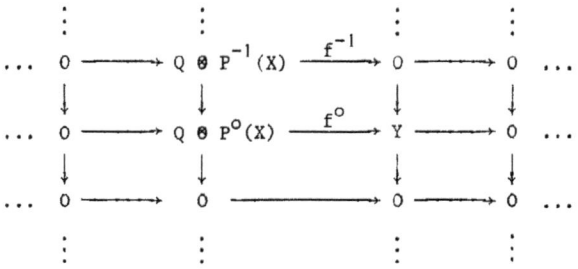

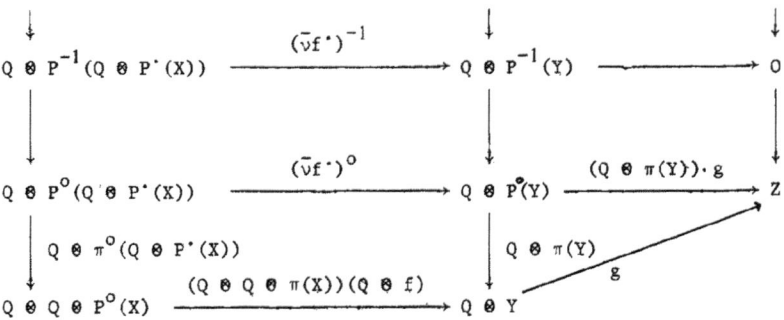

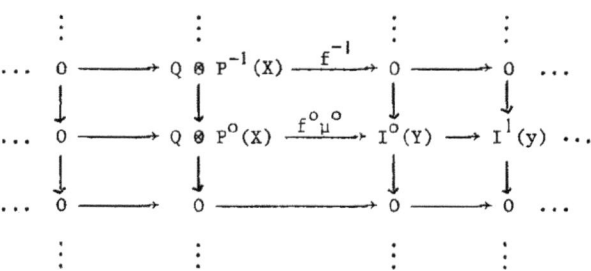

Next let X_1, X_2, Y_1, Y_2 be in mod A, and $h_i : X_i \to Y_i$, $f : Q \otimes X_1 \to Y_1$ and $g : Q \otimes X_2 \to Y_2$ be A-linear maps such that $(1 \otimes h_1)\cdot g = f \cdot h_2$. We have constructed double complexes $\bar{\nu}X_1 \xrightarrow{f^\cdot} Y_1$ and $\bar{\nu}X_2 \xrightarrow{g^\cdot} Y_2$. We claim that $(\bar{\nu}h_1, h_2)$ is a morphism of double complexes. Consider the following diagram of double complexes, where we have only shown the relevant parts:

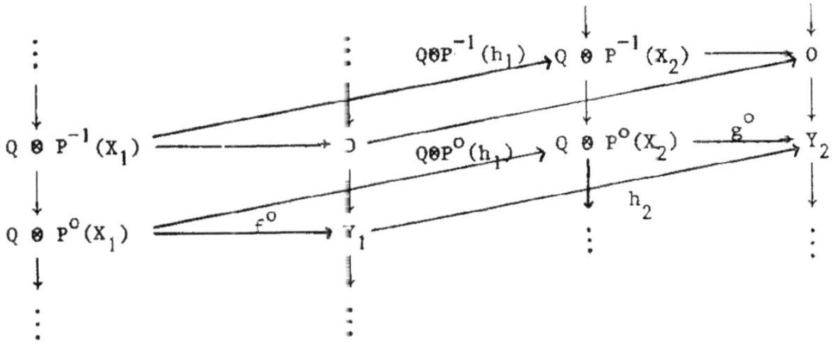

We have to show that $(Q \otimes P^0(h_1))g^0 = f^0 h_2$. All other conditions are trivially satisfied. Indeed, $(Q \otimes P^0(h_1))g^0 = (Q \otimes P^0(h_1))\cdot(Q \otimes \pi(X_2))\cdot g = (Q \otimes \pi(X_1))(Q \otimes h_1)\cdot g = (Q \otimes \pi(X_1))f\cdot h_2 = f^0 h_2$. Here we used that the lower part can be completed to a commutative diagram:

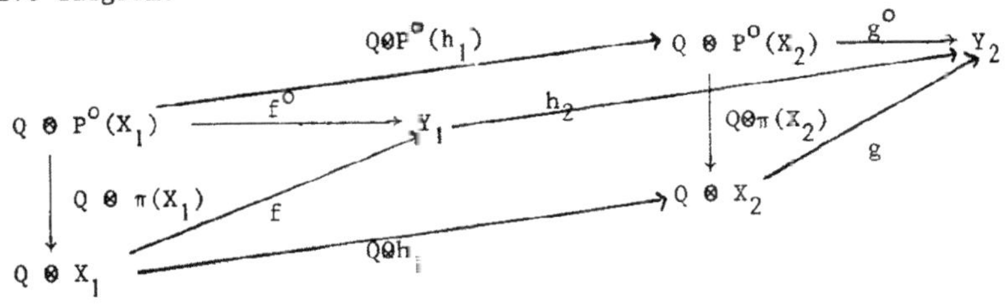

Again, using $I^\cdot$ we consider $(\bar{\nu}h_1, I^\cdot(h_2))$ as a morphism from the double complex $\bar{\nu}X_1 \to I^\cdot(Y_1)$ to the double complex $\bar{\nu}X_2 \to I^\cdot(Y_2)$.

3.4 After these preparations we now turn to the construction. We consider the full subcategory $M^{\leq 0}$ of the repetitive category mod $\hat{A}$ given by the objects $M = (M_i, f_i)$ such that $M_i = 0$ for all $i > 0$. Note that for $M \in$ mod $\hat{A}$ there exists $i(M)$ such that $\eta^{i(M)}M \in M^{\leq 0}$ (compare 1.7).

We will define a functor $G' : M^{\leq 0} \to C^b({}_A I)$ such that $G'\eta = T\bar{\nu}G'$. But first we define a functor $G'' : M^{\leq 0} \to C^{b,b}({}_A I)$, where $C^{b,b}({}_A I)$ denotes the category of bounded double complexes over ${}_A I$.

In contrast to 1.5 it is more convenient to use here the adjoint description of $\hat{A}$-modules. Thus an $\hat{A}$-module can be written in the following way: $M = (M_i, f_i)$, where the M_i are A-modules, all but finitely many being zero, the f_i are A-linear maps $f_i : Q \otimes_A M_i \to M_{i+1}$ such that $(1 \otimes f_i)f_{i+1} = 0$ for all $i \in \mathbb{Z}$. An $\hat{A}$-morphism $h : M = (M_i, f_i) \to N = (N_i, g_i)$ is thus given by a sequence $h = (h_i)$ of A-linear maps $h_i : M_i \to N_i$ such that the following diagram commutes for all $i \in \mathbb{Z}$

$$\begin{array}{ccc} Q \otimes_A M_i & \xrightarrow{f_i} & M_{i+1} \\ \downarrow{\scriptstyle 1 \otimes h_i} & & \downarrow{\scriptstyle h_{i+1}} \\ Q \otimes_A N_i & \xrightarrow{g_i} & N_{i+1} \end{array}$$

Let $M = (M_i, f_i) \in M^{\leq 0}$. Consider the following double complex over ${}_A I$ which is bounded:

$$M^{\cdot\cdot} = (\ldots \to \bar{\nu}^{-i} M_i \xrightarrow{\bar{\nu}^{-i-1} f_i^{\cdot}} \bar{\nu}^{-i-1} M_{i+1} \to \ldots \to \bar{\nu} M_{-1} \xrightarrow{\tilde{f}^{\cdot}_{-1}} I^{\cdot}(M_0) \to 0 \ldots)$$

where $\tilde{f}^o_{-1} = f^o_{-1} \mu^o(M_0)$ and $\tilde{f}^i_{-1} = f^i_{-1}$ for $i \neq 0$.

The hatched part in the following picture indicates the positions of non-zero terms in the double complex $M^{\cdot\cdot}$

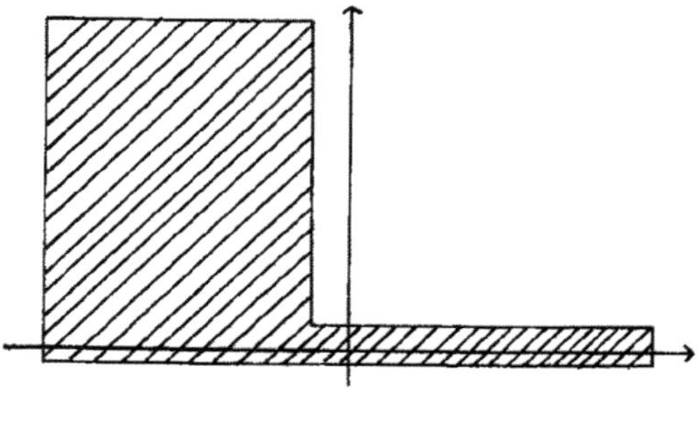

Let $M = (M_i, f_i)$, $N = (N_i, g_i) \in M^{\leq 0}$ and $h = (h_i)$ a morphism from M to N in mod $\hat{A}$, then $h^{\cdot\cdot} = (\ldots \bar{\nu}^{-i} h_i, \bar{\nu}^{-i-1} h_{i+1}, \ldots, \bar{\nu} h_{-1}, I^{\cdot}(h_0))$ is a morphism of double complexes from $M^{\cdot\cdot}$ to $N^{\cdot\cdot}$.

Summarizing the preceding considerations we obtain:

<u>Lemma</u>: $G'' : M^{\leq 0} \to C^{b,b}({}_A I)$ <u>with</u> $G''(M) = M^{\cdot\cdot}$ <u>for</u> $M \in M^{\leq 0}$ <u>and</u> $G''(h) = h^{\cdot\cdot}$ <u>for a morphism</u> h <u>in</u> $M^{\leq 0}$ <u>defines a functor</u>.

Denote by $\text{Tot} : C^{b,b}({}_A I) \to C^{b}({}_A I)$ the functor given by forming the associated total complex. Then let $G' = \text{Tot} \circ G''$. Thus G' is a functor from $M^{\leq 0}$ to $C^b({}_A I)$. It follows immediately from the definition of G'' that $G'\eta = T\bar{\nu}G'$ where T is the shift functor on $C^b({}_A I)$. We obtain a functor, again denoted by G', from $M^{\leq 0}$ to $K^b({}_A I)$ having the property that $G'\eta = T\bar{\nu}G'$, where $\bar{\nu}$ is now considered as an endo-functor on $K^b({}_A I)$. Note that $\bar{\nu}$ is a selfequivalence on $K^b({}_A I)$. We may use this to extend G' to a functor $G : \text{mod}\, \hat{A} \to K^b({}_A I)$ such that $G\eta = T\bar{\nu}G$.

<u>Lemma</u>: G <u>factors over</u> <u>mod</u> $\hat{A}$.

<u>Proof</u>: It is enough to show that $G(\tilde{P}) = 0$ in $K^b({}_A I)$ for an indecomposable projective $\hat{A}$-module $\tilde{P}$. Using $G\eta = T\bar{\nu}G$ we may assume that $\tilde{P}$ is of the form:

$$\tilde{P} = \ldots 0 \sim P \overset{f_{-1}}{\sim} I \sim 0 \ldots$$

where $f_{-1} : Q \otimes_A P \to I$ is the identity and I is an indecomposable injective A-module.

Then it is easily seen that $G(\tilde{P})$ is homotopy-equivalent to the complex $\ldots 0 \longrightarrow I^{-1} \xrightarrow{d^{-1}} I^0 \longrightarrow 0 \ldots$, with $I^0 = I^{-1} = I$ and $d^{-1} = id_I$, which shows the assertion.

We will denote the resulting functor from <u>mod</u> $\hat{A} \to K^b({}_A I)$ again by G. Clearly we still have $G\eta = T\bar{\nu}G$.

Let us briefly indicate the construction of an ivertible natural transformation $GF \to id$. First we make the following observation. Let $M \in M^{\leq 0}$, $M = (M_i, f_i)$ and assume that $M_0 \neq 0$. Consider the exact sequence $0 \longrightarrow M_0 \xrightarrow{v} M \xrightarrow{w} M/M_0 \longrightarrow 0$ in mod $\hat{A}$. Set $M' = M/M_o$. There exists $\underline{u} \in \underline{\mathrm{Hom}}_{\hat{A}}(T^- M', M_o)$ such that $T^- M' \xrightarrow{\underline{u}} M_o \xrightarrow{\underline{v}} M \xrightarrow{\underline{w}} M'$ is a triangle in $\underline{\mathrm{mod}}\ \hat{A}$. It is easily seen that $G(T^- M') \xrightarrow{G(\underline{u})} G(M_o) \xrightarrow{G(\underline{v})} G(M) \xrightarrow{G(\underline{w})} G(M')$ is a triangle in $K^b({}_A I)$. With this we can construct an invertible natural transformation $\mu : GF \to \mathrm{id}$. Clearly we may restrict to the case where $I^{\cdot} \in C^{\leq o}({}_A I)$. If $I^{\cdot} \in C[o,o]$ there is nothing to show. Now suppose that $I^{\cdot} \in C[-i,o]$ (compare 1.6). Then we obtain inductively using the previous observation the following commutative diagram of triangles in $K^b({}_A I)$.

$$\begin{array}{ccccccc}
GF(T^- I'^{\cdot}) & \longrightarrow & GF(I^o) & \longrightarrow & GF(I^{\cdot}) & \longrightarrow & GF(I'^{\cdot}) \\
\downarrow \mu(T^- I'^{\cdot}) & & \downarrow \mu(I^o) & & & & \downarrow \mu(I'^{\cdot}) \\
T^- I'^{\cdot} & \longrightarrow & I^o & \longrightarrow & I^{\cdot} & \longrightarrow & I'^{\cdot}
\end{array}$$

By the axiom (TR3) of a triangulated category (compare § 1.1.1 in [V]) there exists a morphism $\mu(I^{\cdot}) : GF(I^{\cdot}) \to I^{\cdot}$ completing the diagram which is an isomorphism in $K^b({}_A I)$.

Summarizing the considerations of section 3 we obtain:

<u>Theorem</u>: <u>The functor</u> $G : \underline{\mathrm{mod}}\ \hat{A} \to K^b({}_A I)$ <u>is a quasi-inverse to</u> F. <u>Thus</u> G <u>is a triangle equivalence.</u>

References:

[B] Beilinson, A.A.: Coherent sheaves on $\mathbb{P}^n$ and problems of linear algebra. Funct. Anal. Appl. 12(1978), 214-216.

[BBD] Beilinson, A.A.; Bernstein, J.; Deligne, P.: Faisceaux pervers, Astérique (100), 1982.

[BGG] Bernstein, I.N.; Gel'fand, I.M.; Gel'fand, S.I.: Algebraic bundles over $\mathbb{P}^n$ and problems of linear algebra. Funct. Anal. Appl. 12(1978), 212-214.

[BG] Bongartz, K.; Gabriel, P.: Covering spaces in representation theory, Invent. Math. 65(1981), 331-378.

[G] Gabriel, P.: The universal cover of a representation-finite algebra, Proceedings ICRA III, Puebla (1980), Springer Lecture Notes in Mathematics 903, Heidelberg 1981, 68-105.

[H] Happel, D.: On the derived category of a finite-dimensional algebra, to appear in Comment. Math. Helv.

[He] Heller, A.: The loop-space functor in homological algebra, Trans. Amer. Math. Soc. 96(1960), 382-394.

[HW] Hughes, D.; Waschbüsch, J.: Trivial extensions of tilted algebras, Proc. London Math. Soc. (3), 46(1983), 347-364.

[K] Kapranov, M.M.: Derived category of coherent sheaves on Grassman manifolds, Funct. Anal. Appl. 17(1983), 145-146.

[R] Ringel, C.M.: Tame algebras and integral quadratic forms, Springer Lecture Notes in Mathematics 1099, Heidelberg 1984.

[V] Verdier, J.L.: Catégories dérivées, état 0, in SGA $4\frac{1}{2}$, Springer Lecture Notes in Mathematics 569, Heidelberg 1977, 262-311.

Dieter Happel
Fakultät für Mathematik
Universität Bielefeld
D 48 Bielefeld 1
BR Deutschland

ALMOST SPLIT SEQUENCES FOR SOME NON-CLASSICAL LATTICE CATEGORIES

K. W. Roggenkamp

Introduction: Let R be a complete local noetherian domain with field of fractions K. A is a finite dimensional K-algebra, and an R-order Λ in A is a unitary subring of A, which is R-finitely generated and contains a K-basis of A.

We consider the situation of two R-orders $\Lambda \subset \Gamma$ in A with

$$\operatorname{rad} \Gamma \subset \Lambda \subset \Gamma \quad ,$$

and study the full subcategory ${}_\Lambda\mathfrak{M}^o(\Gamma)$ of mod Λ, the objects of which are finitely generated left Λ-modules which are finitely generated and torsionfree over R. Moreover, we require that for such a module M, ΓM is a projective Γ-module. (At this point we should explain our notation:
Since M is R-torsionfree, M is a submodule of $KM = AM$, and so we may form ΓM inside AM. Another interpretation is that ΓM is $\Gamma \otimes_\Lambda M$ modulo the R-torsion submodule.)

The aim of this note is to prove the following result:

Theorem: The category ${}_\Lambda\mathfrak{M}^o(\Gamma)$ has right almost split sequences; i.e., if M'' is not ext-projective in ${}_\Lambda\mathfrak{M}^o(\Gamma)$, then there exists an almost split sequence

$$0 \longrightarrow M' \longrightarrow M \longrightarrow M'' \longrightarrow 0 \quad .$$

Let us look at an example from geometry: We denote by Λ the coordinate ring of three planes in $\mathbb{C}^{(6)}$ intersecting in exactly one point, say the origin. If $R = \mathbb{C}[[X,Y]]$ is the ring of formal power-series in 2 variables, then Λ is the pullback

$$\begin{array}{ccc} \Lambda & \longrightarrow & R \times R \times R \\ \downarrow & & \downarrow \\ \mathbb{C} & \xrightarrow{\Delta} & \mathbb{C} \times \mathbb{C} \times \mathbb{C} \end{array} \quad ,$$

where Δ is the diagonal. Moreover, $R \times R \times R =: \Gamma$ is the normalization. So the category ${}_\Lambda\mathfrak{M}^o(\Gamma)$ consists of Λ-modules which are finitely generated and torsionfree over R and which become projective - in the above sense - over the normalization Γ of Λ. It should be noted that Λ is not Cohen-Macaulay in the sense of

Auslander [A], and so his and Reiten's theory of almost split sequences of CM-modules does not apply here [A]. In this example rad Λ = rad Γ . We let m be the maximal ideal in R . Then obvious modules in ${}_{\Lambda}\mathfrak{M}^{o}(\Gamma)$ are

$$M_1 = (R,0,0) \quad , \quad M_2 = (0,R,0) \quad , \quad M_3 = (0,0,R)$$

$P = \Lambda$; N_1 is the pullback

$$\begin{array}{ccc} N_1 & \longrightarrow & M_2 \oplus M_3 \\ \downarrow & & \downarrow \\ k & \xrightarrow{\Delta_1} & k \oplus k \end{array} \quad , \quad k = R/m \text{ and } \Delta_1 \text{ is the diagonal}$$

Similarly N_2 and N_3 are defined. Moreover, we put $J = N_1 + N_2 + N_3$ in Γ . Then the Auslander-Reiten quiver of ${}_{\Lambda}\mathfrak{M}^{o}(\Gamma)$ is given as

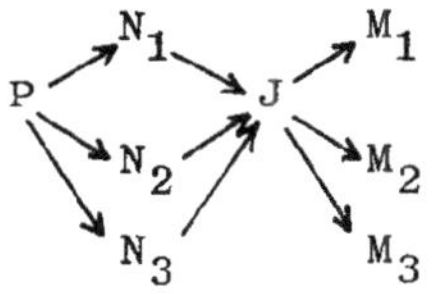

In this example, the modules P, N_1, N_2, N_3 are ext-projective and J, M_1, M_2, M_3 are ext-injective.

We would like to compare this with the situation, where $R = \mathbb{C}[[X]]$ and Λ and Γ are defined as above, similarly the modules. In this case ${}_{\Lambda}\mathfrak{M}^{o}(\Gamma)$ is the category of Λ-lattices. The Auslander-Reiten quiver is

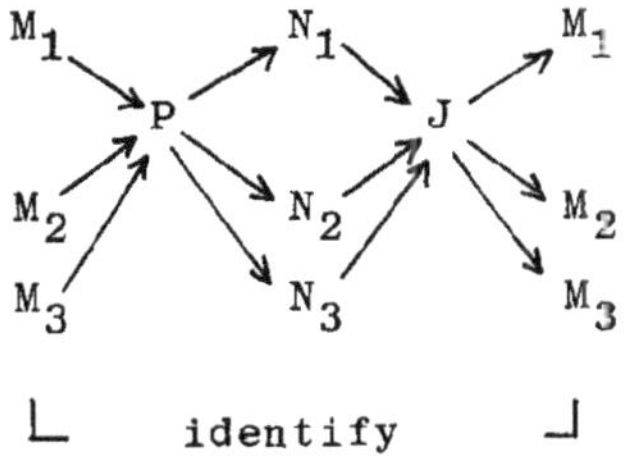

Here only P is ext-projective and J is ext-injective. The reason for the difference is, that in both cases

$$\mathrm{Ker}(P \longrightarrow N_i) = m\, M_i \quad .$$

In the one-dimensional case $m\, M_i \simeq M_i \in {}_{\Lambda}\mathfrak{M}^{o}(\Gamma)$, but in the two-dimensional case, $m\, M_i \notin {}_{\Lambda}\mathfrak{M}^{o}(\Gamma)$.

Proof of the theorem

We put $\mathfrak{A} = \Lambda/\mathrm{rad}\,\Gamma$ and $\mathfrak{B} = \Gamma/\mathrm{rad}\,\Gamma$. Then the $k = R/\max R$ -algebra

$$\mathfrak{D} = \begin{pmatrix} \mathfrak{B} & \mathfrak{B} \\ 0 & \mathfrak{A} \end{pmatrix}$$

has a projective socle, and we denote by $\mathfrak{S}(\mathfrak{D})$ the full subcategory of mod $\mathfrak{D}$, consisting of those modules which have a projective socle. We recall from [RR], that the functor

(1) $\quad \mathbb{F}: {}_{\Lambda}\mathfrak{M}^{o}(\Gamma) \longrightarrow \mathfrak{S}(\mathfrak{D})$,

induced by

$$M \longmapsto \begin{bmatrix} \Gamma M/\mathrm{rad}\,\Gamma\cdot M \\ M/\mathrm{rad}\,\Gamma\cdot M \end{bmatrix},$$

is a representation equivalence between ${}_{\Lambda}\mathfrak{M}^{o}(\Gamma)$ and the subcategory $\mathfrak{C}$ of $\mathfrak{S}(\mathfrak{D})$ generated by the non-simple indecomposables in $\mathfrak{S}(\mathfrak{D})$. $\mathfrak{S}(\mathfrak{D})$ has almost split sequences [S, R1].

Lemma 1: If

$$0 \to X \xrightarrow{\alpha} Y \xrightarrow{\beta} Z \to 0$$

is an almost split sequence in $\mathfrak{S}(\mathfrak{D})$ with X not simple projective, then

$$0 \longrightarrow \mathbb{F}^{-1}(X) \longrightarrow \mathbb{F}^{-1}(Y) \xrightarrow{\mathbb{F}^{-1}(\beta)} \mathbb{F}^{-1}(Z) \longrightarrow 0$$

is an almost split sequence in ${}_{\Lambda}\mathfrak{M}^{o}(\Gamma)$.

Proof: We recall from [R1] Lemma 3:

(2) Let $\varphi: M \longrightarrow N$ be a map between indecomposables in ${}_{\Lambda}\mathfrak{M}^{o}(\Gamma)$. Assume $\mathbb{F}(\varphi) \neq 0$. If $\mathbb{F}(\varphi)$ is irreducible, so is φ .

By (2.4) in [R1], the sequence

(3) $\quad 0 \to \mathrm{Soc}\,X \to \mathrm{Soc}\,Y \to \mathrm{Soc}\,Z \to 0$

is exact, where Soc( ) denotes the socle.
We now put $M = \mathbb{F}^{-1}(Z)$, $E = \mathbb{F}^{-1}(Y)$ and $N = \mathbb{F}^{-1}(X)$. Because of (3) we have

(4) $\quad \Gamma E \simeq \Gamma M \oplus \Gamma N$.

Let now $\tilde{\beta}$ be any preimage of β under $\mathbb{F}$. Then $\tilde{\beta}: E \to M$ is surjective, since $\mathrm{rad}\,\Gamma\cdot M \subset \mathrm{rad}\,\Lambda\cdot M$. We now choose a lifting $\tilde{\alpha}: N \to E$ which induces a split monomorphism $\Gamma N \to \Gamma E$. Then $\tilde{\alpha}\tilde{\beta} = 0$.

Claim 1: The sequence

$$\mathbb{E}: \quad 0 \to N \xrightarrow{\tilde{\alpha}} E \xrightarrow{\tilde{\beta}} M \to 0$$

is exact.

<u>Proof:</u> We only have to show $\operatorname{Im}\tilde{\alpha} \supset \operatorname{Ker}\tilde{\beta}$, noting that $\tilde{\alpha}$ is injective. So, let $e \in \operatorname{Ker}\tilde{\beta}$. Then there exists $n \in N$ such that

$$n\tilde{\alpha} - e \in \operatorname{rad}\Gamma \cdot E \quad ;$$

since the induced sequence

(5) $\quad 0 \to \operatorname{rad}\Gamma \cdot N \to \operatorname{rad}\Gamma \cdot \mathbb{E} \to \operatorname{rad}\Gamma \cdot M \to 0$

is split exact, $n\tilde{\alpha} - e = n'\tilde{\alpha}$ for some $n' \in \operatorname{rad}\Gamma \cdot N \subset N$. Thus $\mathbb{E}$ is an exact sequence in ${}_{\Lambda}\mathfrak{M}^{0}(\Gamma)$.

This proves Claim 1.

<u>Claim 2:</u> $\mathbb{E}$ is right almost split.

<u>Proof:</u> Let $L \in {}_{\Lambda}\mathfrak{M}^{0}(\Gamma)$ and assume that we have a map $\sigma: L \longrightarrow M$ which is not a split epimorphism.

<u>Case 1:</u> $\operatorname{Im}\sigma \in \operatorname{rad}\Gamma \cdot M$. Since the exact sequence is split we get a factorization of σ via $\tilde{\alpha}$.

<u>Case 2:</u> If $\operatorname{Im}\sigma \notin \operatorname{rad}\Gamma \cdot M$, then $\mathbb{F}(\sigma) \neq 0$, and σ is a split epimorphism if and only if $\mathbb{F}(\sigma)$ is a split epimorphism. We then can find a map $\epsilon: L \longrightarrow E$ with $F(\sigma) = \mathbb{F}(\tau\tilde{\beta})$; i.e. $\sigma - \tau\tilde{\beta}: L \longrightarrow M$ factorizes via $\operatorname{rad}\Gamma \cdot M$, and so we can find $\tau': L \longrightarrow E$ with $\sigma = (\tau + \tau')\tilde{\beta}$; i.e. $\mathbb{E}$ is right almost split.

<u>Claim 3:</u> $\mathbb{E}$ is left almost split.

<u>Proof:</u> Let $L \in {}_{\Lambda}\mathfrak{M}^{0}(\Gamma)$ and assume that we have a map $\sigma: N \longrightarrow L$ which is not a split monomorphism.

<u>Case 1:</u> $\operatorname{Im}\sigma \subset \operatorname{rad}\Gamma \cdot L$. Then σ can be extended to a Γ-linear map $\Gamma N \longrightarrow L$, and since the sequence

$$0 \to \Gamma N \to \Gamma E \to \Gamma M \to 0$$

is split exact, we get a factorization of σ via $\tilde{\alpha}$.

<u>Case 2:</u> If $\operatorname{Im}\sigma \not\subset \operatorname{rad}\Gamma \cdot L$, then $\mathbb{F}(\sigma) \neq 0$, and we can find $\tau: E \longrightarrow L$ with $\sigma - \tilde{\alpha}\tau: N \longrightarrow L$ factorizes via $\operatorname{rad}\Gamma \cdot L$. As above we construct with the help of Case 1 a factorization of σ.

This proves Lemma 1.

We now come to the remaining cases. Let S be a simple projective $\mathfrak{D}$-module. Then in $\mathfrak{S}(\mathfrak{D})$ we have an almost split sequence - provided S is not an injective object in $\mathfrak{S}(\mathfrak{D})$:

(6) $\quad 0 \to S \to \overline{P} \xrightarrow{\alpha} X \to 0$.

where $\overline{P}$ is a projective $\mathfrak{D}$-module. This follows from Lemma 6 [R2].

<u>Remark:</u> If the simple projective S is also an injective object in

$\mathfrak{S}(\mathfrak{D})$, then $\mathfrak{D}$ has a direct summand which is simple, and so Λ has a ring direct summand Λ_0, which coincides with a direct summand of Γ (cf. Lemma 2, [R2]).
We lift X to M with $F(M) \simeq X$, $\overline{P}$ to P and α to $\tilde{\alpha}: P \longrightarrow M$; put $Q = \operatorname{Ker} \tilde{\alpha}$.

Lemma 2: In the exact sequence

$\mathbb{E}$: $\quad 0 \to Q \to P \to M \to 0 \quad ,$

Q is Γ-module of the form $Q = \operatorname{rad} \Gamma \cdot \tilde{Q}$ for a projective Γ-module $\tilde{Q}$. If Q is Γ-projective, then $\mathbb{E}$ is a right almost split sequence. If Q is not Γ-projective, then M is ext-projective in ${}_{\Lambda}\mathfrak{M}^o(\Gamma)$.

Proof: Let $\tilde{Q}$ be the kernel of the induced map $\Gamma P \longrightarrow \Gamma M$, which is a split epimorphism, and so $\tilde{Q}$ is Γ-projective. Since $P/\operatorname{rad} \Gamma \cdot P \simeq M/\operatorname{rad} \Gamma \cdot M$, we conclude that $Q \subset \operatorname{rad} \Gamma \cdot P$. The splitting $\Gamma P \longrightarrow \tilde{Q}$ induces a splitting $\operatorname{rad} \Gamma \cdot P \longrightarrow \operatorname{rad} \Gamma \cdot \tilde{Q}$. Since S in (6) is simple, and since $\operatorname{rad} \Gamma \cdot \tilde{Q} \subset Q$, we must have $Q = \operatorname{rad} \Gamma \cdot \tilde{Q}$.

Case 1: Q is Γ-projective; then the exact sequence $\mathbb{E}$ is an exact sequence in ${}_{\Lambda}\mathfrak{M}^o(\Gamma)$, and by similar arguments as in the proof of Lemma 1, $\mathbb{E}$ is right almost split. Note that $\mathbb{E}$ can not be split.

Case 2: Q is not Γ-projective. Then we have to show that M is ext-projective in ${}_{\Lambda}\mathfrak{M}^o(\Gamma)$. Assume we have an exact sequence in ${}_{\Lambda}\mathfrak{M}^o(\Gamma)$,

$$0 \to L' \to L \to M \to 0 \quad ,$$

which is not split exact. Then we can form the pullback

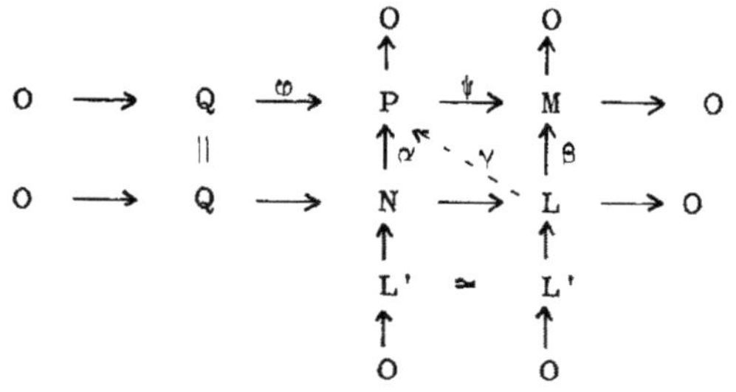

We then have $N \simeq P \oplus L'$. Since β is not a split epimorphism, we also have $N \simeq Q \oplus L$. So $\Gamma N \simeq \Gamma P \oplus \Gamma L' \simeq Q \oplus \Gamma L$. Since L, L' and P lie in ${}_{\Lambda}\mathfrak{M}^o(\Gamma)$, we conclude $Q \in {}_{\Lambda}\mathfrak{M}^o(\Gamma)$; i.e. Q is Γ-projective, a contradiction.

This proves Lemma 2.

<u>Remark:</u> I do not know, whether in Case 1, the sequence

$$0 \to Q \to P \to M \to 0$$

is also left almost split. The obstruction for the proof is as follows: Given a map $\alpha: Q \longrightarrow L$ in ${}_\Lambda\mathfrak{M}^o(\Gamma)$, we form the pushout

$$\begin{array}{ccccccccc} 0 & \longrightarrow & Q & \longrightarrow & P & \longrightarrow & M & \longrightarrow & 0 \\ & & \downarrow & & \downarrow & & \| & & \\ 0 & \longrightarrow & L & \longrightarrow & N & \longrightarrow & M & \longrightarrow & 0 \end{array}$$

Though L and M lie in ${}_\Lambda\mathfrak{M}^o(\Gamma)$, I do not see any reason, why N lies in ${}_\Lambda\mathfrak{M}^o(\Gamma)$. If N were in ${}_\Lambda\mathfrak{M}^o(\Gamma)$, then the classical argument [AR] would work.

Up to now we have shown, that every $M \in {}_\Lambda\mathfrak{M}^o(\Gamma)$, which is not ext-projective, has a right almost split sequence. The situation for the ext-injective modules in ${}_\Lambda\mathfrak{M}^o(\Gamma)$ is more complicated because of the lack of a duality.

<u>Remark:</u> By a recent unpublished result of Smalø it seems likely, that the hypothesis $\mathrm{rad}\,\Gamma \subset \Lambda \subset \Gamma$ in the theorem can be dropped.

References

[A] Auslander, M.: Isolated singularities and almost split sequences.
Representation theory II, Groups and orders, Springer Lecture Notes 1178, 194-242

Finite type implies isolated singularity.
Orders and their applications, Springer Lecture Notes 1142, 1-4

[AR] Auslander, M - I. Reiten: Representation theory of artin algebras IV, V, VI.
Comm. in Algebra 5 (1977), 443-518; 5 (1977), 519-554; 6 (1978), 257-300

[RR] Ringel, C.M. - K.W.Roggenkamp: Diagrammatic methods in the representation theory of orders.
Journal of Algebra 60 (1979), 11-42

[S] Simpson, D.: Vectorspace categories, right peak rings and their socle projective modules.
Journal of Algebra 92, (1985), 532-571

[R1] Roggenkamp, K.W.: Auslander-Reiten species for socle determined categories of hereditary algebras and for generalized Bäckström orders.
Mitt.Math.Sem.Gießen 159 (1983), 1-98

[R2] Roggenkamp, K.W.: Lattices over subhereditary orders and socle projective modules.
Preprint Stuttgart 1985

SPECIAL INSTANTON BUNDLES AND PONCELET CURVES

by

W. Böhmer and G. Trautmann

In [6] A. Hirschowitz and M.S. Narasimhan considered special instanton bundles which are extensions of a line bundle on a quadric in $\mathbb{P}_3$ by a trivial sheaf of rank 2. They satisfy $h^oE(1) = 2$ whereas for the general instanton bundle $h^oE(1) = 0$. It is shown first that $h^oE(1) \leq 2$ for any instanton bundle and that the special bundles of [6] are exactly those with $h^oE(1) = 2$. Second we show that the special instanton bundles are in a one to one correspondence with certain plane Poncelet curves, see definition 4.4, with respect to conics, prop. 4.5, 4.6, which describe the set of jumping lines and the set of unstable planes respectively. Moreover we get as a nice byproduct that the special instanton bundles can be described by persymmetric monad operators.

0. Notations. We fix an algebraically closed field k of characteristic 0, a 4-dimensional k-vector space V, and let $\mathbb{P}_3 = \mathbb{P}V$ be the projective space of 1-dimensional subspaces of it. $\mathcal{O}$ always denotes the structure sheaf of $\mathbb{P}_3$ and $\mathcal{O}(d)$ the invertible sheaf of degree d. For a coherent $\mathcal{O}$-module F we write $F(d) = F \otimes \mathcal{O}(d)$ and $h^iF(d)$ for the dimension of $H^iF(d) = H^i(\mathbb{P}_3, F(d))$. An (algebraic) vector bundle is a locally free $\mathcal{O}$-module of finite type. The sheaves Ω^i of differential i-forms are identified with the images in the canonical Koszul complex $\wedge^{i+1}V^* \otimes \mathcal{O}(-i-1) \longrightarrow \wedge^iV^* \otimes \mathcal{O}(-i)$. We frequently use the canonical isomorphisms $\wedge^iV \simeq \wedge^{4-i}V^*$ based on a fixed isomorphism $\wedge^4V \simeq k$.

1. Instanton bundles and their monads. An n-instanton bundle on $\mathbb{P}_3$ is by definition an (algebraic) vector bundle E of rank 2 with Chern classes $c_1 = 0$, $c_2 = n$, which is stable (i.e. $h^oE = 0$) with $h^1E(-2) = 0$, see [1]. Omitting the trivial case $n = 1$, stability implies $n \geq 2$. We write $H = H^2E(-3)$, $K = H^1E$ and obtain $H^* = H^1E(-1)$ by Serre-duality. It follows from the Hirzebruch-Riemann-Roch theorem and easy vanishing statements, that dim $H = n$ and dim $K = 2n-2$, [1]. It is also wellknown, that any n-instanton E is the cohomology of a

short sequence (called monad)

$$H \otimes \mathcal{O}(-1) \xrightarrow[M]{} H^* \otimes \Omega^1(1) \xrightarrow[B]{} K \otimes \mathcal{O}$$

which is a complex, and in which M is a subbundle and B an epimorphism. The morphisms M , B correspond to operators

$$H \xrightarrow[M]{} H^* \otimes \Lambda^2 V^* \simeq H^* \otimes \Lambda^2 V \qquad H^* \xrightarrow[B]{} K \otimes V$$

which induce the morphisms via

$$\begin{array}{ccc} H \otimes \mathcal{O}(-1) & \longrightarrow & H^* \otimes \Omega^1(1) \\ & {\scriptstyle M}\searrow \quad \nearrow & \\ & H^* \otimes \Lambda^2 V^* \otimes \mathcal{O}(-1) & \end{array} \qquad \begin{array}{ccc} H^* \otimes \Omega^1(1) & \longrightarrow & K \otimes \mathcal{O} \\ \downarrow & & \wr\Vert \\ H^* \otimes \Lambda^3 V \otimes \mathcal{O} & \xrightarrow[\wedge B]{} & K \otimes \Lambda^4 V \otimes \mathcal{O}. \end{array}$$

Moreover M can be assumed symmetric, i.e. the induced operator $\Lambda^2 V^* \longrightarrow \mathrm{Hom}(H,H^*)$ factorizes over $\mathrm{Sym}(H,H^*)$. Note that $E \simeq E'$ iff $(M,B) \sim (M',B')$, i.e. there are $S \in GL(H)$, $R \in GL(K)$ such that

$$M' = (S^t \otimes \mathrm{id}) \circ M \circ S \qquad \text{and } B' = (R \otimes \mathrm{id}) \circ B \circ S^{t,-1}.$$

Choosing bases of K , H and the dual basis in H^* we may express M and B as matrices $M = (a_{ij})$ with $a_{ij} = a_{ji} \in \Lambda^2 V$ and $B = (b_{ij}) \in V$. It is straightforward to verify that the conditions for the monad in addition to the symmetry of M are:

(i) $k^n \otimes V \xrightarrow[\wedge M]{} k^n \otimes \Lambda^3 V \xrightarrow[\wedge B]{} k^{2n-2} \otimes \Lambda^4 V \longrightarrow 0$ is exact

(ii) $k^{2n-2} \xrightarrow[B^t]{} k^n \otimes V$ (identified with the dual $K^* \otimes V^* \longrightarrow H$ of B) satisfies $\mathrm{Im}(B^t) \cap (k^n \otimes v) = 0$ for any nonzero $v \in V$.

If K is the kernel of the morphism B, the exact sequences

$$0 \to K \to H^* \otimes \Omega^1(1) \xrightarrow[B]{} K \otimes \mathcal{O} \to 0$$

and

$$0 \to H \otimes \mathcal{O}(-1) \longrightarrow K \xrightarrow[\pi]{} E \longrightarrow 0$$

are called the first and second display sequences.

<u>1.1 Unstable planes of E:</u> A plane $P = \{v^* = 0\}$ is called unstable for E if $h^0(E|P) \neq 0$. By restricting the second display sequence to P we get $h^0(K|P) = h^0(E|P)$ for any P. Now the first display implies that P is unstable iff the evaluation map $H^* \xrightarrow[B]{} K \otimes V \xrightarrow[v^*]{} K$ has rank $< n$, or in matrix notation rank $v^*(b_{ij}) < n$. Therefore the set of unstable planes of E is the determinantal variety of B in $\mathbb{P}V^*$.

<u>1.2 Jumping lines of E:</u> A line $L \subset \mathbb{P}_3$ is called a jumping line of E if $E|L \not\simeq 2\mathcal{O}_L$ (for generic L $E|L$ is trivial). If L is spanned by

$x, y \in V$, it is a jumping line iff rank $M\wedge x\wedge y < n$, where $M\wedge x\wedge y$ is the composite operator $H \xrightarrow{M} H^* \otimes \Lambda^2 V \xrightarrow{\wedge x\wedge y} H^* \otimes \Lambda^4 V$. This can be derived by restricting the displays to L. More precisely $E|L \simeq \mathcal{O}_L(-i) \oplus \mathcal{O}_L(i)$ iff rank $M\wedge x\wedge y = n-i$.

1.3 Sections of $K(1)$ and $E(1)$: It follows from the Koszul complex and the description of B that any $t \in \Gamma K(1)$ can be identified with an element $a \in H^* \otimes \Lambda^2 V$ satisfying $a\wedge B = 0$. The evaluation at the point $\langle x\rangle$ is then given by $a\wedge x \in H^* \otimes \Lambda^3 V$. Thus, if $a = (a_1,\dots,a_n)$ with $a_i \in \Lambda^2 V$ is the decomposition w.r.t. to a basis of H^*, $\langle x\rangle$ is a zero of t iff $a_i = x\wedge y_i$ with some $y_i \in V$. Therefore the zero scheme $V(t)$ of t is either empty, a point, or a line. It is empty iff $t \in \operatorname{Im} M$, $t \neq 0$.

Lemma: *1) If t is a nonzero section of $K(1)$ and $V(t) \neq \emptyset$, then $V(t)$ is a line which is contained in the zero scheme of the section $\pi(t)$ of $E(1)$.*
2) If s is a nonzero section of $E(1)$ then $V(s)$ is a disjoint union of (not necessarily reduced) lines.

Proof: Let $s \in \Gamma E(1)$ be nonzero. Any component of $V(s)$ has dimension 1, [5]. Let $X \subset V(s)$ be a connected union of components of $V(s)$. By the second display there is a $t \in \Gamma K(1)$ with $\pi(t) = s$ and $\pi(t|X_{red}) = 0$. Since X_{red} is connected, $\Gamma\mathcal{O} \simeq \Gamma\mathcal{O}_{X_{red}}$, and we can find some $u \in H$ s.t. $t' = t-M(u)$ vanishes on X_{red}. By the above $V(t')$ is a line containing X_{red}, and so $X_{red} = V(t')$. This proves 2). If $p \in V(t)$ for some $t \in \Gamma K(1)$ then $V(\pi(t))$ has a component passing through p. Its reduction is a line $L \ni p$. As before $L = V(t')$ with $t' = t-M(u)$ for some u. But now $M(u)(p) = 0$ and hence $u = 0$. Hence $V(t) = V(t')$ is a line.

1.4 Lemma: *The dependency locus of any two independent sections $s, s' \in \Gamma E(1)$ is a quadric surface $Q \subset \mathbb{P}_3$, and for any $p \in Q$ there is a linear combination $s'' = \lambda s + \lambda' s'$ with $p \in V(s'')$.*

Proof: Since $\Lambda^2 E(1) \simeq \mathcal{O}(2)$ the section $s\wedge s'$ defines a quadric equation. Its zero scheme Q satisfies the statement.

2. The estimate $h^0E(1) \leq 2$.

If P is a stable plane, it was shown in [5] that $h^0(E|P)(1) \leq 3$ and hence $h^0E(1) \leq 3$ by the exact sequence $0 \to \Gamma E(1) \to \Gamma(E|P)(1)$. It was also shown in [5] that $h^0E(1) = 2$ for $n = 2$. Let us assume $h^0E(1) = 3$, and that $n \geq 4$ first. If s_o, s_1, s_2 is a basis of $\Gamma E(1)$, the pairs s_o, s_1 and s_o, s_2 define different quadrics Q_1, Q_2 with $V(s_o) \subset Q_1 \cap Q_2$. Since $\deg V(s_o) = n+1$, [5], Bezout's theorem implies that $Q_1 \cap Q_2$ cannot be a curve, and hence $Q_i = P \cup P_i$ with a common plane. By varying s_1, s_2 we can assume that the planes P, P_1, P_2 are all different. Since no $V(s)$ can be contained in a plane (there is an exact sequence $0 \to \mathcal{O} \to E(1) \to \mathcal{I}(2) \to 0$ where $\mathcal{I}$ is the ideal of $V(s)$, which implies $0 = \Gamma E = \Gamma \mathcal{I}(1)$), $V(s_o) = L_o \mathring{\cup} L$ where $L_o \subset P$ and $L = P_1 \cap P_2$, 2) of 1.3). If s is any section of $E(1)$, the dependency quadric of s_o, s must also contain P and L, since s is spanned by s_o, s_1, s_2, and hence $V(s) = L(s) \mathring{\cup} L'(s)$ with $L(s) \subset P$. Let $\{q\} = L \cap P$. We can choose s_1 so that $L(s_1)$ is different from $L_o = L(s_o)$ by 1.4. Let p be the intersection $L(s_o) \cap L(s_1)$. Then $p \neq q$.

<u>Claim:</u> Any line through p is contained in some $V(s)$.

For by the proof of 1.3 we can find sections t_o, t_1 of $K(1)$ s.t. $V(t_i) = L(s_i)$ for $i = 0,1$. If $p = \langle x \rangle$ we can assume $t_i = a_i \otimes (x \wedge y_i)$ with $x, y_o, y_1 \in V$ independent. If a_o, a_1 were independent then $\langle x \rangle$ would be the only zero of $t_o + t_1$, which is excluded by 1.3. Therefore $t_i = a \otimes (x \wedge y_i)$ for some $a \in H^*$. Choosing bases we can write $a = (\alpha_1, \ldots, \alpha_n)$ and $B = (b_{ij})$ and get $x \wedge y_o \wedge \sum_i \alpha_i b_{ij} = 0$ and $x \wedge y_1 \wedge \sum_i \alpha_i b_{ij} = 0$ for any j. Since x, y_o, y_1 are independent, $\sum_i \alpha_i b_{ij} = \lambda_j x$ for any j. But then $t = a \otimes x \wedge y \in H^* \otimes \Lambda^2 V$ is a section of $K(1)$ for any y, which proves the claim.

Now we can choose a line L' with $p \in L' \not\subset P \cup P_1$ = quadric of s_o, s_1 and a section s_2 of $E(1)$ with $L'(s_2) = L'$ (cf. notation above). It follows that $p \notin L(s_2) \subset P$. Now the sections s_o, s_1, s_2 are independent. For otherwise $V(s_2) \subset P \cup P_1$. It follows that any section of $E(1)$ vanishes in p. Let p_o, p_1 be the intersection points of $L(s_2)$ with $L(s_o), L(s_1)$. Repeating the argument just used any section of $E(1)$ also vanishes in p_o, p_1. Since $p_1 \notin L(s_o)$ we must have $p_1 = q \in L = L'(s_o)$. But also for any $p' \in P \setminus \bigcup_{o}^{2} L(s_i)$ there is some $s' \in \Gamma E(1)$ vanishing in p' by 1.4. Since $p, p_o, q \in V(s') = L(s') \cup L'(s')$ the line $L(s')$ must be one of the $L(s_i)$, contradiction.

Now we are left with the case $h^o E(1) = 3$ and $n = 3$. For a basis s_o, s_1, s_2 of $\Gamma E(1)$ we can choose $t_i \in \Gamma K(1)$ with $\pi(t_i) = s_i$ such that $V(t_i)$ are different vanishing lines by using arguments as before. Then $t_i = a_i \wedge x_i \wedge y_i$ with $a_i \in H$.

Case 1) The vectors a_0, a_1, a_2 are linearly independent. Using these as a basis we obtain $x_i \wedge y_i \wedge b_{ij} = 0$ for any i,j. By elementary linear algebra and the monad conditions (i),(ii) we can assume that B can be given the form

$$\begin{pmatrix} x_0 & y_o & 0 & 0 \\ 0 & 0 & x_1 & y_1 \\ x_2 & y_2 & x_2 & y_2 \end{pmatrix} .$$

It follows from this that $\Gamma K(1) \subset k^3 \otimes \Lambda^2 V$ has dimension ≤ 5, by calculating the kernel of B. But then $h^0 E(1) \leq 2$.

Case 2) a_0, a_1 are independent and $a_2 \in \text{span}(a_0, a_1)$. Again we choose a basis s.t. $a_0 = (1,0,0)$, $a_1 = (0\ 1\ 0)$ and $a_2 = (\alpha, \beta, 0)$. Then B can be given the form

$$\begin{pmatrix} x_0 & y_o & 0 & 0 \\ 0 & 0 & x_1 & y_1 \\ \hline & & * & \end{pmatrix}$$

by choosing a suitable basis of K. But now $(\alpha x_2 \wedge y_2, \beta x_2 \wedge y_2, 0) \wedge B = 0$ and this implies that at least two of the lines $V(t_i)$ are identical, contradiction.

Case 3) The a_i span a 1-dimensional subspace for any choice of the t_i. Now we may assume $a_0 = a_1 = a_2 = (1,0,0)$ and the only possibility for B is

$$B = \left(\begin{array}{c|c} \lambda x & 0 \\ \hline B'' & B' \end{array}\right) ,$$

where $x \neq 0 \in V$ and λ is a scalar (all three lines $V(t_i)$ pass through $\langle x \rangle$). If $\lambda = 0$ the matrix B cannot satisfy the monad condition (ii), for by elementary linear algebra we can find a combination of the columns of B, in which the two nonzero components are linearly dependent. Let now $\lambda \neq 0$. Since B' also satisfies the monad condition (ii), we get an exact sequence $0 \to K' \to 2\Omega^1(1) \xrightarrow[B']{} 3\mathcal{O} \to 0$. Using the Eagon-Northcott complex of B', see [4], sequence 0.5, we find $h^0 K'(1) = 0$. Then any $t \in \Gamma K(1)$ is of the form $t = (x \wedge y, 0, 0) \in k^3 \otimes \Lambda^2 V$ for some y and hence $h^0 K(1) = 3$ and $h^0 E(1) = 0$, contradiction.

2.1 Proposition: *If* $h^0 E(1) = 2$ *then the dependency quadric* Q *of* $\Gamma E(1)$ *is nonsingular,* $Q \simeq \mathbb{P}_1 \times \mathbb{P}_1$, *and* $E(1)$ *is an extension* $0 \to 2\mathcal{O} \to E(1) \to \mathcal{O}_Q(2,1-n) \to 0$. *We call this sequence the evalution sequence of* E.

Remark 1: Bundles E which are extensions of the above type are called special 't Hooft bundles in [6].

Remark 2: Using the second display and the behaviour of the zero loci of sections of $K(1)$ and $E(1)$, see 1.3, we also get the evaluation sequence

$$0 \to k^{n+2} \otimes \mathcal{O} \to K(1) \to \mathcal{O}_Q(2,1-n) \to 0 .$$

Proof: 1) By 1.4 any vanishing line $V(t)$, $t \in \Gamma K(1)$, is contained in Q and through any $q \in Q$ there is one such line. Two different $V(t_1), V(t_2)$ are not allowed to meet: For by the proof of the claim we would have $t_i = a \otimes (x \wedge y_i)$ where $\langle x \rangle$ is the intersection and then any line through $\langle x \rangle$ would be a vanishing line, contradicting the fact that it is contained in Q. Now these properties of vanishing lines obviously exclude the cases where Q is a double plane, a pair of planes or a cone, i.e. Q is nonsingular. Moreover all vanishing lines are contained in one ruling of $Q \simeq \mathbb{P}_1 \times \mathbb{P}_1$.

2) Since Q is the dependency locus of two basic sections we immediately have an exact sequence

$$0 \to 2\mathcal{O} \to E(1) \to F \to 0$$

where F as the cokernel is supported by Q and an $\mathcal{O}_Q$-module. By the same sequence prof $F_p = 2$ for any $p \in Q$ and hence F is locally-free on Q. Since no section of $E(1)$ can vanish identically on Q, 1.3, F must be invertible, i.e. $F = \mathcal{O}_Q(a,b)$, since $Q \simeq \mathbb{P}_1 \times \mathbb{P}_1$. The conditions $h^0 F = 0$, $h^1 F(-2) = n$, $h^1 F(-1) = 2n-2$, which follow from known groups of E, imply $(a,b) = (2,1-n)$ or $= (1-n,2)$. The dual of the above sequence and $\mathcal{E}xt^1_{\mathcal{O}}(\mathcal{O}_Q(2,1-n),\mathcal{O}) \simeq \mathcal{O}_Q(n+1,0)$ imply that the ruling of Q, which consists of the vanishing lines of sections of $E(1)$, corresponds to the number $a = 2$, which we choose as the first.

3. Special monads.

The monad (M,B) of E is called <u>special</u> if B has a matrix representation

$$B = \begin{pmatrix} x & x' & & & & \\ y & y' & x & x' & & \\ & & y & y' & \ddots & \\ & & & & x & x' \\ & & & & y & y' \end{pmatrix}$$

with x, x', y, y' a basis of V.

3.1 In this case the sections of $K(1)$ can be computed as follows. Since $\Gamma K(1) \subset k^n \otimes \Lambda^2 V$ is the subspace of all $(\xi_1,\ldots,\xi_n)$, $\xi_i \in \Lambda^2 V$ with $(\xi_1,\ldots,\xi_n) \wedge B = 0$ it can be checked by elementary linear algebra that the rows of the "Hankel" matrix

$$R = \begin{pmatrix} \xi & & & \\ \omega & \xi & & \\ \eta & \omega & \ddots & \\ & \eta & \ddots & \xi \\ & & \ddots & \omega \\ & & & \eta \end{pmatrix}$$

with $\xi = x\wedge x'$, $\omega = x\wedge y' - x'\wedge y$, $\eta = y\wedge y'$, form a basis of $\Gamma K(1)$. Let now $t = (\xi_1,\ldots,\xi_n) = \lambda \circ R$ be any section, $\lambda \in k^{n+2}$, and assume that $V(t) \neq \emptyset$. Then the ξ_i are decomposable, $\xi_i \wedge \xi_i = 0$. By the special shape of R we have

$$(\lambda_i\xi + \lambda_{i+1}\omega + \lambda_{i+2}\eta)^{\wedge 2} = 0 \quad \text{for } i = 1,\ldots,n,$$

which means by the definition of ξ,ω,η, that

$$\lambda_i\lambda_{i+2} - \lambda_{i+1}^2 = 0.$$

Now it is well known that these equations define the rational normal curve in $\mathbb{P}_{n+1}$, i.e. there is a pair $(s,t) \neq 0$ with

$$\lambda = (s^{n+1}, s^n t, \ldots, st^n, t^{n+1}).$$

But now $\xi_i = s^{n-2-i}t^{i-1}(s^2\xi + st\omega + t^2\eta)$ and thus the zero scheme $V(t)$ is the line given by

$$s^2\xi + st\omega + t^2\eta = (sx+ty)\wedge(sx'+ty') \qquad (*)$$

i.e. the line spanned by $sx + ty$, $sx' + ty' \in V$.
Now (*) is the parametrization of a conic $C(B) \subset G_2(V) \subset \mathbb{P}\Lambda^2 V$ contained in the Graßmannian of lines. Let $Q(B)$ be the quadric surface spanned by the system $C(B)$ of lines.

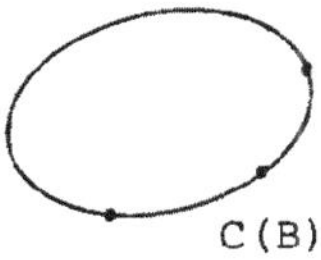
C(B)

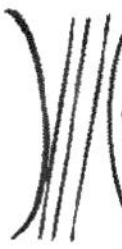
Q(B)

We have shown without using the bundle E that $Q(B)$ is the dependency locus of $\Gamma K(1) \simeq k^{n+2}$. Now as in the proof of 2.1 the cokernel of the evaluation map $k^{n+2} \otimes \mathcal{O} \to K(1)$ can be computed to be $\mathcal{O}_Q(2,1-n)$.

We are now able to prove:

3.2 Proposition: *An n-instanton bundle E has a special monad if and only if* $h^0E(1) = 2$.

Proof: If B is special we have seen that $h^0K(1) = n+2$ and hence $h^0E(1) = 2$. In this case the quadric Q(B) is the evaluation quadric of E. If conversely $h^0E(1) = 2$ the evaluation quadric Q of E is regular by 2.1, and we have the evaluation sequence

$$0 \to k^{n+2} \otimes \mathcal{O} \to K(1) \to \mathcal{O}_Q(2,1-n) \to 0.$$

We can find basis vectors $x,x',y,y' \in V$ such that Q is the regulus defined by $s^2\xi + st\omega + t^2\eta$. Let B_o be the special matrix defined by x,x',y,y' as above, and let K_o be the kernel of $k^n \otimes \Omega^1(1) \xrightarrow[B_o]{} k^{2n-2} \otimes \mathcal{O}$. We then have $Q = Q(B_o)$ and by 3.1 the evaluation sequence

$$0 \to k^{n+2} \otimes \mathcal{O} \to K_o(1) \to \mathcal{O}_Q(2,1-n) \to 0$$

given by the matrix R. This induces an isomorphism

$$\mathrm{Hom}(k^{n+2},k^{n+2}) \xrightarrow{\sim} \mathrm{Ext}^1(\mathcal{O}_Q(2,1-n),k^{n+2} \otimes \mathcal{O}),$$

in which the extension $K_o(1)$ corresponds to id, and the given extension $K(1)$ to a homomorphism φ which defines $K(1)$ as a pushout. Therefore we have a diagram

$$\begin{array}{ccccccccc}
 & & 0 & & 0 & & & & \\
 & & \downarrow & & \downarrow & & & & \\
 & & k^m \otimes \mathcal{O} & \simeq & A & & & & \\
 & & \downarrow & & \downarrow & & & & \\
0 & \longrightarrow & k^{n+2}\otimes\mathcal{O} & \longrightarrow & K_o(1) & \longrightarrow & \mathcal{O}_Q(2,1-n) & \longrightarrow & 0 \\
 & & \downarrow\varphi & & \downarrow\varphi' & & \| & & \\
0 & \longrightarrow & k^{n+2}\otimes\mathcal{O} & \longrightarrow & K(1) & \longrightarrow & \mathcal{O}_Q(2,1-n) & \longrightarrow & 0 \\
 & & \downarrow & & \downarrow & & & & \\
 & & k^m \otimes \mathcal{O} & \simeq & B & & & & \\
 & & \downarrow & & \downarrow & & & & \\
 & & 0 & & 0 & & & &
\end{array}$$

in which A, B are kernel, cokernel respectively. But since $\mathrm{Hom}(K(1),\mathcal{O}) = 0$ we must have $m = 0$, i.e. φ' is an isomorphism. This proves that the matrix B in the monad of E can be chosen as B_o, and 3.2 is proved.

3.3 Remark: The quadric Q(B) defined by a special B is dual to the quadric $xy' - x'y = 0$ in $\mathbb{P}_3^* = \mathbb{P}V^*$ which is the variety of unstable planes of E.

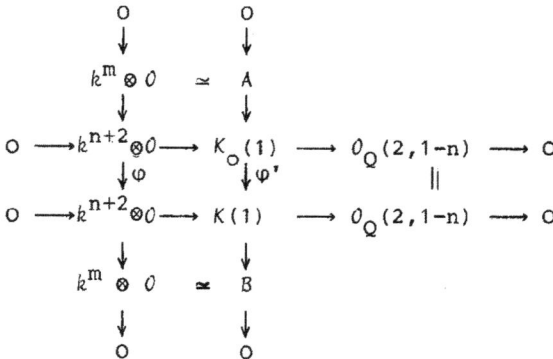

3.4 If the n-instanton bundle has a monad with special B then by the monad condition (i) and since the matrix R represents the kernel of $k^n \otimes \Lambda^2 V \xrightarrow[\Lambda B]{} k^{2n-2} \otimes \Lambda^3 V$ we can write

$$M = A \circ R$$

where A is an $n \times (n+2)$-matrix with coefficients in k. But since M is symmetric, the special "Hankel" form of R implies that A is persymmetric, i.e.

$$A = \begin{pmatrix} \alpha_0 & \alpha_1 & \alpha_2 & \cdots & \alpha_{n+1} \\ \alpha_1 & \alpha_2 & & & \alpha_{n+2} \\ \alpha_2 & & & & \vdots \\ \vdots & & & & \vdots \\ \alpha_{n-1} & \cdots & & \cdots & \alpha_{2n} \end{pmatrix}.$$

But then also M is persymmetric. Hence we have

Corollary: *Any instanton bundle with* $h^0E(1) = 2$ *has a monad* (M,B) *with* B *special and* M *persymmetric.*

Moreover we have the

3.5 Proposition: *Let* B *be a special matrix as above and let* $M = A \circ R$ *with* A *persymmetric. Then* (M,B) *defines an n-instanton bundle (with* $h^0E(1) = 2$) *if and only if* $\det \tilde{A} \neq 0$, *where* $\tilde{A}$ *is the persymmetric matrix*

$$\tilde{A} = \begin{pmatrix} \alpha_0 & \alpha_1 & \cdots & \alpha_n \\ \alpha_1 & & & \vdots \\ \vdots & & & \vdots \\ \alpha_n & \cdots & \cdots & \alpha_{2n} \end{pmatrix}.$$

The elementary but not totally trivial proof can be found in [3].

3.5 Remark: One can describe precisely when two special monads $(A \circ R, B)$ and $(A' \circ R', B')$ define isomorphic bundles. In the special case $B = B'$ the tuples α and α' must be equal up to a scalar multiple.

4. The Poncelet curve of a special 't Hooft bundle.

Let E be an n-instanton bundle with $h^0E(1) = 2$. We can fix a special monad $(A \circ R, B)$ for E as described in 3. We consider the following geometric invariants of E.

4.1 First let $C(E) = C(B)$ be the conic consisting of vanishing lines of sections of $K(1)$ and $E(1)$. It determines a plane $P(E) \subset \mathbb{P}\Lambda^2 V$ in which it is contained such that $C(E) = G_2(V) \cap P(E)$, and it also determines the dependency quadric $Q(E)$ and its dual of unstable planes, see 3.3.

4.2 Any section $s \neq 0$ of $E(1)$ defines a number of points on $C(E)$ which are the components of $V(s)$, 1.3, 2). We give a precise definition of a linear system $L(E) \subset \Gamma\mathcal{O}_{\mathbb{P}_1}(n+1) = \Gamma\mathcal{O}_{C(E)}(n+1)$ including multiplicities, by using the matrix A of the monad: For any $\lambda \in k^{n+2}$ we let $p_\lambda(s,t)$ be the polynomial

$$p_\lambda(s,t) = \det \begin{pmatrix} \lambda_0, \lambda_1, \ldots, \lambda_{n+1} \\ \hline A \\ \hline s^{n+1}, s^n t, \ldots, t^{n+1} \end{pmatrix}$$

and $s_\lambda = \pi(\lambda \circ R) \in \Gamma E(1)$, where $\lambda \circ R \in k^n \otimes \Lambda^2 V$ is considered as a section of $K(1)$. The subspace $L \subset \Gamma\mathcal{O}_{\mathbb{P}_1}(n+1)$ spanned by the p_λ is 2-dimensional (rank $A = n$) and the correspondence $p_\lambda \longleftrightarrow s_\lambda$ is one to one ($p_\lambda = p_{\lambda'}$ iff $\lambda' = a\lambda + b\circ A$ iff $\pi(\lambda\circ R) = \pi(\lambda'\circ R)$).
Moreover (s,t) is a zero of p_λ iff the line determined by $s^2\xi + st\omega + t^2\eta$ is contained in $V(s_\lambda)$, for we have $p_\lambda(s,t) = 0$ iff λ is a linear combination $\lambda = b\circ A + c(s^{n+1},\ldots,t^{n+1})$ which in turn is equivalent to

$$\pi(\lambda\circ R) = c\pi((s^{n+1},\ldots,t^{n+1})\circ R) = c(s^{n-1},\ldots,t^{n-1}) \otimes (s^2\xi + st\omega + t^2\eta)$$

Thus $L = L(E)$ is independent of A and linearly isomorphic to $\Gamma E(1)$.

4.3 Remark: It is elementary to prove that the system $\mathbb{P}L$ is base point free iff $\det \tilde{A} \neq 0$, which is exactly the condition for A to define a bundle, 3.5.

4.4 Third we denote by $\tilde{\Delta}(E) \subset \mathbb{P}\Lambda^2 V$ the hypersurface of degree n given by $\det M\wedge\zeta = 0$ where we use $\Lambda^4 V \simeq k$. This depends only on E and $\Delta(E) = \mathbb{G}_2(V) \cap \tilde{\Delta}(E)$ is the divisor of jumping lines of E, 1.2. The curve $S(E) = P(E) \cap \tilde{\Delta}(E)$ is a curve of degree n which already determines $\tilde{\Delta}(E)$, since the entries of $M = A\circ R$ are linear combinations of ξ, ω, η, and $P(E)$ is spanned by these vectors.

4.5 Definition: Let C be a conic in $\mathbb{P}_2$, $\mathbb{P}L \subset \mathbb{P}\Gamma\mathcal{O}_C(n+2)$ a pencil of divisors on C of degree n+1. A curve S of degree n in $\mathbb{P}_2$ is called Poncelet related to C with respect to $\mathbb{P}L$ if for any two points of any one of the divisors of $\mathbb{P}L$ the tangents to C meet on S.

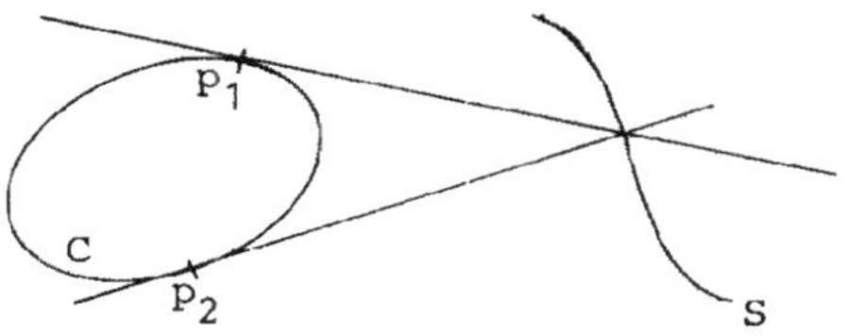

Remark: Note that $\mathbb{P}L$ is already determined by S, for starting with any point on S and drawing tangents to C one gets a divisor on C and a closed polygon of lines, [10].

4.6 Proposition: *If E is an n-instanton bundle with $h^0E(1) = 2$ then the curve S(E) is Poncelet related to C(E) w.r.t. $\mathbb{P}L(E)$.*

Proof: If $s_o^2\xi + s_o t_o \omega + t_o^2 \eta$ is a point on C(E) = C(B) its tangent line in P(E) is given by the equation

$$s_o^2\xi(p) + s_o t_o \omega(p) + t_o^2 \eta(p) = 0$$

where we put $\xi(p) \in k$ for $\xi \wedge p \in \Lambda^4 V$. If we denote $\sigma_o = (s_o^{n+1}, s_o^n t_o, \dots, t_o^{n+1})$ this implies by the special shape of the matrix R that

$$\sigma_o \circ R(p) = 0$$

for any point p on the tangent line. Now let $s_1^2\xi + s_1 t_1 \omega + t_1^2 \omega$ be another point on C(E) s.t. $(s_o, t_o), (s_1, t_1)$ are zeros of the same polynomial p_λ. Then the intersection point p_{o1} of the tangents satisfies

$$\sigma_o \circ R(p_{o1}) = 0 \ , \ \sigma_1 \circ R(p_{o1}) = 0.$$

Now by the definition of p_λ and by $p_\lambda(s_i, t_i) = 0$ we conclude (eliminate λ) that there is a non-trivial linear combination

$$b \circ A = c_o \sigma_o + c_1 \sigma_1.$$

This implies $b \circ A \circ R(p_{o1}) = 0$, i.e. rank $A \circ R(p_{o1}) < n$ or $p_{o1} \in S(E)$.

4.7 Corollary: *There is a bijection between the moduli space of special 't Hooft bundles E on $\mathbb{P}_3$ with $c_2 = n$ and the set of pairs (P,L) where $P \subset \mathbb{P}\Lambda^2 V$ is a plane s.t. $P \cap \mathbb{G}_2(V) = C$ is a regular conic and $L \subset \Gamma O_C(n+1)$ is a 2-dimensional linear system without base points.*

Proof: An n-instanton bundle E defines P(E), L(E) as above. If P,L are given, we can find a special matrix B as in 3. s.t. C = C(B). Now it is easy to see that any 2-dimensional subspace $L \subset \Gamma\mathcal{O}_{\mathbb{P}_1}(n+1)$ can be obtained as the span of polynomials p_λ defined by a rank-n matrix A as above, which is not necessarily persymmetric. Nevertheless the condition that L is base point free can easily be verified to be equivalent to the condition that the pair $(A\circ R, B)$ defines a monad and hence an n-instanton bundle E with $h^0E(1) = 2$. Then A can be replaced by a persymmetric A which is unique up to a scalar multiple.

4.8 Remark: One can give the space of pairs (P,L) directly an algebraic structure as a fibration over the open part of the Hilbert scheme of conics in $\mathbb{G}_2(V)$, consisting of regular conics, with fibre the open subset $G_2^0\Gamma\mathcal{O}_{\mathbb{P}_1}(n+1)$ consisting of systems without base points, and show that this is the coarse moduli scheme of special instanton bundles, see [6]. This can be achieved here directly using the matrices B and A with parameters.

4.9 Remark: A detailed characterization of the different types of jumping lines of a special instanton bundle, which also generalizes the case $c_2 = 2$ of [5], can be found in [3].

References.

[1] W. Barth, K. Hulek, Monads and Moduli of vector bundles, Manuscripta math. 25, 323-347 (1978)

[2] W. Böhmer, Monaden und Matrizen für Vektorbündel über $\mathbb{P}_n$, Dissertation,Kaiserslautern 1984

[3] W. Böhmer, Geometric and computational interpretation of instanton bundles on $\mathbb{P}_3$ with $h^0E(1) = 2$, Preprint

[4] L. Gruson, R. Lazarsfeld, C. Peskine, On a theorem of Castelnuovo and the equations defining space curves, Preprint

[5] R. Hartshorne, Stable vector bundles of rank 2 on $\mathbb{P}_3$, Math. Ann. 238, 229-280 (1978)

[6] A. Hirschowitz, M.S. Narasimhan, Fibrés de 't Hooft speciaux et applications, Proc. Nice Conf. 1981, Birkhäuser 1982

[7] M.S. Narasimhan, G. Trautmann, The compactification of the moduli scheme of stable rank 2 bundles on $\mathbb{P}_3$ with $c_1 = 0$, $c_2 = 2$, Prepr.

[8] Ch. Okonek, M. Schneider, H. Spindler, Vector bundles on complex projective spaces, Birkhäuser 1980

[9] G. Trautmann, Moduli for vector bundles on $\mathbb{P}_n(\mathbb{C})$, Math. Ann. 237, 167-186 (1978)

[10] G. Trautmann, Poncelet curves and associated theta characteristics, to appear in Expositiones Mathematicae.

Jean-Marc DREZET

GROUPE DE PICARD DES VARIÉTÉS DE MODULES DE FAISCEAUX SEMI-STABLES SUR $\mathbb{P}_2$

1 - INTRODUCTION

Soient $r \geq 2$, c_1, c_2 des entiers. On note $M(r,c_1,c_2) = M$ la variété de module des faisceaux semi-stables (au sens de Gieseker-Maruyama) sur $\mathbb{P}_2$ (= $\mathbb{P}_2(\mathbb{C})$), de rang r et de classes de Chern c_1, c_2. C'est une variété projective, intègre et normale.

On se propose d'étudier son groupe de Picard. Le premier résultat obtenu est le

THEOREME 1 : *La variété* $M(r,c_1,c_2)$ *est localement factorielle.*

Soient Cl(M) le groupe des classes d'équivalence linéaire de diviseurs de Weil de M, Pic(M) son groupe de Picard. La normalité de M implique que le morphisme canonique

$$\mathrm{Pic}(M) \longrightarrow \mathrm{Cl}(M)$$

est injectif. Le théorème 1 équivaut à sa surjectivité.

Nous verrons cependant (§9), suivant une idée de J. Le Potier, que le complèté de l'anneau local d'un point singulier de M peut très bien ne pas être factoriel.

Calculons maintenant Pic(M). Le résultat fait intervenir les conditions d'existence des fibrés stables ([8] , [6]). Il existe une unique fonction

$$\delta : \mathbb{Q} \longrightarrow \mathbb{Q}$$

possèdant la propriété suivante : pour tous entiers $r \geq 1$, c_1, c_2, on a $\dim(M(r,c_1,c_2)) > 0$ si et seulement si

$$\Delta = \frac{1}{r}(c_2 - \frac{r-1}{2r}c_1^2) \geq \delta(\frac{c_1}{r}) .$$

Posons $\mu = \frac{c_1}{r}$. On obtient alors le

THEOREME 2 : *Si* $\Delta = \delta(\mu)$, *on a* $\mathrm{Pic}(M) \simeq \mathbb{Z}$.

Si $\Delta > \delta(\mu)$, *on a* $\mathrm{Pic}(M) \simeq \mathbb{Z} \oplus \mathbb{Z}$.

Ce résultat a été prouvé par Strømme dans le cas où $r = 2$ et M est lisse ([20]). On peut ensuite décrire les éléments de Pic(M). Ils sont ici définis par une propriété universelle (dans [20], des générateurs de Pic(M) sont décrits sous forme de diviseurs premiers de M, correspondant à des faisceaux stables possèdant certaines propriétés géométriques).

Soit S une variété algébrique lisse et irréductible, F un faisceau cohérent sur $S \times \mathbb{P}_2$, tel que pour tout point s de S le faisceau F_s sur $\mathbb{P}_2$ soit semi-stable de rang r et de classes de Chern c_1, c_2. On déduit de F un morphisme

$$f_F : S \longrightarrow M .$$

Soient p_S, p_2 : $S \times \mathbb{P}_2 \longrightarrow S$, $S \times \mathbb{P}_2 \longrightarrow \mathbb{P}_2$ les projections. On définit successivement

$E_i = p_{S!}(F \otimes p_2^* \mathcal{O}(-i))$ pour $i = 0,1,2$ (ce sont des éléments du groupe de Grothendieck K(S) de S),

$l_i = \det(E_i)$, dans Pic(S) ,

et si $\alpha = (a_0, a_1, a_2)$ est un triplet d'entiers,

$L_{F,\alpha} = \bigotimes_{i=0}^{2} l_i^{\otimes a_i}$, qui est aussi un élément de Pic(S).

Posons

$\chi_i = rg(E_i)$ ($= \chi(E(-i))$ si E est un faisceau semi-stable sur $\mathbb{P}_2$, de rang r et de classes de Chern c_1, c_2). On a alors le

THEOREME 3 : *Si* $a_0\chi_0 + a_1\chi_1 + a_2\chi_2 = 0$, *il existe un unique élément* L_α *de* Pic(M) *tel que pour tout* F , *on ait*

$$f_F^* L_\alpha \simeq L_{F,\alpha} .$$

On obtient ainsi tous les éléments de Pic(M):

THEOREME 4 : *Posons* $H = \{ (a_0, a_1, a_2) \in \mathbb{Z}^3,\ a_0\chi_0 + a_1\chi_1 + a_2\chi_2 = 0 \}$. *Alors le morphisme*

$$\gamma : H \longrightarrow Pic(M)$$
$$\alpha \longmapsto L_\alpha$$

est surjectif. Si $\Delta > \delta(\mu)$, *c'est un isomorphisme.*

Le noyau de γ sera décrit au §6 dans le cas où $\Delta = \delta(\mu)$.

Les résultats précédents sont démontrés en détail dans [7] , et reposent en partie sur ceux de [8], [5], et [6].

On se contentera ici de donner les grandes lignes des démonstrations, en insistant sur les points les plus importants.

<u>Contenu des chapitres suivants</u>

Dans le §2 on donne les définitions des faisceaux stables et semi-stables et de leurs variétés de modules, ainsi que les propriétés les plus importantes de ces variétés.

Les démonstrations des théorèmes 1 à 4 sont esquissées dans le §3. Elles se ramènent en gros à celles de trois assertions qui seront éxaminées dans les chapitres suivants.

Le §4 donne une méthode de construction de fibrés en droites sur un bon quotient d'une variété lisse X par un groupe réductif G, à partir de caractères de G.

Cette méthode est est appliquée dans le §5 au cas où X est un espace de monades. Dans ce cas le quotient est justement une variété de modules de faisceaux semi-stables sur $\mathbb{P}_2$.

Le §6 traite le cas spécial des *variétés de modules de hauteur nulle.* Par définition, ce sont les variétés de modules telles que $\Delta = \delta(\mu)$.On peut donner une description assez précise de ces variétés.

Le §7 est consacré à l'étude de certaines variétés de modules pour lesquelles $\Delta > \delta(\mu)$.

Dans le §8 on démontre le résultat suivant : la sous-variété de M correspondant aux faisceaux semi-stables non stables est de codimension au moins 2, sauf dans certains cas particuliers, où M est isomorphe à $\mathbb{P}_5$. Ce résultat est essentiel à la démonstration du théorème 1.

Dans le §9, suivant une idée de J.Le Potier, on montre que le complété $\widehat{\mathcal{O}}_x$ de l'anneau local d'un point x de M peut très bien ne pas être factoriel, bien que $\mathcal{O}_x$ le soit d'après le théorème 1.

2 - FAISCEAUX SEMI-STABLES ET VARIETES DE MODULES

a - Notations

Soit E un faisceau cohérent sur $\mathbb{P}_2$, de rang $r > 0$ et de classes de Chern c_1, c_2 . On pose

$$\mu(E) = \frac{c_1}{r} \quad , \text{ qu'on appelle la } \textit{pente} \text{ de } E ,$$

$$\Delta(E) = \frac{1}{r}\left(c_2 - \frac{r-1}{2r} c_1^2\right) \quad , \text{ qu'on appelle le } \textit{discriminant} \text{ de } E.$$

Si $\chi(E)$ désigne la caractéristique d'Euler-Poincaré de E, la formule de Riemann-Roch s'écrit

$$\chi(E) = r(P(\mu(E)) - \Delta(E)),$$

où P est le polynôme

$$P(X) = \frac{X^2}{2} + \frac{3X}{2} + 1 .$$

On notera Q le fibré quotient canonique sur $\mathbb{P}_2$, c'est à dire le co-noyau du morphisme canonique

$$\mathcal{O}(-1) \longrightarrow \mathcal{O} \otimes H^0(\mathcal{O}(1))^* .$$

b - Faisceaux semi-stables

Un faisceau cohérent E sur $\mathbb{P}_2$ est dit *semi-stable* (resp. *stable*) s'il est sans torsion et si pour tout sous-faisceau cohérent propre F de E on a

$$\mu(F) \leqq \mu(E) ,$$

et en cas d'égalité

$$\Delta(F) \geqq \Delta(E) \qquad (\text{ resp. } >) .$$

Cette notion de stabilité est celle de Gieseker et Maruyama ([11] ,[18]).

c - Variétés de modules de faisceaux semi-stables

La *variété de modules* $M(r,c_1,c_2)$ est l'unique variété algébrique (à isomorphisme près) possèdant les propriétés suivantes :

Soit Φ le foncteur de la catégorie des variétés algébriques dans celle des ensembles associant à la variété S l'ensemble des classes d'équivalence de faisceaux cohérents F sur $S \times \mathbb{P}_2$ tels que pour tout point fermé s de S, F_s soit semi-stable, de rang r et de classes de Chern c_1, c_2. Deux tels faisceaux F,F' sont dits *équivalents* s'il existe un fibré en droites L sur S tel que $F' \simeq F \otimes p_S^* L$, p_S désignant la projection sur S. Alors

(i) Il existe un morphisme de foncteurs

$$\Phi \longrightarrow \mathrm{Hom}(-,M) .$$

(ii) Si $\Phi \longrightarrow \mathrm{Hom}(-,M')$ est un autre morphisme de foncteurs, il existe un unique morphisme $f : M \longrightarrow M'$ tel qu'on ait un diagramme commutatif

$$\begin{array}{ccc} & \nearrow & \mathrm{Hom}(-,M) \\ \Phi & & \big\downarrow \mathrm{Hom}(-,f) \\ & \searrow & \mathrm{Hom}(-,M') \end{array}$$

En particulier, d'après (i), d'un faisceau F induisant un élément de $\Phi(S)$ on déduit un morphisme

$$f_F : S \longrightarrow M .$$

La variété M est projective et normale ([11],[18]), et irréductible ([10],[8]).

d - Points fermés de $M(r,c_1,c_2)$

Soit E un faisceau semi-stable. Il existe une filtration

$$E_0 = 0 \subset E_1 \subset \dots \subset E_m = E$$

de E telle que pour $1 \leq i \leq m$, E_i/E_{i-1} soit stable, de mêmes pente et discriminant que E. Une telle filtration, pas obligatoirement unique, est dite de Jordan-Hölder. La classe d'isomorphisme du gradué $\bigoplus_{1 \leq i \leq m} E_i/E_{i-1}$ est indépendante de la filtration choisie, on la note Gr(E). Deux faisceaux semi-stables E,E' sont dits *équivalents* si Gr(E) = Gr(E').

Les points fermés de $M(r,c_1,c_2)$ s'identifient naturellement aux classes d'équivalence de faisceaux semi-stables de rang r et de classes de Chern c_1, c_2. Si F induit un élément de $\Phi(S)$, et si s est un point fermé de S, $f_F(s)$ n'est autre que $Gr(F_s)$.

L'ensemble des points de M correspondant aux faisceaux stables est un ouvert lisse, noté M_s, qui est non vide si dim(M) > 0. Les points fermés de M_s sont donc les classes d'isomorphisme de faisceaux stables de rang r et de classes de Chern c_1, c_2 .

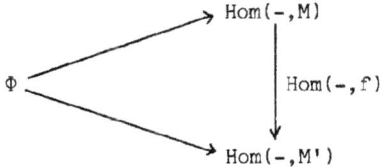

3 - PLAN DES DEMONSTRATIONS

a - Faisceaux semi-stables et monades

Pour démontrer les résultats du §1 on se ramène facilement au cas où $-r+1 < c_1 \leq 0$. Alors la suite spectrale de Beilinson montre que tout faisceau semi-stable E de rang r et de classes de Chern c_1, c_2 est isomorphe à la cohomologie en degré -1 d'un complexe du type

$$\Lambda^2 Q^* \otimes H_{-2} \xrightarrow{A} Q^* \otimes H_{-1} \xrightarrow{B} \mathcal{O} \otimes H_0 ,$$

H_0, H_{-1}, H_{-2} étant des $\mathbb{C}$-espaces vectoriels de dimension finie, le morphisme de faisceaux A étant injectif et B surjectif. De tels complexes sont appelés *monades* dans [2] , voir aussi [8], [14] . Un isomorphisme $E \simeq \mathrm{Ker}(B)/\mathrm{Im}(A)$ induit des isomorphismes

$$H_i \simeq H^1(E(i)) , \text{ pour } i = 0, -1, -2,$$

et par conséquent $\dim(H_i) = -\chi_{-i}$, qui ne dépend que de r, c_1, c_2. Soit $\mathcal{M}$ la variété des complexes du type précédent, avec A injectif, B surjectif, tels que le faisceau Ker(B)/Im(A) soit semi-stable. D'après [8] , c'est une variété lisse irréductible. Sur $\mathcal{M}$ agit de façon évidente le groupe réductif

$$G = (\prod_{i=0,-1,-2} GL(H_i))/\mathbb{C}^* .$$

Sur $\mathcal{M} \times \mathbb{P}_2$ existe une *monade universelle* dont la cohomologie V en degré -1 est une famille de faisceaux semi-stables de rang r et de classes de Chern c_1, c_2 paramétrée par $\mathcal{M}$. Le morphisme $f_V : \mathcal{M} \longrightarrow M$ est G-invariant et en fait on peut même montrer que c'est un bon quotient de $\mathcal{M}$ par G ([8]). Si $\mathcal{M}_s = f_V^{-1}(M_s)$, la restriction de $f_V : \mathcal{M}_s \longrightarrow M_s$ est un quotient géométrique et G agit librement sur $\mathcal{M}_s$. On en déduit comme dans [15] un morphisme de groupes

$$\pi_M : \mathrm{Char}(G) \longrightarrow \mathrm{Pic}(M_s) ,$$

Char(G) désignant le groupe des caractères de G. Si λ est un tel caractère, on a $\pi_M(\lambda) = (\mathcal{M}_s \times \mathbb{C})/G$, l'action de G sur $\mathcal{M}_s \times \mathbb{C}$ étant définie par

$$(g,(m,t)) \longmapsto (gm,\lambda(g)t) .$$

Les démonstrations des résultats du §1 reposent sur les assertions suivantes :

(i) Le morphisme π_M se factorise par Pic(M), c'est à dire qu'on a un diagramme commutatif

$$\begin{array}{ccc} \mathrm{Char}(G) & \xrightarrow{\ \pi_M\ } & \mathrm{Pic}(M_s) \\ & \searrow & \big\downarrow \text{restriction} \\ & & \mathrm{Pic}(M) \end{array} \quad .$$

(ii) Le morphisme π_M est surjectif.

(iii) Si M n'est pas isomorphe à $\mathbb{P}_5$, on a $\mathrm{codim}_M(M \setminus M_s) \geq 2$.

Dans le reste du §3 on supposera ces assertions démontrées.

b - Démonstration du théorème 1

Il s'agit de montrer que le morphisme canonique $\mathrm{Pic}(M) \longrightarrow \mathrm{Cl}(M)$ est surjectif. On peut supposer que M n'est pas isomorphe à $\mathbb{P}_5$. On a un diagramme commutatif

$$\begin{array}{ccc} \mathrm{Pic}(M) & \xrightarrow{\ \beta\ } & \mathrm{Cl}(M) \\ \big\downarrow r_1 & & \big\downarrow r_2 \\ \mathrm{Pic}(M_s) & \xrightarrow{\ \beta'\ } & \mathrm{Cl}(M_s) \end{array} \quad ,$$

où les flèches verticales sont les restrictions. D'après (i) et (ii), r_1 est surjectif, et β' est un isomorphisme car M_s est lisse. Donc $r_2 \circ \beta$ est surjectif. D'autre part, r_2 est un isomorphisme d'après (iii), donc β est surjectif, ce qui démontre le théorème 1.

c - Démonstration du théorème 2

Dans le cas où $\Delta = \delta(\mu)$, on montrera dans le §6 que si M_0 désigne l'ouvert des points lisses de M, on a $\mathrm{Pic}(M_0) \simeq \mathbb{Z}$. Compte tenu du théorème 1 on a $\mathrm{Pic}(M) \simeq \mathbb{Z}$.

Supposons maintenant que $\Delta > \delta(\mu)$. On a alors $\mathrm{Char}(G) \simeq \mathbb{Z}^2$. On a donc d'après (ii) un morphisme surjectif $\mathbb{Z}^2 \longrightarrow \mathrm{Pic}(M)$. Il suffit alors de trouver deux fibrés en droites indépendants sur M_s. Cela sera fait dans un cas particulier dans le §7.

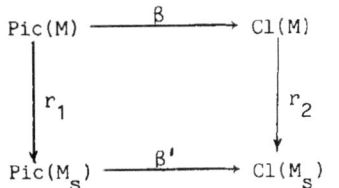

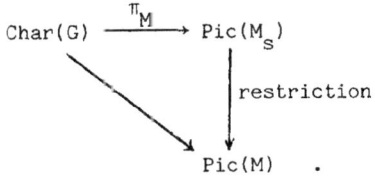

d - Démonstrations des théorèmes 3 et 4

Soient $\alpha = (a_0, a_1, a_2)$ un élément de H et λ le caractère de G défini par α, c'est à dire que

$$\lambda(\mathbb{C}^*(g_{-2}, g_{-1}, g_0)) = \prod_{i=0,1,2} \det(g_{-i})^{a_i} .$$

Soit D_λ le fibré en droites sur M déduit de λ, à l'aide de (i). Le G-fibré $f_V^* D_\lambda$ est isomorphe à

$$\prod_{i=0,1,2} (\Lambda^{-\chi_i} H_{-i})^{\otimes a_i} ,$$

muni de l'action évidente de G.

On va montrer que pour toute variété lisse irréductible S, et tout faisceau F sur $S \times \mathbb{P}_2$ tel que pour tout point s de S, F_s soit semi-stable de rang r et de classes de Chern c_1, c_2, on a

$$f_F^* D_\lambda \simeq L_{F,\alpha} ,$$

ce qui prouvera l'existence de L_α. Soient p_1, p_2 les projections $\mathcal{M} \times \mathbb{P}_2 \longrightarrow \mathcal{M}$, $\mathcal{M} \times \mathbb{P}_2 \longrightarrow \mathbb{P}_2$. On a des isomorphismes canoniques

$$R^1 p_{1*}(V \otimes p_2^* \mathcal{O}(-i)) \simeq \mathcal{O}_{\mathcal{M}} \otimes H_{-i} ,$$

pour $i=0,1,2$. D'après la définition de D_λ, on en déduit un isomorphisme canonique

$$f_V^* D_\lambda \simeq L_{V,\alpha} .$$

Soit s un point de S. Il existe des trivialisations sur un voisinage U(s) de s

$$(*) \qquad R^1 p_{S*}(F \otimes p_2^* \mathcal{O}(-i))_{|U(s)} \simeq \mathcal{O}_{U(s)} \otimes H_{-i} ,$$

d'où on déduit à l'aide de la suite spectrale de Beilinson un morphisme

$$\varphi_s : U(s) \longrightarrow \mathcal{M}$$

tel qu'on ait un diagramme commutatif

$$\begin{array}{ccc} U(s) & \xrightarrow{\varphi_s} & \mathcal{M} \\ & {\scriptstyle f_{F|U(s)}} \searrow \quad \swarrow {\scriptstyle f_V} & \\ & M & \end{array}$$

On déduit alors de φ_s un isomorphisme

$$(f_{F|U(s)})^* D_\lambda = \varphi_s^* f_V^* D_\lambda \simeq L_{F,\alpha|U(s)} .$$

Si on choisit une autre trivialisation que (*), le nouveau morphisme φ_s obtenu est le produit de l'ancien par un morphisme $U(s) \longrightarrow G$. Il en découle que l'isomorphisme $(f_{F|U(s)})^* D_\lambda \simeq L_{F,\alpha|U(s)}$ reste le même. Donc tous ces

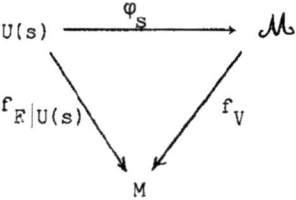

isomorphismes locaux se recollent pour donner un isomorphisme $f_V^*D_\lambda \simeq L_{F,\alpha}$.
Ceci prouve l'existence de L_α.

Pour montrer que L_α est unique il suffit de trouver un faisceau F comme précédemment, tel que $f_F^* : \mathrm{Pic}(M) \longrightarrow \mathrm{Pic}(S)$ soit injectif. On considère pour cela la famille de faisceaux F utilisée pour construire M ([11]) : S est une variété lisse sur laquelle opère un groupe du type PGL(N), N > 0, et f_F est PGL(N)-invariant. L'injectivité de f_F découle alors du fait que PGL(N) ne possède pas de caractère non trivial.

4 - FIBRES EN DROITES SUR UN QUOTIENT

Le résultat énoncé ici servira à construire des fibrés en droites sur M, sachant qu'on a un bon quotient $f_V : \mathcal{M} \longrightarrow M$ par G.

Plus généralement, soit X une variété algébrique lisse irréductible sur laquelle opère un groupe algébrique réductif connexe G_o. On suppose qu'il existe un bon quotient $\pi : X \longrightarrow N$. Soit λ un caractère de G_o. On en déduit une action de G_o sur $X \times \mathbb{C}$: $g(x,t) = (gx, \lambda(g)t)$. Notons $L_\lambda^!$ le G_o-fibré en droites sur X obtenu. Dans [15], on montre que si l'action de G_o sur X est libre et si π est un quotient géométrique, alors $L_\lambda^!/G_o$ est un fibré en droites sur N. On va voir qu'on peut déduire de λ un fibré en droites sur N sous des conditions plus générales que les précédentes.

Soit V_λ la réunion des ouverts V de N tels qu'il existe une fonction régulière λ^{-1}-*invariante* $f : \pi^{-1}(V) \longrightarrow \mathbb{C}^*$, c'est à dire telle que $f(gx) = \lambda(g)^{-1}f(x)$ pour tous x dans $\pi^{-1}(V)$ et g dans G_o. Soit $(f_i : \pi^{-1}(V_i) \longrightarrow \mathbb{C}^*)$ la famille de toutes ces fonctions λ^{-1}-invariantes. Pour tous i,j, la fonction $f_i/f_j : \pi^{-1}(V_i \cap V_j) \longrightarrow \mathbb{C}^*$ est G_o-invariante, donc définit une application régulière $V_i \cap V_j \longrightarrow \mathbb{C}^*$, aussi notée f_{ij}. Alors (f_{ij}) est une famille de cocycles définissant un fibré en droites L_λ sur V_λ. On a un isomorphisme canonique de G_o-fibrés en droites

$$L_\lambda^!|_{V_\lambda} \simeq (\pi|_{V_\lambda})^* L_\lambda .$$

Si l'action de G_o est libre et si π est un quotient géométrique, on a $V_\lambda = N$, et le fibré L_λ n'est autre que $L_\lambda^!/G_o$.

Dans le cas général on cherche sous quelles conditions on a $V_\lambda = N$. On voit aisément qu'il est nécessaire que pour tout point x_o de X, si G_{ox_o} désigne le stabilisateur de x_o, on ait $\lambda(G_{ox_o}) = \{1\}$. Cette condition ne semble pas suffisante, mais il est facile de voir qu'elle découle des conditions énoncées dans le théorème 5 ci-dessous.

Notons que si x_o est un point fermé de X, il existe une unique orbite fermée de X contenue dans $\overline{G_o x_o}$, c'est l'orbite de dimension minimale contenue dans $\overline{G_o x_o}$. On démontre alors le

<u>THEOREME 5</u> : *On a* $V_\lambda = N$ *si la condition suivante est réalisée : pour tout point* x_o *de* X, *soit* $G_o x$ *l'unique orbite fermée de* X *contenue dans* $\overline{G_o x_o}$. *Soit*

$$g : D - \{P\} \longrightarrow G_o$$
$$t \longmapsto g_t$$

un morphisme ((D,P) *étant un germe de courbe lisse*), *tel que*

$$\beta : D - \{P\} \longrightarrow \overline{G_o x_o}$$
$$t \longmapsto g_t x_o$$

admette un prolongement $\overline{\beta}$ *à* D *tel que* $\overline{\beta}(P) = x$. *Alors le morphisme*

$$\varphi = \lambda \circ g : D - \{P\} \longrightarrow \mathbb{C}^*$$

admet un prolongement $\overline{\varphi}$ *à* D , *et l'élément* $\overline{\varphi}(P)$ *de* $\mathbb{C}^*$ *est indépendant de* g .

Plus simplement on peut exprimer la condition du théorème 5 en disant que si $\lim\limits_{t \to P} g_t x_o = x$, alors $\lim\limits_{t \to P} \lambda(g_t)$ est un point de $\mathbb{C}^*$ indépendant de g.

5 - APPLICATION

On va appliquer le théorème 5 à l'espace des monades $\mathcal{M}$ du §3, sur lequel agit le groupe réductif $G = (GL(H_{-2}) \times GL(H_{-1}) \times GL(H_0))/\mathbb{C}^*$. Il faut montrer que tout caractère λ de G vérifie la condition du théorème 5. Un tel caractère est défini par un triplet (a_0, a_1, a_2) d'entiers tel que

$$a_0\chi_0 + a_1\chi_1 + a_2\chi_2 = 0 .$$

On a

$$\lambda(\mathbb{C}^*(g_{-2}, g_{-1}, g_0)) = \prod_{i=0,1,2} \det(g_{-i})^{a_i} .$$

Soient x_o un point de $\mathcal{M}$, Gx l'unique orbite fermée de $\mathcal{M}$ contenue dans $\overline{G_o x_o}$, (D,P) un germe de courbe lisse et

$$g : D - \{P\} \longrightarrow G$$
$$t \longmapsto g_t$$

un morphisme tel que

$$\beta : D - \{P\} \longrightarrow \mathcal{M}$$
$$t \longmapsto g_t x_o$$

admette un prolongement $\overline{\beta}$ à D, avec $\overline{\beta}(P) = x$. Soit $\varphi = \lambda \circ g : D - \{P\} \longrightarrow \mathbb{C}^*$. Soit

$$0 = z_0 \subset z_1 \subset \dots \subset z_m = x_o$$

une filtration de x_o par des sous-monades, correspondant à une filtration de

Jordan-Hölder de la cohomologie en degré -1 de x_0,

$$z_j : \Lambda^2 Q^* \otimes H^j_{-2} \longrightarrow Q^* \otimes H^j_{-1} \longrightarrow \mathcal{O} \otimes H^j_0 \quad .$$

Soit Γ_i la variété de drapeaux de H_i contenant $(H^1_i, \dots, H^m_i)$, pour $i=-2,-1,0$. Le groupe G agit sur $\Gamma = \Gamma_{-2} \times \Gamma_{-1} \times \Gamma_0$ de telle sorte que si $h \in G$, $h((H^j_i))$ soit le triplet de drapeaux associé à la filtration

$$0 = hz_0 \subset hz_1 \subset \dots \subset hz_m = hx_0 \; .$$

Puisque Γ est une variété complète, il existe un prolongement $\overline{\vartheta}$ à D du morphisme

$$\vartheta : D - \{P\} \longrightarrow \Gamma$$
$$t \longmapsto g_t((H^j_i)) \quad .$$

Il existe un morphisme $\sigma : D \longrightarrow G$ tel que pour tout $t \in D$ on ait

$$\sigma(t)\overline{\vartheta}(t) = ((H^j_i)) \quad ,$$

cela découle du fait que Γ s'identifie au quotient $G/\mathrm{Stab}((H_i))$. Alors, pour tout $t \in D$, $\sigma(t)\overline{\vartheta}(t)$ possède une sous-monade

$$z_j(t) : \Lambda^2 Q^* \otimes H_{-2} \longrightarrow Q^* \otimes H_{-1} \longrightarrow \mathcal{O} \otimes H_0 \quad .$$

Pour tout $t \in D - \{P\}$, $z_j(t)/z_{j-1}(t)$ est isomorphe à z_j/z_{j-1}. C'est à dire que si G_j désigne le groupe opérant sur l'espace de monades $\mathcal{M}_j$ contenant z_j/z_{j-1}, $z_j(t)/z_{j-1}(t)$ est un élément de $G_j z_j/z_{j-1}$. Il en est de même de $z_j(P)/z_{j-1}(P)$, car la cohomologie de z_j/z_{j-1} étant un faisceau stable, l'orbite de z_j/z_{j-1} est fermée dans $\mathcal{M}_j$. On a en fait $x \simeq \bigoplus_{1 \leq j \leq m} z_j/z_{j-1}$.

Le triplet (a_0,a_1,a_2) définit aussi un caractère λ_j de G_j, car on a

$$\frac{\dim(H^j_i/H^{j-1}_i)}{\dim(H^j_k/H^{j-1}_k)} = \frac{\dim(H_i)}{\dim(H_k)} \quad , \text{ pour } -2 \leq i,k \leq 0 \; .$$

Si $g_j(t)$ est l'élément de G_j induit par $\sigma(t)g_t$ on a

$$\varphi(t) = \lambda(g_t) = \lambda(\sigma(t))^{-1} \prod_{1 \leq j \leq m} \lambda_j(g_j(t)) \; .$$

Les morphismes g_j s'étendent en fait à D tout entier, car z_j/z_{j-1} ayant une cohomologie en degré -1 stable, on peut montrer que le morphisme

$$G_j \longrightarrow \mathcal{M}_j$$
$$g \longmapsto gz_j/z_{j-1}$$

est propre. Par conséquent on obtient une extension $\overline{\varphi}$ de φ à D, avec

$$\overline{\varphi}(P) = \lambda(\sigma(P))^{-1} \prod_{1 \leq j \leq m} \lambda_j(g_j(P)) \; .$$

Il reste à montrer que $\overline{\varphi}(P)$ ne dépend pas du choix de g. On remarque d'abord que $\overline{\varphi}(P)$ ne dépend que du drapeau $\overline{\vartheta}(P)$. Supposons qu'on ait un autre morphisme $g' : D - \{P\} \longrightarrow G$, possèdant la même propriété que g , et d'où on déduit φ', ϑ'. Il faut montrer que $\overline{\varphi}'(P) = \overline{\varphi}(P)$. La famille de drapeaux $\overline{\vartheta}'(P)$ est compatible avec une filtration de Jordan-Hölder de la cohomologie de x dont les gradués sont les mêmes que ceux déduits de $\overline{\vartheta}(P)$. Puisque x est une somme directe de monades dont les cohomologies sont stables, il existe un automorphisme de x (c'est à dire un automorphisme de la cohomologie de x), envoyant la première filtration sur la seconde. Cet automorphisme provient d'un élément g_o du stabilisateur de x. On a alors $g_o\overline{\vartheta}'(P) = \overline{\vartheta}(P)$. En remplaçant g' par $g_o g'$, on voit que

$$\overline{\varphi}'(P) = \lambda(g_o)^{-1}\overline{\varphi}(P) .$$

Il reste à montrer que $\lambda(g_o) = 1$.

Plus généralement, on prouvera que si g_o est un élément du stabilisateur d'un point quelconque y de $\mathcal{M}$, on a $\lambda(g_o) = 1$. Il suffit pour cela de montrer que si f_o est l'automorphisme de la cohomologie E de y déduit de g_o, il existe une filtration de Jordan-Hölder de E invariante par f_o. En effet, supposons cela établi. Soit $0 = z_o \subset z_1 \subset \ldots \subset z_m = y$ la filtration de y induite par la filtration précédente de E. De g_o on déduit un élément g_j du stabilisateur de z_j/z_{j-1} et de (a_0,a_1,a_2) un caractère λ_j du groupe opérant sur l'espace des monades contenant z_j/z_{j-1}. On a alors $\lambda(g_o) = \prod_{1\leq j\leq m}\lambda_j(g_j)$. Mais la cohomologie de z_j/z_{j-1} étant stable, tous les g_j sont égaux à l'identité, et par conséquent $\lambda(g_o) = 1$.

Il reste à montrer qu'il existe une filtration de Jordan-Hölder de E invariante par f_o. Considérons la filtration suivante de E :

$$0 = E_o \subset E_1 \subset \ldots\ldots \subset E_n = E ,$$

où pour $1\leq i\leq n$, E_i/E_{i-1} est la somme de tous les sous-faisceaux stables de E/E_{i-1} de mêmes pente et discriminant que E. Cette filtration est invariante par f_o. On cherche une filtration de Jordan-Hölder de E plus fine que la précédente. On est donc ramené au cas où E est une somme directe de faisceaux stables. On peut alors écrire

$$E = \bigoplus_{1\leq i\leq p} E_i' \otimes \mathbb{C}^{p_i} ,$$

où les E_i' sont des faisceaux stables de mêmes pente et discriminant que E, non isomorphes deux à deux. Les $E_i' \otimes \mathbb{C}^{p_i}$ sont stables par f_o, et la restriction de f_o à $E_i' \otimes \mathbb{C}^{p_i}$ est donnée par une matrice carrée $n_i \times n_i$. Il suffit donc de trigonaliser toutes ces matrices pour obtenir une filtration de Jordan-Hölder de E invariante par f_o.

6 - VARIETES DE MODULES DE HAUTEUR NULLE

Soit $M(\cdot,c_1,c_2)$ une variété de modules de hauteur nulle, c'est à dire telle que $\Delta = \delta(\mu)$. On va montrer que si M_o désigne l'ouvert des points lisses de M, on a $\mathrm{Pic}(M_o) \simeq \mathbb{Z}$. Pour cela, on utilise une description de M qui possède son intérêt propre.

a - Définition de l'application δ

Un fibré vectoriel algébrique F sur $\mathbb{P}_2$ est dit *exceptionnel* s'il est stable et si on a $\mathrm{Ext}^1(F,F) = 0$. En général, la dimension de la variété de modules contenant un faisceau stable E est égale à $\dim(\mathrm{Ext}^1(E,E))$. On voit donc que si F est un fibré exceptionnel, la variété de modules correspondant à F est réduite à un point. On en déduit qu'un fibré exceptionnel est homogène, c'est à dire invariant par PGL(2). En fait on peut montrer qu'un faisceau cohérent E sur $\mathbb{P}_2$ est un fibré exceptionnel si et seulement si E est simple et rigide (E est *simple* signifie que ses seuls endomorphismes sont les homothéties, et E est *rigide* que $\mathrm{Ext}^1(E,E) = 0$). Les exemples les plus simples de fibrés exceptionnels sont les fibrés $\mathcal{O}(k)$ et $Q(k)$, $k \in \mathbb{Z}$.

Un fibré exceptionnel est entièrement déterminé, à isomorphisme près, par sa pente. Dans [8] , on donne une description de l'ensemble des pentes des fibrés exceptionnels et on calcule aussi le rang et le discriminant d'un fibré exceptionnel, connaissant sa pente. Dans [5], on donne un procédé de récurrence pour construire tous les fibrés exceptionnels.

Soit μ un nombre rationnel. Il existe alors un fibré exceptionnel, unique à isomorphisme près, tel que $|\mu - \mu(F)| < x_F$, x_F étant la plus petite solution de l'équation

$$X^2 - 3X + 1/\mathrm{rg}(F)^2 = 0 .$$

Dans l'intervalle $]\mu(F) - x_F, \mu(F) + x_F[$, l'application δ est donnée par les formules suivantes :

$$\text{Si } \mu \leq \mu(F), \quad \delta(\mu) = P(\mu - \mu(F)) - \Delta(F) ,$$

$$\text{si } \mu \geq \mu(F), \quad \delta(\mu) = P(\mu(F) - \mu) - \Delta(F) ,$$

où P est le polynôme défini dans le §2.

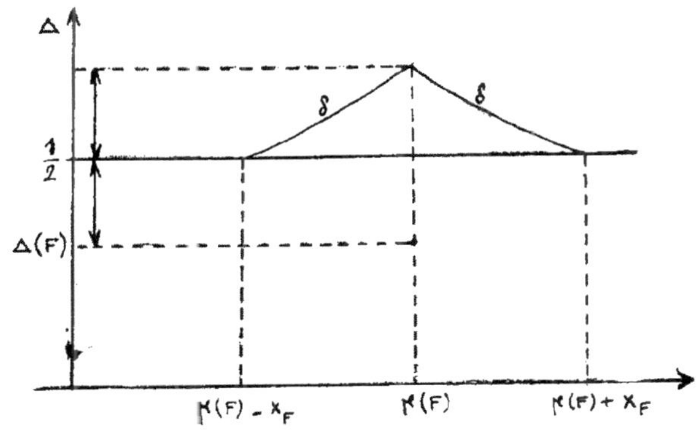

Pour montrer que ces formules sont plausibles, on se contentera de vérifier que si E est un fibré stable de pente μ, avec $\mu(F) - x_F < \mu < \mu(F)$, on a $\Delta(E) \geq \delta(\mu)$. D'après l'expression de la formule de Riemann-Roch du §2, on a

$$\chi(F^* \otimes E) = rg(E)rg(F)(P(\mu - \mu(F)) - \Delta(E) - \Delta(F)) ,$$

et il suffit donc de vérifier que $\chi(F^* \otimes E) \leq 0$. Pour cela il suffit de montrer que $h^0(F^* \otimes E) = 0$ et $h^2(F^* \otimes E) = 0$. La première égalité découle de la stabilité de E et F et du fait que $\mu(E) < \mu(F)$. Pour démontrer la seconde, notons qu'on a par dualité de Serre $h^2(F^* \otimes E) = h^0(F(-3) \otimes E^*)$, et $h^0(F(-3) \otimes E^*) = \dim(\mathrm{Hom}(E,F(-3))) = 0$, car E et F sont stables et $\mu(E) > \mu(F(-3))$ car $x_F < 1$.

b - Variétés de modules de hauteur nulle et modules de Kronecker

Soit $M(r,c_1,c_2)$ une variété de modules de hauteur nulle et F le fibré exceptionnel associé à $M(r,c_1,c_2)$ dans a-. On définit dans [5] des triplets particuliers de fibrés exceptionnels, appelés *triades*. Si (U,V,F) est une triade, et E un faisceau semi-stable de rang r et de classes de Chern c_1,c_2, une généralisation de la suite spectrale de Beilinson permet de prouver le résultat suivant : si $\mu \leq \mu(F)$, il existe des entiers m,n, dépendant uniquement de r,c_1,c_2 et un morphisme injectif de faisceaux

$$\varphi : U \otimes \mathbb{C}^m \longrightarrow V \otimes \mathbb{C}^n$$

dont le conoyau soit isomorphe à E. La donnée d'un morphisme tel que φ équivaut à celle d'une application linéaire

$$\psi : \mathbb{C}^m \otimes \mathrm{Hom}(U,V)^* \longrightarrow \mathbb{C}^n \quad .$$

De telles applications linéaires sont appelées *modules de Kronecker* dans [13].

Posons $q = \dim(\mathrm{Hom}(U,V))$ (en fait on a $q = 3rg(F)$). Sur l'espace projectif $\mathbb{P}$ des droites de $L(\mathbb{C}^m \otimes \mathrm{Hom}(U,V)^*,\mathbb{C}^n)$ agit le groupe réductif $G_0 = SL(m) \times SL(n)$ de la façon suivante :

$$((g_1,g_2),\psi) \longmapsto (g_2 \circ \psi \circ (g_1^{-1} \otimes I)) \quad .$$

Cette action se linéarise de manière évidente et on peut donc définir une notion de (semi-)stabilité pour les points de $\mathbb{P}$ (Cf.[19]). Soit $\mathbb{P}^{ss}$ l'ouvert des points semi-stables de $\mathbb{P}$, et $N(q,m,n) = \mathbb{P}^{ss}/G_0$. C'est une variété algébrique projective irréductible et normale.

En associant à la classe d'équivalence de E la G_0-orbite de ψ, on définit un isomorphisme

$$M(r,c_1,c_2) \simeq N(q,m,n) \ .$$

C'est cet isomorphisme qui permet de calculer $\mathrm{Pic}(M_0)$. Le cas où $\mu > \mu(F)$ est analogue.

c - Périodicité

Soient m,n des entiers tels que $\dim(N(q,m,n)) > 0$. On a alors $qn - m > 0$ et $N(q,n,qn-m)$ est isomorphe à $N(q,m,n)$. L'isomorphisme s'obtient de la façon suivante : soit

$$\varphi : \mathbb{C}^m \otimes \mathrm{Hom}(U,V)^* \longrightarrow \mathbb{C}^n$$

un module de Kronecker semi-stable. On en déduit une application linéaire

$$\omega : \mathbb{C}^m \longrightarrow \mathbb{C}^n \otimes \mathrm{Hom}(U,V) ,$$

qui est injective. Son conoyau est donc de dimension $qn - m$. Soit β un isomorphisme $\mathrm{Coker}(\omega) \simeq \mathbb{C}^{qn-m}$. Alors à l'image de φ on associe celle de

$$\beta \circ p : \mathbb{C}^n \otimes \mathrm{Hom}(U,V) \longrightarrow \mathbb{C}^{qn-m}$$

dans $N(q,n,qn-m)$, p étant l'application quotient (une fois prouvée bien entendu la semi-stabilité de $\beta \circ p$).

Supposons que $N(q,m,n)$ corresponde à $M(r,c_1,c_2)$ comme dans b-. Alors $N(q,n,qn-m)$ correspond à une autre variété de modules de hauteur nulle $M(r',c_1',c_2')$ avec

$$\mu(F) - x_F < \frac{c_1'}{r'} < \frac{c_1}{r} .$$

On peut itérer ce procédé et on obtient ainsi une infinité de variétés de modules isomorphes à $M(r,c_1,c_2)$.

Exemples : En prenant $F = \mathcal{O}$ et en considérant $N(3,1,1)$ on obtient

$M(1,0,1) \simeq M(3,-1,2) \simeq M(8,-3,8) \simeq \ldots . \simeq M(377,-144,10529) \simeq \ldots . \simeq \mathbb{P}_2$.

En prenant $F = \mathcal{O}$ et en considérant $N(3,2,2)$ on obtient

$M(2,0,2) \simeq M(6,-2,5) \simeq M(16,-6,25) \simeq \ldots \simeq \mathbb{P}_5$.

En prenant $F = Q^*$ et en considérant $N(6,1,1)$ on obtient

$M(4,-2,4) \simeq M(24,-10,60) \simeq \ldots . \simeq M(4756,-1970,1942420) \simeq \mathbb{P}_5$.

d - Groupe de Picard

Pour montrer que $\mathrm{Pic}(M_o) \simeq \mathbb{Z}$, il suffit d'après b- de prouver que si m,n sont des entiers tels que $\dim(N(q,m,n)) > 0$, et si N_o désigne l'ouvert des points lisses de $N(q,m,n)$, on a $\mathrm{Pic}(N_o) \simeq \mathbb{Z}$.

Soit $N_s(q,m,n)$ l'ouvert de $N(q,m,n)$ correspondant aux modules de Kronecker stables. Posons $W = L(\mathbb{C}^m \otimes \mathbb{C}^q, \mathbb{C}^n)$, et soit W^s l'ouvert des points stables de W. Le groupe réductif $G_1 = (GL(m) \times GL(n))/\mathbb{C}^*$ agit sur W, induisant l'action de G_o sur $\mathbb{P}$. Le morphisme canonique $W^s \longrightarrow N_s(q,m,n)$ est un quotient géométrique, et G_1 agit librement sur W^s. On en déduit comme dans [15] et le §3 une suite exacte de groupes abéliens

$$\mathcal{O}^*(W^s) \longrightarrow \mathrm{Char}(G_1) \longrightarrow \mathrm{Pic}(N_s(q,m,n)) \longrightarrow \mathrm{Pic}(W^s) .$$

Puisque $\mathrm{Char}(G_1) \simeq \mathbb{Z}$ et $\mathrm{Pic}(W^s) = 0$, il reste à montrer que $\mathcal{O}^*(W^s) = \mathbb{C}^*$. Ceci est vrai si et seulement si $\mathrm{codim}_W(W \backslash W^s) \geq 2$.

Soient z,z' les solutions de l'équation $2X^2 - qX + 1 = 0$, avec $z < z'$. On peut montrer que si

$$z \leq \frac{m}{n} \leq z' ,$$

et si $m \neq 2$ dans le cas où $z' = m/n$, on a $\mathrm{codim}_W(W \backslash W^s) \geq 2$. D'après c-, on peut toujours se ramener au cas où les inégalités précédentes sont vérifiées, car il est facile de voir que

$$z \leq \frac{1}{q - z'} \leq z' .$$

L'exception $m = 2$, $z' = m/n$ ne se produit que dans le cas où $q = 3$ et $(m,n) = (2,2)$ (et alors F est un fibré en droites). Les variétés de modules rencontrées dans ces cas sont isomorphes à $\mathbb{P}_5$, par exemple si F est le fibré trivial, $N(3,2,2)$ correspond à la variété de modules $M(2,0,2)$. On a dans ces cas $M_0 \simeq M(r,c_1,c_2)$, et bien évidemment, $\mathrm{Pic}(M_0) \simeq \mathbb{Z}$, mais on ne peut parvenir à ce résultat en calculant $\mathrm{Pic}(M_s)$, car $M \backslash M_s$ est une hypersurface.

<u>Remarque</u> : L'isomorphisme $M(r,c_1,c_2) \simeq N(q,m,n)$ permet aussi de calculer la cohomologie entière de $M(r,c_1,c_2)$ (si cette variété est lisse), en utilisant la la méthode de Atiyah-Bott ([1],[16]). On démontre ainsi que $H^\cdot(M(r,c_1,c_2),\mathbb{Z})$ est sans torsion, nulle en degré impair, et on obtient des formules de récurrence pour calculer les nombres de Betti de $M(r,c_1,c_2)$. Ainsi, la variété de modules de hauteur nulle lisse la plus simple non isomorphe à une grasmannienne est $M(4,-1,3)$, qui est de dimension 6. On obtient pour cette variété

$$b_2 = 1, \quad b_4 = 3, \quad b_6 = 3 .$$

e - <u>Détermination du noyau de γ</u>

Rappelons que γ désigne le morphisme $H \longrightarrow \mathrm{Pic}(M)$ défini au §1 et au §3,d-, H désignant le sous-groupe de $\mathbb{Z}^3$ constitué des triplets (a_0,a_1,a_2) tels que $a_0\chi_0 + a_1\chi_1 + a_2\chi_2 = 0$. On a en général $H \simeq \mathbb{Z}^2$ (et $H \simeq \mathbb{Z}$ si χ_0 ou χ_2 est nul). Dans notre cas, M est de hauteur nulle et si $H \simeq \mathbb{Z}^2$, le noyau de γ est isomorphe à $\mathbb{Z}$. On va trouver un générateur de ce noyau.

On utilise une version à paramètres de la suite spectrale de Beilinson. Soit S une variété algébrique lisse et irréductible, E un faisceau cohérent sur $S \times \mathbb{P}_2$, tel que pour tout point s de S, E_s soit semi-stable de rang r eet de classes de Chern c_1, c_2. Il existe une suite spectrale de faisceaux cohérents sur $S \times \mathbb{P}_2$, convergeant vers E en degré 0, vers 0 en les autres degrés, de terme E_1

$$E_1^{p,q} = p_S^*(R^c p_{S*}(E \otimes p_2^* \mathcal{O}(-p))) \otimes p_2^* \Lambda^p Q^* .$$

Supposons que $\mu(F) - x_F < \mu < \mu(F)$ (l'autre cas est analogue). D'après a-, on a alors, pour tout faisceau semi-stable E_o sur $\mathbb{P}_2$, de rang r et de classes de Chern c_1, c_2, $h^i(F^* \otimes E_o) = 0$ pour tout $i \geq 0$. Il en découle que dans K(S) on a $p_{S!}(E \otimes p_2^* F^*) = 0$. Le foncteur $p_{S!}$ étant additif on obtient

$$\sum (-1)^{p+q} p_{S!}(E_1^{p,q} \otimes p_2^* F^*) = 0$$

dans K(S), c'est à dire

$$0 = \sum (-1)^{p+q} (F^* \otimes \Lambda^p Q^*) R^q p_{S*}(E \otimes p_2^* \mathcal{O}(-p)) ,$$

$$\text{(A)} \qquad = \sum (-1)^p (F^* \otimes \Lambda^p Q^*) p_{S!}(E \otimes p_2^* \mathcal{O}(-p)) .$$

Posons

$$\alpha_o = (\chi(F^*), -\chi(F^* \otimes Q^*), \chi(F^*(-1))) .$$

En prenant le rang de (A), on voit que α_o est un élément de H, et en considérant son déterminant, que $\gamma(\alpha_o) = 0$.

Pour montrer que α_o engendre Ker(γ), il suffit de vérifier que $\chi(F^*)$, $\chi(F^* \otimes Q^*)$ et $\chi(F^*(-1))$ sont premiers entre eux. Considérons la suite spectrale de Beilinson $E_r^{p,q}$ de faisceaux cohérents sur $\mathbb{P}_2$, convergeant vers F^* en degré 0 et vers 0 en les autres degrés, de terme E_1

$$E_1^{p,q} = \mathcal{O}(p) \otimes H^q(F^* \otimes \Lambda^{-p} Q^*) .$$

On en déduit immédiatement que

$$rg(F^*) = -\chi(F^* \otimes Q^*) + \chi(F^*) + \chi(F^*(-1)) ,$$

$$c_1(F^*) = \chi(F^* \otimes Q^*) - 2\chi(F^*(-1)) .$$

Puisque $rg(F^*)$ et $c_1(F^*)$ sont premiers entre eux d'après [8], il en est de même de $\chi(F^*)$, $\chi(F^* \otimes Q^*)$ et $\chi(F^*(-1))$.

7 - ETUDE D'UN CAS PARTICULIER

On démontre ici que le morphisme canonique

$$\pi_M : \mathrm{Char}(G) \longrightarrow \mathrm{Pic}(M(r,c_1,c_2))$$

est un isomorphisme dans les cas où $-r+1 < c_1 < -1$, et où c_1, $r+c_1$ ne sont pas des diviseurs de $-c_2 + c_1(c_1+1)/2$. Pour cela, on utilise la description d'Ellingsrud de certains ouverts de M_s (cf. [10]).

a - La construction d'Ellingsrud

Si E est un fibré vectoriel sur $\mathbb{P}_1$, il existe une unique suite d'entiers $(k_1,\dots,k_r)$ telle que $E \simeq \mathcal{O}(k_1) \oplus \dots \oplus \mathcal{O}(k_r)$, et que $k_1 \geq \dots \geq k_r$. On l'appelle le *type de décomposition* de E (cf. Grothendieck [12]). On dit que E est *rigide* si $\mathrm{Ext}^1(E,E) = 0$. Cela revient à dire que si $1 \leq i,j \leq r$, on a $|k_i - k_j| \leq 1$.

Soit x un point de $\mathbb{P}_2$, U_x l'ouvert de M_s constitué des classes d'isomorphisme de fibrés vectoriels stables de rang r et de classes de Chern c_1, c_2 dont la restriction à chaque droite de $\mathbb{P}_2$ passant par x est rigide. D'après Brun-Hirschowitz ([4]), on a $\mathrm{codim}_{M_s}(M_s \setminus U_x) \geq 2$.

Soit P_x l'éclatement de $\mathbb{P}_2$ en x, c'est la sous-variété de $\mathbb{P}_2 \times \mathbb{P}_2^*$ constituée des couples (y,l), où l est une droite de $\mathbb{P}_2$ contenant x et y. Soient $p : P_x \longrightarrow \mathbb{P}_2$ et $q : P_x \longrightarrow L$ les projections, L désignant la droite de $\mathbb{P}_2^*$ définie par x.

Soit E un fibré vectoriel stable sur $\mathbb{P}_2$, de rang r et de classes de Chern c_1, c_2, tel que le type de décomposition de la restriction de E à chaque droite passant par x soit $(0,\ldots,0,-1,\ldots,-1)$. On a alors, en posant

$$A = q_*p^*E, \quad B = (q_*p^*(E^*(-1)))^* \quad ,$$

une suite exacte sur P_x

$$0 \longrightarrow q^*A \longrightarrow p^*E \longrightarrow q^*B \otimes p^*\mathcal{O}(-1) \longrightarrow 0 \quad .$$

On a

$$rg(A) = r + c_1 \quad , \quad rg(B) = -c_1 \ ,$$
$$c_1(A) = -c_2 + c_1(c_1+1)/2 = -c_1(B) \quad .$$

De plus, le type de décomposition de A (resp. B) est constitué d'entiers négatifs (resp. positifs).

Réciproquement, soient A,B des fibrés vectoriels sur L possèdant les propriétés précédentes. Les extensions

$$0 \longrightarrow q^*A \longrightarrow E_0 \longrightarrow q^*B \otimes p^*\mathcal{O}(-1) \longrightarrow 0$$

sont classifiées par $W = \mathrm{Ext}^1_{\mathcal{O}_{P_x}}(q^*B \otimes p^*\mathcal{O}(-1), q^*A)$. Sur $W \times P_x$ existe une *extension universelle* :

$$0 \longrightarrow p_1^*q^*A \longrightarrow F \longrightarrow p_1^*(q^*B \otimes p^*\mathcal{O}(-1)) \longrightarrow 0 \ ,$$

p_1 désignant la projection $W \times P_x \longrightarrow P_x$. Sur W agit le groupe $G_1 = (\mathrm{Aut}(A) \times \mathrm{Aut}(B))/\mathbb{C}^*$, et pour tous w,w' dans W, on a $F_w \simeq F_{w'}$ si et seulement si w et w' sont dans la même orbite. Soit W^s l'ouvert de W des points w tels que F_w provienne d'un fibré vectoriel sur $\mathbb{P}_2$ (c'est à dire que $F_w|_{p^{-1}(x)}$ soit trivial), et que ce fibré soit stable. De F on déduit alors un fibré vectoriel F' sur $W^s \times \mathbb{P}_2$. Les fibres du morphisme canonique $f_{F'} : W^s \longrightarrow M_s$ sont précisément les orbites de l'action de G_1. Ceci permet de démontrer l'égalité

$$\dim(\overline{\mathrm{Im}(f_{F'})}) = \dim(M) - \dim(\mathrm{Ext}^1(A,A)) - \dim(\mathrm{Ext}^1(B,B)),$$

pourvu que W^s ne soit pas vide. Si A et B sont rigides, $\mathrm{Im}(f_{F'})$ est un ouvert non vide U_o de M_s. Si un au moins des fibrés A,B n'est pas rigide, puisque rg(A) ne divise pas $c_1(A)$ et rg(B) ne divise pas $c_1(B)$, on a

$$\dim(\mathrm{Ext}^1(A,A)) + \dim(\mathrm{Ext}^1(B,B)) \geq 2 ,$$

donc, U_x étant la réunion de U_o et d'un nombre fini de sous-variétés du type $\overline{\mathrm{Im}(f_{F'})}$, avec A ou B non rigide, on a $\mathrm{codim}_{M_s}(U_x \setminus U_o) \geq 2$, et par conséquent $\mathrm{codim}_{M_s}(M_s \setminus U_o) \geq 2$. Il suffit donc de montrer que le morphisme canonique

$$\pi' : \mathrm{Char}(G) \longrightarrow \mathrm{Pic}(U_o)$$

est un isomorphisme. Ce morphisme s'obtient en considérant l'action de G sur $f_V^{-1}(U_o) \subset \mathcal{M}$.

Dans tout ce qui suit on supposera que A et B sont rigides.

b - Etude de W^s

Soit H la sous-variété fermée de W constituée des points w tels que F_w ne provienne pas de $\mathbb{P}_2$, c'est à dire que $F_{w|p^{-1}(x)}$ ne soit pas trivial.

LEMME 6 : *La sous-variété* H *est une hypersurface irréductible de* W .

Soit w un point de W. La fibre $p^{-1}(x)$ est isomorphe à L, et la restriction à $p^{-1}(x)$ de l'extension définie par w donne une suite exacte

$$0 \longrightarrow A \longrightarrow F_{w/p^{-1}(x)} \longrightarrow B \longrightarrow 0 .$$

Le fibré $F_{w/p^{-1}(x)}$ est trivial si et seulement si l'application

$$D(w) : H^0(B(-1)) \longrightarrow H^1(A(-1))$$

déduite de la suite exacte précédente est bijective. Donc H est l'ensemble des points w de W vérifiant l'équation $\det(D(w)) = 0$. C'est par conséquent une sous-variété équidimensionnelle de codimension 1 de W.

Pour montrer qu'elle est irréductible, il suffit de prouver qu'elle est lisse en dehors d'une sous-variété de H de codimension au moins 3 dans W. Pour cela, on remarque d'abord que le morphisme de Kodaïra-Spencer de F en chaque point w de W

$$\varphi : W \longrightarrow \mathrm{Ext}^1(F_w,F_w)$$

est surjectif (en général, c'est à dire si A et B ne sont plus nécessairement rigides, on peut montrer que $\mathrm{Coker}(\varphi) \simeq \mathrm{Ext}^1(A,A) \oplus \mathrm{Ext}^1(B,B)$, ce qui permet de retrouver la formule donnant la dimension de $\overline{\mathrm{Im}(f_{F'})}$). D'autre part, le morphisme canonique

$$\mathrm{Ext}^1(F_w,F_w) \longrightarrow \mathrm{Ext}^1(F_{w|p^{-1}(x)},F_{w|p^{-1}(x)})$$

est aussi surjectif. En effet, il suffit de prouver que si J désigne le faisceau d'idéaux de $p^{-1}(x)$, on a

$$\mathrm{Ext}^2(F_w, F_w \otimes J) = 0 .$$

Par dualité de Serre on a

$$\mathrm{Ext}^2(F_w, F_w \otimes J) \simeq \mathrm{Hom}(F_w, F_w \otimes J^* \otimes \omega_{P_x})^* ,$$

et ce dernier est nul, à cause du type de décomposition des restrictions de F_w aux droites de $\mathbb{P}_2$ ne passant pas par x (on a $J^* \otimes \omega_{P_x} \simeq p^*\mathcal{O}(-1) \otimes q^*\mathcal{O}(-2)$). En définitive, le morphisme de Kodaïra-Spencer de la restriction de F_w à $W \times p^{-1}(x)$ est surjectif, autrement dit cette restriction est une famille complète de fibrés sur $p^{-1}(x)$. Posons $E_o = \mathcal{O}_L(1) \oplus (r-2)\mathcal{O}_L \oplus \mathcal{O}_L(-1)$. Il résulte de la théorie des déformations des fibrés sur $\mathbb{P}_1$ (Brieskhorn [3]), que l'ensemble des points w de H tels que $F_w \simeq E_o$ est un ouvert lisse de H. Il suffit donc de montrer que le complémentaire de cet ouvert est de codimension au moins 3 dans W. Cela découle du fait que si $F_{w|p^{-1}(x)}$ n'est ni trivial ni isomorphe à E_o, on a

$$\dim(\mathrm{Ext}^1(F_{w|p^{-1}(x)}, F_{w|p^{-1}(x)})) \geq 3 .$$

Ceci achève la démonstration du lemme 6.

<u>LEMME 7</u> : *On a* $\mathrm{codim}_W((W\backslash H) - W^s) \geq 2$.

Cela découle d'un résultat plus général dû à J.Le Potier : soit E une famille complète de faisceaux cohérents sur $\mathbb{P}_2$, de rang r et de classes de Chern c_1, c_2, paramétrée par une variété algébrique lisse irréductible T. On suppose que pour tout point t de T on ait $\mathrm{Ext}^2(E_t, E_t) = 0$, et qu'il existe une droite de $\mathbb{P}_2$ sur laquelle E_t soit rigide. Alors, si T^{ss} désigne l'ouvert de T des points t tels que E_t soit semi-stable, on a $\mathrm{codim}_T(T\backslash T^{ss}) \geq 2$. Pour ce résultat l'hypothèse $\Delta > \delta(\mu)$ est essentielle.

On déduit des lemmes précédents un isomorphisme $\mathcal{O}^*(W^s)/\mathbb{C}^* \simeq \mathbb{Z}$, un générateur étant donné par une équation irréductible de H. Posons

$$A = a\mathcal{O}(k) \oplus a_o\mathcal{O}(k+1) , \quad B = b\mathcal{O}(k') \oplus b_o\mathcal{O}(k'+1) ,$$

avec a,b,a_o,b_o strictement positifs. Tout automorphisme de A provient d'une matrice

$$\begin{bmatrix} X & Y \\ 0 & Z \end{bmatrix} ,$$

avec X dans GL(a), Y dans $GL(a_o)$, Z étant un morphisme $a\mathcal{O}(k) \longrightarrow a_o\mathcal{O}(k+1)$. On en déduit aisément que $\mathrm{Char}(\mathrm{Aut}(A)) = \mathrm{Char}(GL(a) \times GL(a_o))$, et de même, $\mathrm{Char}(\mathrm{Aut}(B)) = \mathrm{Char}(GL(b) \times GL(b_o))$. Donc $\mathrm{Char}(G_1)$ est constitué des morphismes

$$\varphi_{s,t,u,v} : (\mathrm{Aut}(A) \times \mathrm{Aut}(B))/\mathbb{C}^* \longrightarrow \mathbb{C}^* ,$$

$$\mathbb{C}^*\left(\begin{bmatrix} X & Z \\ 0 & Y \end{bmatrix}, \begin{bmatrix} X' & Z' \\ 0 & Y' \end{bmatrix}\right) \longmapsto \det(X)^s\det(Y)^t\det(X')^u\det(Y')^v ,$$

s,t,u,v étant des entiers tels que $sa + ta_o + ub + vb_o = 0$. Donc $Char(G_1)$ est isomorphe à $\mathbb{Z}^3$. A toute fonction inversible $f : W^s \longrightarrow \mathbb{C}^*$ correspond un caractère de G : $\lambda(g) = f(gw)/f(w)$, qui ne dépend pas du point w de W^s. Le caractère associé à l'équation $\det(D(w)) = 0$ de H rencontrée dans la démonstration du lemme 6 est $\varphi_{-k,-k-1,-k',-k'-1}$, qui n'est pas une puissance non triviale d'un autre caractère. Il en découle que $\det(D(w)) = 0$ est une équation irréductible de H. On notera plus simplement $\chi = \varphi_{-k,-k-1,-k',-k'-1}$.

c - Groupe de Picard de U_o

Le groupe G_1 agit librement sur W^s. Il est donc tentant d'en déduire une suite exacte

$$\mathcal{O}^*(W^s) \longrightarrow Char(G_1) \xrightarrow{\vartheta} Pic(U_o) \xrightarrow{f^*_{F'}} Pic(W^s) = 0 \quad .$$

Malheureusement, G_1 n'étant pas réductif, le morphisme ϑ ne peut pas être défini a priori. Pour le construire effectivement, on établit un lien entre W^s et l'espace $\mathcal{M}$ des monades du §3.

Soit w un point de W^s. De la suite exacte

$$0 \longrightarrow q^*A \longrightarrow p^*F'_w \longrightarrow q^*B \otimes p^*\mathcal{O}(-1) \longrightarrow 0$$

on déduit des isomorphismes

(1) $H^1(F'_w(-2)) \simeq H^0(B(-2))$, $H^1(F'_w) \simeq H^1(A)$, $H^1(F'_w(-1)) \simeq H^0(B(-1))$.

Supposons fixés des isomorphismes

(2) $H_{-2} \simeq H^0(B(-2))$, $H_{-1} \simeq H^0(B(-1))$, $H_o \simeq H^1(A)$.

On en déduit un morphisme $\Phi : W^s \longrightarrow \mathcal{M}$. Si w est un point de W^s, la monade $\Phi(w)$ est obtenue en considérant d'abord la monade canonique associée à F'_w par la suite spectrale de Beilinson, d'où on déduit un élément de $\mathcal{M}$ en composant avec les isomorphismes (1) et (2). Le morphisme Φ est compatible avec un morphisme de groupes algébriques $G_1 \longrightarrow G$ induisant $\overline{\Psi} : Char(G) \longrightarrow Char(G_1)/\mathbb{Z}\chi$.

Un calcul aisé montre que $\overline{\Psi}$ est surjectif. En utilisant le fait que le morphisme $Char(G) \longrightarrow Pic(U_o)$ est lui parfaitement défini, on peut alors montrer que ϑ l'est. On a alors un diagramme commutatif

$$\begin{array}{ccc} \mathbb{Z}^2 \simeq Char(G) & \xrightarrow{\pi'} & Pic(U_o) \\ \downarrow & & \| \\ \mathbb{Z}^2 \simeq Char(G_1)/\mathbb{Z}\chi & \longrightarrow & Pic(U_o) \end{array} \quad .$$

Il en découle que π' est un isomorphisme.

8 - FAISCEAUX SEMI-STABLES NON STABLES

On donne ici des éclaircissements sur la démonstration de l'assertion (iii) du §3,a- : si M n'est pas isomorphe à $\mathbb{P}_5$, on a $\mathrm{codim}_M(M\setminus M_s) \geq 2$. C'est une conséquence du résultat suivant :

PROPOSITION 8 : *Soit* F *une famille complète de faisceaux semi-stables sur* $\mathbb{P}_2$, *de rang* r *et de classes de Chern* c_1, c_2, *paramétrée par une variété algébrique lisse irréductible* X. *Soit* X^s *l'ouvert de* X *des points* x *tels que* F_x *soit stable. Alors, si* $M(r,c_1,c_2)$ *n'est pas isomorphe à* $\mathbb{P}_5$, *on a* $\mathrm{codim}_X(X\setminus X^s) \geq 2$.

Rappelons que "F est *complète* " signifie qu'en tout point x de X, le morphisme de Kodaïra-Spencer de F

$$\varphi : T_x X \longrightarrow \mathrm{Ext}^1(F_x, F_x)$$

est surjectif.

Le théorème (4.10) de [8] assure que M_s est dense dans M. Dans la démonstration de ce théorème est prouvée l'inégalité

$$\mathrm{codim}_X(X\setminus X^s) \geq \mathrm{Inf}_{(r_i)}\Big(\sum_{i<j} r_i r_j (2\Delta - 1) \Big) = \varepsilon ,$$

(r_i) parcourant l'ensemble S des suites $(r_1,\ldots,r_k)$, $k > 1$, d'entiers positifs de somme r, telles que pour $1 \leq i \leq r$, $r_i\mu$ et $r_i(P(\mu) - \Delta)$ soient des entiers. En fait, $(r_1,\ldots,r_k)$ est un élément de S si et seulement si il existe un faisceau semi-stable E de rang r et de classes de Chern c_1, c_2 tel qu'on ait $\mathrm{Gr}(E) = E_1 \oplus \ldots \oplus E_k$, avec $\mathrm{rg}(E_i) = r_i$ pour $1 \leq i \leq k$. On a alors $r_i^2(2\Delta - 1) \geq 1$, car cet entier est égal à $d - 1$, d étant la dimension de la variété de modules contenant E_i, et on a $d \geq 2$ d'après [8]. On a donc, si $i < j$, $r_i r_j(2\Delta - 1) \geq 1$.

Supposons que $\varepsilon \leq 1$. On doit alors avoir $k = 2$ et $r_1 r_2(2\Delta - 1) = 1$. Ceci entraine $r_1 = r_2$ et $r_1^2(2\Delta - 1) = 1$.

Soit M' la variété de modules contenant E_1. On a $\dim(M') = 2$. Il existe un procédé permettant de trouver toutes les variétés de modules de dimension donnée. Supposons que $-1/2 < \mu < 0$ (le cas $-1 < \mu < 1/2$ est analogue). On trouve alors que M', puisqu'elle est de dimension 2, est une variété de modules de hauteur nulle, le fibré exceptionnel qui lui est associé est $\mathcal{O}$ (cf. §6), et M' fait partie de la suite infinie de variétés de modules isomorphes à $\mathbb{P}_2$, dont le premier terme est M(1,0,1). Il en découle que M est de hauteur nulle, le fibré exceptionnel qui lui est associé est aussi $\mathcal{O}$, et M fait partie de la suite infinie de variétés de modules isomorphes à $\mathbb{P}_5$, dont le premier terme est M(2,0,2). Ceci démontre la proposition 8.

Si $\mathrm{codim}_M(M\setminus M_s) = 1$, on a donc $M \simeq \mathbb{P}_5$, et $M\setminus M_s$ est une hypersurface de degré 3. Il existe une infinité de couples (M,M'). Le plus simple est $(M(2,0,2),M(1,0,1))$, ensuite vient $(M(5,-2,5),M(3,-1,2))$. Il ne faut pas croire cependant que si $M \simeq \mathbb{P}_5$, on ait $\mathrm{codim}_M(M\setminus M_s) = 1$. En fait il existe une autre série de variétés de modules de hauteur nulle isomorphes à $\mathbb{P}_5$, dont le fibré exceptionnel associé est Q^*. La plus simple d'entre elles est $M(4,-2,4)$ qui contient les classes d'isomorphisme de fibrés uniformes non homogènes construits par Elencwajg [9]. Pour ces variétés on a $M = M_s$.

9 - LE COMPLETE DE L'ANNEAU LOCAL D'UN POINT DE $M\setminus M_s$

Soient X,Y des éléments distincts de $\mathrm{Hilb}^2(\mathbb{P}_2) \simeq M(1,0,2)$, I_X, I_Y les faisceaux d'idéaux de X,Y respectivement. Le faisceau $E = I_X \oplus I_Y$ est semi-stable de rang 2 et de classes de Chern 0,4. Soit x le point de $M(2,0,4) = M$ correspondant à E. Le résultat suivant est dû à J.Le Potier :

PROPOSITION 9 : *Le complété de l'anneau local* $\mathcal{O}_{M,x}$ *n'est pas factoriel.*

Cependant, d'après le théorème 1, $\mathcal{O}_{M,x}$ est factoriel. Je pense que la proposition 9 est valable pour tout point d'une variété de modules non isomorphe à $\mathbb{P}_5$ correspondant à un faisceau semi-stable non stable.

Soit m une monade de $\mathcal{M}$ dont la cohomologie est isomorphe à E. L'orbite Gm est fermée dans $\mathcal{M}$. Le stabilisateur de m s'identifie à $\mathrm{Aut}(E)$, lui-même isomorphe à $\mathbb{C}^* \times \mathbb{C}^*$. C'est un groupe réductif. Il opère aussi de manière évidente sur $T_{\mathcal{M},m}/T_{Gm,m} = T_0$. Soient N le quotient $T_0/(\mathbb{C}^* \times \mathbb{C}^*)$ (bien défini, car T est une variété affine et $\mathbb{C}^* \times \mathbb{C}^*$ un groupe réductif) ν l'image de 0 dans N. D'après le théorème de Luna ([17],[19]), $\hat{\mathcal{O}}_{M,x}$ est isomorphe à $\hat{\mathcal{O}}_{N,\nu}$. Il suffit donc de montrer que $\hat{\mathcal{O}}_{N,\nu}$ n'est pas factoriel. Pour cela il suffit de montrer que $\mathcal{O}_{N,\nu}$ ne l'est pas. D'après [8], on a un isomorphisme canonique $T_0 \simeq \mathrm{Ext}^1(E,E)$, et l'action de $\mathbb{C}^* \times \mathbb{C}^*$ sur $\mathrm{Ext}^1(E,E)$ est l'action de $\mathrm{Aut}(E)$ "par conjugaison". On a

$$\mathrm{Ext}^1(E,E) \simeq \mathrm{Ext}^1(I_X,I_Y) \oplus \mathrm{Ext}^1(I_Y,I_X) \oplus \mathrm{Ext}^1(I_X,I_X) \oplus \mathrm{Ext}^1(I_Y,I_Y).$$

Le groupe $\mathbb{C}^* \times \mathbb{C}^*$ laisse invariante cette décomposition et agit trivialement sur les deux derniers facteurs. Donc N est le produit d'un espace affine et de $N' = (\mathrm{Ext}^1(I_X,I_Y) \oplus \mathrm{Ext}^1(I_Y,I_X))/(\mathbb{C}^* \times \mathbb{C}^*)$. Soit ν' l'image de 0 dans N'. Il suffit de montrer que $\mathcal{O}_{N',\nu'}$ n'est pas factoriel.

On a $\mathrm{Ext}^1(I_X,I_Y) \simeq \mathrm{Ext}^1(I_Y,I_X) \simeq \mathbb{C}^3$, et l'action de $\mathbb{C}^* \times \mathbb{C}^*$ sur $N' \simeq \mathbb{C}^6$ est la suivante :

$$(\lambda,\mu)(a_o,a_1,a_2,b_o,b_1,b_2) = (\frac{\lambda}{\mu}a_o,\frac{\lambda}{\mu}a_1,\frac{\lambda}{\mu}a_2,\frac{\mu}{\lambda}b_o,\frac{\mu}{\lambda}b_1,\frac{\mu}{\lambda}b_2) \quad .$$

Considérons les fonctions $(\mathbb{C}^* \times \mathbb{C}^*)$-invariantes

$$u = a_o b_o \ , \ v = a_o b_1 \ , \ w = a_1 b_o \ , \ t = a_1 b_1 \ .$$

On en déduit des fonctions régulières $\overline{u}$, $\overline{v}$, $\overline{w}$, $\overline{t}$ sur N'. On a

$$\overline{u}\,\overline{t} = \overline{v}\,\overline{w} \ .$$

Supposons que $\mathcal{O}_{N',\nu'}$ soit factoriel. Il existe alors des fonctions $(\mathbb{C}^* \times \mathbb{C}^*)$-invariantes définies sur un voisinage de 0 dans $\mathbb{C}^6$, α_o, α_1, β_o, β_1, telles que

$$u = \alpha_o\beta_o \ , \ v = \alpha_o\beta_1 \ , \ w = \alpha_1\beta_o \ , \ t = \alpha_1\beta_1 \quad .$$

On a alors

$$\frac{u}{w} = \frac{a_o}{a_1} = \frac{\alpha_o}{\alpha_1} \ ,$$

donc on peut écrire $\alpha_o = a_o\alpha'_o$, α'_o étant une fonction régulière inversible définie sur un voisinage de 0 dans $\mathbb{C}^6$. Dans le développement en série entière de α_o existe donc un terme de la forme ca_o, avec c dans $\mathbb{C}^*$. Ceci est impossible car les éléments $(\mathbb{C}^* \times \mathbb{C}^*)$-invariants de $\mathbb{C}[[a_o,a_1,a_2,b_o,b_1,b_2]]$ n'ont dans leur développement que des termes de degré total pair. Ceci démontre la proposition 9.

BIBLIOGRAPHIE

1 - ATIYAH, M.F., BOTT, R.: The Yang-Mills equations over Riemann Surfaces. Phil. Trans. Roy. Soc. London. A 308, 523-615 (1982)

2 - BARTH, W.: Moduli of vector bundles on the projective plane. Invent. Math. 42, 63-91 (1977)

3 - BRIESKORN, E.: Uber holomorphe $\mathbb{P}_n$-Bündel über $\mathbb{P}_1$. Math. Ann. 157, 343-357 (1967)

4 - BRUN, J., HIRSCHOWITZ, A.: Droites de saut des fibrés de rang élevé sur $\mathbb{P}_2$. Math. Zeits. 181, 171-178 (1982)

5 - DREZET, J.M.: Fibrés exceptionnels et suite spectrale de Beilinson généralisée sur $\mathbb{P}_2(\mathbb{C})$. Preprint Paris (1985)

6 - DREZET, J.M.: Fibrés exceptionnels et variétés de modules de faisceaux semi-stables sur $\mathbb{P}_2(\mathbb{C})$. Preprint Paris (1985)

7 - DREZET, J.M.: Groupe de Picard des variétés de modules de faisceaux semi-stables sur $\mathbb{P}_2(\mathbb{C})$. Preprint Paris (1985)

8 - DREZET, J.M., LE POTIER, J.: Fibrés stables et fibrés exceptionnels sur $\mathbb{P}_2$. Ann. Scient. Ec. Norm. Sup. 18, 193-244 (1985)

9 - ELENCWAJG, G.: Des fibrés uniformes non homogènes. Math. Ann. 239, 185-192 (1979)

10 - ELLINGSRUD, G.: Sur l'irréductibilité du module des fibrés stables sur $\mathbb{P}_2$. Math. Zeits. 182, 189-192 (1983)

11 - GIESEKER, D.: On the moduli of vector bundles on an algebraic surface. Ann. of Math. 106, 45-60 (1977)

12 - GROTHENDIECK, A.: Sur la classification des fibrés holomorphes sur la sphère de Riemann. Amer. J. of Math. 79, 121-138 (1957)

13 - HULEK, K.: On the classification of stable rank-r vector bundles over the projective plane. In: Vector bundles and differential equations (A. Hirschowitz ed.). Proceedings (Nice 1979). Progress in Math. 7, Birkhäuser (1980)

14 - LE POTIER, J.: Fibrés stables de rang 2 sur $\mathbb{P}_2(\mathbb{C})$. Math. Ann. 241, 217-256 (1979)

15 - LE POTIER, J.: Sur le groupe de Picard de l'espace de modules de fibrés stables sur $\mathbb{P}_2$. Ann. Scient. Ec. Norm. Sup. 14, 141-155 (1981)

16 - LE POTIER, J., VERDIER, J.L.(Ed.): Module Des Fibrés Stables Sur Les Courbes Algébriques. Progress in Math. 54. Birkhäuser (1985)

17 - LUNA, D.: Slices Etales. Bull. de la Soc. Math. de France 33 (1973)

18 - MARUYAMA, M.: Moduli of stable sheaves II. J. Math. Kyoto Univ. 18, 557-614 (1978)

19 - MUMFORD, D., FOGARTY, J.: Geometric Invariant Theory. Erg. der Math. und ihre Grenzg. 34. Springer Verlag (1982)

20 - STRØMME, S.A.: Ample divisors on fine moduli spaces on the projective plane. Math. Zeits. 187, 405-423 (1984)

ON THE RATIONALITY OF THE MODULI SPACE FOR STABLE RANK-2 VECTOR BUNDLES ON P^2

G. Ellingsrud[1] and S. A. Strømme[2]

[1]Matematisk Institutt, Universitetet i Oslo
P.O.Box 1053, N - 0316 OSLO, Norway
[2]Matematisk Institutt, Universitetet i Bergen
N - 5014 BERGEN-U, Norway

INTRODUCTION.

(1) Denote by $M(c_1, c_2)$ the moduli space for stable rank-2 vector bundles on the projective plane P over an algebraically closed field k of arbitrary characteristic, with the given Chern classes c_1 and c_2. The fact that $M(c_1, c_2)$ is a nonsingular, irreducible variety of dimension $= 4c_2 - c_1^2 - 3$ has been known for a long time, cf. Maruyama's 1978 paper [10]. In the present note we prove the following

THEOREM 1. (a) <u>If</u> $c_1^2 - 4c_2 \equiv 0 \pmod 8$, <u>then</u> $M(c_1, c_2)$ <u>is rational</u>.
(b) <u>If</u> $c_1^2 - 4c_2 \equiv 0 \pmod 8$, <u>then there exists a rational variety which is a</u> P^1- <u>bundle over an open dense subset of</u> $M(c_1, c_2)$.

The proof presented here dates from 1979, when it appeared in the Oslo Preprint Series [3]. That paper was never published, since we learned that Hulek [9] had treated the case of odd first Chern class, and Barth's paper [2], treating the case c_1 even, is even older. Now, however, Maruyama [11] has established that there is a gap in Barth's proof that is not easily filled in. In particular, Barth's claim that $M(0, 2m)$ is rational is still open for $m \geq 2$, so theorem 1 is still the best proven result on the rationality question. Even though Maruyama gave a different proof of theorem 1 in [11], we feel that the original proof still carries an independent interest. For example, the ideas of this proof have later been applied to the study of moduli spaces over ruled surfaces [8], as well as to the study of the Picard group of $M(c_1, c_2)$, see [4], [13].

(2) Notation. If V is a vector space, the set of one-dimensional subspaces will be denoted by PV. In contrast, if A is a locally free sheaf on a variety X, then $\mathbb{P}(A)$ will be the projectivized bundle in the sense of E.G.A., i. e. $\mathbb{P}(A) = \underline{\mathrm{Proj}}(\mathrm{Sym}(A))$. In particular, PV is the set of k-rational points of the projective space $\mathbb{P}(V^*)$.

THE PROOF.

(3) Fix, once and for all, a closed point $p \in P$. Let L be the projective line parametrizing the pencil of lines in P passing through p, and let $F \subset P \times L$ be the incidence correspondence, with projections $f: F \to P$ and $g: F \to L$. Then f is the blowing up of P in p, and g can be identified with the structure map of the ruled surface $\mathbb{P}(O_L \oplus O_L(1))$. The Picard group of F is freely generated by the two elements

$$\sigma := c_1(g^* O_L(1)) \quad \text{and}$$
$$\tau := c_1(f^* O_P(1)).$$

If n,m are integers, we shall use the notation

$$O_F(n\tau + m\sigma) := f^* O_P(1) \otimes g^* O_L(1).$$

The Grothendieck-Riemann-Roch theorem for the morphism g implies the following:

LEMMA. *Let* D *be a torsion-free sheaf on* F *of rank* r *and with Chern classes* $c_1(D) = c\tau$ *and* $c_2(D) = n$. *Assume that* $R^1 g_* D = 0$. *Then* $g_* D$ *is locally free on* L *of rank* = $r + c$ *and degree* = $-n+c(c+1)/2$.

(4) As usual when dealing with stable rank-2 bundles on a projective space, we restrict our attention to the particular cases $c_1 = -1$ and $c_1 = 0$. A 2-bundle with this restriction is stable if and only if it has no global sections. Since we are only interested in the question of rationality, the exact choice of stability definition is of minor importance. Indeed, for the purpose of this proof, we shall introduce a very strong stability condition called well-behavedness (see (8) below).

(5) Let c and n be integers such that $-1 \leq c \leq 0$ and $n \geq 2 + c$. In the case $c = 0$ we shall write $n = 2m + i$, where $0 \leq i \leq 1$. Let E be a stable rank-2 vector bundle on P with $c_1(E) = c$, $c_2(E) = n$. For a line K in P, the restriction of E to K satisfies

$$E_K \approx O_K(d) \oplus O_K(c-d)$$

where $d = d(K)$ is a non-negative integer.

(6) We say that E <u>has the property (J(p))</u> if the following condition is satisfied:

If $c = -1$: $d(K) = 0$ for all lines K containing p.
If $c = 0$: $d(K) \leq 1$ for all lines K containing p, with equality for only a finite number of such lines.

(7) Assume that E has the property (J(p)). Then, since the fibers of $g\colon F \to L$ are identified with the lines in P through p, standard base change theory [7,III-12] shows that $R^1 g_* f^* E$ vanishes. Thus the lemma in (3) above applies to give that $g_* f^* E$ is locally free of rank $= 2 + c$ and degree $= -n$. If $c = -1$, this leaves us with only one possibility: $g_* f^* E \approx O_L(-n)$. But if $c = 0$, the rank is 2, so $g_* f^* E$ is of the form $O_L(a) \oplus O_L(b)$, where $a + b = -n$. The most "generic" behaviour occurs when $|a - b| \leq 1$, which motivates the following

(8) <u>Definition.</u> E is <u>well-behaved (with respect to p)</u> if it has the property (J(p)) and in addition, if $c = 0$,

$$g_* f^* E \approx O_L(-m) \oplus O_L(-m-i),$$

where $n = 2m + i$, $0 \leq i \leq 1$.

(9) If $c = -1$ and E is well-behaved, base-change theory shows that the natural map $g^* g_* f^* E \to f^* E$ is pointwize injective. Consideration of the Chern classes shows that the cokernel is the linebundle $O_F(n\sigma-\tau)$. Hence there is induced a short exact sequence on F

$$(*1) \qquad 0 \longrightarrow O_F(-n\sigma) \xrightarrow{r} f^* E \longrightarrow O_F(n\sigma-\tau) \longrightarrow 0$$

If $c = 0$ and E is well-behaved, then $g_* f^* E \otimes O_L(m) \approx O_L \oplus O_L(-i)$, so $h^0(F, f^* E(m\sigma)) = 2 - i \geq 1$. Let r be a non-zero section of $f^* E(m\sigma)$. Then r is pointwize injective on each fiber K of g which satisfies $d(E) = 0$. Furthermore, r vanishes on no full fiber of g, since $H^0(L, g_* f^* E(m-1)) = 0$. It follows that r vanishes in codimension 2, and gives rise to a short exact sequence on F

$$(*2) \qquad 0 \longrightarrow O_F(-m\sigma) \xrightarrow{r} f^*E \longrightarrow I_Z(m\sigma) \longrightarrow 0$$

where Z is a local complete intersection subscheme of F of length = $c_2(f^*E(m\sigma)) = n$.

(10) For a well-behaved bundle E, let S(E) be the vector space

$$S(E) := \begin{cases} H^0(F, f^*E(n\sigma)) & (\text{if } c = 0) \\ H^0(F, f^*E(m\sigma)) & (\text{if } c = -1) \end{cases}$$

Consider the set X of $\sim$ -equivalence classes of pairs (E,r) where E is a well-behaved bundle on P and r is a non-zero element of S(E). Two pairs are equivalent, $(E,r) \sim (E',r')$, if there exists an isomorphism $E \to E'$ carrying r to r'. In particular, since stable bundles are simple, $(E,r) \sim (E,r')$ if and only if r and r' are proportional elements of S(E). Note that the dimension of S(E) is 2 if c = 0 and n is even, and 1 otherwize, corresponding to the cases (b) and (a) of theorem 1, respectively.

(11) If M_0 is the set of isomorphism classes of well-behaved vector bundles, there is a natural forgetful map

$$b: X \longrightarrow M_0$$

whose fiber over a bundle E is simply PS(E) (the set of one-dimensional subspaces of S(E)). In particular, b is a bijection in case (a) of theorem 1, and a $\mathbb{P}^1$-bundle (at least set-theoretically) in case (b). In order to prove the theorem, it suffices to show that X can be given the structure of a rational variety in such a way that the map b is in fact a morphism. But let us work set-theoretically for a little while still.

(12) Consider now the set W defined as follows:

If c = -1, W is the set of isomorphism classes of short non-split exact sequences of coherent sheaves on F of the form

$$(*3) \qquad 0 \longrightarrow O_F(-n\sigma) \longrightarrow D \longrightarrow O_F(n\sigma-\tau) \longrightarrow 0$$

where "isomorphism" means the existence of vertical isomorphisms making everything commute:

$$\begin{array}{ccccccccc} 0 & \longrightarrow & O_F(-n\sigma) & \longrightarrow & D & \longrightarrow & O_F(n\sigma-\tau) & \longrightarrow & 0 \\ & & \downarrow & & \downarrow & & \downarrow & & \\ 0 & \longrightarrow & O_F(-n\sigma) & \longrightarrow & D' & \longrightarrow & O_F(n\sigma-\tau) & \longrightarrow & 0 \end{array}$$

The set W can then be canonically identified with the set $PExt^1(O_F(n\sigma-\tau),O_F(-n\sigma))$. If $w \in W$, let $D(w)$ be the middle term, and let $r(w)$ be the induced global section of $D(w) \otimes O_F(n\sigma)$.

If $c = 0$, W is to be the set of isomorphism classes of short non-split exact sequences of the form

$$(*4) \qquad 0 \longrightarrow O_F(-m\sigma) \longrightarrow D \longrightarrow I_Z(m\sigma) \longrightarrow 0$$

where Z is a local complete intersection subscheme of F of length = n. Here isomorphism is defined exactly as above. For an element $w \in W$, let $Z(w)$ and $D(w)$ be the closed subscheme and the middle term, respectively. Also, let $r(w)$ be the induced global section of $D(w) \otimes O_F(m\sigma)$, this is well defined up to a non-zero scalar multiple. There is a natural forgetful map

$$a: W \longrightarrow H$$

given by $a(w) = Z(w)$, where $H \subset Hilb(F)$ is an open subvariety of the Hilbert scheme of F. The fiber of a over a point $\{Z\} \in H$ is then canonically identified with $PExt^1(I_Z(m\sigma),O_F(-m\sigma))$.

(13) Let Y be the set of those $w \in W$ such that $D(w)$ is locally free and restricts to the trivial 2-bundle on the exceptional fiber $B := f^{-1}(p)$. For $w \in Y$, put $E(w) := f_*D(w)$. A theorem of Schwarzenberger ensures that $E(w)$ is then locally free and that the natural map $f^*E(w) \to D(w)$ is an isomorphism [12,thm.5]. It is straightforward to check that $E(w)$ is well-behaved with respect to p, and that the pair $u(w) := (E(w),r(w))$ is an element of X, giving rise to a map of sets

$$u: Y \longrightarrow X.$$

(14) On the other hand, a pair (E,r) in X gives rise to a sequence (*1) or (*2), respectively, thus giving a natural map of sets

$$v\colon X \longrightarrow Y$$

which is clearly an inverse to the map u above.

(15) Returning to the program outlined in (11), we proceed now to show that X has a natural structure of a variety. Using the bijections u and v above, it suffices to give such a structure to Y. This can be done as follows:

If $c = -1$, W has a natural structure of a projective space

$$a\colon W = \mathbb{P}(A^*) \longrightarrow \mathrm{Spec}(k)$$

where A is the vector space $A := \mathrm{Ext}^1(O_F(n\sigma-\tau), O_F(-n\sigma))$, and Y is an open subvariety, as can be seen from the semicontinuity of cohomology.

If $c = 0$, let $J \subset O_{F\times H}$ be the universal ideal, where H is the open subvariety of Hilb(F) parametrizing local complete intersections of length = n. Consider the relative Ext-sheaf on H

$$A := \mathrm{Ext}^1(pr_H; J(m\sigma), O_{F\times H}(-m\sigma)).$$

Base-change theory for relative Ext [1, Satz 1] implies that A is locally free on H and induces the vector spaces

$$A(Z) := \mathrm{Ext}^1(I_Z(m\sigma), O_F(-m\sigma))$$

on the fibers. It is now easy to see that W can be given the scheme structure of the projective bundle

$$a\colon W = \mathbb{P}(A^*) \longrightarrow H$$

Again, $Y \subset W$ is naturally an open subvariety.

(16) We have given Y, and thereby also X, a scheme structure. There remains to show that the composed map

$$\varphi := bu\colon Y \longrightarrow M_0$$

is indeed a morphism of varieties. By the definition of a coarse moduli space, this will be the case once we construct a vector bundle $\underline{E}$ on $P \times Y$ with the property that for all $y \in Y$, the fiber $\underline{E}(y)$ represents the isomorphism class $\varphi(y)$ in M_0.

In both cases, W is of the form $\mathbb{P}(A^*)$, hence comes equipped with a tautological linebundle quotient

$$a^* A^* \longrightarrow O_W(\lambda)$$

The induced global section of $a^* A(\lambda)$ gives rise to a "tautological" extension of sheaves on $F \times W$

(*5) $$0 \longrightarrow O_{F \times W}(\lambda - n\sigma) \longrightarrow \underline{D} \longrightarrow O_{F \times W}(n\sigma - \tau) \longrightarrow 0$$

(*6) $$0 \longrightarrow O_{F \times W}(\lambda - m\sigma) \longrightarrow \underline{D} \longrightarrow J(m\sigma) \longrightarrow 0$$

(according to whether $c = -1$ (*5) or $c = 0$ (*6)). Let $\underline{D}'$ be the restriction to $F \times Y \subset F \times W$ of $\underline{D}$ and put

$$\underline{E} := (f \times 1)_* \underline{D}'$$

Schwarzenberger's theorem mentioned in (13) above is easily generalized to families [5,1.13], so $\underline{E}$ is indeed a bundle with the required properties.

(17) We have proved that there exists a variety Y and a vector bundle $\underline{E}$ on $P \times Y$, well-behaved on the fibers, such that the induced morphism

$$\varphi: Y \longrightarrow M(c,n)$$

has as image the open subset M_0 of $M(c,n)$. Furthermore, φ is an imbedding if either c or n is odd, and a $\mathbb{P}^1$-bundle if both c and n are even, corresponding to the two cases of theorem 1. Also, Y is an open subset of the variety W, which is rational, since it is a projective bundle over either a point ($c = -1$) or a certain Hilbert scheme ($c = 0$), which is birational to the n-th symmetric power of the plane and therefore rational by [6]. The only thing that remains to be checked is that Y, or equivalently, M_0, is nonempty. We distinguish between the cases $c = -1$ and $c = 0$.

<u>$c = -1$:</u> In this case, all the $D(w)$ are extensions of linebundles and hence locally free. The restriction of (*3) to the exceptional divisor B looks like this:

(*3') $$0 \longrightarrow O_B(-n) \longrightarrow D_B \longrightarrow O_B(n) \longrightarrow 0$$

Therefore, to show that $D(w)_B$ is trivial for general w, it suffices to show that the natural restriction map

$$\mathrm{Ext}^1(O_F(n\sigma-\tau),O_F(-n\sigma)) \longrightarrow \mathrm{Ext}^1(O_B(n),O_B(-n))$$

is surjective. This map is identified with the restriction map of cohomology

$$H^1(F,O_F(\tau-2n\sigma)) \longrightarrow H^1(B,O_B(-2n))$$

Since the divisor class of B in F is $(\tau-\sigma)$, the cokernel of this map is contained in the second cohomology group $H^2(F,O_F(-(2n+1)\sigma)) = 0$, hence it is surjective.

<u>c = 0:</u> In this case, take Z to be n points of F disjoint from B. Then a similar argument completes the proof also in this case. Indeed, the spectral sequence relating local and global Ext gives rise to the vertical maps in the following exact commutative diagram, where the horizontal maps are the natural restriction maps:

$$\begin{array}{ccc}
0 & & 0 \\
\downarrow & & \downarrow \\
H^1(F,\underline{Hom}(I_Z(m\sigma),O_F(-m\sigma))) & \xrightarrow{\gamma} & H^1(B,\underline{Hom}(O_B(m),O_B(-m))) \\
\downarrow & & \downarrow \\
\mathrm{Ext}^1(I_Z(m\sigma),O_F(-m\sigma)) & \xrightarrow{\beta} & \mathrm{Ext}^1(O_B(m),O_B(-m)) \\
\downarrow{\scriptstyle\alpha} & & \downarrow \\
H^0(F,\underline{Ext}^1(I_Z(m\sigma),O_F(-m\sigma))) & \longrightarrow & 0 \\
\downarrow & & \\
H^2(F,\underline{Hom}(I_Z(m\sigma),O_F(-m\sigma))) & &
\end{array}$$

If we can show that α and β are surjective, we are done. Now α is surjective since $H^2(F,\underline{Hom}(I_Z(m\sigma),O_F(-m\sigma))) = H^2(F,O_F(-2m\sigma)) = 0$. To show that β is surjective, it suffices to show that γ is surjective. γ can be identified with the natural map

$$H^1(F,O_F(-2m\sigma)) \longrightarrow H^1(B,O_B(-2m))$$

which is surjective since $H^2(F,O_F(-2m\sigma-(\tau-\sigma))) = 0$.

This completes the proof.

References.

1 Banica,C.,Putinar,M. ,Schumacher,G.: Variation der globalen Ext in Deformationen kompakter komplexer Räume. Math. Ann. 250, 135-155(1980)

2 Barth, W.: Moduli of vector bundles on the projective plane. Invent. Math. 42,63-91(1977)

3 Ellingsrud,G., Strømme,S.A.: On the moduli space for stable rank-2 vector bundles on $\mathbb{P}^2$. Preprint, Oslo (1979)

4 Ellingsrud,G., Strømme,S.A.: The Picard group of the moduli space for stable rank-2 vector bundles on $\mathbb{P}^2$ with odd first Chern class. Preprint, Oslo (1979)

5 Ellingsrud,G., Strømme,S.A.: Stable Rank-2 Vector Bundles on $\mathbb{P}^3$ with $c_1=0$ and $c_2=3$. Math. Ann. 255,123-135(1981)

6 Fogarty,J.: Algebraic families on an algebraic surface. Am. J. Math. 90,511-521(1968)

7 Hartshorne,R.: Algebraic Geometry. Graduate Texts in Math. 52, Berlin, Heidelberg, New York: Springer 1977

8 Hoppe,H.J.,Spindler,H.: Modulräume stabiler 2-Bündel auf Regelflächen. Math. Ann. 249,127-140(1980)

9 Hulek,K.: Stable Rank-2 Vector Bundles on $\mathbb{P}_2$ with c_1 odd. Math. Ann. 242,241-266(1979)

10 Maruyama,M.: Moduli of stable sheaves II. J.Math. Kyoto Univ. 18,557-614(1978)

11 Maruyama,M.: On rationality of the moduli spaces of vector bundles of rank 2 on $\mathbb{P}^2$. Proc. Sendai Conf. (1985)

12 Schwarzenberger,R.L.E.: Vector Bundles on Algebraic Surfaces. Proc. Lond. Math. Soc. 11,601-622(1961)

13 Strømme,S.A.: Ample Divisors on Fine Moduli Spaces on the Projective Plane. Math. Z. 187,405-423(1984)

A THEOREM ON ZERO SCHEMES OF SECTIONS IN TWO-BUNDLES OVER AFFINE SCHEMES WITH APPLICATIONS TO SET THEORETIC INTERSECTIONS

O. Forster and K. Wolffhardt

We consider the following problem. Let E be a rank 2 vector bundle over an affine scheme X and f a section of E with zero scheme $Z \subset X$. If codim Z = 2 and there exists a reasonable theory of Chern classes on X, then Z represents the second Chern class $c_2(E)$. Since the second Chern classes of a vector bundle and of its dual coincide, one may ask whether E^* admits a section φ with the same zero scheme Z.

We prove that this is true if X is an affine algebraic surface over an algebraically closed field (Proposition 1.3). The proof uses Serre's extension theory for codimension 2 ideals and the cancellation theorem of Murthy-Swan. In an elementary way we then prove the existence of φ in a more general situation: X is an arbitrary affine scheme and the only condition is that $\det(E)|Z$ be trivial (Theorem 1.5).

We apply these results to prove generalizations of the theorem of Storch [St] and Eisenbud-Evans [EE] on the minimal number of equations for the set theoretic description of closed subschemes of an affine scheme. By other methods, similar results have been obtained by Boratyński [B], Lyubeznik [L], and Mandal [M]. In Theorem 2.6 we prove: Let $Y \subset X = \operatorname{Spec} R$ be a subscheme. If Y is defined by a locally principal ideal $I \subset R$ such that the conormal module I/I^2 is generated by m elements ($m \geq 2$), then Y can be set theoretically defined by m functions. For arbitrary codimension we derive the following result: Y can be set theoretically defined by $n := \dim X$ functions if Y is a locally complete intersection without zero-dimensional components. In fact n functions suffice in a more general case. The conditions on the ideal I of Y are as follows. For $k \geq 1$ let Y_k be the set of all points $y \in Y$ such that I_y requires at least k generators. We suppose $\dim Y_k \leq n-k$ for $1 \leq k \leq n-1$, and $Y_n = \emptyset$. Then Y can be set theoretically defined by n functions (cf. Theorem 3.6).

§ 1. Zero schemes of sections in 2-bundles

(1.1) Let E be a *vector bundle* over a locally ringed space $(X, \mathcal{O}_X)$. By this we mean a locally free $\mathcal{O}_X$-module of finite type. We denote its dual bundle by E^*. A section $f \in \Gamma(X,E)$ defines a morphism of $\mathcal{O}_X$-modules

$$E^* \longrightarrow \mathcal{O}_X, \quad \varphi \longmapsto \langle\varphi, f\rangle,$$

which we identify with f. The ringed subspace Z of X with structure sheaf

$$\mathcal{O}_Z := \text{Coker}\,(E^* \xrightarrow{f} \mathcal{O}_X)$$

is called the *zero scheme* of f and denoted by $Sch_E(f)$ or briefly by Sch(f). Its underlying topological space is

$$V(f) = V_E(f) := \{x \in X : f(x) = 0\}.$$

Here f(x) denotes the element induced by f in the vector space $E(x) := E_x/m_xE_x$.

<u>(1.2)</u> Suppose now that the vector bundle E on X has constant rank 2 and that the zero scheme Z = Sch(f) of a section f in E has codimension 2. If for example X is an non-singular variety over an algebraically closed field, Z represents the Chern class $c_2(E)$, which is equal to $c_2(E^*)$. So the question arises if the dual bundle E* admits a section with the same zero scheme Z.

Of course, this is not always true. Assume for instance that X is Cohen-Macaulay in every point of Z. Then a simple necessary condition can be formulated as follows: If both E and E* admit sections with zero scheme Z, then $\det(E)^2|Z$ is trivial. To see this, we consider the conormal bundle $\nu_Z = \mathcal{I}_Z/\mathcal{I}_Z^2$ of Z. The epimorphism

$$E^* \xrightarrow{f} \mathcal{I}_Z \longrightarrow 0$$

induces an isomorphism $E^*|Z \xrightarrow{\sim} \nu_Z$. Analogously, we have an isomorphism $E|Z \xrightarrow{\sim} \nu_Z$. This implies that $\det(E)^2|Z \simeq \mathcal{O}_Z$. This necessary condition is evidently fulfilled if Z consists of finitely many points. This assumption is sufficient, as the following proposition shows.

<u>(1.3) Proposition.</u> Let X be an affine algebraic surface over an algebraically closed field and E an algebraic vector bundle of rank 2 over X. Let $f \in \Gamma(X,E)$ be a section such that Sch(f) is zero-dimensional and consists of Cohen-Macaulay points of X. Then there exists a section $\varphi \in \Gamma(X,E^*)$ in the dual bundle with $Sch(\varphi) = Sch(f)$.

Remark. Later we will prove a theorem which contains Proposition 1.3 as a special case. Nevertheless we will bring a separate proof of 1.3 because it is of independent interest.

<u>Proof.</u> Let Z = Sch(f), and $\mathcal{I}_Z := \text{Im}(f: E^* \longrightarrow \mathcal{O}_X)$, the ideal sheaf of Z. Since X is Cohen-Macaulay in every $x \in Z$, we have an exact sequence (Koszul complex)

$$0 \longrightarrow L^* \longrightarrow E^* \xrightarrow{f} \mathcal{I}_Z \longrightarrow 0,$$

where $L = \det(E)$. This exact sequence defines an element $\xi \in \mathrm{Ext}^1(I_Z,L^*) = \Gamma(X,\mathcal{E}xt^1(I_Z,L^*))$. Now

$$\mathcal{E}xt^1(I_Z,L^*) \cong \mathcal{E}xt^2(\mathcal{O}_Z,L^*) \cong \det(\nu_Z)^* \otimes L^*$$
$$= \det(E) \otimes L^* \otimes \mathcal{O}_Z \cong \mathcal{O}_Z.$$

Since E^* is locally free, we have by Serre theory: ξ_x is a generator of $\mathcal{E}xt^1(I_Z,L^*)_x$ for all $x \in Z$. On the other hand

$$\mathcal{E}xt^1(I_Z,L) \cong \det(E) \otimes L \otimes \mathcal{O}_Z.$$

Since Z is zero-dimensional, we have a (noncanonical) isomorphism $\mathcal{E}xt^1(I_Z,L) \cong \mathcal{E}xt^1(I_Z,L^*)$. Let $\tilde{\xi} \in \mathrm{Ext}^1(I_Z,L)$ be the element which corresponds to ξ under this isomorphisms and let

$$0 \longrightarrow L \longrightarrow V \longrightarrow I_Z \longrightarrow 0$$

be the extension corresponding to $\tilde{\xi}$. Again by Serre, V is locally free of rank 2. We will prove $V \cong E$. First, by Schanuel's lemma,

$$V \oplus L^* \cong E^* \oplus L.$$

We have to use the following

(1.4) Lemma. Let W be a vector bundle over a two-dimensional affine scheme X with $\det(W) \cong \mathcal{O}_X$. Then $W \cong W^*$.

Proof of the lemma. We may assume that W has constant rank m. The assertion is clear if $m \leq 2$, since for a vector bundle E of constant rank 2 one has

$$E^* \cong E \otimes \det E^*.$$

If $m > 2$, by a well known theorem of Serre, we can write $W = W' \oplus \mathcal{O}_X^{m-2}$, where W' is a vector bundle of rank 2, and the assertion follows.

We return to the proof of Proposition 1.3. Applying Lemma 1.4 we obtain

$$V \oplus L^* \cong E^* \oplus L \cong E \oplus L^*.$$

By the cancellation theorem of Murthy and Swan [MS] this implies $V \cong E$, and we have an exact sequence

$$0 \longrightarrow L \longrightarrow E \xrightarrow{\varphi} I_Z \longrightarrow 0,$$

which proves Proposition 1.3.

Remark. For the application of Murthy-Swan's cancellation theorem we had to suppose that X is an affine algebraic surface over an algebraically closed field. Actually the assertion holds in a much more general situation.

<u>(1.5) Theorem.</u> Let X be an affine scheme, E a vector bundle of rank 2 over X and $f \in \Gamma(X,E)$ a section with zero scheme $Z = \mathrm{Sch}(f)$. Suppose that the restriction of the line bundle $L := \det(E)$ to Z is trivial. Then there exists a section $\varphi \in \Gamma(X,E^*)$ with zero scheme Z.

Note that we do not require that X is Cohen-Macaulay in the points of Z nor that Z is of codimension 2. The condition that $\det(E)|Z$ is trivial is automatically fulfilled if Z consists of finitely many points.

<u>Proof.</u> Since $L|Z$ is trivial there exists a section $h \in \Gamma(X,L)$ such that $h|Z$ has no zeroes. Therefore $(f,h) \in \Gamma(X,E \oplus L)$ is *unimodular* (i.e. a section without zeroes). Hence there exists a section $(\psi,\lambda) \in \Gamma(X,E^*\oplus L^*)$ such that

$$(*) \qquad \langle\psi,f\rangle + \langle\lambda,h\rangle = 1.$$

Define

$$\Phi := \psi \otimes \psi + i(\lambda) : E \longrightarrow E^*,$$

where $i(\lambda) : E \longrightarrow E^*$ is defined by

$$\langle i(\lambda)v,w\rangle := \langle\lambda, v \wedge w\rangle$$

for sections v,w of E. Let $\varphi := f\circ\Phi \in \Gamma(X,E^*)$ be the composition of the maps

$$E \xrightarrow{\Phi} E^* \xrightarrow{f} \mathcal{O}_X,$$

i.e.

$$\langle\varphi,v\rangle = \langle\Phi(v),f\rangle = \langle\psi,v\rangle\langle\psi,f\rangle + \langle\lambda,v\wedge f\rangle.$$

It remains to show that

$$\mathrm{Im}(E \xrightarrow{\varphi} \mathcal{O}_X) = \mathrm{Im}(E^* \xrightarrow{f} \mathcal{O}_X) =: \mathcal{I}_Z.$$

We prove the equality $\mathrm{Im}\varphi_x = \mathcal{I}_{Z,x}$ first for $x \in V(\lambda)$. By defnition, $\mathrm{Im}\,\varphi \subset \mathcal{I}_Z$. From (*) it follows that $\langle\psi,f\rangle(x) = 1$. Now

$$\langle\varphi,f\rangle = \langle\psi,f\rangle^2,$$

hence $\varphi_x(f)$ is invertible, so $\mathrm{Im}\,\varphi_x = \mathcal{O}_{X,x} \supset \mathcal{I}_{Z,x}$. The equality $\mathrm{Im}\,\varphi_x = \mathcal{I}_{Z,x}$ for $x \notin V(\lambda)$ follows immediately from the fact that $\Phi|X \setminus V(\lambda)$ is an isomorphism. This will be shown using the following funny formula.

<u>(1.6) Proposition.</u> Let E be a rank 2 vector bundle and let $S,A : E \longrightarrow E^*$ be morphisms, S symmetric and A antisymmetric. Then

$$\det(S+A) = \det(S) + \det(A).$$

(These determinants are sections of the line bundle $\det(E^*)^2$.) Since the assertion is local, the formula can be verified by simple matrix calculus.

We apply this proposition to Φ and get

$$\det\Phi = \det(\psi\otimes\psi) + \det(i(\lambda)) = 0 + \lambda^2,$$

hence $\det\Phi$ is invertible on $X \setminus V(\lambda)$, q.e.d.

§ 2. Set theoretic description of hypersurfaces

For the proof of our theorem on the set theoretic description of hypersurfaces in affine schemes we need some preparations.

(2.1) Let $X = \mathrm{Spec}(R)$ be the spectrum of a ring R, and $\Omega = \mathrm{Specm}(R) \subset X$ its maximal spectrum. For subsets $Z \subset Y \subset X$, where Z is closed in Y, we have the notions of combinatorial (Krull) *dimension* $\dim Y$ and $\mathrm{codim}_Y Z$. We will also use the following notations:

$$\mathrm{dimm}\, Y := \dim(Y\cap\Omega),$$
$$\mathrm{Codim}_Y Z := \min\{\mathrm{codim}_Y Z, \mathrm{codim}_{Y\cap\Omega}(Z\cap\Omega)\}.$$

While always $\dim(Y\cap\Omega) \leq \dim Y$, examples show that $\mathrm{codim}_{Y\cap\Omega}(Z\cap\Omega)$ may be less, equal or bigger that $\mathrm{codim}_Y Z$.

(2.2) Lemma. Let Y be an affine scheme whose underlying topological space is noetherian. Let $L_1,\dots,L_r$ be line bundles on Y such that $L_1\oplus\dots\oplus L_r$ admits a unimodular section. Then there exists a unimodular section $(f_1,\dots,f_r) \in \Gamma(Y,L_1\oplus\dots\oplus L_r)$ such that

$$\mathrm{Codim}_Y V(f_1,\dots,f_k) \geq k$$

for all $k = 1,\dots,r$.

Proof. Let $(g_1,\dots,g_r) \in \Gamma(X,L_1\oplus\dots\oplus L_r)$ be unimodular. Then $f_1,\dots,f_r$ are constructed by induction in such a way that $(f_1,\dots,f_k,g_{k+1},\dots,g_r)$ is unimodular and the above inequalities hold.

(2.3) Proposition. Let L be a line bundle on an affine scheme X and $\varphi \in \Gamma(X,L^*)$. Set $Y := \mathrm{Sch}(\varphi)$. Suppose that $L|Y$ is generated by m global sections, $m \geq 2$. Then there exist $f_1,\dots,f_m \in \Gamma(X,L)$ such that

$$\mathrm{Sch}(f_1,\dots,f_m) \subset Y.$$

If Y has a noetherian topology, the sections $f_1,\dots,f_m$ can be chosen in such a way that in addition

$$\mathrm{Codim}_Y \mathrm{Sch}(f_1,\dots,f_m) \geq m - 1.$$

<u>Proof.</u> Choose $g_1,\ldots,g_m \in \Gamma(X,L)$ that generate $L|Y$. Then g_1 has no zeroes on $V(g_2,\ldots,g_m) \cap Y$. Therefore there exists also a $\varphi_1 \in \Gamma(X,L^*)$ which has no zeroes on $V(g_2,\ldots,g_m) \cap Y$. Then $(\varphi_1,g_2,\ldots,g_m)|Y$ is a unimodular section of $L^* \oplus L^{\oplus(m-1)}|Y$. If Y is a noetherian topological space, we may assume by Lemma (2.2) that

$$\operatorname{Codim}_Y V(\varphi_1,g_2,\ldots,g_{m-1}) \cap Y \geq m - 1.$$

Set

$$Z := \operatorname{Sch}(\varphi,\varphi_1,g_2,\ldots,g_{m-1}) \subset Y$$

and

$$X' := \operatorname{Sch}(g_2,\ldots,g_{m-1}).$$

Since $g_m|Z$ has no zeroes, $L|Z$ is trivial. Application of Theorem 1.5 to the bundle $L^* \oplus L^*|X'$ and its section $(\varphi,\varphi_1)|X'$ yields $(f_1,f_2) \in \Gamma(X,L\oplus L)$ such that

$$Z = \operatorname{Sch}(f_1,f_2) \cap X' = \operatorname{Sch}(f_1,f_2,g_2,\ldots,g_{m-1}).$$

Now

$$(f_1,f_2,\ldots,f_m) := (f_1,f_2,g_2,\ldots,g_{m-1})$$

satisfies the assertion of (2.3).

<u>(2.4)</u> In the sequel we will use the following notations. For a module M over a ring R we denote by $\mu(M)$ the minimal number of generators. We say that an ideal $I \subset R$ is *generated up to radical by* m *elements*, if there exists an ideal $J \subset I$ with $\sqrt{J} = \sqrt{I}$ and $\mu(J) \leq m$.

<u>(2.5)</u> We will need the following fact: If F is a finitely generated $\mathcal{O}_Y$-module on a reduced scheme Y such that $\mu(F_y)$ is constant, then F is locally free.

The next theorem gives an estimate of the number of generators up to radical of a hypersurface ideal I by the number of generators of the conormal module I/I^2.

<u>(2.6) Theorem.</u> Let R be a ring and $I \subset R$ a finitely generated locally principal ideal with $\mu(I/I^2) \leq m$ for some $m \geq 2$. Then I is generated up to radical by m elements. If $\operatorname{Supp}(I/I^2)$ is noetherian, the following more precise statement holds: There exists an ideal $J \subset I$ with $\sqrt{J} = \sqrt{I}$, $\mu(J) \leq m$, and

$$\operatorname{Codim}_{\operatorname{Supp}(I/I^2)} \operatorname{Supp}(I/J) \geq m - 1.$$

Proof. Set $\mathfrak{a} := \sqrt{\operatorname{Ann} I}$, $R' := R/\mathfrak{a}$ and let $X' := \operatorname{Spec} R'$ be the affine scheme of R'. The underlying space of X' is $V(\mathfrak{a}) = \operatorname{Supp} I$. Since $\mu((I/\mathfrak{a}I)_x) = 1$ for all $x \in X'$, and R' is reduced, by (2.5), the R'-module $I/\mathfrak{a}I$ is locally free of rank 1. We denote by L the line bundle on X' associated to $I/\mathfrak{a}I$. The inclusion $I \longrightarrow R$ induces a morphism $\varphi : L \to \mathcal{O}_X'$ with

$$V_{L^*}(\varphi) = V(I) \cap X' = \operatorname{Supp}(I/I)^2;$$

this is the underlying topological space of $Y := \operatorname{Sch}(\varphi)$. We have $\Gamma(Y, L|Y) \cong I/(I+\mathfrak{a})I$, hence $\mu(\Gamma(Y, L|Y)) \leq \mu(I/I^2) \leq m$. By Proposition 2.3 there exist sections

$$f_1, \ldots, f_m \in \Gamma(X', L) = I/\mathfrak{a}I$$

with

$$Z := V(f_1, \ldots, f_m) \subset Y,$$

and, if $\operatorname{Supp}(I/I^2)$ is noetherian, $\operatorname{Codim}_Y Z \geq m - 1$. Let $F_1, \ldots, F_m \in I$ be representatives of $f_1, \ldots, f_m$, and $J \subset R$ the ideal generated by $F_1, \ldots F_m$. By construction

$$\operatorname{Supp}(I/J) = Z \subset \operatorname{Supp}(I/I^2),$$

hence $V(I) = V(J)$. This proves Theorem 2.6.

(2.7). Corollary. Let I be a finitely generated, locally principal ideal in a ring R such that $\operatorname{Specm}(R/I)$ is noetherian and satisfies

$$\dim \operatorname{Specm}(R/I) \leq n - 1$$

for some $n \geq 2$. Then I is generated up to radical by n elements.

Proof. Since $Y = \operatorname{Specm}(R/I)$ has dimension $\leq n - 1$ and $\mu((I/I^2)_y) \leq 1$ for all $y \in Y$, it follows that I/I^2 is generated by n elements [F], [Sw].

Remark. Corollary 2.7 says in particular: Let R be an n-dimensional noetherian ring, $n \geq 2$. Then every locally principal ideal $I \subset R$ can be generated up to radical by n elements. This has been proved by Boratyński [B], for R a 2-dimensional affine algebra over an algebraically closed field and by Murthy for n-dimensional regular affine algebras over algebraically closed fields (mentioned in [L]). Mandal proved it for arbitrary n-dimensional noetherian Cohen-Macaulay rings [M].

(2.8) Corollary. Let $Y \subset X$ be an effective Cartier divisor in an n-dimensional Stein space X, $n \geq 3$. Then the ideal $I(Y)$ of Y is generated up to radical by $[\frac{n+1}{2}]$ holomorpic functions.

Remark. On an n-dimensional Stein space any vector bundle of rank d can be generated by d + [n/2] global sections. (In [FR] this is proved over Stein manifolds; the proof is valid for arbitrary Stein spaces by the results of Hamm [H], [H'], on the topology of Stein spaces with singularities). This implies that I(Y) can be generated by 1 + [n/2] holomorphic functions (without restriction on n).

Proof of (2.8). By the above remark, $I(Y)/I(Y)^2$ can be generated by $1 + [\frac{n-1}{2}] = [\frac{n+1}{2}]$ elements.

§ 3. Set theoretic description of subschemes

(3.1) Lemma. Let M be a finitely generated module over a ring R. We denote by X the affine scheme of R and by $\mathcal{M}$ the $\mathcal{O}_X$-module associated to M. Suppose that $Y_0 := \mathrm{Supp}(\mathcal{M})$ is noetherian. Then there exist $\alpha_1,\ldots,\alpha_m \in R$ such that for

$$Y_j := V(\alpha_1,\ldots,\alpha_j) \cap Y_0$$

we have

i) $\mathcal{M}|Y_{j-1} \setminus Y_j$ is free for $j = 1,\ldots,m$,

ii) $Y_m = \emptyset$.

Here $Y_{j-1} \setminus Y_j$ is considered as a reduced subscheme of X. For any locally closed subscheme $Z \subset X$ the restriction $\mathcal{M}|Z$ denotes the sheaf $\mathcal{M} \otimes \mathcal{O}_Z$ on Z.

Proof. The α_j are constructed by induction. To find α_{j+1} let $y \in Y_j$ be a point such that $\mu(\mathcal{M}_y)$ is minimal on Y_j. Then by (2.5), the sheaf $\mathcal{M}|Y_j$ is free in a neighbourhood of y in Y_j which can be chosen as $Y_j \setminus V(\alpha_{j+1})$.

(3.2) Lemma. Let P be a module over a ring R, and $\alpha \in R$ such that P_α is a free R_α-module of rank r and $D(\alpha) := \mathrm{Spec}(R) \setminus V(\alpha)$ is a noetherian topological space. Then for every $g \in P$ there exists an $f \in P$ such that

i) $f \equiv g \mod \alpha P$.

ii) $\mathrm{Codim}_{D(\alpha)} V(f|D(\alpha)) \geq r$.

Proof. There exist $e_1,\ldots,e_r \in \alpha P$ such that their images $\bar{e}_j := e_j|D(\alpha) \in P_\alpha$ form a basis of P_α. Define $g_j \in R_\alpha$ by

$$g|D(\alpha) = \sum_{j=1}^{r} g_j \bar{e}_j .$$

By induction on j choose $a_j \in R$, such that the sets $Y_o := D(\alpha)$ and

$$Y_j := \{x \in Y_{j-1} : g_j(x) = a_j(x)\},$$

satisfy

$$\operatorname{Codim}_{Y_{j-1}} Y_j \geq 1$$

for $j = 1,\dots,r$.
For this it suffices that $g_j(x_\mu) \neq a_j(x_\mu)$, $\mu = 1,\dots,m$, where $\{x_1,\dots,x_m\}$ meets all irreducible components of Y_{j-1} and of $Y_{j-1} \cap \operatorname{Specm}(R)$. For

$$f := g - \sum_{j=1}^{r} a_j e_j$$

we have $V(f|V(\alpha)) = Y_r$, which implies the assertion.

<u>(3.3)</u> Let M be a finitely generated module over a ring R. For $k \in \mathbb{N}$ we define subsets $X_k(M)$ of $X := \operatorname{Spec}(R)$ as

$$X_k(M) := \{x \in X : \mu(M_x) \geq k\}.$$

All $X_k(M)$ are closed sets. We have $X_o(M) = X$, $X_1(M) = \operatorname{Supp}(M)$, and $X_k(M) = \emptyset$ for large k. We will apply this concept especially to the conormal module I/I^2 of a finitely generated ideal I. Note that $X_1(I/I^2) = \operatorname{Supp}(I) \cap V(I)$ and $X_k(I/I^2) = X_k(I)$ for $k \geq 2$.

To estimate the minimal number of generators of a module we define the invariant

$$b(M) := \begin{cases} \sup\{k + \operatorname{dimm} X_k(M) : k \geq 1 \text{ and } X_k(M) \neq \emptyset\}, & \text{if } M \neq 0, \\ 0, & \text{if } M = 0. \end{cases}$$

If Specm(R) is noetherian, we have $\mu(M) \leq b(M)$ (cf. [F], [Sw]).

<u>(3.4) Proposition.</u> Let M be a finitely generated R-module such that Supp(M) is noetherian. For $k \in \mathbb{N}$ let $X'_k := X_k(M) \setminus X_{k+1}(M)$. There exists an $f \in M$ such that

$$\operatorname{Codim}_{X'_k} V(f|X'_k) \geq k$$

for all k.
(Note that, by definition, the empty subset of any topological space has codimension $+\infty$.)

Proof. Let $\mathrm{Supp}(M) = Y_0 \supset Y_1 \supset \ldots \supset Y_m = \emptyset$ be a stratification as in Lemma 3.1. We find $f \in M$ by constructing $f_j = f|Y_j$ for $j = m, m-1, \ldots, 0$ inductively with the aid of Lemma 3.2.

Remark. Proposition 3.4 contains as a special case the following well known result [S]: Let P be a finitely generated projective module of rank r over a ring with noetherian spectrum. Then there exists an $f \in P$ such that $\mathrm{Codim}\, V(f) \geq r$. If, in particular, $\dim \mathrm{Specm}(R) < r$, the module P has a direct summand isomorphic to R.

(3.5) Corollary. Let M be a finitely generated R-module such that Supp(M) is noetherian. Suppose that for some $m \geq 2$ we have

$$b(M) \leq m, \qquad X_m(M) = \emptyset.$$

Then there exist elements $f_1, \ldots, f_{m-2} \in M$ such that for $j = 1, \ldots, m-2$ the module $M_j := M/(f_1, \ldots, f_j)$ satisfies

$$b(M_j) \leq m - j, \qquad X_{m-j}(M_j) = \emptyset.$$

Proof by induction on j, using Proposition 3.4.

In particular M_{m-2} has a support $Y := \mathrm{Supp}(M_{m-2})$ with $\dim m\, Y \leq 1$, and M_{m-2} induces by (2.5) a line bundle on the reduced subscheme Y of Spec R.

(3.6) Theorem. Let I be a finitely generated ideal of a ring R such that $\mathrm{Supp}(I/I^2)$ is noetherian. Suppose that for some m we have

$$b(I/I^2) \leq m \quad \text{and} \quad X_m(I/I^2) = \emptyset.$$

Then there exists an ideal $J \subset I$ with $\mu(J) \leq m$, $\sqrt{J} = \sqrt{I}$, and $\dim m\, \mathrm{Supp}(I/J) \leq 0$.

Proof. For $m = 1$, we have $I/I^2 = 0$ and the assertion is trivial. Therefore suppose $m \geq 2$. By Corollary 3.5 there exist $f_1, \ldots, f_{m-2} \in I$ such that the ideal

$$I' := I/(f_1, \ldots, f_{m-2})$$

of the ring

$$R' := R/(f_1, \ldots, f_{m-2})$$

satisfies

$$b(I'/I'^2) \leq 2 \quad \text{and} \quad X_2(I'/I'^2) = \emptyset.$$

Identifying Spec(R') with $V(f_1, \ldots, f_{m-2}) \subset \mathrm{Spec}(R)$ we have $V(I') = V(I)$. By Theorem 2.6 there exists an ideal $J' \subset I'$ generated by two elements f'_{m-1}, f'_m, such that

$$V(J') = V(I') \qquad \text{and} \qquad \dim \operatorname{Supp}(I'/J') \leqq 0$$

Let $f_{m-1}, f_m \in I$ be representatives of f'_{m-1}, f'_m and $J := (f_1, \ldots, f_m)$. Since $V(J) = V(J')$ and $I/J \cong I'/J'$, the assertion follow.

(3.7) Remark. The assumptions on $b(I/I^2)$ and $X_m(I/I^2)$ in (3.6) are for $m \geqq 2$ equivalent to

$$\dim (V(I) \cap \operatorname{Supp}(I)) \leqq m-1,$$
$$\dim X_k(I) \leqq m-k \text{ for } k = 2,\ldots,m-1,$$
$$X_m(I) = \emptyset.$$

Therefore (3.6) applies in particular to locally complete intersections. By a *locally complete intersection ideal* we mean an ideal I in a ring R such that $\mu(I_x) \leqq$ height (I_x) for all $x \in V(I)$. Note that, by this definition, I_x need not be generated by a regular sequence in R_x (which would be automatically the case if R were supposed to be Cohen-Macaulay). Further we do not require V(I) to be of pure codimension. For a finitely generated locally complete intersection ideal I in an n-dimensional ring we have

$$\dim X_k(I) \leqq n-k \text{ for } k \geqq 2.$$

Therefore Theorem 3.6 implies

(3.8) Corollary. Let R be an n-dimensional noetherian ring and $I \subset R$ a locally complete intersection ideal such that V(I) has no zero-dimensional components. Then I can be generated up to radical by n elements.

In the case of Cohen-Macaulay rings this result has been obtained by other methods by Lyubeznik [L] for height $(I) \geqq 2$, and by Mandal [M] also for height one.

In general, Corollary 3.8 is not correct if V(I) has zero-dimensional components. For example let R be the coordinate ring of a smooth n-dimensional affine algebraic variety X over an algebraically closed field, and I the ideal of a single point $x \in X$. If I is generated up to radical by n elements, then the class of $\{x\}$ in the Chow group $A^n(X)$ of codimension n cycles is a torsion element. This is not always the case (see eg [MM] or [R]).

References

[B] Boratyński, M.: Every curve on a nonsingular surface can be defined by two equations. Proc. AMS 96 (1986) 391-393.

[EE] Eisenbud, D. and E.-G. Evans jr.: Every algebraic set in n-space is the intersection of n hypersurfaces. Invent. math. 19 (1973) 107-112.

[F] Forster, O.: Über die Anzahl der Erzeugenden eines Ideals in einem Noetherschen Ring. Math. Zeitschr. 84 (1964) 80-87.

[FR] Forster, O. und K.J. Ramspott: Analytische Modulgarben und Endromisbündel. Invent. Math. 2 (1966) 145-170.

[H] Hamm, H.A.: Zum Homotopietyp Steinscher Räume. Journ. f. d. r. u. a. Math. 338 (1983) 121-135.

[H'] Hamm, H.A.: Zum Homotopietyp q-vollständiger Räume. Journ. f. d. r. u. a. Math. 364 (1986) 1-9.

[L] Lyubeznik, G.: Some theorems on set theoretic intersections. Preprint 1985.

[M] Mandal, S.: On set theoretic intersection in affine spaces. Preprint 1985.

[MM] Mohan Kumar, N. and M.P. Murthy: Algebraic cycles and vector bundles over affine three-folds. Ann. Math. 116 (1982).

[MS] Murthy, M.P. and R.G. Swan: Vector bundles over affine surfaces. Inv. math. 36 (1976) 125-165.

[R] Rojtman, A.A.: The torsion of the group of O-cycles modulo rational equivalence. Ann. Math. 111 (1980) 553-569.

[S] Serre, J.-P.: Modules projectifs et espaces fibrés à fibre vectorielle. Sém. Dubreil-Pisot 1957/58, exp. 23 (1958).

[St] Storch, U.: Bemerkung zu einem Satz von M. Kneser. Arch. Math. 23 (1972) 403-404.

[Sw] Swan, R.G.: The number of generators of a module. Math. Zeitschr. 102 (1967) 318-322.

Mathematisches Institut der
Ludwig-Maximilians-Universität
Theresienstraße 39
D - 8000 München 2
West Germany

GPSR Compliance
The European Union's (EU) General Product Safety Regulation (GPSR) is a set of rules that requires consumer products to be safe and our obligations to ensure this.

If you have any concerns about our products, you can contact us on

ProductSafety@springernature.com

In case Publisher is established outside the EU, the EU authorized representative is:

Springer Nature Customer Service Center GmbH
Europaplatz 3
69115 Heidelberg, Germany

www.ingramcontent.com/pod-product-compliance
Ingram Content Group UK Ltd.
Pitfield, Milton Keynes, MK11 3LW, UK
UKHW021316180426
11947UKWH00015B/1263

* 9 7 8 3 5 4 0 1 8 2 6 3 4 *